Current Topics in Microbiology and Immunology

Volume 362

Current Topics in Microbiology and Immunology

Previously published volumes

Further volumes can be found at www.springer.com

Vol. 329: **Griffin, Diane E.**; **Oldstone, Michael B. A.** (**Eds.**): Measles. 2009. ISBN 978-3-540-70522-2

Vol. 330: **Griffin, Diane E.**; **Oldstone, Michael B. A.** (**Eds.**): Measles. 2009. ISBN 978-3-540-70616-8

Vol. 331: **Villiers, E. M. de** (**Ed.**): TT Viruses. 2009. ISBN 978-3-540-70917-8

Vol. 332: **Karasev A.** (**Ed.**): Plant produced Microbial Vaccines. 2009. ISBN 978-3-540-70857-5

Vol. 333: **Compans, Richard W.**; **Orenstein, Walter A.** (**Eds.**): Vaccines for Pandemic Influenza. 2009. ISBN 978-3-540-92164-6

Vol. 334: **McGavern, Dorian**; **Dustin, Micheal** (**Eds.**): Visualizing Immunity. 2009. ISBN 978-3-540-93862-0

Vol. 335: **Levine, Beth**; **Yoshimori, Tamotsu**; **Deretic, Vojo** (**Eds.**): Autophagy in Infection and Immunity. 2009. ISBN 978-3-642-00301-1

Vol. 336: **Kielian, Tammy** (**Ed.**): Toll-like Receptors: Roles in Infection and Neuropathology. 2009. ISBN 978-3-642-00548-0

Vol. 337: **Sasakawa, Chihiro** (**Ed.**): Molecular Mechanisms of Bacterial Infection via the Gut. 2009. ISBN 978-3-642-01845-9

Vol. 338: **Rothman, Alan L.** (**Ed.**): Dengue Virus. 2009. ISBN 978-3-642-02214-2

Vol. 339: **Spearman, Paul**; **Freed, Eric O.** (**Eds.**): HIV Interactions with Host Cell Proteins. 2009. ISBN 978-3-642-02174-9

Vol. 340: **Saito, Takashi**; **Batista, Facundo D.** (**Eds.**): Immunological Synapse. 2010. ISBN 978-3-642-03857-0

Vol. 341: **Bruserud, Øystein** (**Ed.**): The Chemokine System in Clinical and Experimental Hematology. 2010. ISBN 978-3-642-12638-3

Vol. 342: **Arvin, Ann M.** (**Ed.**): Varicella-zoster Virus. 2010. ISBN 978-3-642-12727-4

Vol. 343: **Johnson, John E.** (**Ed.**): Cell Entry by Non-Enveloped Viruses. 2010. ISBN 978-3-642-13331-2

Vol. 344: **Dranoff, Glenn** (**Ed.**): Cancer Immunology and Immunotherapy. 2011. ISBN 978-3-642-14135-5

Vol. 345: **Simon, M. Celeste** (**Ed.**): Diverse Effects of Hypoxia on Tumor Progression. 2010. ISBN 978-3-642-13328-2

Vol. 346: **Christian Rommel**; **Bart Vanhaesebroeck**; **Peter K. Vogt** (**Ed.**): Phosphoinositide 3-kinase in Health and Disease. 2010. ISBN 978-3-642-13662-7

Vol. 347: **Christian Rommel**; **Bart Vanhaesebroeck**; **Peter K. Vogt** (**Ed.**): Phosphoinositide 3-kinase in Health and Disease. 2010. ISBN 978-3-642-14815-6

Vol. 348: **Lyubomir Vassilev**; **David Fry** (**Eds.**): Small-Molecule Inhibitors of Protein-Protein Interactions. 2011. ISBN 978-3-642-17082-9

Vol. 349: **Michael Kartin** (**Eds.**): NF-kB in Health and Disease. 2011. ISBN 978-3-642-16017-2

Vol. 350: **Rafi Ahmed**; **Tasuku Honjo** (**Eds.**): Negative Co-receptors and Ligands. 2011. ISBN 978-3-642-19545-7

Vol. 351: **Marcel, B. M. Teunissen** (**Ed.**): Intradermal Immunization. 2011. ISBN 978-3-642-23689-1

Vol. 352: **Rudolf Valenta**; **Robert L. Coffman** (**Eds.**) Vaccines against Allergies. 2011. ISBN 978-3-642-20054-0

Vol. 353: **Charles E. Samuel** (**Ed.**): Adenosine Deaminases Acting on RNA (ADARs) and A-to-I Editing. 2011. ISBN 978-3-642-22800-1

Vol. 354: **Pamela A. Kozlowski** (**Ed.**): Mucosal Vaccines. 2012. ISBN 978-3-642-23692-1

Vol. 355: **Inko K. Mellinghoff; Charles L. Sawyers** (**Eds.**): Therapeutic Kinase inhibitors, 2012. ISBN 978-3-642-28295-9

Vol. 356: **Cornelis Murre** (**Ed.**): Epigenetic Regulation of Lymphocyte Development. 2012. ISBN 978-3-642-24102-4

Vol. 357: **Nicholas J. Mantis** (**Ed.**): Ricin and Shiga Toxins. 2012. ISBN 978-3-642-27469-5

Vol. 359: **Benhur Lee, Paul Rota** (**Eds.**): Henipavirus. 2012. ISBN 978-3-642-29818-9

Marco Falasca
Editor

Phosphoinositides and Disease

Responsible series editor: Peter K. Vogt

Springer

Editor
Marco Falasca
Inositide Signalling Group
The Blizard Institute
Queen Mary University of London
London
UK

ISSN 0070-217X
ISBN 978-94-007-5024-1 ISBN 978-94-007-5025-8 (eBook)
DOI 10.1007/978-94-007-5025-8
Springer Dordrecht Heidelberg New York London

Library of Congress Control Number: 2012950382

Printed on acid-free paper

Springer is part of Springer Science+Business Media (www.springer.com)

Contents

Chapter 1
An Introduction to Phosphoinositides

Tania Maffucci

Abstract Phosphoinositides (PIs) are minor components of cellular membranes that play critical regulatory roles in several intracellular functions. This chapter describes the main enzymes regulating the turnover of each of the seven PIs in mammalian cells and introduces to some of their intracellular functions and to some evidences of their involvement in human diseases. Due to the complex interrelation between the distinct PIs and the plethora of functions that they can regulate inside a cell, this chapter is not meant to be a comprehensive coverage of all aspects of PI signalling but rather an introduction to this complex signalling field. For more details of their regulation/functions and extensive description of their intracellular roles, more detailed reviews are suggested on each single topic.

Abbreviations

ArPIKfyve	Associated regulator of PIKfyve
CML	Chronic myelogenous leukemia
CMT	Charcot-Marie-Tooth
DAG	Diacylglycerol
ER	Endoplasmic reticulum
FERM	Four-point one, Ezrin, Radixin, Moesin
FYVE	Fab1/YOTB/Vac1/EEA1
GPCRs	G-protein coupled receptors
GRAM	Glucosyltransferases, Rab-like GTPase activators and Myotubularins
GSK3	Glycogen synthase kinase 3
HIF	Hypoxia inducible factor
LPA	Lysophosphatidic acid

T. Maffucci (✉)
Centre for Diabetes, Blizard Institute, Inositide Signalling Group, Queen Mary University of London, Barts and The London School of Medicine and Dentistry, 4 Newark Street, London, E1 2AT, UK
e-mail: t.maffucci@qmul.ac.uk

M. Falasca (ed.), *Phosphoinositides and Disease*, Current Topics in Microbiology and Immunology 362, DOI: 10.1007/978-94-007-5025-8_1,

LPI	Lysophosphatidylinositol
Ins(1,4,5)P_3	Inositol 1,4,5-trisphosphate
MIPS	Myo-inositol-3-phosphate synthase
MTM	Myotubularin
MTMR	Myotubularin-related
mTOR	Mechanistic target of rapamycin
PDK1	3-phosphoinositide-dependent protein kinase 1
PH	Pleckstrin homology
PHD	Plant HomeoDomain
PIKfyve	PhosphoInositide Kinase for five position containing a Fyve finger
PIPP	Proline-rich inositol polyphosphate 5-phosphatase
PIs	phosphoinositides
PI3K	phosphoinositide 3-kinase
PLA	phospholipase A
PLC	Phospholipase C
PLD	Phospholipase D
PtdIns3*P*	Phosphatidylinositol 3-phosphate
PtdIns4*P*	Phosphatidylinositol 4-phosphate
PtdIns5*P*	Phosphatidylinositol 5-phosphate
PtdIns(3,4)P_2	Phosphatidylinositol 3,4-bisphosphate
PtdIns(4,5)P_2	Phosphatidylinositol 4,5-bisphosphate
PtdIns(3,5)P_2	Phosphatidylinositol 3,5-bisphosphate
PtdIns(3,4,5)P_3	Phosphatidylinositol 3,4,5-trisphosphate
PIP4 Ks	PtdIns5P 4-kinases
PIP5Ks	PtdIns4P 5-kinases
PTEN	Phosphatase and tensin homolog
PX	Phox homology
RTK	Receptor tyrosine kinase
SHIP	Src homology 2-domain-containing inositol phosphatase
SKIP	Skeletal muscle and kidney enriched 5-phosphatase
TAPP	Tandem PH domain-containing Protein
TGN	Trans-Golgi network
Vps34	Vacuolar protein sorting 34

The Phosphoinositide Family

Phosphoinositides (PIs) are phospholipids comprising two fatty acid chains linked by a glycerol moiety to a water-soluble inositol (Ins) head group. Ins is a cyclohexanehexol that can exist in nine isomeric forms, of which the *myo*-inositol form is the most used in biology (Michell 2008).

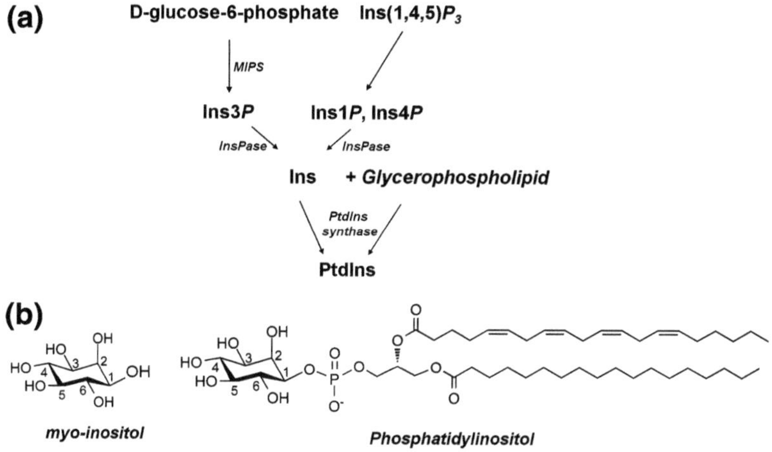

Fig. 1.1 **a** Schematic representation of PtdIns synthesis. **b** Structures of *myo*-inositol and PtdIns

myo-inositol (Fig. 1.1) is synthesised by a NAD^+-dependent *myo*-inositol-3-phosphate synthase (MIPS) that catalyses the cyclisation of D-glucose-6-phosphate to D-*myo*-inositol-3-phosphate (Ins3*P*). Alternatively. *myo*-inositol can derive from Ins1*P* or Ins4*P*, products of dephosphorylation of the Ins(1,4,5)P_3 Ins3*P*, Ins1P_1 or Ins4*P* are dephosphorylated by the inositol monophosphatase (InsPase) to generate Ins (Michell 2008). The synthesis of phosphatidylinositol (PtdIns) is catalysed by the enzyme PtdIns synthase that links the 1-position of the *myo*-inositol to the diester phosphate of a glycerophospholipid (Fig. 1.1). PtdIns is primarily synthesised in the endoplasmic reticulum (ER) and then delivered to other membranes by vesicular transport or via cytosolic PtdIns transfer protein (Di Paolo and De Camilli 2006). Differential phosphorylation of the hydroxyls at the 3-, 4- , and 5-position within the *myo*-inositol headgroup of PtdIns generates seven distinct derivatives, named PIs (Fig. 1.2). The majority of PIs possesses the same fatty acids [a saturated C_{18} residue (stearoyl) in the 1-position and a tetra-unsaturated C_{20} residue (arachidonoyl) in the 2-position]. All seven PIs are naturally occurring in the cell membranes of all higher eukaryotes and they can be inter-converted into each other by the action of specific kinases or phosphatases. Intracellular localisation of PIs-regulating enzymes and/or their relocation upon cellular stimulation is critical for localised modulation of PIs levels.

Mechanisms of PIs-dependent Cellular Signalling

Because of their lipid tail, PIs are obligatory membrane-bound; therefore, they can mark specific membrane compartments, or subdomains within a membrane. Through this property, PIs can act as components of cellular membranes and in some cases they can define the membrane or subdomain of the membranes. PIs can

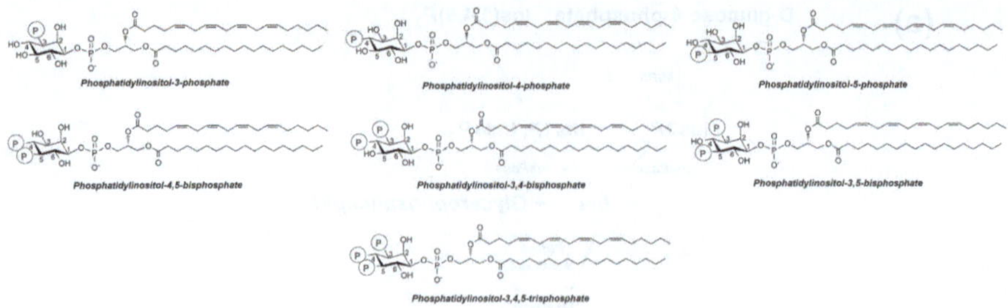

Fig. 1.2 Structures of the seven PIs

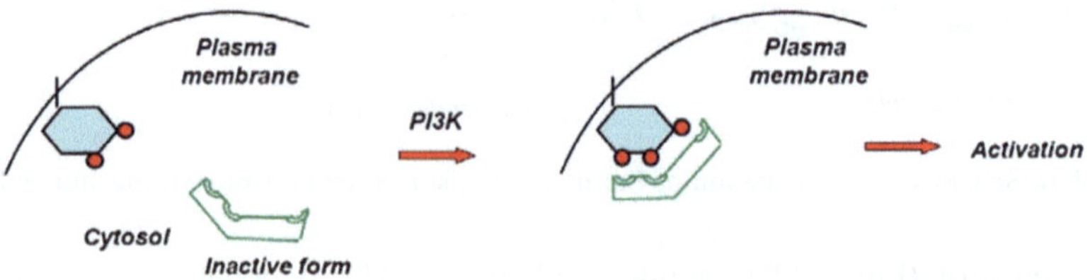

Fig. 1.3 PI-dependent activation of a target protein through recruitment to the plasma membrane. Depicted example refers specifically to the PtdIns(3,4,5)P_3-mediated mechanism of activation

also act as signalling molecules. Their cytosolic soluble headgroup allow PIs to bind to cytosolic proteins or to cytosolic domains of membrane proteins. Binding is mediated by specific PI binding domain(s) within the target proteins and, as a consequence of this interaction, proteins can be activated through conformational changes or through association to a specific cellular membrane (spatial regulation). Moreover, because some PIs are specifically synthesised upon cellular stimulation in normal cells, their regulated synthesis can mediate activation of the target protein not only spatially but also temporally. The rapid synthesis/turnover of some of them is critical to activate signals within the cells only when they are requested to act. Indeed, PIs can rapidly be converted into each other and in particular phosphatases can dephosphorylate specific PIs, often switching off intracellular signals. One of the most studied examples is activation of protein kinase B/Akt that requires translocation of the enzyme to the plasma membrane, a process that is mediated by the interaction of Akt pleckstrin homology (PH) domain and phosphatidylinositol 3,4,5-trisphosphate [PtdIns(3,4,5)P_3], (Fig. 1.3).

PIs can also act as precursors of other "non PIs" signalling molecules. The paradigmatic example of this is the hydrolysis of phosphatidylinositol 4,5-bisphosphate [PtdIns(4,5)P_2] by phospholipase C (PLC) isoforms. These enzymes convert PtdIns(4,5)P_2 to inositol 1,4,5-trisphosphate and diacylglycerol (Fig. 1.4a), that in turn can mediate release of intracellular calcium and activation of protein kinases Cs, respectively. Similarly, PIs can be hydrolysed by members of the family of PLA (Fig. 1.4b). In particular, the action of PLA_2 on PtdIns releases arachidonic acid, whose role as second messenger has been well studied.

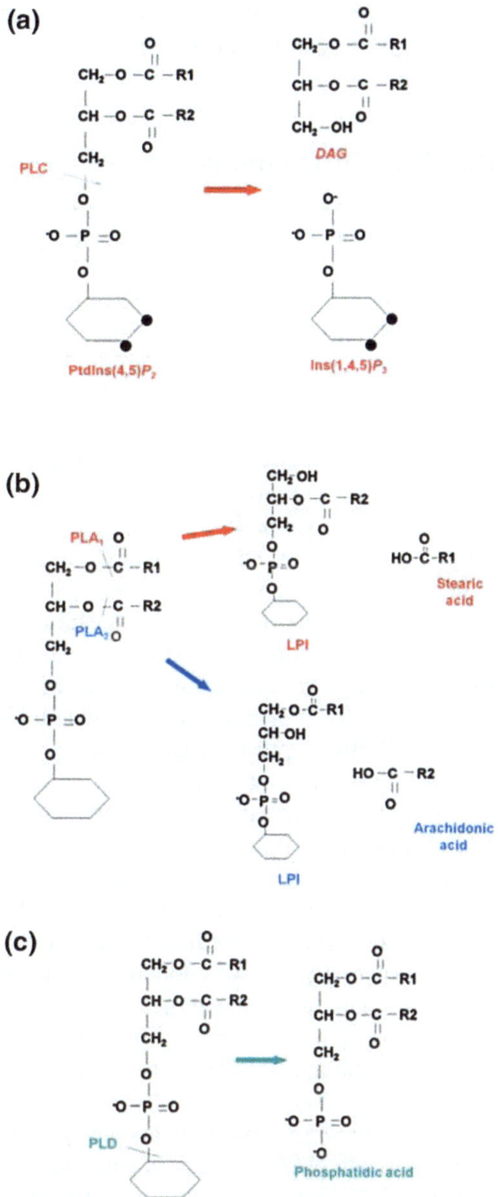

Fig. 1.4 Synthesis of second messengers from PIs: **a** hydrolysis of PtdIns(4,5)P_2 into Ins(1,4,5)P_3 and diacylglycerol (*DAG*) by phospholipase C (*PLC*), **b** hydrolysis of PIs into lysophosphatidylinositol (*LPI*) and stearic or arachidonic acid by phospholipases A, **c** hydrolysis of PIs into phosphatidic acid by phospholipase D (*PLD*)

More recently, an important role in signalling has also been established for the other product of PLA action, lysophosphatidylinositol (LPI) (Piñeiro et al. 2011). LPI can be generated by PLA_1 or PLA_2 (Fig. 1.4b) and it seems that the 2-arachidonoyl species is the most important in signalling.

Finally, PLD can release phosphatidic acid (Fig. 1.4c) which can have a role in signalling or it can be further converted into the second messenger lysophosphatidic acid (LPA).

PIs-Binding Domains

Proteins can bind to PIs through specific protein domains, each showing a distinct affinity and selectivity. The interaction PI–protein domain alone is usually not sufficient to determine the intracellular localisation or re-localisation of the effector protein. Binding of other domains within the protein to other components of the membrane, simultaneous binding of the same domain to a PI and protein(s) (Maffucci and Falasca 2001) or other cooperative mechanisms (Lemmon 2008) guarantee a highly regulated recruitment and activation of signalling molecules. Some of the domains that have been shown to bind PIs are listed below.

PH domains are modules of about 100 amino acids, first identified in pleckstrin, the major protein kinase C substrate in platelets (Lemmon 2008). Although very different in their primary structure, all PH domains possess a similar tertiary structure, consisting of a 7-stranded β-sandwich structure formed by two near-orthogonal β sheets (Lemmon 2008). It has been estimated that only $\sim$10 % of all PH domains bind strongly and specifically to PIs (Lemmon 2008). The majority of PH domains show either low specificity or low affinity (or both) for PIs, and require additional mechanisms to guarantee the specific targeting of the host protein (Maffucci and Falasca 2001).

***Fab**1/**YOTB**/**Vac**1/**EEA**1* (*FYVE*) domains are zinc finger modules of about 60–70 amino acids which specifically bind the monophosphate phosphatidylinositol 3-phosphate (PtdIns3*P*) (Stenmark and Aasland 1999). They can be found as a single finger or as a tandem repeat in proteins and consist of two double stranded antiparallel β sheets and a small C-terminal α-helix. The structure is held together by two tetrahedrally coordinated Zn^{2+} ions (Kutateladze 2006; Lemmon 2008). In contrast to PH, FYVE domains bind more strongly to membrane-embedded PtdIns3*P* than the isolated headgroup Ins(1,3)P_2 (Kutateladze 2006; Lemmon 2008). Endosomal targeting of most FYVE domains requires dimerisation of the domains although this may be crucial for a subset not all FYVE domains (Kutateladze 2006).

Phox homology (*PX*) domains are regions of 130 amino acids named after the two phagocyte NADPH oxidase subunits p40^{phox} and p47^{phox} (Kutateladze 2007). Despite little sequence similarity, PX domains show a highly conserved 3D structure consisting of three stranded β-sheet and a subdomain of three to four α-helices (Kutateladze 2007). All PX domains found in *S. cerevisiae* bind PtdIns3*P* whereas PX domains able to bind phosphatidylinositol 3,5-bisphosphate [PtdIns(3,5)P_2] and phosphatidylinositol 3,4-bisphosphate [PtdIns(3,4)P_2] have also been found in mammals (Lemmon 2008). PX domain binding to the membrane requires a combined action of headgroup binding, electrostatic attraction and membrane insertion, as FYVE and in contrast to PH domain. There are only few examples of PX domains able to bind PtdIns3*P* with high affinity and whose interaction with this PI is sufficient to target them to membranes. The majority of them bind PtdIns3*P* weakly and are recruited to the membrane only as part of a multicomplex.

***P**lant **H**omeo**D**omain* (*PHD*) consists of two atypical zinc fingers, characterised by a Cys4-His-Cys3 structure that coordinates two Zn^{2+} ions. Several chromatin-regulating proteins contain PHD fingers, including members of the ING family which can modulate the activity of histone acetyl transferase and histone deacetylase (Gozani et al. 2003). ING2 has been reported to possess a PHD finger which specifically binds phosphatidylinositol 5-phosphate [PtdIns5*P*] and PtdIns3*P* in vitro and possibly PtdIns5*P* in vivo (Gozani et al. 2003).

***F**our-point one, **E**zrin, **R**adixin, **M**oesin* (*FERM*) domains are modules of approximately 300 amino acids found in a family of peripheral membrane proteins that are able to link membrane to cytoskeleton (Chishti et al. 1998). The FERM domain is composed of three subdomains which can interact with numerous protein-binding partners and can bind to PtdIns(4,5)P_2 (Tepass 2009).

***G**lucosyltransferases, **Ra**b-like GTPase activators and **M**yotubularins* (*GRAM*) domains consist generally of 70 amino acids (Doerks et al. 2000). Data have suggested that at least GRAM domains from myotubularins can bind or contribute to binding to PtdIns(3,5)P_2 (Berger et al. 2003; Tsujita et al. 2004). Binding to phosphatidylinositol 4-phosphate [PtdIns4*P*] has also been reported for one GRAM domain (Yamashita et al. 2006).

ENTH and ANTH domains form a superhelical solenoid of α-helices which can bind PIs with relatively little stereospecificity (Lemmon 2008). Oligomerisation of multiple low-affinity PtdIns(4,5)P_2-binding sites seems to be critical for membrane association of the ANTH domain while a well defined pocket within the ENTH domain forms the PtdIns(4,5)P_2-binding site (Lemmon 2008).

Monophosphates: Phosphatidylinositol 3-phosphate

Kinases

PtdIns3*P* (Fig. 1.2) is one of the lipid products of the family of enzymes phosphoinositide 3-kinase (PI3K) which catalyse phosphorylation at position 3 of the inositol ring of some PIs (Falasca and Maffucci 2006; Falasca and Maffucci 2009). Of the three classes of PI3Ks (Vanhaesebroeck et al. 2010), PtdIns3*P* can derive from direct phosphorylation of PtdIns by class III PI3 K (vacuolar protein sorting 34, Vps34) or members of class II PI3Ks (Fig. 1.5). These isoforms will be discussed in this paragraph. Class I PI3Ks will be discussed within the paragraph on PtdIns(3,4,5)P_3.

Class III PI3K (*hVps*34) is ubiquitously expressed in mammalian cells and only catalyses the synthesis of PtdIns3*P* (Backer 2008). It is generally considered responsible for the synthesis of a constitutive, mostly endosomal-associated, pool

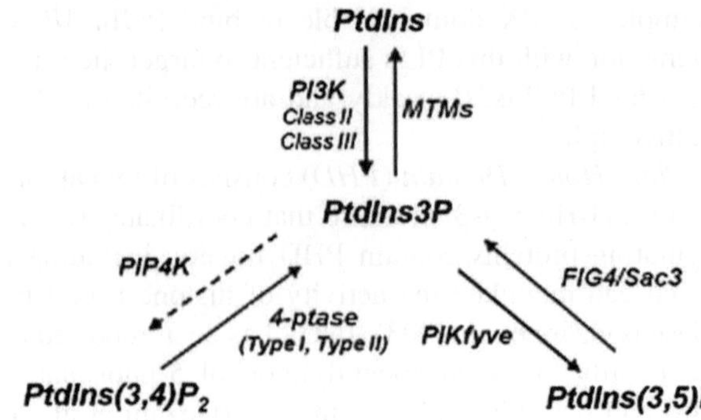

Fig. 1.5 Schematic representation of PtdIns3*P* turnover

of PtdIns3*P*. Vps34 is closely associated with the protein kinase Vps15 which has been described as a Vps34 regulatory protein although the precise role of Vps15 in Vps34 regulation is still unclear (Backer 2008). Questions still remain of whether the activity of Vps34 is modulated by cellular stimulation (Vanhaesebroeck et al. 2010).

Class II PI3Ks are monomers of high molecular weight and exist in three isoforms, PI3K-C2α, PI3K-C2β and PI3K-C2γ (Falasca and Maffucci 2007, 2012). Class II PI3Ks can phosphorylate both PtdIns and PtdIns4*P* but not PtdIns(4,5)P_2 in vitro, with a preferential activity for PtdIns (Arcaro et al. 1998). Work from our laboratory has demonstrated that PI3K-C2α and PI3K-C2β specifically generate PtdIns3*P* at the plasma membrane of muscle and cancer cells upon insulin and LPA stimulation, respectively, (Maffucci et al. 2003; 2005a; Falasca et al. 2007) indicating that class II PI3Ks are mostly responsible for the synthesis of the stimulated pool of this PI. A PI3K-C2α-dependent synthesis of PtdIns3*P* has also been detected in neurosecretory granules (Meunier et al. 2005). No evidence of a role for PI3K-C2γ in modulating PtdIns3*P* levels in vivo has been provided so far. Evidence is also emerging indicating that there is a pool of PtdIns3*P*, likely synthesised by class II PI3Ks, and directly regulated by the phosphatases myotubularins (MTMs). For instance, down-regulation of MTMs in *C. elegans* is able to rescue the phenotype of a Vps34 null mutant (Xue et al. 2003). Similarly, down-regulation of PI3K-C2β specifically rescues the inhibition of Akt phosphorylation induced by down-regulation of MTM1 (Razidlo et al. 2011). The mechanism of activation of class II PI3Ks is still not clear although data suggest that it may involve translocation of the enzymes to the plasma membrane (Maffucci et al. 2005; Falasca et al. 2007; Falasca and Maffucci 2012).

Phosphatases

PtdIns3*P* can be dephosphorylated by several phosphatases, mostly MTMs which are members of the protein tyrosine phosphatases superfamily (Robinson and Dixon 2006) (Fig. 1.5). Originally thought to be protein phosphatases, it was soon observed that recombinant myotubularin MTM1 and myotubularin-related (MTMR) 1, 2, 3, 4, 6 and 7 were able to selectively dephosphorylate PtdIns3*P*. Data later revealed that some MTM/MTMR (such as MTM1, MTMR 2, 3, 6) were also able to dephosphorylate PtdIns(3,5)P_2 (see below). MTMRs are conserved among eukaryotes and 14 family members have been detected in humans. Interestingly, nearly half of them are predicted to be catalytically inactive. Although overexpressed MTMs are usually found diffusely distributed into the cytosol, localisation in other intracellular compartments has also been reported. For instance, endogenous MTMR7 has been described as partially localised to Golgi-like structures in a neuroblastoma cell line; MTM1 can partially localise to the plasma membrane (Robinson and Dixon 2006) and it has also been observed in Rab5-positive early endosomes (Cao et al. 2007) where it critically regulates PtdIns3*P* levels (Cao et al. 2008). MTMR2 specifically localizes to Rab7-positive late endosomes (Cao et al. 2007) and can regulate late endosomal PtdIns3*P* levels (Cao et al. 2008). Overexpressed MTMR3 is largely cytosolic but also present in punctuate elements and in the ER.

The mechanism of regulation of MTMs is not completely understood. It has been suggested that intracellular relocation may represent one way of regulating their activity or that the several non-catalytic domains may have a role in regulation of their enzymatic activity (Choudhury et al. 2006). For instance, binding of PtdIns5*P* seems to activate MTM1, MTMR3 and MTMR6 at least in vitro (Clague and Lorenzo 2005). Moreover evidence seems to suggest that binding of inactive MTMs to active family members can represent some sort of mechanism of regulation (Zou et al. 2009). A recent hypothesis has suggested that MTM specificity can be regulated by their recruitment to different protein complexes (Lecompte et al. 2008).

Other Routes of PtdIns3P Turnover

PtdIns3*P* levels are also regulated by kinases/phosphatases that can convert it into other PIs (Fig. 1.5) and these routes will be discussed in the paragraphs related to the respective PI. In particular, interconversion of PtdIns3*P* and PtdIns(3,5)P_2 is regulated by the enzyme PIKfyve and the 5-phosphatase FIG4/Sac3. Direct synthesis of PtdIns3*P* and PtdIns(3,4)P_2 can be potentially mediated by PIP4 Ks, detected in vitro and possibly occurring in vivo (Divecha and Halstead 2004). Finally, dephosphorylation of PtdIns(3,4)P_2 to PtdIns3*P* has also been described.

Intracellular Roles

Membrane trafficking. PtdIns3*P* intracellular localisation has been studied by using specific PtdIns3*P*-binding domains fused to the GFP, mostly tandem FYVE domains. Despite several limitations and drawbacks of these tools, these experiments clearly revealed a strong accumulation of this PI in the endosomal system in resting mammalian cells (Gillooly et al. 2000) where it regulates membrane transport and membrane dynamics by recruiting proteins containing FYVE, PX and PH domains (Lindmo and Stenmark 2006). It is generally accepted that hVPs34 is responsible for the synthesis of the endosomal PtdIns3*P* and indeed inhibition of hVps34 blocks homotypic early endosomes fusion, the formation of internal vesicles in multivesicular bodies and it delays trafficking of receptors (Backer 2008). Vps34 also controls protein sorting to the lysosome and membrane homeostasis. Emerging data indicate that the endosomal pool of PtdIns3*P* can also contribute to growth factor signalling by critically regulating maturation of an early endocytic intermediate, known as APPL endosomes (Zoncu et al. 2009). Evidence suggests that other PI3Ks can also control the endosomal pool of PtdIns3*P* (Falasca and Maffucci 2009).

Autophagy. hVps34 forms a multiprotein complex with the proautophagic Beclin1/Atg6, Bif-1 and UVRAG that initiates autophagosome formation (Backer 2008; Vanhaesebroeck et al. 2010) and it is critical for autophagosome–lysosome fusion during late steps of autophagy. These data support the hypothesis of a key role for PtdIns3*P* in this process, as also suggested by data indicating that knockdown of MTMR14 and, possibly MTMR6, increases autophagy and the total numbers of autophagic organelles (Vergne et al. 2009); that overexpression of a dominant negative mutant of MTMR3 as well as down-regulation of MTMR3 induce autophagosome formation (Taguchi-Atarashi et al. 2010) and that the local pools of PtdIns3*P* are critical for both autophagy initiation and regulation of autophagosome membrane structure (Taguchi-Atarashi et al. 2010). A direct involvement of PtdIns3*P* in autophagosome formation has been proposed (Axe et al. 2008) as well as a role in the recruitment of Atg18, a protein involved in autophagosome formation (Proikas-Cezanne et al. 2007).

Exocytosis. Evidence indicates that PtdIns3*P* has a role in exocytosis, possibly in different cellular contexts. A PI3K-C2α-dependent pool of PtdIns3*P* plays a critical role in exocytosis of neurosecretory granules by regulating the ATP-dependent priming of the vesicles (Meunier et al. 2005; Wen et al. 2008). Similarly, PI3K-C2α is involved in glucose-induced insulin secretion (Leibiger et al. 2010) and insulin granules' exocytosis (Dominguez et al. 2011). Although a role for PtdIns(3,4)P_2 in this process has been suggested (Liebiger et al. 2010) it has been reported that PtdIns3*P* can directly support fusion in particular experimental conditions using reconstituted proteoliposomes (Mima and Wickner 2009a) and this phosphoinositide is part of a minimal set of lipids required for fusion (Mima and Wickner 2009b). The possibility, therefore, exists that PI3K-C2α may regulate insulin granule fusion directly by maintaining a pool of PtdIns3*P* necessary for this event. A role for an insulin-dependent PtdIns3*P* in regulation of glucose disposal into fat and muscle cells through modulation of the translocation of the glucose transporter protein GLUT4 to the plasma membrane has also been demonstrated (Maffucci et al. 2003a). Synthesis of this stimulated pool of PtdIns3*P* occurs at the plasma membrane of muscle cells and adipocytes (Maffucci et al. 2003a) and it is mediated by PI3K-C2α (Falasca et al. 2007). Indeed, down-regulation of PI3K-C2α inhibits glucose transport (Falasca et al. 2007). Consistent with a role for PtdIns3*P*, overexpression of MTM impairs insulin-induced GLUT4 translocation (Chaussade et al. 2003). More recently, it has been reported that a Rab5-dependent

PtdIns3*P* synthesis is important for GLUT4 translocation (Lodhi et al. 2008). The precise role of PtdIns3*P* in GLUT4 translocation remains to be addressed (Ishiki et al. 2005; Kanda et al. 2005; Kong et al. 2006). Whether PtdIns3*P* itself is directly involved in this process or whether the de novo synthesis of this PI through PI3K-C2α activation is required for its further conversion into PtdIns(3,5)P_2, whose role in GLUT4 translocation has also been reported (Shisheva 2008a), is still a matter of investigation.

Cell migration. LPA activates PI3K-C2β which in turn catalyses the synthesis of PtdIns3*P* at the plasma membrane of ovarian and cervical cancer cells (Maffucci et al. 2005a). Translocation to the plasma membrane of exogenous PI3K-C2β upon LPA stimulation has been observed (Maffucci et al. 2005a), suggesting that a stimulated targeting of the class II enzymes can be responsible for temporally and spatially regulated synthesis of PtdIns3*P*. Data indicate that the LPA-dependent pool of PtdIns3*P* regulates cell migration (Maffucci et al. 2005a), later confirmed in other cellular systems (Domin et al. 2005).

Regulation of ion channels. PtdIns3*P* indirectly activates the Ca^{2+}-activated K^+ channel, KCa3.1 (Srivastava et al. 2006a). A role for MTMR6-dependent modulation of PtdIns3*P* levels has been suggested in this context (Srivastava et al. 2005, 2006b). Involvement of the class II PI3K isoform PI3K-C2β in this process has also been reported (Srivastava et al. 2009).

Studies using specific PtdIns3*P* binding PH domains tagged to GFP have revealed the presence of a nuclear pool of this PI (Safi et al. 2004; Maffucci et al. 2003b), possibly specifically restricted to nuclear speckles (Maffucci et al. 2003b). In this respect, it is noteworthy that PI3K-C2α has also been found confined in this specific intranuclear structures (Didichenko and Thelen 2001). Nuclear localisation was also reported for PI3K-C2β (Sindić et al. 2001; Visnjić et al. 2002). It remains to be established what the role of nuclear PtdIns3*P* is.

Human Disease

Since PtdIns3*P* plays a key role in autophagy, it would be interesting to determine whether it might be involved in human pathologies characterised by deregulation of this process. Autophagy is altered in several diseases, including cancer, some myopathies (where autophagic vacuoles accumulate) and several neurodegenerative disorders (where the accumulation of misfolded proteins can derive from defective autophagy). No mutation in PIK3C3 has been found in this kind of disease (Nicot and Laporte 2008) although mutation or deletion of components of the VPS34 complexes, such as beclin 1 and ultraviolet radiation resistance-associated gene protein have been found in breast, ovarian and colon cancer (Levine and Kroemer 2008). Data are also now revealing a key role for autophagy in pancreatic islets of Langerhans (Ebato et al. 2008) and β cells homeostasis (Jung et al. 2008) and altered autophagy has been detected in β cells from Type 2 diabetic individuals (Masini et al. 2009; Marchetti and Masini 2009).

A rare variant in the promoter of PIK3C3 gene has been reported to be associated with bipolar disorder and schizophrenia in a candidate gene study, (Nicot and Laporte 2008). A potential role for PtdIns3*P* in cancer is suggested by data indicating a key role for PI3K-C2β in migration of some cancer cells (Maffucci et al. 2005a; Katso et al. 2006) and data suggesting a role for PI3K-C2β in regulation of apoptosis in some cell lines (Elis et al. 2008). PI3K-C2α and PI3K-C2β expression is elevated in a large number of human small cell lung cancer cell lines compared with normal lung epithelial cells (Arcaro et al. 2002). The potential contribution of PtdIns3*P* to cancer still needs to be properly investigated.

Mutations in the MTMs have been associated with distinct forms of Charcot-Marie-Tooth (CMT) diseases, a heterogeneous group of genetic peripheral neuropathies affecting motor and sensory nerves and characterised by progressive distal muscle atrophy and weakness (McCrea and De Camilli 2009; Nicot and Laporte 2008). Mutations all over the gene encoding for MTMR2 were detected in the autosomal recessive demyelinating neuropathy CMT type 4B (CMT4B1) and five distinct mutations in the *MTMR*13 gene have been reported to cause CMT4B2. About 200 mutations in the gene encoding MTM1, resulting in reduced protein levels or loss of protein, have been found in patients affected by X-linked centronuclear myopathy (Nicot and Laporte 2008), a very severe congenital myopathy.

Monophosphates: Phosphatidylinositol 4-phosphate

Kinases

PtdIns4*P* (Fig. 1.2) is generated from the precursor PtdIns through the action of PtdIns 4-kinases (Balla and Balla 2006; D'Angelo et al. 2008). Four isoforms of PtdIns 4-kinases have been identified in mammals, originally classified as Types II and III PI4Ks (Fig. 1.6).

Type II PI4Ks, PI4KIIα and PI4KIIβ, are characterised by their insensitivity to the inhibitor wortmannin (Balla and Balla 2006). Because of their palmitoylation, Type II PI4Ks are tightly membrane-bound proteins, although a significant larger fraction of PI4KIIβ than PI4KIIα is cytosolic. PI4KIIα was cloned from membranes of large dense core vesicles in neurons and although it has been suggested that both isoforms may be associated with synaptic vesicles, it seems that the main isoform in these vesicles is PI4KIIα (Hammond and Schiavo 2007). Both PI4KIIα and PI4KIIβ localised to intracellular membranes, mostly TGN and endosomes (Balla and Balla 2006) but they are also present in compartments that contain the AP-3 adaptor complex and in GLUT4-containing vesicles (D'Angelo et al. 2008). A plasma membrane localisation has also been reported for both Type II PI4Ks either in basal conditions (PI4KIIα) or upon stimulation of cells with platelet derived growth factor (PI4KIIβ) in a Rac-dependent manner (D'Angelo et al.

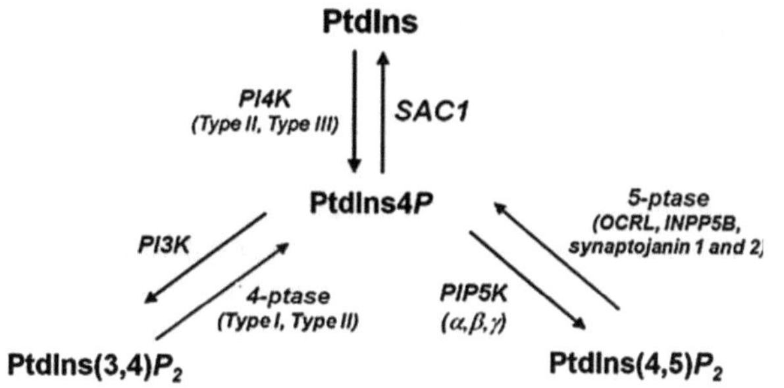

Fig. 1.6 Schematic representation of PtdIns4*P* turnover

2008). Regulation of Type II PI4Ks is not completely clear but it has been reported that calcium inhibits both isoforms whereas membrane association increases PI4KIIβ activity (Balla and Balla 2006).

Type III PI4Ks, PI4KIIIα and PI4KIIIβ, are sensitive to wortmannin (Balla and Balla 2006). In mammalian cells, PI4KIIIα is mainly localised in the ER, in a perinuclear compartment and in the nucleolus (Balla and Balla 2006). PI4KIIIβ is mainly present in the Golgi complex (D'Angelo et al. 2008) although it has also been reported in the nucleus (Balla and Balla 2006). Both Arf1 and neuronal calcium sensor 1 regulate PI4KIIIβ activity which can be further activated through phosphorylation by protein kinases D1 and D2 (D'Angelo et al. 2008).

Phosphatases

PtdIns4*P* can be dephosphorylated by the phosphatase Sac1 and this process is responsible for its turnover at the ER and the Golgi (Fig. 1.6). Original data on mammalian Sac1 revealed that this enzyme was able to dephosphorylate PtdIns4*P* and PtdIns3*P* in vitro (Rohde et al. 2003); however, its specific localisation within the ER and the Golgi complex supports a main role for this enzyme in modulation of PtdIns4*P* levels in vivo. It has been suggested that redistribution of Sac1 between the Golgi complex and the ER upon growth factor stimulation can induce an accumulation of PtdIns4*P* within the Golgi, which in turn can enhance stimulated secretion in proliferating cells (Blagoveshchenskaya et al. 2008).

Other Routes of PtdIns4P Turnover

PtdIns4*P* can be converted into the bisphosphates PtdIns(4,5)P_2 and PtdIns(3,4)P_2 by the action of PtdIns4*P* 5-kinases (PIP5Ks, Type I PIPK) and PI3K respectively (Fig. 1.6). Similarly, dephosphorylation of both PIs by specific 5- and 4-phosphatases has been reported to generate PtdIns4*P* (Fig. 1.6).

Intracellular Roles

Consistent with its subcellular localisation, PI4KIIα has a role in trans-Golgi network (TGN)-to-endosome and TGN-to-plasma membrane transport, in the association of AP-3 with endosomal compartments and in degradation of the epidermal growth factor receptor (D'Angelo et al. 2008). Similarly, PI4KIIIβ can regulate the structural architecture of the Golgi complex, TGN-to-plasma membrane transport and the sphingolipid synthetic pathway (D'Angelo et al. 2008).

The intracellular roles of PtdIns4*P* have been mostly ascribed to its subsequent conversion to PtdIns(4,5)P_2 and possibly PtdIns(3,4,5)P_3. In this respect, it must be noted that the pool of PtdIns4*P* to be converted into PtdIns(4,5)P_2 can be either delivered to the plasma membrane by membrane carriers derived from the Golgi complex and from recycling organelles or it can be produced locally at the plasma membrane by PI4Ks (Di Paolo and De Camilli 2006). For instance, it seems that de novo synthesis of PtdIns4*P*, followed by action of PIP5K, is required for generation of synaptic PtdIns(4,5)P_2 rather than stored PtdIns4*P* (Hammond and Schiavo 2007). Synthesis of this PI, therefore, represents a key step in modulation of the levels of PtdIns(4,5)P_2.

Beside its role as precursor of PtdIns(4,5)P_2, the identification of selective PtdIns4*P* effectors within the cell has now suggested that PtdIns4*P* itself may regulate some intracellular functions. In particular, it has been shown that PtdIns4*P* can participate in recruitment of adaptor and coat proteins to specific domains of the TGN. Examples include AP-1, which promotes clathrin-dependent TGN-to-endosome trafficking and epsinR which can induce membrane curvature. PtdIns4*P* seems also to regulate assembly of COPII at the ER exit sites. Some lipid transport proteins can also bind PtdIns4*P*, including oxysterol-binding proteins, which binds cholesterol and oxysterols, and ceramide transport protein, which transfers ceramide between the ER and the TGN (Balla and Balla 2006). Recently, it has been reported that inhibition of PtdIns4*P* production results in impairment of glycosphingolipid synthesis through its role in regulation of the glucosylceramide-transfer protein FAPP2 (D'Angelo et al. 2007). Some evidence of a role for PI4KIIIβ in insulin secretion has also been provided (Balla and Balla 2006).

Although the roles of PI4KIIIα in the ER are not completely known it is likely that this enzyme is responsible for the synthesis of a pool of PtdIns4*P* at the level of the ER-plasma membrane sites. Indeed, it has been reported that the production of a plasma membrane pool of PtdIns4*P* is mediated by PI4KIIIα (Balla and Balla 2006). The importance of this is highlighted by mass measurements of PtdIns(4,5)P_2 revealing that almost 50 % of the total cellular pool of this PI is synthesised via a wortmannin-sensitive PI4K (Balla and Balla 2006).

Human Disease

Accumulation of PI4KIIα has been observed in some cancer types and specific down-regulation of this enzyme reduces growth of different xenografts in mice in a mechanism involving regulation of angiogenesis through control of hypoxia inducible factor (HIF)-1α expression (Li et al. 2010). Interestingly, down-regulation of HIF-1α expression induced by knockdown of PI4KIIα is rescued by exogenous addition of PtdIns4*P*, confirming that the lipid kinase activity of the enzyme is crucial for its role (Li et al. 2010). Although expression of PI4KIIβ appears to be up regulated in some but not all analysed xenografts, overexpression of this specific isoform does not increase HIF-1α expression, as observed in the case of PI4KIIα overexpression (Li et al. 2010).

Monophosphates: Phosphatidylinositol 5-phosphate

Kinases

PIKfyve/MTMs. The main route of PtdIns5*P* synthesis in mammalian cells seems to involve synthesis of PtdIns(3,5)P_2 (which is catalysed by Type III PIPK, PIKfyve, as described below) and subsequent dephosphorylation by MTMs (Fig. 1.7). MTM1, MTMR1, MTMR2 and MTMR3 were all able to dephosphorylate PtdIns(3,5)P_2 and the product of the reaction was identified as PtdIns5*P* (Tronchere et al. 2004). Moreover, overexpression of MTM1 in Jurkat cells was able to increase PtdIns5*P* levels (Tronchere et al. 2004), indicating that this enzyme can induce PtdIns5*P* synthesis in mammalian cells.

The direct phosphorylation of PtdIns by PIKfyve has also been suggested (Shisheva 2008b). For instance, an increase in PtdIns5*P* levels has been detected in Ba/F3 and NIH3T3 cells expressing the oncogenic tyrosine kinase nucleophosmin anaplastic lympohoma kinase compared to control cells. Treatment with siRNA specifically targeting PIKfyve reduces PtdIns5*P* levels (Coronas et al. 2008). Consistent with this, overexpression of PIKfyve seems to induce an accumulation of PtdIns5*P* in some cell lines (Sbrissa et al. 2002). However, whether this is due to a direct PIKfyve-dependent phosphorylation of PtdIns or rather to PIKfyve-mediated generation of PtdIns(3,5)P_2 and subsequent dephosphorylation by MTMs is difficult to be assessed.

PIP4Ks. Levels of PtdIns5*P* can also be regulated by PtdIns5*P* 4-kinases (PIP4Ks, Type II PIPKs) which can reduce PtdIns5*P* levels by phosphorylating it at position 4, leading to PtdIns(4,5)P_2 (see below). Three isoforms of PIP4Ks exists, namely PIP4Kα, β, and γ, with the β showing the highest expression of all isoforms and the γ being highly expressed in the kidney and in neurons (Clarke et al. 2009). Although studies have reported that overexpression of PIP4Kα (Roberts et al. 2005) and β (Roberts et al. 2005; Jones et al. 2006) does not affect

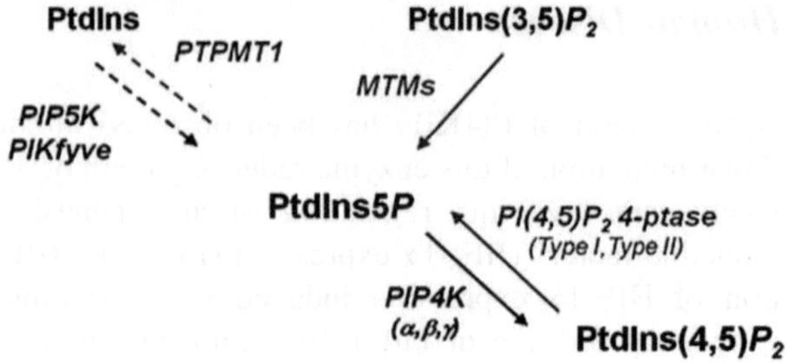

Fig. 1.7 Schematic representation of PtdIns5*P* turnover

the basal levels of PtdIns5*P*, UV irradiation induces phosphorylation of PIP4Kβ in a mechanism involving p38 activation, resulting in inhibition of the enzyme and subsequent accumulation of nuclear PtdIns5*P* (Jones et al. 2006). PIP4Kβ possesses an almost exclusive nuclear localisation (Divecha et al. 1993; Richardson et al. 2007) where it can specifically localise in nuclear speckles (Bunce et al. 2008), therefore it was generally accepted that this isoform was specifically involved in modulation of a nuclear pool of PtdIns5*P*. By contrast, it has been suggested that PIP4Kα can modulate increases in a cytoplasmic pool of PtdIns5*P* in HeLa cells upon inhibition of the phospho-tyrosine phosphatase (Wilcox and Hinchliffe 2008), consistent with a preferred cytosolic intracellular localisation. Nevertheless, it has been recently demonstrated that PIP4Kβ can interact with PIP4Kα in vitro and in vivo and modulate the nuclear localisation of PIP4Kα (Bultsma et al. 2010), consistent with original observations (Divecha et al. 1993). PIP4Kγ appears to be associated with vesicles in kidney and brain. PIP4Kα, first purified and cloned from erythrocytes (Boronenkov and Anderson 1995) and platelets (Divecha et al. 1995), possesses a much higher enzymatic activity than the other two isoforms (Clarke et al. 2009; Bultsma et al. 2010).

PIP5Ks (PtdIns4*P* 5-kinase, Type I PIPK) can also directly phosphorylate PtdIns in vitro to generate PtdIns5*P* (Tolias et al. 1998) but whether this occurs in vivo remain to be addressed, with one study showing that transfection of PIP5Kα does not increase the levels of PtdIns5*P* in COS7 (Roberts et al. 2005). No study so far has described a PI3K-dependent phosphorylation of PtdIns5*P* to generate PtdIns(3,5)P_2.

Phosphatases

A phosphatase with a highly specific 5-phosphatase activity towards PtdIns5*P* in vitro has been described in mammals (Merlot et al. 2003; Pagliarini et al. 2004). This enzyme, originally named phospholipid-inositol phosphatase and later renamed PTPMT1 (Blero et al. 2007), can potentially convert PtdIns5*P* to PtdIns but whether this enzyme has a role in reducing the levels of PtdIns5*P* in vivo is not clear.

By contrast, the role of some phosphatases in promoting an increase in PtdIns5*P* levels has been demonstrated. For instance, MTMs are key regulators of

PtdIns5P levels, as already discussed. Accumulation of PtdIns5*P* can also derive from dephosphorylation of PtdIns(4,5)P_2 by the action of 4-phosphatases. Actually the identification of a 4-phosphatase, invasion plasmid gene D (IpgD), in *Shigella flexneri* able to convert PtdIns(4,5)P_2 into PtdIns5*P* (Niebuhr et al. 2002) represented an important step towards the identification of some of the intracellular functions of PtdIn5*P* (see below). Two mammalian PtdIns(4,5)P_2 4-phosphatases have been identified (Ungewickell et al. 2005) which can generate PtdIns5*P* through dephosphorylation of PtdIns(4,5)P_2 (see below). Some increase in PtdIns5*P* levels have been detected in HEK293 cells overexpressing Type I but not Type II PtdIns(4,5)P_2 4-phosphatase (Zou et al. 2007).

Intracellular Roles

Levels of PtdIns5*P* increase 20-fold during G1 suggesting that this PI may have a role in regulation of cell cycle progression (Clarke et al. 2001). As discussed above, accumulation of a PIP4Kβ-dependent pool of PtdIns5*P* has been detected in the nucleus upon stress signals and it has been shown that nuclear PtdIns5*P* can modulate ING2 binding to chromatin (Jones et al. 2006) and this in turn can regulate the ability of ING2 to activate p53 (Gozani et al. 2003). This role of nuclear PtdIns5*P* seems to be supported by data indicating that overexpression of Type I PtdIns(4,5)P_2 4-phosphatase promotes p53 acetylation and cell death (Zou et al. 2007). Increased cellular levels of PtdIns5*P* stimulate the complex Cul3-SPOP which can regulate ubiquitylation of several substrates, including PIP4KIβ in a mechanism involving p38 MAPK (Bunce et al. 2008).

A role for a cytosolic pool of PtdIns5*P* has been discovered in cells during *S. flexneri* infection. Accumulation of this PI at the entry foci of the bacteria has been reported as a result of release of the 4-phosphatase IpgD (Pendaries et al. 2006). Moreover, it has been reported that expression of IpgD but not its inactive mutant is able to induce Akt phosphorylation in host cells and this is inhibited by expression of PIP4Kβ (Pendaries et al. 2006; Ramel et al. 2009). Moreover, incubation of cells with exogenous PtdIns5*P* is also specifically able to induce Akt phosphorylation and this was suggested to occur through activation of class I PI3Ks (Pendaries et al. 2006). More recently, it has been suggested that PtdIns5*P* has a role in protecting Akt from dephosphorylation (Ramel et al. 2009).

Data have also suggested that PtdIns5*P* may have a regulatory role in the osmotic response pathway, since its levels are decreased upon hypo-osmotic shock in 3T3-L1 fibroblasts and adipocytes (Sbrissa et al. 2002). Using an indirect assay to measure PtdIns5*P* accumulation, it was also reported that insulin increases the levels of this PI in 3T3-L1 adipocytes and CHO-IR cells (Sbrissa et al. 2004) and data also suggested a role for this PI in GLUT4 translocation (Sbrissa et al. 2004). Accumulation of PtdIns5*P* has been detected in platelets upon thrombin stimulation (Morris et al. 2000).

Bisphosphates: Phosphatidylinositol 4,5-bisphosphate

Phosphatidylinositol 4,5-bisphosphate [PtdIns(4,5)P_2, Fig. 1.2) was originally mainly studied as a precursor of key second messengers Ins(1,4,5)P_3/diacylglycerol (Fig. 1.3) or PtdIns(3,4,5)P_3. Nevertheless in the last years, it has become increasingly evident that PtdIns(4,5)P_2 can also act as a second messenger itself.

Kinases

PIP4Ks. PtdIns(4,5)P_2 can be synthesised through the action of PtdIns5*P* 4-kinases that phosphorylates PtdIns5*P*, as already discussed (Fig. 1.8). However, this route is mostly studied for its role in modulation of PtdIns5*P* levels rather than for its contribution to PtdIns(4,5)P_2 synthesis.

PIP5Ks. PtdIns(4,5)P_2 is mainly synthesised through the phosphorylation of PtdIns4*P* at the 5-position by PIP5Ks (van den Bout and Divecha 2009), (Fig. 1.8). Ectopic expression of PIP5Ks but not PIP4Ks increases the synthesis of PtdIns(4,5)P_2 in cells (Halstead et al. 2005). Three isoforms of PIP5Ks exist (PIP5Kα, PIP5Kβ and PIP5Kγ) and data indicate that the three isoforms are not redundant. PIP5Kα localises at the plasma membrane, the Golgi complex, in nuclear speckles and in membrane ruffles. PIP5Kβ localises at the plasma membrane and it has also been detected in perinuclear region. Mouse PIP5Kγ661 can localise to focal adhesions and at adherens junctions in epithelial cells (van den Bout and Divecha 2009). Several different mechanisms of regulation of PIP5Ks have been reported, including regulation through the small GTP-binding proteins Rac and Rho, the ADP-ribosylation factors, the protein talin or through phospholipase D and phosphatidic acid (van den Bout and Divecha 2009).

Phosphatases

5-*phosphatases*. PtdIns(4,5)P_2 can be dephosphorylated by 5-phosphatases to generate PtdIns4*P* (Fig. 1.8). The first identified enzyme of this family, Type I, does not appear to dephosphorylate PIs (Blero et al. 2007). All the other 5-phosphatases are Type II enzymes and nine different forms have been described in humans. Some of them are listed in this paragraph. *OCRL* exists in two isoforms, termed a and b (Lowe 2005) and it preferentially dephosphorylates PtdIns(4,5)P_2 although it can also dephosphorylate PtdIns(3,5)P_2 and PtdIns(3,4,5)P_3 (Ooms et al. 2009). OCRL1 localises to the TGN, lysosome and endosomes in particular in clathrin-coated buds at the TGN and clathrin-coated vesicles between the TGN and endosomes (Lowe 2005). OCRL can also translocate to membrane ruffles upon cellular stimulation, in a mechanism dependent on Rac1 (Ooms et al. 2009). *INPP5B* possesses a domain organisation very similar to OCRL and it preferentially dephosphorylates

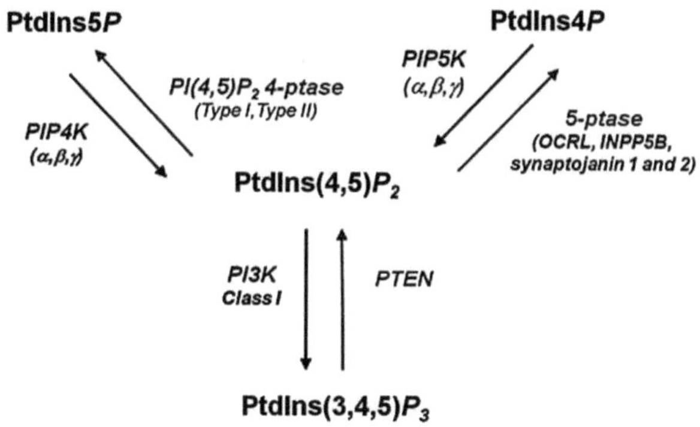

Fig. 1.8 Schematic representation of PtdIns(4,5)P_2 turnover

PtdIns(4,5)P_2 and PtdIns(3,4,5)P_3 and localises to the Golgi and endocytic pathway. It is also able to translocate to lamellipodia upon cellular stimulation, where it can co-localise with Rab5 and actin. *Synaptojanin* 1 is a nerve terminal protein of 145 kDa crucial for synaptic vesicle trafficking and recycling (Blero et al. 2007). Mice deficient for this phosphatase die shortly after birth (Cremona et al. 1999). *Synaptojanin* 2 was first identified by polymerase chain reaction and it is necessary for the formation of clathrin-coated pits and for lamellipodia formation during cell migration. The proline rich domains of the two synaptojanins can associate with different proteins, locating the enzymes in different cellular compartments (Blero et al. 2007). *SKIP* is a 5-phosphatase which can dephosphorylate PtdIns(3,4,5)P_3 (see below). Nevertheless, in vitro kinetic analysis has revealed a preference for PtdIns(4,5)P_2 compared to PtdIns(3,4,5)P_3 but whether this can occur in vivo is still not clear. Similarly, SHIP2, INPP5E and PIPP have been shown to be able to dephosphorylate PtdIns(4,5)P_2 in vitro (see below).

4-*phosphatases*. Dephosphorylation of PtdIns(4,5)P_2 by a 4-phosphatase was first observed in studies on IpgD. Two mammalian PtdIns(4,5)P_2 4-phosphatases have been identified [named PtdIns(4,5)P_2 4-phosphatase type I and type II] which lead to accumulation of PtdIns5P. Both exogenous and endogenous proteins appear to localise in the late endosomes/lysosomal membranes with data suggesting that they may be lysosomal transmembrane proteins (Ungewickell et al. 2005).

Intracellular Roles

Channel activity. Activity of several ion channels has been reported to depend on PIs (a list can be found in Logothetis et al. 2010). In particular, PtdIns(4,5)P_2 is involved in regulation of inward rectifier and voltage-gated potassium channels, calcium channels and pumps, transient receptor potential channels, epithelial sodium channels, ion exchangers (Di Paolo and De Camilli 2006; Logothetis et al. 2010). PtdIns(4,5)P_2 can regulate channel function through direct binding (Halstead et al. 2005). On the other hand, several channels appear to be dependent on PtdIns(4,5)P_2 even if they lack specific PtdIns(4,5)P_2 binding sites (van Rossum et al. 2005).

Actin modulation. One of the most studied roles of PtdIns(4,5)P_2 is regulation of the actin cytoskeleton, often in concert with PtdIns(3,4,5)P_3 or small GTPases. PtdIns(4,5)P_2 can regulate actin filament assembly through different mechanisms including: (1) inhibition of actin monomer sequestering; (2) filament depolymerisation; (3) filament barbed ends capping; (4) activation of the Arp2/3 complex (Saarikangas et al. 2010). On the other hand, PtdIns(4,5)P_2 can positively regulate actin-membrane linkage by releasing proteins containing FERM domain from their autoinhibitory state (Saarikangas et al. 2010). Through these actions, PtdIns(4,5)P_2 participates in regulation of cell shape, motility, cytokinesis and several other functions (Di Paolo and De Camilli 2006). PtdIns(4,5)P_2 is also critical for regulating local polymerisation of actin during phagocytosis.

Trafficking. PtdIns(4,5)P_2 is directly involved in exocytosis, acting both on plasma membrane proteins and on vesicle proteins and it can probably cooperate with SNARE interactions (Di Paolo and De Camilli 2006). PtdIns(4,5)P_2 appears to be required for all forms of endocytosis by recruiting and regulating critical endocytic proteins at the plasma membrane (Ooms et al. 2009). A role for OCRL in regulation of transport from early endosomes to TGN and for INPP5B in the early secretory pathways has also been suggested (Ooms et al. 2009). Whether PtdIns(4,5)P_2 is directly involved in these processes is not clear.

Apoptosis. As precursor of PtdIns(3,4,5)P_3, PtdIns(4,5)P_2 has been indirectly implicated in regulation of apoptosis. Evidence that PtdIns(4,5)P_2 itself can have a role in regulation of apoptosis has been provided by studies indicating that this PI can directly inhibit purified caspase 3 and 8 activity in vitro and that overexpression of PIP5Kα can suppress apoptosis in a mechanism that does not involves PtdIns(3,4,5)P_3 synthesis or Akt activation (Mejillano et al. 2001). PtdIns(4,5)P_2 levels are reduced during apoptosis induced by oxidative stress and UV irradiation in a mechanism independent of caspase activation and involving down-regulation of PIP5Kα (Halstead et al. 2006).

Human Disease

Lowe Syndrome. Mutations in OCRL causes Lowe Syndrome (also known as **O**culo**C**erebro**R**enal syndrome of **L**owe), an X-linked disorder which induces bilateral congenital cataracts, mental retardation, neonatal hypotonia and renal Fanconi syndrome (McCrea and De Camilli 2009). The mechanisms by which OCRL loss determine the clinical onset of Lowe syndrome is not clear. Magnetic resonance imaging brain scans of Lowe Syndrome's patients show cystic abnormalities in the white matter. High levels of lysosomal enzymes are also found in the plasma, suggesting a defect in the endosomal trafficking (Ooms et al. 2009). Fibroblasts from Lowe syndome's exhibit reduced actin stress fibres, increased F-actin puncta, enhanced sensitivity to actin-depolymerising agents and an altered localisation of actin-binding proteins such as gelsolin and α-actinin (Ooms et al. 2009).

Dent disease. Mutations of OCRL have been observed also in a subset of cases of Dent disease, an X-linked disease with renal disorder similar but not identical to the defects detected in Lowe Syndromes (McCrea and De Camilli 2009).

Channelopathies. Mutations that appear to impair the interactions between channels and PtdIns(4,5)P_2, as established by in vitro studies, and resulting in decreased open probability have been detected in Andersen–Tawil syndrome and in hyperprostaglandin E syndrome (HPS), the antenatal form of Bartter syndrome (Halstead et al. 2005; Logothetis et al. 2010). Similarly, disruption of the interaction between the channel and PtdIns(4,5)P_2 seems to result from few mutations linked to LQT syndrome (Logothetis et al. 2010). Few of the mutations in Kir6.2 that have been linked to congenital hyperinsulinism can be mapped to the putative PtdIns(4,5)P_2 binding site (Logothetis et al. 2010). Finally, few mutations identified in patients with thyrotoxic hypokalemic periodic paralysis were also shown to alter the interaction between PtdIns(4,5)P_2 and Kir2.6 (Logothetis et al. 2010).

Lethal contractural syndrome type 3. An inactivating mutation in the gene encoding for PIP5Kγ has been found to be responsible for this autosomal recessive lethal congenital contractural syndrome (Narkis et al. 2007).

Bisphosphates: Phosphatidylinositol 3,4-bisphosphate

Kinases

PI3K. The direct route of phosphatidylinositol 3,4-bisphosphate [PtdIns(3,4)P_2] synthesis occurs through phosphorylation of PtdIns4*P* by a member of the PI3K family (Fig. 1.9). While in vitro studies have revealed that both class I and class II PI3K isoforms can generate PtdIns(3,4)P_2, the relative contribution of each isoform to in vivo synthesis of PtdIns(3,4)P_2 is still not clear. In fact, demonstration that accumulation of PtdIns(3,4)P_2 can occur in vivo through direct action of a PI3K-dependent on PtdIns4*P*, as opposed to PtdIns(3,4,5)P_3 synthesis followed by the action of a 5-phosphatase has proven to be very difficult because several stimuli are able to simultaneously generate PtdIns(3,4)P_2 and PtdIns(3,4,5)P_3 making difficult to discriminate which effect are due to each PI and to identify intracellular functions specifically dependent on PtdIns(3,4)P_2. The relative levels of PtdIns(3,4)P_2 compared to PtdIns(3,4,5)P_3 seem to depend on the cell types, receptor type activated, relative levels of phosphatase and tensin homolog (PTEN), tissue- and cell-specific expression of 5- and 4-phosphatases (Ooms et al. 2009).

PIP4K. PIP4Ks have also been shown to be able to phosphorylate PtdIns3*P* to generate PtdIns(3,4)P_2 in vitro (Rameh et al. 1997) and this observation, together with data showing that PtdIns(3,4)P_2 can be converted to PtdIns(3,4,5)P_3 by PIP5Ks during oxidative stress (Halstead et al. 2001) has led to the hypothesis of the existence of this alternative pathway to generate PtdIns(3,4,5)P_3 which for instance may have been developed before evolution of class I PI3Ks (Divecha and Halstead 2004).

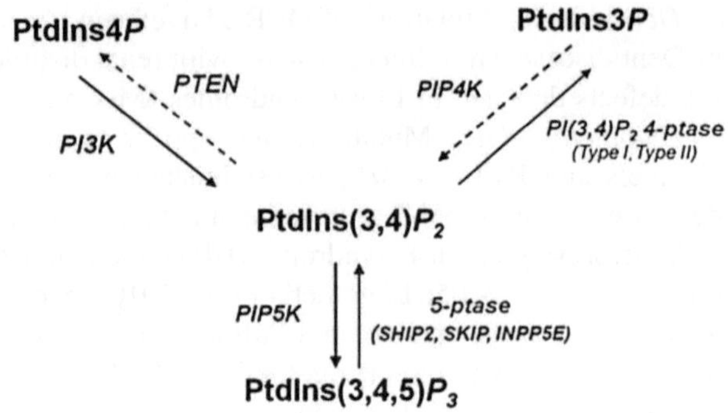

Fig. 1.9 Schematic representation of PtdIns(3,4)P_2 turnover

Phosphatases

4-phosphatases. Dephosphorylation of PtdIns(3,4)P_2 by inositol polyphosphate 4-phosphatases (4-ptases) leads to PtdIns3P (Fig. 1.9). Two PtdIns(3,4)P_2 4-ptases have been identified, Type I and II, that exist as two alternative splice mutants (α and β). Recombinant 4-ptase I is able to hydrolyse PtdIns(3,4)P_2 in vitro and its overexpression has been shown to decrease total levels of PtdIns(3,4)P_2 but not PtdIns(3,4,5)P_3 (Kisseleva et al. 2002). Moreover, Weeble cells (which lacks 4-ptase mRNA and protein) show an increase in PtdIns(3,4)P_2 but not PtdIns(3,4,5)P_3 (Shin et al. 2005). The inositol polyphosphate 4-phosphatase Type II (INPP4B) preferentially dephosphorylates PtdIns(3,4)P_2 both in vitro and in vivo with no effect on PtdIns(3,4,5)P_3 (Gewinner et al. 2009).

PTEN. In vitro studies have suggested that PTEN can dephosphorylate PtdIns(3,4)P_2 but it is not clear whether this occurs in vivo.

Other Routes of PtdIns(3,4)P_2 Turnover

PtdIns(3,4)P_2 can accumulate upon dephosphorylation of PtdIns(3,4,5)P_3 by 5-phosphatases (Fig. 1.9) and this route is still believed to be the main pathway for PtdIns(3,4)P_2 accumulation upon cellular stimulation.

Intracellular Roles

The identification of intracellular functions specifically regulated by PtdIns(3,4)P_2 has proven to be extremely challenging because of the simultaneous synthesis of PtdIns(3,4,5)P_3 that almost always occurs upon cellular stimulation. Furthermore, some PH domains able to bind PtdIns(3,4)P_2 can also bind PtdIns(3,4,5)P_3 (including the highly characterised Akt PH domain), making difficult to assess the contribution of each PI in activation of the target protein. A specific role for PtdIns(3,4)P_2 in Akt activation has been suggested by data reporting that mouse embryonic fibroblasts lacking 4-ptase I display an increased Akt activation in serum-starved conditions and enhanced activation upon growth factor stimulation which results in increased cell proliferation and colonies in soft agar, reduced apoptosis, and show an enhanced ability to form tumours in vivo (Ivetac et al. 2009).

The identification of PH domains specifically able to bind PtdIns(3,4)P_2 such as the C-terminal domains of TAPP1 and TAPP2 (**Ta**ndem **PH** domain-containing **P**rotein) (Dowler et al. 2000) has been useful to investigate the intracellular localisation of PtdIns(3,4)P_2 and to analyse stimulus-dependent increase. These data suggest a specific role for PtdIns(3,4)P_2 during oxidative stress. Different pools of PtdIns(3,4)P_2 exist within the cells and can be modulated upon cellular stimulation. In particular, an increase in PtdIns(3,4)P_2 was detected at the plasma membrane upon certain cellular stimulation but also detected in ER, MVBs, nuclear envelope and mitochondria and data suggested that the pool in the ER may be regulated by a different mechanism compared to the plasma membrane and endosome pool (Watt et al. 2004). Type Iα 4-ptase has been involved in regulation of megakaryocyte and fibroblast differentiation (Blero et al. 2007).

Human Disease

INPP4B acts as a tumour suppressor in epithelial cancers and a positive correlation between loss of INPP4B protein expression and reduced overall survival has been reported for breast cancer and ovarian cancer patients (Gewinner et al. 2009). INPP4B down-regulation increases anchorage-independent growth, cell proliferation and motility whereas INPP4B overexpression reduces tumour growth. Since INPP4B does not appear to dephosphorylate PtdIns(3,4,5)P_3, a key role for PtdIns(3,4)P_2 in these processes is suggested. Further evidence of a role for INP44B in cancer comes from the observation that INPP4B is frequently deleted in breast cancers and it is silenced in malignant proerythroblasts with high levels of phosphorylated Akt (Bunney and Katan 2010). More recently, down-regulation of INPP4B has been detected in prostate cancer and it has been shown that the expression levels of the phosphatase are regulated by activation of the androgen receptor (Hodgson et al. 2011).

A model of spontaneous mutant mouse (weeble mutant mice) has revealed that mutation in the type I 4-ptase gene is associated with neuronal loss in the cerebellum, ataxia and neonatal death (Nystuen et al. 2001).

Bisphosphates: Phosphatidylinositol 3,5-bisphosphate

Kinases

PtdIns(3,5)P_2 is synthesised by a member of the family of Type III PIPKs through PtdIns3P phosphorylation (Dove et al. 2009) (Fig. 1.10). The first PtdIns(3,5)P_2-synthesising enzyme to be identified in higher eukaryotes was the mouse

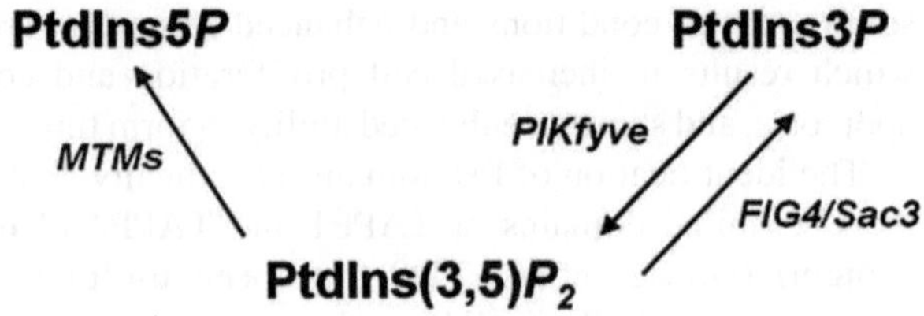

Fig. 1.10 Schematic representation of PtdIns(3,5)P_2 turnover

PIKfyve (**P**hospho**I**nositide **K**inase for **f**ive position containing a **Fyve** finger). Three alternatively spliced forms of this enzyme have been described so far (Shisheva 2008b). Beside PtdIns3*P*, PIKfyve appears also to be able to phosphorylate PtdIns to generate PtdIns5*P*, although the physiological relevance of this is still controversial. Interestingly, PIKfyve possesses also protein kinase activity and both kinase activities are catalysed by the same domain (Shisheva 2008b). The mechanism of PIKfyve activation involves formation of a complex with the activator ArPIKfyve (**A**ssociated **r**egulator of PIKfyve) which is facilitated by association with the PtdIns(3,5)P_2 phosphatase Sac3. In this respect, Sac3 can act both as an antagonist and as a promoter of PIKfyve activity. Indeed, this represents one of the examples of a multiprotein complex made of a kinase/phosphatase/adaptor necessary for tight control of a PI synthesis (Lecompte et al. 2008).

Phosphatases

MTMs. PtdIns(3,5)P_2 can be dephosphorylated by MTMs (Fig. 1.10) and recent hypothesis suggest that inactive MTMs can dimerise with active MTMs and change their specificity towards this PI rather than PtdIns3*P* (Lecompte et al. 2008).

FIG4/Sac3. PtdIns(3,5)P_2 can be dephosphorylated by the 5-phosphatase FIG4/Sac3 and the scaffolding protein VAC14 (Fig. 1.10). Consistent with this, a slight increase in the levels of PtdIns(3,5)P_2 has been detected in 3T3-L1 adipocytes upon down-regulation of Sac3 using siRNA (Ikonomov et al. 2009a). Consistent with this, mutations of FIG4 or Sac14 in mouse models resulted in 50–75 % reduction in the levels of PtdIns(3,5)P_2 in cultured fibroblasts.

5 *phosphatases*. A potential role for the the 5-phosphatase 72 kDa-5 phosphatase/INPP5E (described below) in dephosphorylating PtdIns(3,5)P_2 has been suggested by data revealing that overexpression of this enzyme increases the levels of PtdIns3*P* (Kong et al. 2006) (Fig. 1.10). Whether the endogenous enzyme has a role in modulating the levels of PtdIns(3,5)P_2 remains to be addressed. Similarly, recent data have revealed that SHIP2 can dephoshorylate PtdIns(3,5)P_2 in vitro and it actually shows a similar affinity for this PI as for PtdIns(3,4,5)P_3 (Ooms et al. 2009) but it is not known whether this activity occurs also in vivo.

Intracellular Roles

Overexpressed wild type PIKfyve localises on early endosomes, MVB, late endosomes and TGN although the localisation is strongly dependent on the cell type, the level of protein expressed and on the ratio between PtdIns(3,5)P_2 and PtdIns3P (Shisheva 2008a).

Membrane trafficking and dynamics. PtdIns(3,5)P_2 has a well- established role in membrane trafficking and it is currently believed that specific cargo molecules can stimulate synthesis of this PI to regulate their correct endosomes–lysosomes trafficking (Dove et al. 2009). Defects in PtdIns(3,5)P_2 synthesis results in accumulation of swollen intracellular vacuoles, enlarged late endosomes and abnormal multivesicular bodies which have been observed both in cultured cells, in animal models and human tissues (see below).

Autophagy. Recent data have revealed a critical role for PtdIns(3,5)P_2 in autophagy in the central nervous system (Ferguson et al. 2010). Indeed, increase of p62 and LC3-II has been specifically detected in the brain and spinal cord of *Fig*$4^{-/-}$ mice (Ferguson et al. 2009). Accumulation of p62 and LAMP-2 has been detected in astrocytes from *Fig*$4^{-/-}$ mice and impaired autophagy appears to underlie neurodegeneration in mice. Abnormal autophagy was also detected in the mouse mutant ingls in which a mutation in Vac14 impairs its interaction with FAB1 reducing PtdIns(3,5)P_2 levels and inducing a neurodegeneration similar to that observed in *Fig*$4^{-/-}$ mice.

Glucose transport. PtdIns(3,5)P_2 may play a role in glucose transport in adipocytes through regulation of GLUT4 translocation to the plasma membrane. Inhibition of PtdIns(3,5)P_2 synthesis by either overexpression of kinase dead mutants of PIKfyve (Ikonomov et al. 2002), down-regulation of PIKfyve using siRNA (Ikonomov et al. 2007) or pharmacological inhibition of PIKfyve (Ikonomov et al. 2009b) has been shown to inhibit GLUT4 translocation and glucose transport upon insulin stimulation. Similar effects were detected upon loss of ArPIKfyve (Ikonomov et al. 2007). Down-regulation of Sac3 by siRNA results in a slight increase in GLUT4 translocation to the plasma membrane and glucose transport whereas overexpression of Sac3 inhibits the insulin-induced GLUT4 translocation (Ikonomov et al. 2009a). Data also suggest that insulin can inhibit the phosphatase activity of Sac3, further promoting PtdIns(3,5)P_2 accumulation. The complex regulation of PtdIns(3,5)P_2 synthesis, together with the fact that PIKfyve substrate (and Sac3 product), is PtdIns3P, whose role in glucose transport is also well-established (Falasca and Maffucci 2009), makes difficult to discriminate the relative contribution of each PI in this process.

Human Disease

François-Neetens Mouchetée Fleck Corneal Dystrophy. Mutations in the gene encoding PIKfyve have been detected in patients with this autosomal dominant disease. Affected patients exhibit small white flecks in the stroma of the cornea, possibly swollen keratocytes filled with vesicles containing lipids and mucopolysaccarides (Nicot and Laporte 2008; McCrea and De Camilli 2009). CMT4J Recessive mutations of human FIG4 was detected on chromosome 6q21 in four unrelated patients with hereditary motor and sensory neuropathy and this novel form of autosomal recessive Charcot Marie-Tooth was designed Type 4J (Chow et al. 2007). *ALS*. Heterozygous deleterious mutations of FIG4 have also been observed in ~1 % of patients with amyotrophic lateral sclerosis (Chow et al. 2009). An autosomal recessive mutation designed "pale tremor" was detected in mice with severe tremor, diluted pigmentation and neuronal degeneration in the central nervous system (Chow et al. 2007). Cultured fibroblasts from these mice revealed altered PtdIns(3,5)P_2 levels as a consequence of an insertion into Fig4. Large vacuoles containing LAMP-2 were detected in the cytoplasm of these fibroblasts, indicating a defect in endosomes-lysosomes (Chow et al. 2007). Mutations of *Fig*4 and *Sac*14 in mice which alter PtdIns(3,5)P_2 levels result in progressive neurodegeneration starting in fetal life and induce lethality between birth and 6 weeks of age. Data suggest that neuronal cell death can occur because of extensive vacuolisation and this is then followed by defective astrocyte response and spongiform degeneration (Ferguson et al. 2009; Ferguson et al. 2010).

Phosphatidylinositol 3,4,5-trisphosphate

PtdIns(3,4,5)P_3 is the "paradigmatic" second messenger since it is barely detectable in normal, resting cells. A plethora of different stimuli, including (but not limited to) growth factors, hormones and several stress signals, can rapidly and transiently increase its levels and the tightly regulated PtdIns(3,4,5)P_3 synthesis can in turn regulate activation of several molecules. The list of cellular functions activated through PtdIns(3,4,5)P_3 synthesis is constantly growing. Interest on PtdIns(3,4,5)P_3 has been fuelled by the discovery that enzymes regulating its turnover are often mutated in several forms of cancer. This evidence has made the PtdIns(3,4,5)P_3-dependent signalling pathways amongst the most studied in signal transduction. Several excellent reviews have been published on PI3Ks (Engelman et al. 2006; Kok et al. 2009; Vanhaesebroeck et al. 2010) or the PI3K/Akt pathways (Vivanco and Sawyers 2002; Manning 2004; Manning and Cantley 2007), to mention only few of them.

Kinases

PI3K. PtdIns(3,4,5)P_3 derives from phosphorylation of PtdIns(4,5)P_2 by the action of class I PI3Ks (Fig. 1.11). Class I PI3Ks were classically divided in two subfamilies based on the receptors able to activate them, with class IA originally thought to be activated only downstream of receptor tyrosine kinases (RTKs) and class IB downstream of G-protein coupled receptors (GPCRs) (Vanhaesebroeck et al. 2001). More recently, activation of the class IA p110β downstream of GPCRs has been reported and, similarly, it has been shown that class IB p110γ can be activated downstream of RTKs through Ras activation (Vanhaesebroeck et al. 2010). Class IA PI3Ks are heterodimers comprising a regulatory and a catalytic subunit. Three main regulatory subunits exist (encoded by three distinct genes), named p85α (which also exists in two shorter isoforms, p55α and p50α), p85β and p55γ. Class IA catalytic subunits are p110α, p11β and p110δ. The regulatory subunits can bind to phosphorylated residues of activated RTKs or adaptor proteins through their SH2 domain and such a binding recruits the dimer to the plasma membrane. In addition, binding to the receptor/adaptor relieves the catalytic subunit from the inhibitory effect of p85. Class IB comprises only one isoform of catalytic subunit (p110γ) and the regulatory subunits p101 and p87. Activation of class IB involves their direct interaction with G$\beta\gamma$ subunits of G-proteins.

PIP5K. Despite possessing only one PI3K isoform (Vps34) able to catalyse only the synthesis of PtdIns3P, fission yeast *S. pombe* can accumulate PtdIns(3,4,5)P_3 and PtdIns(3,4)P_2 upon deletion of *ptn*1 (its orthologue of PTEN) (Mitra et al. 2004). This has suggested the possibility that a mechanism to generate PtdIns(3,4,5)P_3 has evolved even before the appearance of class I PI3K isoforms through sequential phosphorylation of PtdIns3P by PIP4Ks and PIP5Ks (Divecha and Halstead 2004). The observation that conversion of PtdIns(3,4)P_2 to PtdIns(3,4,5)P_3 by PIP5Ks can occur upon oxidative stress (Halstead et al. 2001) suggests that this alternative route may be conserved in mammalian cells.

Phosphatases

PTEN. In normal, resting cells PtdIns(3,4,5)P_3 is barely detectable, mostly because of the action of the enzyme (PTEN, Fig. 1.11). PTEN was discovered in 1997 and identified as a tumour suppressor gene located on chromosome 10q23, a genomic region that suffers loss of heterozygosity in many cancers (Maehama and Dixon 1999; Di Cristofano and Pandolfi 2000). By specifically dephosphorylating PtdIns(3,4,5)P_3 at its 3-position, PTEN switches off the PI3K-dependent signalling, leading to accumulation of PtdIns(4,5)P_2. The *Pten* knockout mouse is embryonic lethal (death between day 6.5 and 9.5 post coitum), revealing the critical role of this enzyme in embryogenesis. *Pten*$^{-/-}$ embryonic stem cells have an impaired ability to differentiate into embryoid bodies in vitro and in vivo (Di Cristofano and Pandolfi 2000). The heterozygous mice have an increased incidence of cancers (Suzuki et al. 1998; Podsypanina et al. 1999), consistent with the observation that loss of PTEN functions is a feature of many human cancers (see below). Several mouse models of loss of PTEN in different organs have been generated, revealing the key role of this enzyme in physiology and pathology (Knobbe et al. 2008). Studies have also suggested roles for the protein phosphatase activity of PTEN (Leslie et al. 2009).

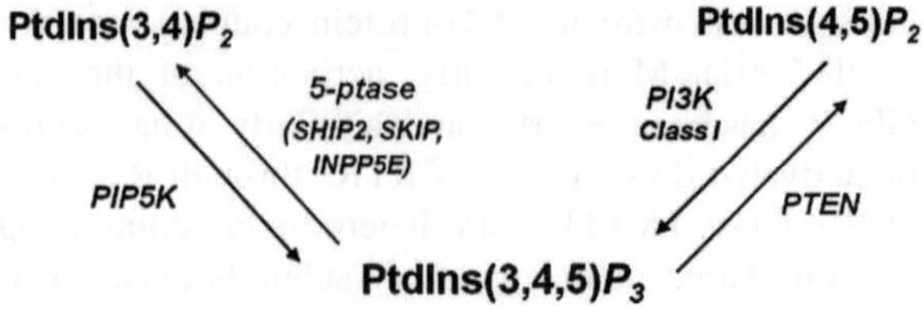

Fig. 1.11 Schematic representation of PtdIns(3,4,5)P_3 turnover

5-phosphatases. Dephosphorylation of PtdIns(3,4,5)P_3 by 5-phosphatases leads to accumulation of PtdIns(3,4)P_2 (Fig. 1.11), therefore, in contrast to PTEN, which is clearly able to switch off the PI3K-dependent pathways, it is not clear if the 5-phosphatases can actually act as a "switch-off" signal since PtdIns(3,4)P_2 is potentially able to activate cellular signals either overlapping or distinct from PtdIns(3,4,5)P_3.

Src homology 2-*domain-containing inositol phosphatase* 1 (SHIP1) is expressed in haematopoietic cells. SHIP2 is expressed in both haemopoietic and non-haemopoietic tissues (Ooms et al. 2009). SHIP2 was originally believed to be able to specifically dephosphorylate PtdIns(3,4,5)P_3 among the PIs but a more recent analysis of the enzyme kinetics in vitro using the recombinant SHIP2 catalytic domain has indicated an almost similar specificity for PtdIns(3,4,5)P_3 and PtdIns(3,5)P_2 among the PIs, with higher specificity for soluble inositols (Ooms et al. 2009). Data also indicate that SHIP2 can dephosphorylate PtdIns(4,5)P_2 at least in vitro. The role of SHIP2 as switch off or propagator of cellular signals may be cell specific. $SHIP1^{-/-}$ mice exhibits a shortened life span, overproduction of granulocytes and macrophages which infiltrate the spleen and the lungs (Helgason et al. 1998) and are osteoporotic (Takeshita et al. 2002). The first $SHIP2^{-/-}$ mice reported showed a fatal neonatal hypoglycaemia as a consequence of increased insulin sensitivity (Clement et al. 2001). However, the third exon of the *Phox2a* gene was also inadvertently deleted in this mouse model and in another SHIP2 knockout model, mice appeared viable and showed a normal glucose and insulin tolerance and no changes in peripheral glucose uptake (Sleeman et al. 2005). Interestingly, these mice appeared highly resistant to weight gain when fed a high-fat diet with increased metabolic rate and no increase in serum lipids, insulin or glucose levels (Astle et al. 2006).

Skeletal muscle and kidney enriched 5-*phosphatase* (*SKIP*) is a 5-phosphatase which is expressed in the heart, skeletal muscle and kidney (Astle et al. 2006). Although in vitro kinetic analysis indicates a preference for PtdIns(4,5)P_2 over PtdIns(3,4,5)P_3, this latter PI seems to be the preferred substrate in vivo (Ooms et al. 2009). Homozygous deletion of SKIP is embrionycally lethal and $Skip^{+/-}$ mice showed increased glucose tolerance and insulin sensitivity on a standard chow diet. On a high-fat diet $Skip^{+/-}$ mice showed increased insulin sensitivity and glucose tolerance and a reduced weight gain and hyperglycemia compared to wild-type mice, suggesting a potential role for SKIP in protection against diet-induced obesity (Ijuin et al. 2008).

72-kDa 5PTASE/TYPE IV/INPP5E is expressed in various tissues, including brain, testis, breast and haemopoietic cells. Original study indicated that it can dephosphorylate PtdIns(4,5)P_2 and PtdIns(3,4,5)P_3 (Kisseleva et al. 2002). Kinetic analysis have now revealed that this is the most potent PtdIns(3,4,5)P_3 5-phosphatase, with an affinity for this PI approximately 10-fold greater than any other 5-ptases able to dephosphorylate PtdIns(3,4,5)P_3. Overexpression of 5-ptase IV results in inhibition of PDGF-induced Akt phosphorylation, inhibition of cell growth and increased apoptosis (Kisseleva et al. 2002). Antisense-mediated reduction of 72-kDa 5-ptase/type IV/INPP5E in the hypothalamus increased PtdIns(3,4,5)P_3 levels and results in reduced food intake and weight loss and reduced serum insulin, leptin and glucose levels, suggesting that this enzyme may also have a role in regulation of glucose homoeostasis and energy metabolism (Ooms et al. 2009). Data have revealed a role for INPP5E in some diseases of the group of ciliopathies (Bielas et al. 2009).

Proline-rich inositol polyphosphate 5-phosphatase (PIPP) is localised to plasma membrane ruffles (Astle et al. 2006). Although the enzyme appears to use PtdIns(4,5)P_2 as its substrate, it can also dephosphorylate PtdIns(3,4,5)P_3 to regulate neurite elongation (Blero et al. 2007).

Intracellular Roles

The list of intracellular functions regulated by PtdIns(3,4,5)P_3 is too long to be fully discussed in this chapter; therefore, only few examples will be mentioned. Furthermore for simplicity, most of the functions described here are related to PtdIns(3,4,5)P_3-dependent Akt activation, the most studied of the PtdIns(3,4,5)P_3-dependent targets (Manning 2004; Manning and Cantley 2007).

The mechanism of PI3K-dependent Akt activation is a paradigm of PI-dependent activation of signalling cascades. Akt possesses a PH domain which is able to bind to PtdIns(3,4)P_2 or PtdIns(3,4,5)P_3. The binding of Akt PH domain to the newly generated PIs allows its translocation to the plasma membrane and induces conformational changes that lead to Akt activation. Activation of Akt is mediated by phosphorylation of its residue Thr308 by 3-phosphoinositide-dependent protein kinase 1 (PDK1) which itself possesses a PH domain able to bind PtdIns(3,4,5)P_3 (Komander et al. 2004). Mutations in PDK1 PH domain that impair PtdIns(3,4,5)P_3 binding strongly inhibits Akt activation in homozygous knock-in embryonic stem cell and knock-in mice indicating that binding of domain to PtdIns(3,4,5)P_3 is critical for Akt activation (McManus et al. 2004; Bayascas et al. 2008). Furthermore, binding of Akt PH domain to PtdIns(3,4,5)P_3 induces a conformational change which is critical for the PDK1-dependent phosphorylation (Calleja et al. 2007).

It must, however, be remembered that it is becoming increasingly evident that not all Pi3K-dependent functions occurs through activation of Akt and several more processes may be regulated through PtdIns(3,4,5)P_3-mediated activation of distinct targets, including small GTP-binding proteins, members of the protein kinase C family or of the PLC family (Falasca et al. 1998). Furthermore, recent evidence has also revealed the existence of PDK1-dependent, Akt-independent cellular signalling activated downstream of oncogenic p110α (Vasudevan et al. 2009) further indicating that the role of PtdIns(3,4,5)P_3 in physiology and pathology is not limited to Akt activation.

Brief summary of data from transgenic mice. Knockout and knock-in mice for the distinct class I PI3K isoforms have been generated and helped elucidating the critical roles of PI3Ks [and possibly PtdIns(3,4,5)P_3] in mammals (a summary of the main phenotypes and references to the relative original papers is presented in Vanhaesebroeck et al. 2010). Both p110α and p110β knockout mice die early during embryogenesis (E9.5 and E3.5 respectively) and knock-in strategies have highlighted a critical role for p110α in vascular development and in metabolism. Growth retardation and mild insulin resistance with age has been described in mice expressing a catalytic inactive p110β mutant. p110γ and p110δ knockout and knock-in mice are viable with immunological (and cardiac in the case of p110γ knockout mice) defects. Results from gene targeted mice for the class IA regulatory subunits have been less clear. While homozygous knockout mice lacking all p85α variants show perinatal lethality, immunological and metabolic defects, heterozygous mice are viable and show increased insulin sensitivity and glucose tolerance (Vanhaesebroeck et al. 2005). Mice deficient in all p85 isoforms in either muscle or the liver show altered insulin signalling in these tissues (Engelman et al. 2006).

Insulin signalling. Class I PI3Ks play a well-established role in regulation of metabolism and in modulating signalling pathways downstream of insulin receptor activation (Engelman et al. 2006). Several insulin-dependent processes are regulated by Akt activation, including glucose uptake in muscle and adipocytes (through regulation of translocation of the glucose transporter GLUT4 to the plasma membrane), glycogen synthesis [through phosphorylation of glycogen synthase kinase 3 (GSK3)] inhibition of gluconeogenesis in the liver [through inhibition of the forkhead (FOXO) family of transcription factors]. Akt is also involved in regulation of fatty acid synthesis. Other PtdIns(3,4,5)P_3 targets may also have a role in these processes. One such example is atypical protein kinase C, whose activation is also dependent on PtdIns(3,4,5)P_3 (directly and indirectly through PDK1) and which has also a role in mediating insulin functions (Farese and Sajan 2010).

Cell proliferation. PtdIns(3,4,5)P_3 can control cell cycle progression through different mechanisms and in particular it is involved in regulation of the G_1/S cell cycle progression (Liang and Slingerland 2003). Most of these functions occurs downstream of Akt. Some examples are listed here. Akt can directly phosphorylate (and inactivate) the cell cycle inhibitor p27Kip. On the other hand, the Akt-dependent phosphorylation of FOXO (and subsequent inactivation) can inhibit the FOXOs-mediated p27Kip transcription. Similarly, the Akt-dependent FOXOs inactivation inhibits transcription of retinoblastoma-like 2, another key cell cycle inhibitor, allowing G_1-S cell-cycle transition. The Akt-dependent inhibition of GSK3β stabilises cyclin D1 and c-Myc (Liang and Slingerland 2003). Furthermore, Akt can promote transcriptional activation of cyclin D and c-Myc genes and can regulate translation of cyclin D1 mRNA.

Cell survival. Akt has a pivotal role in promoting survival and activation of anti-apoptotic pathways (Franke et al. 1997; Marte and Downward 1997; Downward 1998). The role of Akt in survival is again both direct and indirect. For instance, Akt can directly phosphorylate the pro-apoptotic proteins BAD, inducing its cytosolic sequestering, and MDM2 which in turn leads to p53 degradation and inhibition of p53-dependent apoptosis. Furthermore, by inactivating FOXOs, Akt can indirectly block the transcription of pro-apoptotic proteins such as FasL and Bim. Akt is also involved in activation of XIAP that can inhibit apoptosis.

Cell growth. Akt is a key regulator of the mechanistic target of rapamycin (mTOR), a kinase that integrates signals from growth factors, nutrients and stresses to positively regulate anabolic processes (transcription, protein synthesis, ribosome biogenesis, nutrient transport and mitochondrial metabolism) and negatively regulate catabolic processes (mRNA degradation, ubiquitin-dependent proteolysis, autophagy and apoptosis) (Wullschleger et al. 2006; Sarbassov et al. 2005a; Ma and Blenis 2009). The interplay between Akt and mTOR is very complicated. Akt can activate mTOR in a multiprotein complex, mTORC1, whereas a second complex, mTORC2, contributes to Akt full activation (Sarbassov et al. 2005b; Huang and Manning 2009). In addition, hyperactivation of mTOR/S6K1 pathway can downregulate Akt activation in some cellular systems through a negative feedback loop whose existence has been highlighted by the use of mTOR inhibitors (Mavrommati and Maffucci 2011).

Cell migration. PI3Ks have been implicated in several events coordinating cell migration such as regulation of cell polarisation, formation of cell protrusions and cell adhesion (Cain and Ridley 2009). Although some events seems to be directly regulated by PtdIns(3,4,5)P_3 others involve cross-talks with other signalling molecules, including Rho GTPases and integrins. One of the downstream target of PtdIns(3,4,5)P_3 in the context of cell migration is PLCγ1, whose N-terminal PH domain can bind to this PI (Falasca et al. 1998). A role for a PI3K/PLCγ1 pathway has been reported as regulator of growth factor-induced motility of breast cancer (Piccolo et al. 2002) and endothelial cells (Maffucci et al. 2009).

Human Disease

Diabetes. Due to the central role of PI3Ks in regulation of insulin signalling, it is not surprising that these enzymes have been extensively investigated for their potential role in insulin resistance development, a feature of Type 2 diabetes. For instance, it is well-established that serine phosphorylation of the adaptor proteins insulin receptor substrates can occur in oxidative stress (often associated with obesity) and it results in attenuation of PI3K-dependent functions, including glucose transport (Engelman et al. 2006). Evidence suggested that, in the context of nutrients overload, such as in obesity, hyperactivation of mTOR and S6K1 can also contribute to the inhibition of PI3K-dependent functions (Um et al. 2006). Polymorphisms in the *PIK3R1* gene are associated with increased risk of Type 2 diabetes and elevated levels of p85 have been found in muscle of a mouse model of pregnancy-induced insulin resistance, in women with pregnancy-induced insulin resistance and in muscle of Type 2 diabetic individuals, with no changes in the p110 levels (Engelman et al. 2006). A cohort study of Caucasian residents from the UK and Belgium identified a 16 bp deletion located in the proximal region of the *SHIP*2 gene 3′ untranslated region significantly associated with individuals with Type 2 diabetes, compared with healthy individuals (Marion et al. 2002) and data in vitro suggested that this deletion results in increased expression of SHIP2 (Ooms et al. 2009). Finally, studies have suggested that some *SHIP*2 polymorphisms (found in control subjects) may protect individuals from Type 2 diabetes, whereas others (found in diabetic individuals) are strongly associated with metabolic phenotypes such as hypertension and obesity (Ooms et al. 2009).

Cancer. Deregulation of PI3K-dependent signalling pathways is linked to the development of cancer (Maehama and Dixon 1999; Shayesteh et al. 1999; Vivanco and Sawyers 2002; Bader et al. 2005; Shaw and Cantley 2006; Vogt et al. 2007) and to increased resistance to treatment with chemotherapeutics (Clark et al. 2002; Liang et al. 2003; She et al. 2003).

Deregulation of PI3K. Amplification of PIK3CA, the gene encoding p110α, has been detected in several cancers, including head and neck, cervical, gastric and lung cancers (Samuels et al. 2004; Zhao and Vogt 2008). Point mutations have been detected throughout the p110α coding region, mostly clustered in two hot spots (E545 in the helical PIK domain and H1047 near the end of the catalytic domain). These mutations, responsible for hyperactivation of the enzyme, have been found in brain, colon, breast and hepatocellular cancers (Vanhaesebroeck et al. 2010). Small deletions and insertions in *PIK3CA* have also been described. Different mutations can increase p110α activity through different mechanisms. Mutations of *PIK3R1* (which encodes p85α, p55α and p50α) have also been found and some of them may be involved in abrogating the inhibitory action of p85α on the catalytic subunits (Vanhaesebroeck et al. 2010). Increased expression of p110β and p110δ has been detected in some colon and bladder tumours and in glioblastoma (Benistant et al. 2000; Knobbe and Reifenberger 2003). Similarly, we have reported a specific increase in p110γ levels in pancreatic cancer (Edling et al. 2010) and this isoform has been involved in tumour angiogenesis (Hamada et al. 2005) and drug resistance of chronic myeloid leukemia cells (Hickey and Cotter 2006).

Deregulation of phosphatases. PTEN. Somatic mutations, gene deletion and gene inactivation of *PTEN* have been detected in several types of cancer, including glioblastomas, prostate cancers, breast cancers and melanoma (Engelman et al. 2006; Carracedo and Pandolfi 2008). Epigenetic mechanisms can also be responsible for down-regulation of PTEN protein expression. Loss of heterozygosity at 10q23 is a very common event in most primary tumours (25–50 %) but the complete loss of PTEN function in early stage tumours has been reported only in endometrial and ovarian cancer (Di Cristofano and Pandolfi 2000) whereas the complete inactivation of PTEN is usually a feature of late-stage, more aggressive and usually metastatic tumours. Germline mutations in the PTEN gene result in Cowden's, Bannayan-Riley-Ruvalcaba syndromes and Lhermitte-Duclos disease, which are characterised by multiple benign tumours called hamartomas, high risk of cancer and increased incidence of macrocephaly (McCrea and De Camilli 2009; Blero et al. 2007).

Other 5-phosphatases. The involvement of other 5-phosphatases in cancer has been much less investigated. Nevertheless, evidence of their contribution is now emerging and these data are elegantly described in a recent review (Bunney and Katan 2010). For instance, mutations in *INPP5D* (the gene encoding SHIP1) have been detected in acute myeloid leukemia and in acute lymphoblastic leukemia. Furthermore, it has been shown that BCR/ABL, the oncogene responsible for chronic myelogenous leukemia (CML), downregulates the expression of SHIP1 and reduced levels of SHIP1 have been found in primary cells from patients affected by CML. SHIP2 might have a role in promoting breast cancer and increased expression of this enzyme has been detected in a panel of breast cancer cells. A more complex scenario appears from INP55E, with data suggesting an up-regulation in some cancers (such as cervical cancer) and down-regulation in others (such as stomach cancer). Interestingly, data suggest an increase in INP55E expression following treatment, for instance in the case of non-Hodgkin's lymphoma or in pancreatic cancer cell lines after treatment with gemcitabine. Data on the potential role of PIPP are also emerging.

Hyperactivation of PI3K or loss of PTEN results in accumulation of PtdIns(3,4,5)P_3 and hyperactivation of PtdIns(3,4,5)P_3-dependent targets, including Akt. The PI3K/Akt pathway is currently considered an attractive target for therapeutic intervention and several compounds targeting the different components of the pathway have been developed or are in development (Vivanco and Sawyers 2002; Luo et al. 2003; Hennessy et al. 2005; Guertin and Sabatini 2007; Liu et al. 2009; Falasca 2010), with some of them currently in clinical trials for cancer treatment (Liu et al. 2009). Although a description of the different strategies used to inhibit the PI3K/Akt pathway and of the several inhibitors developed is beyond the purpose of this chapter, it is worth mentioning that pro-apoptotic activity in vitro and anti-cancer activity in vivo has been detected for compounds that likely act by interfering with the binding of Akt PH domain to PtdIns(3,4,5)P_3, including inositol 1,3,4,5,6-pentakisphosphate (Razzini et al. 2000; Piccolo et al. 2004; Maffucci et al. 2005b), ether lipid analogues and PH domain-targeting inhibitors (Kozikowski et al. 2003; Gills et al. 2006; Crowell et al. 2007) such as perifosine, the most developed Akt inhibitor currently available (Kondapaka et al. 2003). Blocking the Akt PH/PtdIns(3,4,5)P_3 interaction is an interesting and useful strategy since data have demonstrated that Akt PH domain is critical for Akt-driven tumourigenesis. The most advanced and promising drugs targeting the PI3K/Akt pathway are derivatives of the mTORC1 inhibitor rapamycin which have shown considerable anti-tumour activity against specific tumours in clinical trials (Faivre et al. 2006). Indeed, some of them have been approved for use against specific cancer types (Mavrommati and Maffucci 2011). However, overall, rapamycin clinical trials resulted less efficient than predicted, partly because of the existence of negative feedback loops leading to up-regulation of Akt (Manning 2004; O'Reilly et al. 2006; Zhang et al. 2007) and/or ERK/MAPK pathway (Carracedo et al. 2008; Mavrommati and Maffucci, 2011).

Conclusion

PIs are very versatile and are responsible for spatial and temporal activation of key signalling molecules, ultimately regulating several cellular functions. On the other hand, it is well-established that deregulated turnover of PIs, due to mutations/ alteration in the kinases/phosphatases controlling their synthesis/dephosphorylation, is associated with several human diseases. All these data clearly support a dominant role for PIs in both physiological and pathological conditions.

Acknowledgments I would like to thank Diabetes UK, Diabetes Research & Wellness Foundation and Barts and The London Charity for their support. I thank Prof Marco Falasca for introducing me to the "PIs' world", for our long, useful discussions on PIs and for critical reading of this manuscript. I also thank Dr Andrew Riley for the structure of the different PIs.

References

Arcaro A, Volinia S, Zvelebil MJ et al (1998) Human phosphoinositide 3-kinase C2beta, the role of calcium and the C2 domain in enzyme activity. J Biol Chem 273:33082–33090

Arcaro A, Khanzada UK, Vanhaesebroeck B et al (2002) Two distinct phosphoinositide 3-kinases mediate polypeptide growth factor-stimulated PKB activation. EMBO J 21:5097–5108

Astle MV, Seaton G, Davies EM et al (2006) Regulation of phosphoinositide signaling by the inositol polyphosphate 5-phosphatases. IUBMB Life 58:451–456

Axe EL, Walker SA, Manifava M et al (2008) Autophagosome formation from membrane compartments enriched in phosphatidylinositol 3-phosphate and dynamically connected to the endoplasmic reticulum. J Cell Biol 182:685–701

Backer JM (2008) The regulation and function of class III PI3Ks: novel roles for Vps34. Biochem J 410:1–17

Bader AG, Kang S, Zhao L et al (2005) Oncogenic PI3K deregulates transcription and translation. Nat Rev Cancer 5:921–929

Balla A, Balla T (2006) Phosphatidylinositol 4-kinases: old enzymes with emerging functions. Trends Cell Biol 17:351–361

Bayascas JR, Wullschleger S, Sakamoto K et al (2008) Mutation of the PDK1 PH domain inhibits protein kinase B/Akt, leading to small size and insulin resistance. Mol Cell Biol 28:3258–3272

Benistant C, Chapuis H, Roche S (2000) A specific function for phosphatidylinositol 3-kinase α (p85α-p110α) in cell survival and for phosphatidylinositol 3-kinase β (p85α-p110β) in de novo DNA synthesis of human colon carcinoma cells. Oncogene 19:5083–5090

Berger P, Schaffitzel C, Berger I et al (2003) Membrane association of myotubularin-related protein 2 is mediated by a pleckstrin homology-GRAM domain and a coiled-coil dimerization module. Proc Natl Acad Sci U S A 100:12177–12182

Bielas SL, Silhavy JL, Brancati F et al (2009) Mutations in INPP5E, encoding inositol polyphosphate-5-phosphatase E, link phosphatidyl inositol signaling to the ciliopathies. Nat Genet 41:1032–1036

Blagoveshchenskaya A, Cheong FY, Rohde HM et al (2008) Integration of Golgi trafficking and growth factor signaling by the lipid phosphatase SAC1. J Cell Biol 180:803–812

Blero D, Payrastre B, Schurmans S et al (2007) Phosphoinositide phosphatases in a network of signalling reactions. Pflugers Arch 455:31–44

Boronenkov IV, Anderson RA (1995) The sequence of phosphatidylinositol-4-phosphate 5-kinase defines a novel family of lipid kinases. J Biol Chem 270:2881–2884

Bultsma Y, Keune WJ, Divecha N (2010) PIP4Kbeta interacts with and modulates nuclear localization of the high-activity PtdIns5P-4-kinase isoform PIP4Kalpha. Biochem J 430:223–235
Bunce MW, Boronenkov IV, Anderson RA (2008) Coordinated activation of nuclear ubiquitin ligase Cul3-SPOP by the generation of phosphatidylinositol 5-phosphate. J Biol Chem 283:8678–8686
Bunney TD, Katan M (2010) Phosphoinositide signalling in cancer: beyond PI3K and PTEN. Nat Rev Cancer 10:342–352
Cain RJ, Ridley AJ (2009) Phosphoinositide 3-kinases in cell migration. Biol Cell 101:13–29
Calleja V, Alcor D, Laguerre M et al (2007) Intramolecular and intermolecular interactions of protein kinase B define its activation in vivo. PLoS Biol 5:e95
Cao C, Laporte J, Backer JM et al (2007) Myotubularin lipid phosphatase binds the hVps15/hVps34 lipid kinase complex on endosomes. Traffic 8:1052–1067
Cao C, Backer JM, Laporte J et al (2008) Sequential actions of myotubularin lipid phosphatases regulate endosomal PI(3)P and growth factor receptor trafficking. Mol Biol Cell 19:3334–3346
Carracedo A, Pandolfi PP (2008) The PTEN-PI3K pathway: of feedbacks and cross-talks. Oncogene 27:5527–5541
Carracedo A, Ma L, Teruya-Feldstein J et al (2008) Inhibition of mTORC1 leads to MAPK pathway activation through a PI3K-dependent feedback loop in human cancer. J Clin Invest 118:3065–3074
Chaussade C, Pirola L, Bonnafous S et al (2003) Expression of myotubularin by an adenoviral vector demonstrates its function as a phosphatidylinositol 3-phosphate [PtdIns(3)P] phosphatase in muscle cell lines: involvement of PtdIns(3)P in insulin-stimulated glucose transport. Mol Endocrinol 17:2448–2460
Chishti AH, Kim AC, Marfatia SM et al (1998) The FERM domain: a unique module involved in the linkage of cytoplasmic proteins to the membrane. Trends Biochem Sci 23:281–282
Choudhury P, Srivastava S, Li Z et al (2006) Specificity of the myotubularin family of phosphatidylinositol-3-phosphatase is determined by the PH/GRAM domain. J Biol Chem 281:31762–31769
Chow CY, Zhang Y, Dowling JJ et al (2007) Mutation of FIG 4 causes neurodegeneration in the pale tremor mouse and patients with CMT4J. Nature 448:68–72
Chow CY, Landers JE, Bergren SK et al (2009) Deleterious variants of FIG 4, a phosphoinositide phosphatase, in patients with ALS. Am J Hum Genet 84:85–88
Clague MJ, Lorenzo O (2005) The myotubularin family of lipid phosphatases. Traffic 6:1063–1069
Clark AS, West K, Streicher S et al (2002) Constitutive and inducible Akt activity promotes resistance to chemotherapy, trastuzumab, or tamoxifen in breast cancer cells. Mol Cancer Ther 1:707–717
Clarke JH, Letcher AJ, D'santos CS et al (2001) Inositol lipids are regulated during cell cycle progression in the nuclei of murine erythroleukaemia cells. Biochem J 357:905–910
Clarke JH, Wang M, Irvine RF (2009) Phosphatidylinositol 5-phosphate 4-kinases localization, regulation and function of Type II phosphatidylinositol 5-phosphate 4-kinases. Adv Enzyme Regul 2009 Nov 6
Clement S, Krause U, Desmedt F et al (2001) The lipid phosphatase SHIP2 controls insulin sensitivity. Nature 409:92–97
Coronas S, Lagarrigue F, Ramel D et al (2008) Elevated levels of PtdIns5P in NPM-ALK transformed cells: implication of PIKfyve. Biochem Biophys Res Commun 372:351–355
Cremona O, Di Paolo G, Wenk MR et al (1999) Essential role of phosphoinositide metabolism in synaptic vesicle recycling. Cell 99:179–188
Crowell JA, Steele VE, Fay JR (2007) Targeting the AKT protein kinase for cancer chemoprevention. Mol Cancer Ther 6:2139–2148
D'Angelo G, Polishchuk E, Di Tullio G et al (2007) Glycosphingolipid synthesis requires FAPP2 transfer of glucosylceramide. Nature 449:62–67
D'Angelo G, Vicinanza M, Di Campli A et al (2008) The multiple roles of PtdIns(4)*P*—not just the precursor of PtdIns(4,5)P_2. J Cell Sci 121:1955–1963

Di Cristofano A, Pandolfi PP (2000) The multiple roles of PTEN in tumor suppression. Cell 100:387–390

Didichenko SA, Thelen M (2001) Phosphatidylinositol 3-kinase c2alpha contains a nuclear localization sequence and associates with nuclear speckles. J Biol Chem 276:48135–48142

Di Paolo G, De Camilli P (2006) Phosphoinositides in cell regulation and membrane dynamics. Nature 443:651–657

Divecha N, Rhee SG, Letcher AJ et al (1993) Phosphoinositide signalling enzymes in rat liver nuclei: phosphoinositidase C isoform beta 1 is specifically, but not predominantly, located in the nucleus. Biochem J 289:617–620

Divecha N, Truong O, Hsuan JJ et al (1995) The cloning and sequence of the C isoform of PtdIns4P 5-kinase. Biochem J 309:715–719

Divecha N, Halstead JR (2004) Of yeast and men. The evolution of PtdIns(3,4,5)P(3) synthesis. EMBO Rep 5:865–866

Doerks T, Strauss M, Brendel M et al (2000) GRAM, a novel domain in glucosyltransferases, myotubularins and other putative membrane-associated proteins. Trends Biochem Sci 25:483–485

Domin J, Harper L, Aubyn D et al (2005) The class II phosphoinositide 3-kinase PI3K-C2beta regulates cell migration by a PtdIns3P dependent mechanism. J Cell Physiol 205:452–462

Dominguez V, Raimondi C, Somanath S et al (2011) Class II phosphoinositide 3-kinase regulates exocytosis of insulin granules in pancreatic beta cells. J Biol Chem 286:4216–4225

Downward J (1998) Mechanisms and consequences of activation of protein kinase B/Akt. Curr Opin Cell Biol 10:262–267

Dove SK, Dong K, Kobayashi T et al (2009) Phosphatidylinositol 3,5-bisphosphate and Fab1p/PIKfyve underPPIn endo-lysosome function. Biochem J 419:1–13

Dowler S, Currie RA, Campbell DG et al (2000) Identification of pleckstrin-homology-domain-containing proteins with novel phosphoinositide-binding specificities. Biochem J 351:19–31

Ebato C, Uchida T, Arakawa M et al (2008) Autophagy is important in islet homeostasis and compensatory increase of beta cell mass in response to high-fat diet. Cell Metab 8:325–332

Edling CE, Selvaggi F, Buus R et al (2010) Key role of phosphoinositide 3-kinase class IB in pancreatic cancer. Clin Cancer Res 16:4928–4937

Elis W, Triantafellow E, Wolters NM et al (2008) Down-regulation of class II phosphoinositide 3-kinase alpha expression below a critical threshold induces apoptotic cell death. Mol Cancer Res 6:614–623

Engelman JA, Luo J, Cantley LC (2006) The evolution of phosphatidylinositol 3-kinases as regulators of growth and metabolism. Nat Rev Gen 7:606–619

Faivre S, Kroemer G, Raymond E (2006) Current development of mTOR inhibitors as anticancer agents. Nat Rev Drug Discov 5:671–688

Falasca M (2010) PI3K/Akt signalling pathway specific inhibitors: a novel strategy to sensitize cancer cells to anti-cancer drugs. Curr Pharm Des 16:1410–1416

Falasca M, Logan SK, Lehto VP et al (1998) Activation of phospholipase C gamma by PI 3-kinase-induced PH domain-mediated membrane targeting. EMBO J 17:414–422

Falasca M, Maffucci T (2006) Emerging roles of phosphatidylinositol-3-monophosphate as a dynamic lipid second messenger. Arch Physiol Biochem 112:274–284

Falasca M, Maffucci T (2007) Role of class II phosphoinositide 3-kinase in cell signalling. Biochem Soc Trans 35:211–214

Falasca M, Maffucci T (2012) Regulation and cellular functions of class II phosphoinositide 3-kinases. Biochem J 443:587–601

Falasca M, Hughes WE, Dominguez V et al (2007) The role of phosphoinositide 3-kinase C2alpha in insulin signalling. J Biol Chem 282:28226–28236

Falasca M, Maffucci T (2009) Rethinking phosphatidylinositol 3-monophosphate. Biochim Biophys Acta 1793:1795–1803

Farese RV, Sajan MP (2010) Metabolic functions of atypical protein kinase C: "good" and "bad" as defined by nutritional status. Am J Physiol Endocrinol Metab 298:E385–E394

Ferguson CJ, Lenk GM, Meisler MH (2009) Defective autophagy in neurons and astrocytes from mice deficient in PI(3,5)P2. Hum Mol Genet 18:4868–4878

Ferguson CJ, Lenk GM, Meisler MH (2010) PtdIns(3,5)P2 and autophagy in mouse models of neurodegeneration. Autophagy 6:170–171
Franke TF, Kaplan DR, Cantley LC (1997) PI3 K: downstream AKTion blocks apoptosis. Cell 88:435–437
Gewinner C, Wang ZC, Richardson A et al (2009) Evidence that inositol polyphosphate 4-phosphatase type II is a tumor suppressor that inhibits PI3K signaling. Cancer Cell 16:115–125
Gillooly DJ, Morrow IC, Lindsay M et al (2000) Localization of phosphatidylinositol 3-phosphate in yeast and mammalian cells. EMBO J 19:4577–4588
Gills JJ, Holbeck S, Hollingshead M et al (2006) Spectrum of activity and molecular correlates of response to phosphatidylinositol ether lipid analogues, novel lipid-based inhibitors of Akt. Mol Cancer Ther 5:713–722
Gozani O, Karuman P, Jones DR et al (2003) The PHD finger of the chromatin-associated protein ING2 functions as a nuclear phosphoinositide receptor. Cell 114:99–111
Guertin DA, Sabatini DM (2007) Defining the role of mTOR in cancer. Cancer Cell 12:9–22
Halstead JR, Roefs M, Ellson CD et al (2001) A novel pathway of cellular phosphatidylinositol(3,4,5)-trisphosphate synthesis is regulated by oxidative stress. Curr Biol 11:386–395
Halstead JR, Jalink K, Divecha N (2005) An emerging role for PtdIns(4,5)P2-mediated signalling in human disease. Trends Pharmacol Sci 26:654–660
Halstead JR, van Rheenen J, Snel MHJ et al (2006) A role for PtdIns(4,5)P2 and PIP5Kα in regulating stress-induced apoptosis. Curr Biol 16:1850–1856
Hamada K, Sasaki T, Koni PA et al (2005) The PTEN/PI3K pathway governs normal vascular development and tumor angiogenesis. Genes Dev 19:2054–2065
Hammond GR, Schiavo G (2007) Polyphosphoinositol lipids: under-PPInning synaptic function in health and disease. Develop Neurobiol 67:1232–1247
Helgason CD, Damen JE, Rosten P et al (1998) Targeted disruption of SHIP leads to hemopoietic perturbations, lung pathology, and a shortened life span. Genes Dev 12:1610–1620
Hennessy BT, Smith DL, Ram PT et al (2005) Exploiting the PI3K/AKT pathway for cancer drug discovery. Nat Rev Drug Discov 4:988–1004
Hickey FB, Cotter TG (2006) BCR-ABL regulates phosphatidylinositol 3-kinase-p110gamma transcription and activation and is required for proliferation and drug resistance. J Biol Chem 281:2441–2450
Hodgson MC, Shao LJ, Frolov A et al (2011) Decreased expression and androgen regulation of the tumor suppressor gene INPP4B in prostate cancer. Cancer Res 71:572–582
Huang J, Manning BD (2009) A complex interplay between Akt, TSC2 and the two mTOR complexes. Biochem Soc Trans 37:217–222
Ijuin T, Yu YE, Mizutani K et al (2008) Increased insulin action in SKIP heterozygous knockout mice. Mol Cell Biol 28:5184–5195
Ikonomov OC, Sbrissa D, Mlak K et al (2002) Requirement for PIKfyve enzymatic activity in acute and long-term insulin cellular effects. Endocrinology 143:4742–4754
Ikonomov OC, Sbrissa D, Dondapati R et al (2007) ArPIKfyve-PIKfyve interaction and role in insulin-regulated GLUT4 translocation and glucose transport in 3T3-L1 adipocytes. Exp Cell Res 313:2404–2416
Ikonomov OC, Sbrissa D, Ijuin T et al (2009a) Sac3 is an insulin-regulated phosphatidylinositol 3,5-bisphosphate phosphatase. J Biol Chem 284:23961–23971
Ikonomov OC, Sbrissa D, Shisheva A (2009b) YM201636, an inhibitor of retroviral budding and PIKfyve-catalyzed PtdIns(3,5)P2 synthesis, halts glucose entry by insulin in adipocytes. Biochem Biophys Res Commun 382:566–570
Ishiki M, Randhawa VK, Poon V et al (2005) Insulin regulates the membrane arrival, fusion, and C-terminal unmasking of glucose transporter-4 via distinct phosphoinositides. J Biol Chem 280:28792–28802
Ivetac I, Gurung R, Hakim S et al (2009) Regulation of PI(3)K/Akt signalling and cellular transformation by inositol polyphosphate 4-phosphatase-1. EMBO Rep 10:487–493
Jones DR, Bultsma Y, Keune WJ et al (2006) Nuclear PtdIns5P as a transducer of stress signaling: an in vivo role for PIP4Kbeta. Mol Cell 23:685–695

Jung HS, Chung KW, Won Kim J et al (2008) Loss of autophagy diminishes pancreatic beta cell mass and function with resultant hyperglycemia. Cell Metab 8:318–324

Kanda H, Tamori Y, Shinoda H et al (2005) Adipocytes from Munc18c-null mice show increased sensitivity to insulin-stimulated GLUT4 externalization. J Clin Invest 115:291–301

Katso RM, Pardo OE, Palamidessi A et al (2006) Phosphoinositide 3-Kinase C2beta regulates cytoskeletal organization and cell migration via Rac-dependent mechanisms. Mol Biol Cell 17:3729–3744

Kisseleva MV, Cao L, Majerus PW (2002) Phosphoinositide-specific inositol polyphosphate 5-phosphatase IV inhibits Akt/protein kinase B phosphorylation and leads to apoptotic cell death. J Biol Chem 227:6266–6272

Knobbe CB, Reifenberger G (2003) Genetic alterations and aberrant expression of genes related to the phosphatidyl-inositol-3'-kinase/protein kinase B (Akt) signal transduction pathway in glioblastomas. Brain Pathol 13:507–518

Knobbe CB, Lapin V, Suzuki A et al (2008) The roles of PTEN in development, physiology and tumorigenesis in mouse models: a tissue-by-tissue survey. Oncogene 27:5398–5415

Kok K, Geering B, Vanhaesebroeck B (2009) Regulation of phosphoinositide 3-kinase expression in health and disease. Trends Biochem Sci 34:115–127

Komander D, Fairservice A, Deak M et al (2004) Structural insights into the regulation of PDK1 by phosphoinositides and inositol phosphates. EMBO J 23:918–928

Kondapaka SB, Singh SS, Dasmahapatra GP et al (2003) Perifosine, a novel alkylphospholipid, inhibits protein kinase B activation. Mol Cancer Ther 2:1093–1103

Kong AM, Horan KA, Sriratana A et al (2006) Phosphatidylinositol 3-phosphate [PtdIns3P] is generated at the plasma membrane by an inositol polyphosphate 5-phosphatase: endogenous PtdIns3P can promote GLUT4 translocation to the plasma membrane. Mol Cell Biol 26: 6065–6081

Kozikowski AP, Sun H, Brognard J et al (2003) Novel PI analogues selectively block activation of the pro-survival serine/threonine kinase Akt. J Am Chem Soc 125:1144–1145

Kutateladze TG (2006) Phosphatidylinositol 3-phosphate recognition and membrane docking by the FYVE domain. Bioch Biophys Acta 1761:868–877

Kutateladze TG (2007) Mechanistic similarities in docking of the FYVE and PX domains to phosphatidylinositol 3-phosphate containing membranes. Prog Lip Res 46:315–327

Lecompte O, Poch O, Laporte J (2008) PtdIns5P regulation through evolution: roles in membrane trafficking? Trends Biochem Sci 33:453–460

Leibiger B, Moede T, Uhles S et al (2010) Insulin-feedback via PI3K-C2α activated PKBα/Akt1 is required for glucose-stimulated insulin secretion. FASEB J 24:1824–1837

Lemmon MA (2008) Membrane recognition by phospholipid-binding domains. Nat Rev Mol Cell Biol 9:99–111

Leslie NR, Maccario H, Spinelli L et al (2009) The significance of PTEN's protein phosphatase activity. Adv Enzyme Regul 49:190–196

Levine B, Kroemer G (2008) Autophagy in the pathogenesis of disease. Cell 132:27–42

Li J, Lu Y, Zhang J et al (2010) PI4KIIα is a novel regulator of tumor growth by its action on angiogenesis and HIF-1α regulation. Oncogene 29:2550–2559

Liang J, Slingerland JM (2003) Multiple roles of the PI3K/PKB (Akt) pathway in cell cycle progression. Cell Cycle 2:339–345

Liang K, Jin W, Knuefermann C et al (2003) Targeting the phosphatidylinositol 3-kinase/Akt pathway for enhancing breast cancer cells to radiotherapy. Mol Cancer Ther 2:353–360

Lindmo K, Stenmark H (2006) Regulation of membrane traffic by phosphoinositide 3-kinases. J Cell Sci 119:605–614

Liu P, Cheng H, Roberts TM et al (2009) Targeting the phosphoinositide 3-kinase pathway in cancer. Nat Rev Drug Discov 8:627–644

Lodhi IJ, Bridges D, Chiang SH et al (2008) Insulin stimulates phosphatidylinositol 3-phosphate production via the activation of Rab5. Mol Biol Cell 19:2718–2728

Logothetis DE, Petrou VI, Adney SK et al (2010) Channelopathies linked to plasma membrane phosphoinositides. Pflugers Arch 2010 Apr 16. [Epub ahead of print]

Lowe M (2005) Structure and function of the Lowe syndrome protein OCRL1. Traffic 6:711–719

Luo J, Manning BD, Cantley LC (2003) Targeting the PI3 K-Akt pathway in human cancer: rationale and promise. Cancer Cell 4:257–262

Ma XM, Blenis J (2009) Molecular mechanisms of mTOR-mediated translational control. Nat Rev Mol Cell Biol 10:307–318

Maehama T, Dixon JE (1999) PTEN: a tumour suppressor that functions as a phospholipid phosphatase. Trends Cell Biol 9:125–128

Maffucci T, Falasca M (2001) Specificity in pleckstrin homology (PH) domain membrane targeting: a role for a phosphoinositide-protein co-operative mechanism. FEBS Lett 506:173–179

Maffucci T, Brancaccio A, Piccolo E et al (2003a) Insulin induces phosphatidylinositol-3-phosphate formation through TC10 activation. EMBO J 22:4178–4189

Maffucci T, Razzini G, Ingrosso A et al (2003b) Role of pleckstrin homology domain in regulating membrane targeting and metabolic function of insulin receptor substrate 3. Mol Endocrinol 17:1568–1579

Maffucci T, Cooke FT, Foster FM et al (2005a) Class II phosphoinositide 3-kinase defines a novel signaling pathway in cell migration. J Cell Biol 169:789–799

Maffucci T, Piccolo E, Cumashi A et al (2005b) Inhibition of the phosphatidylinositol 3-kinase/Akt pathway by inositol pentakisphosphate results in antiangiogenic and antitumor effects. Cancer Res 65:8339–8349

Maffucci T, Raimondi C, Abu-Hayyeh S et al (2009) A phosphoinositide 3-kinase/phospholipase Cgamma1 pathway regulates fibroblast growth factor-induced capillary tube formation. PLoS ONE 4:e8285

Manning BD (2004) Balancing Akt with S6K: implications for both metabolic diseases and tumorigenesis. J Cell Biol 167:399–403

Manning BD, Cantley LC (2007) AKT/PKB signaling: navigating downstream. Cell 129: 1261–1274

Marchetti P, Masini M (2009) Autophagy and the pancreatic beta-cell in human type 2 diabetes. Autophagy 5:1055–1056

Marion E, Kaisaki PJ, Pouillon V et al (2002) The gene INPPL1, encoding the lipid phosphatase SHIP2, is a candidate for type 2 diabetes in rat and man. Diabetes 51:2012–2017

Marte BM, Downward J (1997) PKB/Akt: connecting phosphoinositide 3-kinase to cell survival and beyond. Trends Biochem Sci 22:355–358

Masini M, Bugliani M, Lupi R et al (2009) Autophagy in human type 2 diabetes pancreatic beta cells. Diabetologia 52:1083–1086

Mavrommati I, Maffucci T (2011) mTOR inhibitors: facing new challenges ahead. Curr Med Chem 18:2743–2762

McCrea HJ, De Camilli P (2009) Mutations in phosphoinositide metabolizing enzymes and human disease. Physiology (Bethesda) 24:8–16

McManus EJ, Collins BJ, Ashby PR, Prescott AR, Murray-Tait V, Armit LJ, Arthur JS, Alessi DR (2004) The in vivo role of PtdIns(3,4,5)P3 binding to PDK1 PH domain defined by knockin mutation. EMBO J 23:2071–2082

Merlot S, Meili R, Pagliarini DJ et al (2003) A PTEN-related 5-phosphatidylinositol phosphatase localized in the Golgi. J Biol Chem 278:39866–39873

Mejillano M, Yamamoto M, Rozelle AL et al (2001) Regulation of apoptosis by phosphatidylinositol 4,5-bisphosphate inhibition of caspases and caspase inactivation of phosphatidylinositol phosphate 5-kinase. J Biol Chem 276:1865–1872

Meunier FA, Osborne SL, Hammond GR et al (2005) Phosphatidylinositol 3-kinase C2alpha is essential for ATP-dependent priming of neurosecretory granule exocytosis. Mol Biol Cell 16:4841–4851

Michell RH (2008) Inositol derivatives: evolution and functions. Nat Rev Mol Cell Biol 9: 151–161

Mima J, Wickner W (2009a) Phosphoinositides and SNARE chaperones synergistically assemble and remodel SNARE complexes for membrane fusion. Proc Natl Acad Sci U S A 106: 16191–16196

Mima J, Wickner W (2009b) Complex lipid requirements for SNARE- and SNARE chaperone-dependent membrane fusion. J Biol Chem 284:27114–27122

Mitra P, Zhang Y, Rameh LE et al (2004) A novel phosphatidylinositol(3,4,5)P3 pathway in fission yeast. J Cell Biol 166:205–211
Morris JB, Hinchliffe KA, Ciruela A et al (2000) Thrombin stimulation of platelets causes an increase in phosphatidylinositol 5-phosphate revealed by mass assay. FEBS Lett 475:57–60
Narkis G, Ofir R, Landau D et al (2007) Lethal contractural syndrome type 3 (LCCS3) is caused by a mutation in PIP5K1C, which encodes PIPKI gamma of the phophatidylinsitol pathway. Am J Hum Genet 81:530–539
Nicot AS, Laporte J (2008) Endosomal phosphoinositides and human diseases. Traffic 9: 1240–1249
Niebuhr K, Giuriato S, Pedron T et al (2002) Conversion of PtdIns(4,5)P(2) into PtdIns(5)P by the *S. flexneri* effector IpgD reorganizes host cell morphology. EMBO J 21:5069–5078
Nystuen A, Legare ME, Schultz LD et al (2001) A null mutation in inositol polyphosphate 4-phosphatase type I causes selective neuronal loss in weeble mutant mice. Neuron 32:203–212
Ooms LM, Horan KA, Rahman P et al (2009) The role of the inositol polyphosphate 5-phosphatases in cellular function and human disease. Biochem J 419:29–49
O'Reilly KE, Rojo F, She QB et al (2006) mTOR inhibition induces upstream receptor tyrosine kinase signaling and activates Akt. Cancer Res 66:1500–1508
Pagliarini DJ, Worby CA, Dixon JE (2004) A PTEN-like phosphatase with a novel substrate specificity. J Biol Chem 279:38590–38596
Pendaries C, Tronchère H, Arbibe L et al (2006) PtdIns5P activates the host cell PI3-kinase/Akt pathway during Shigella flexneri infection. EMBO J 25:1024–1034
Piccolo E, Innominato PF, Mariggio MA et al (2002) The mechanism involved in the regulation of phospholipase Cgamma1 activity in cell migration. Oncogene 21:6520–6529
Piccolo E, Vignati S, Maffucci T et al (2004) Inositol pentakisphosphate promotes apoptosis through the PI 3-K/Akt pathway. Oncogene 23:1754–1765
Piñeiro R, Maffucci T, Falasca M (2011) The putative cannabinoid receptor GPR55 defines a novel autocrine loop in cancer cell proliferation. Oncogene 30:142–152
Podsypanina K, Ellenson LH, Nemes A et al (1999) Mutation of Pten/Mmaca in mice causes neoplasia in multiple organ systems. Proc Natl Acad Sci U S A 96:1563–1568
Proikas-Cezanne T, Ruckerbauer S, Stierhof YD et al (2007) Human WIPI-1 puncta-formation: a novel assay to assess mammalian autophagy. FEBS Lett 581:3396–3404
Razidlo GL, Katafiasz D, Taylor GS (2011) Myotubularin regulates Akt-dependent survival signaling via phosphatidylinositol 3-phosphate. J Biol Chem 286:20005–20019
Rameh LE, Tolias KF, Duckworth BC et al (1997) A new pathway for synthesis of phosphatidylinositol-4,5-bisphosphate. Nature 390:192–196
Ramel D, Lagarrigue F, Dupuis-Coronas S et al (2009) PtdIns5P protects Akt from dephosphorylation through PP2A inhibition. Biochem Biophys Res Commun 387:127–131
Razzini G, Berrie CP, Vignati S et al (2000) Novel functional PI 3-kinase antagonists inhibit cell growth and tumorigenicity in human cancer cell lines. FASEB J 14:1179–1187
Richardson JP, Wang M, Clarke JK et al (2007) Genomic tagging of endogenous type IIbeta phosphatidylinositol 5-phosphate 4-kinase in DT40 cells reveals a nuclear localization. Cell Signal 19:1309–1314
Roberts HF, Clarke JH, Letcher AJ et al (2005) Effects of lipid kinase expression and cellular stimuli on phosphatidylinositol 5-phosphate levels in mammalian cell lines. FEBS Lett 579:2868–2872
Robinson FL, Dixon JE (2006) Myotubularin phosphatases: policing 3-phosphoinositides. Trends Cell Biol 16:403–412
Rohde HM, Cheong FY, Konrad G et al (2003) The human phosphatidylinositol phosphatase SAC1 interacts with the coatomer I complex. J Biol Chem 278:52689–52699
Saarikangas J, Zhao H, Lappalainen P (2010) Regulation of the actin cytoskeleton-plasma membrane interplay by phosphoinositides. Physiol Rev 90:259–289
Safi S, Vandromme M, Caussanel S et al (2004) Role for the pleckstrin homology domain-containing protein CKIP-1 in phosphatidylinositol 3-kinase-regulated muscle differentiation. Mol Cell Biol 24:1245–1255

Samuels Y, Wang Z, Bardelli A et al (2004) High frequency of mutations of the PIK3CA gene in human cancers. Science 304:554

Sarbassov DD, Ali SM, Sabatini DM (2005a) Growing roles for the mTOR pathway. Curr Opin Cell Biol 17:596–603

Sarbassov DD, Guertin DA, Ali SM et al (2005b) Phosphorylation and regulation of Akt/PKB by the rictor-mTOR complex. Science 307:1098–1101

Sbrissa D, Ikonomov OC, Deeb R et al (2002) Phosphatidylinositol 5-phosphate biosynthesis is linked to PIKfyve and is involved in osmotic response pathway in mammalian cells. J Biol Chem 277:47276–47284

Sbrissa D, Ikonomov OC, Strakova J et al (2004) Role for a novel signaling intermediate, phosphatidylinositol 5-phosphate, in insulin-regulated F-actin stress fiber breakdown and GLUT4 translocation. Endocrinology 145:4853–4865

Shaw RJ, Cantley LC (2006) Ras, PI(3)K and mTOR signalling controls tumour cell growth. Nature 441:424–430

Shayesteh L, Lu Y, Kuo WL et al (1999) PIK3CA is implicated as an oncogene in ovarian cancer. Nat Genet 21:99–102

She QB, Solit D, Basso A et al (2003) Resistance to gefitinib in PTEN-null HER-overexpressing tumor cells can be overcome through restoration of PTEN function or pharmacologic modulation of constitutive phosphatidylinositol 3'-kinase/Akt pathway signaling. Clin Cancer Res 9:4340–4346

Shisheva A (2008a) Phosphoinositides in insulin action on GLUT4 dynamics: not just PtdIns(3,4,5)P3. Am J Physiol Endocrinol Metab 295:E536–E544

Shisheva A (2008b) PIKfyve: partners, significance, debates and paradoxes. Cell Biol Int 32: 591–604

Sindić A, Aleksandrova A, Fields AP et al (2001) Presence and activation of nuclear phosphoinositide 3-kinase C2beta during compensatory liver growth. J Biol Chem 276:17754–17761

Sleeman MW, Wortley KE, Lai KM et al (2005) Absence of the lipid phosphatase SHIP2 confers resistance to diary obesity. Nat Med 11:199–205

Srivastava S, Li Z, Lin L et al (2005) The phosphatidylinositol 3-phosphate phosphatase myotubularin—related protein 6 (MTMR6) is a negative regulator of the Ca2+ -activated K+ channel KCa3.1. Mol Cell Biol 25:3630–3638

Srivastava S, Choudhury P, Li Z et al (2006a) Phosphatidylinositol 3-phosphate indirectly activates KCa3.1 via 14 amino acids in the carboxy terminus of KCa3.1. Mol Biol Cell 17:146–154

Srivastava S, Ko K, Choudhury P et al (2006b) Phosphatidylinositol-3 phosphatase myotubularin-related protein 6 negatively regulates CD4 T cells. Mol Cell Biol 26:5595–5602

Srivastava S, Di L, Zhdanova O et al (2009) The class II phosphatidylinositol 3 kinase C2beta is required for the activation of the K+ channel KCa3.1 and CD4 T-cells. Mol Biol Cell 20:3783–3791

Stenmark H, Aasland R (1999) FYVE-finger proteins—effectors of inositol lipid. J Cell Sci 112:4175–4183

Suzuki A, de la Pompa JL, Stambolic V et al (1998) High cancer susceptibility and embryonic lethality associated with mutation of the PTEN tumour suppressor gene in mice. Curr Biol 8:1169–1178

Taguchi-Atarashi N, Hamasaki M, Matsunaga K et al (2010) Modulation of local PtdIns3P levels by the PI phosphatase MTMR3 regulates constitutive autophagy. Traffic 11:468–478

Takeshita S, Namba N, Zhao JJ et al (2002) SHIP-deficient mice are severely osteoporotic due to increased numbers of hyper-resorptive osteoclasts. Nat Med 8:943–949

Tepass U (2009) FERM proteins in animal morphogenesis. Curr Opin Genet Dev 19:357–367

Tolias KF, Rameh LE, Ishihara H et al (1998) Type I phosphatidylinositol-4-phosphate 5-kinases synthesize the novel lipids phosphatidylinositol 3,5-bisphosphate and phosphatidylinositol 5-phosphate. J Biol Chem 273:18040–18046

Tronchere H, Laporte J, Pendaries C et al (2004) Production of phosphatidylinositol 5-phosphate by the phosphoinositide 3-phosphatase myotubularin in mammalian cells. J Biol Chem 279:7304–7312

Tsujita K, Itoh T, Ijuin T et al (2004) Myotubularin regulates the function of the late endosome through the GRAM domain-phosphatidylinositol 3, 5-bisphosphate interaction. J Biol Chem 279:13817–13824

Um SH, D'Alessio D, Thomas G (2006) Nutrient overload, insulin resistance, and ribosomal protein S6 kinase 1, S6K1. Cell Metab 3:393–402

Ungewickell A, Hugge C, Kisseleva M et al (2005) The identification and characterization of two phosphatidylinositol-4, 5-bisphosphate 4-phosphatases. Proc Natl Acad Sci U S A 102: 18854–18859

van den Bout I, Divecha N (2009) PIP5K-driven PtdIns(4,5)P_2 synthesis: regulation and cellular functions. J Cell Sci 122:3837–3850

Vanhaesebroeck B, Leevers SJ, Ahmadi K et al (2001) Synthesis and function of 3-phosphorylated inositol lipids. Annu Rev Biochem 70:535–602

Vanhaesebroeck B, Ali K, Bilancio A et al (2005) Signalling by PI3K isoforms: insights from gene-targeted mice. Trends Biochem Sci 30:194–204

Vanhaesebroeck B, Guillermet-Guibert J, Graupera M et al (2010) The emerging mechanisms of isoform-specific PI3K signalling. Nat Rev Mol Cell Biol 11:329–341

van Rossum DB, Patterson RL, Sharma S et al (2005) Phospholipase Cgamma1 controls surface expression of TRPC3 through an intermolecular PH domain. Nature 434:99–104

Vasudevan KM, Barbie DA, Davies MA et al (2009) AKT-independent signaling downstream of oncogenic PIK3CA mutations in human cancer. Cancer Cell 16:21–32

Vergne I, Roberts E, Elmaoued RA et al (2009) Control of autophagy initiation by phosphoinositide 3-phosphatase jumpy. EMBO J 28:2244–2258

Visnjić D, Crljen V, Curić J et al (2002) The activation of nuclear phosphoinositide 3-kinase C2beta in all-trans-retinoic acid-differentiated HL-60 cells. FEBS Lett 529:268–274

Vivanco I, Sawyers CL (2002) The phosphatidylinositol 3-Kinase AKT pathway in human cancer. Nat Rev Cancer 2:489–501

Vogt PK, Kang S, Elsliger MA et al (2007) Cancer-specific mutations in phosphatidylinositol 3-kinase. Trends Biochem Sci 32:342–349

Watt SA, Kimber WA, Fleming IN et al (2004) Detection of novel intracellular agonist responsive pools of phosphatidylinositol 3, 4-bisphosphate using the TAPP1 pleckstrin homology domain in immunoelectron microscopy. Biochem J 377:653–663

Wen PJ, Osborne SL, Morrow IC et al (2008) Ca2+ -regulated pool of phosphatidylinositol-3-phosphate produced by phosphatidylinositol 3-kinase C2alpha on neurosecretory vesicles. Mol Biol Cell 12:5593–5603

Wilcox A, Hinchliffe KA (2008) Regulation of extranuclear PtdIns5*P* production by phosphatidylinositol phosphate 4-kinase 2α. FEBS Lett 582:1391–1394

Wullschleger S, Loewith R, Hall MN (2006) TOR signaling in growth and metabolism. Cell 124:471–484

Xue Y, Fares H, Grant B et al (2003) Genetic analysis of the myotubularin family of phosphatases in Caenorhabditis elegans. J Biol Chem 278:34380–34386

Yamashita S, Oku M, Wasada Y et al (2006) PI4P-signaling pathway for the synthesis of a nascent membrane structure in selective autophagy. J Cell Biol 173:709–717

Zhang H, Bajraszewski N, Wu E et al (2007) PDGFRs are critical for PI3K/Akt activation and negatively regulated by mTOR. J Clin Invest 117:730–738

Zhao L, Vogt PK (2008) Class I PI3K in oncogenic cellular transformation. Oncogene 27: 5486–5496

Zoncu R, Perera RM, Balkin DM et al (2009) A phosphoinositide switch controls the maturation and signaling properties of APPL endosomes. Cell 136:1110–1121

Zou J, Marjanovic J, Kisseleva MV et al (2007) Type I phosphatidylinositol-4,5-bisphosphate 4-phosphatase regulates stress-induced apoptosis Proc Natl Acad Sci U S A 104:16834–16839

Zou J, Chang SC, Marjanovic J et al (2009) MTMR9 increases MTMR6 enzyme activity, stability, and role in apoptosis. J Biol Chem 284:2064–2071

Chapter 2
Phosphoinositides and Cardiovascular Diseases

Alessandra Ghigo, Alessia Perino and Emilio Hirsch

Abstract Phosphoinositides (PIs), a family of phosphorylated derivatives of the membrane lipid phosphatidylinositol, are established regulators of multiple cellular functions. An increasing amount of evidence has highlighted potential links between PI-mediated signaling pathways and the etiology of many human diseases, including cardiovascular pathologies. This chapter will provide a detailed overview of the peculiar functions of the major cardiovascular PIs in the pathogenesis of atherosclerosis, heart failure, and arrhythmias.

Introduction

Inositol phospholipids or phosphoinositides (PIs) are a family of phosphorylated derivatives of the membrane lipid phosphatidylinositol (Sasaki et al. 2007). Among these, phosphatidylinositol 4′5′-bisphosphate (PIP2) and phosphatidylinositol 3′4′5′-trisphosphate (PIP3) represent the two major plasma membrane PIs (Czech 2000). PIP2 is synthesized from the membrane phosphatidylinositol via sequential activation of two phosphoinositide kinases, PI4K and PI5K that catalyze the addition of a phosphate group on the inositol D4 and D5 positions, respectively (Lee and Rhee 1995). In the heart, PIP2 can function as a key second messenger, controlling the activity of a wide plethora of ion channels and thus contributing to heart rhythm modulation. Alternatively, PIP2 plays an important role as a substrate for the phospholipase C (PLC), which hydrolyses PIP2 leading to inositol trisphosphate (IP3) and

A. Ghigo · A. Perino · E. Hirsch (✉)
Department of Genetics, Biology and Biochemistry, University of Torino, Molecular Biotechnology Center, Via Nizza, 52, 10126 Torino, Italy
e-mail: emilio.hirsch@unito.it

M. Falasca (ed.), *Phosphoinositides and Disease*, Current Topics in Microbiology and Immunology 362, DOI: 10.1007/978-94-007-5025-8_2,

diacyl-glycerol (DAG) second messenger formation. Among these, IP3 has emerged as a key modulator of cardiac excitation–contraction coupling as well as of cardiac hypertrophy. In addition, PIP2 constitutes the preferential substrate of phosphatidylinositol 3 kinases (PI3Ks) that, by adding a phosphate group on the inositol D3 position, generates the second messenger PIP3 (Cantley 2002). In the cardiovascular system, PIP3 is mainly involved in the control of cardiomyocyte apoptosis, hypertrophy, and contractility.

Deregulation of PI metabolism underlies the onset and progression of several cardiovascular pathologies, ranging from atherosclerosis to heart failure, with different PIs playing specific non-redundant roles in defined pathological contexts. This chapter will provide a detailed overview of the pathophysiological function of the major cardiovascular PIs, including PIP2, and its derivative products, IP3, and PIP3. We will first focus on their involvement in vascular diseases, such as atherosclerosis, where PIs are key regulators of multiple cell types involved in the onset and progression of the disease. Atherosclerosis-associated vascular damage can in turn trigger cardiac dysfunction, ultimately leading to heart failure, a complex multi-factorial disease, whose progression can be affected by PI level modulation. A dedicated section will describe the role of PIs in the crucial phases of heart failure evolution. Finally, we will report the impact of PI activity on heart rhythm disturbances that often occur in association with heart failure, further triggering life-threatening complications.

Atherosclerosis

Atherosclerosis is a systemic disease of the vessel wall occurring principally in large and medium-size elastic and muscular arteries, such as aorta, carotids, coronaries, and peripheral arteries. The pathology consists in an altered permeability of the endothelium that permits the accumulation of lipoproteins within the intima, resulting in endothelial cell activation and leukocyte recruitment (Ross 1995). This process leads to the formation of the atherosclerotic plaque that can be subjected to rupture, leading to thrombosis and consequent occlusion of the vessel. Among cardiovascular PIs, PIP3 represents the major regulator of the atherosclerotic process and is critically involved in the function of key cellular players such as leukocytes and platelets. An additional contribution to the atherosclerotic vascular remodeling is provided by IP3, which primarily regulates the activity of endothelial and vascular smooth muscle cells.

PIP3

Inflammation represents a key element of the atherosclerotic process and involves the migration of leukocytes into the atherosclerotic lesion (Ross 1999). Once recruited to the inflammatory site, leukocytes are able to secrete inflammatory mediators, to proteolytically cleave the extracellular matrix and to cross-talk with local vascular cells (Libby 2002). PIP3 signaling is a key participant in each of these events. PIP3 is produced by all class I PI3Ks, but the inflammatory aspect of atherosclerosis is dominated by the pool of PIP3 produced by PI3Kγ, an enzyme highly expressed in the hematopoietic cell lineage (Fougerat et al. 2009). In agreement, the inhibition of the PI3Kγ-PIP3-Akt signaling pathway in murine models of atherosclerosis (ApoE- or LDLR-deficient mice) significantly reduces the development of early and advanced atherosclerotic lesions (Fougerat et al. 2008). Similar findings have been reported in PI3Kγ-ApoE double knock-out mice, which exhibit smaller lesions than single ApoE-knockout controls (Chang et al. 2007). The reduced inflammation present in these mice is linked to the impaired migration of PI3Kγ-null neutrophils and macrophages toward different chemokine stimuli and to their defective oxidative burst (Hirsch et al. 2000; Patrucco et al. 2004). Furthermore, transgenic mice carrying a hyperactivation of the PI3Kγ-PIP3 pathway are not able to correctly localize PIP3 at the leading edge of the migrating cell, thus resulting in decreased directional migration in response to selective stimuli (Costa et al. 2007). Indeed, the formation of a proper leading edge is strictly required for the directional migration of leukocytes to the site of inflammation (Ridley et al. 2003).

Critical events following exacerbated plaque inflammation and rupture are platelet aggregation and thrombus formation, the main causes of infarction and stroke. Several factors regulate these processes, including the functional state of cellular enzymes and membrane receptors. Platelet adhesion and aggregation at sites of vascular injury are dependent on the interaction between von Willebrand factor (vWF) and two major platelet adhesion receptors, glycoprotein (GP) Ib-V-IX and integrin $\alpha_{\mathrm{II}}\beta_{\mathrm{b3}}$. Several lines of evidence have implicated PI3K-PIP3 in such signaling pathways (Yap et al. 2002; Kasirer-Friede et al. 2004; Kim et al. 2009). Indeed, PIP3 is produced in platelets upon the interaction of the GpIb-V-IX complex with vWF and is required for the activation of a cascade involving NO/cGMP/PKG, leading to p38 MAP-kinase and ERK signaling, calcium mobilization, and, in turn, to integrin $\alpha_{\mathrm{II}}\beta_{\mathrm{b3}}$-dependent stable adhesion and aggregation (Li et al. 2006; Stojanovic et al. 2006; Yin et al. 2008). Several evidences demonstrate that the PIP3 produced by PI3Kβ is essential in mediating these effects (Jackson et al. 2005). In vivo, platelet tethering is followed by stable adhesion and spreading, a process involving collagen receptors. Several studies have demonstrated a critical role for PIP3 in the organization of a signal transduction network initiated by the collagen receptor GpVI (Pasquet et al. 1999). In this context, PIP3 is essential for the stable interaction of PLCγ2 with the plasma membrane, thereby promoting PIP2 hydrolysis (Ragab et al. 2007). Blokade of PI3Kβ-PIP3 signaling pathway results in an impaired activation of downstream signaling molecules, resulting in defective aggregation responses (Canobbio et al. 2009; Martin et al. 2010; Consonni et al. 2012).

IP3

Monocyte-derived macrophages play a key role in the pathogenesis of atherosclerosis since they are activated within the atheromatous lesions (Ross 1986). Macrophages are able to accumulate lipids mainly through LDL receptor- and scavenger receptor-dependent mechanisms. Binding of LDL to these receptors stimulates PI breakdown, thus inducing a rapid IP3 production and subsequent Ca^{2+} mobilization from intracellular stores (Ishikawa et al. 1989). LDL plays an important role in the process of atherogenesis, since this lipid is able to alter the function not only of macrophages but also of other cell types involved in the pathophysiology of the disease. Indeed, LDL stimulates the production of IP3 in endothelial cells (Myers et al. 1992; Hamilton et al. 1994) and in vascular smooth muscle cells (Resink et al. 1992). In addition to humoral factors, physical or mechanical stresses also contribute to vascular remodeling. Indeed, pressure by itself triggers IP3 production and intracellular calcium release in rat vascular smooth muscle cells, resulting in cell proliferation and DNA synthesis, two processes that are closely related to the smooth muscle cell expansion and plaque formation (Hishikawa et al. 1994).

Hypertrophy and Heart Failure

Coronary atherosclerosis is the most important primary etiologic factor in the industrialized world predisposing to ischemic heart disease and the subsequent development of heart failure (Redfield 2002). Heart failure is a complex, multifactorial disease that is characterized by the inability of the heart to pump sufficient blood to meet the metabolic needs of the body. An essential precursor of heart failure is represented by cardiac hypertrophy, which is an adaptive response of the heart to stress conditions, such as volume overload and pressure overload, resulting in an abnormal increase in cardiac mass. Although at the beginning hypertrophy has beneficial effects, resulting in normalized wall stress, chronic hypertrophy has maladaptive features and is associated with a significant increase in the risk of heart failure. In recent years, PIP3 has emerged as a critical switch of cardiac hypertrophy and heart failure development, with selected pools of PIP3 playing distinct non-redundant roles. Also, IP3 and its immediate precursor PIP2 have been shown to participate to these processes, with IP3 regulating cardiac hypertrophy in response to selected stimuli and PIP2 functioning as a key regulator of cardiomyocyte apoptosis and the subsequent transition to heart failure.

PIP3

In the healthy heart, the pool of PIP3 produced by PI3Kα is a master regulator of cardiomyocyte metabolism and survival (Damilano et al. 2010). Selective cardiac overexpression of a dominant-negative PI3Kα results in smaller cardiomyocytes and reduced heart size, with normal tissue architecture and contractility (Shioi et al. 2000; Crackower et al. 2002; McMullen et al. 2003). On the other hand, selective cardiac overexpression of a constitutively active PI3Kα leads to myocardial hypertrophy, without progression to cardiac dysfunction (Shioi et al. 2000). These findings phenocopy the genetic loss of myocardial PTEN, that leads to PIP3 accumulation (Crackower et al. 2002). All these results support the conclusion that PI3Kα boosts myocardial growth through the activation of canonical PIP3-dependent anabolic pathways. Beyond cardiomyocyte size control, PIP3 produced by PI3Kα represents a master switch of physiological but not of pathological hypertrophy. Mice expressing a dominant-negative PI3Kα are resistant to left ventricular hypertrophy induced by exercise training but they develop compensatory hypertrophy following pressure overload (McMullen et al. 2003). Importantly, the pool of PIP3 produced by PI3Kα is functionally relevant in conditions of myocardial damage, where it protects cardiomyocytes from cell death and dysfunction caused by various pathological noxae, including dilative myocardiopathy, myocardial infarction, chronic adrenergic stimulation, and pressure overload (McMullen et al. 2007; Lin et al. 2010). Taken together, cardiac PI3Kα-PIP3 signaling is required for the protection of cardiac function.

On the other hand, the pool of PIP3 produced by PI3Kγ drives the heart toward the development of heart failure. A critical event in the natural history of heart failure is represented by immune cell recruitment to the myocardium and their subsequent activation (Frangogiannis 2008). Similar to the migration of white blood cells into the atherosclerotic plaque, PIP3 produced by PI3Kγ is a crucial component of signal transduction controlling leukocyte migration to the heart (Hirsch et al. 2006). PI3Kγ-PIP3 signaling pathway is also active within the cardiomyocytes (Ghigo et al. 2011). To discern the specific contribution of PI3Kγ-PIP3 signaling in leukocytes and cardiomyocytes, the major cell types involved in the process of cardiac maladaptive remodelling, a model of pressure overload-induced heart failure in bone marrow-transplanted chimeric mice has been used (Damilano et al. 2011). Selective inhibition of PIP3 production within the leukocyte compartment causes a reduced leukocyte infiltration in the myocardium soon after transverse aortic constriction (TAC) that correlates with reduced fibrosis and preserved diastolic function. Conversely, selective inhibition of PIP3 production within cardiomyocytes is required to counteract TAC-induced cardiac dysfunction at later time points. This maladaptive role of cardiac PIP3 appears to be linked to its impact on β-adrenergic receptor (β-AR) signaling. Deregulation of β-AR responses is a key pathophysiological feature of heart failure (Rockman et al. 2002). Indeed, uncontrolled catecholamine stimulation of β-ARs in a failing myocardium leads both to a reduction in the number of ligand-accessible β-ARs and to a diminished response to stimulation of remaining receptors (two processes called downregulation and desensitization, respectively) (Bristow et al. 1982; Bristow 1998). Due to desensitization and downregulation, adrenergic signaling is therefore progressively impaired in the failing myocardium, which loses tonic and phasic contractile responses to catecholamine stimulation. Within this scenario, PIP3 produced by PI3Kγ is an essential regulator of β-AR endocytosis (Naga Prasad et al. 2001). In particular, PI3Kγ constitutes a cytoplasmic complex with G-protein coupled receptor kinase 2 (GRK-2) that, upon β-AR stimulation, mediates translocation of PI3Kγ to the activated receptor (Naga Prasad et al. 2000). Herein, a local pool of PIP3 produced by PI3Kγ is required for the recruitment of phosphoinositide-binding endocytic proteins, such as β-arrestin and the clathrin adaptor AP-2, which leads to the consequent organization of clathrin-coated pits orchestrating the internalization of the receptor (Laporte et al. 2000; Naga Prasad et al. 2002). In vivo studies have confirmed these mechanistic insights. Inhibition of PIP3 localized at the β-AR, through transgenic overexpression of an inactive form of PI3Kγ, prevents the development of β-AR dysfunction in response to chronic catecholamine stimulation and protects from heart failure (Nienaber et al. 2003; Perrino et al. 2005; Perrino et al. 2006). Furthermore, β-AR density remains unchanged after pressure overload in mice expressing a PI3Kγ defective in PIP3 production. Similarly, the administration of a PI3Kγ-specific pharmacological inhibitor in wild-type mice suffering from pressure overload-induced heart failure can significantly improve both β-AR density and left ventricular contractility (Perino et al. 2011). Accordingly, the mechanical unloading of the failing human heart is associated with significant reduction in GRK2-associated PI3Kγ activity initially enhanced by the disease (Perrino et al. 2007). Taken together, these findings picture a scenario where PI3Kγ-PIP3 deregulation leads to maladaptive β-adrenergic perturbation during heart failure. In conclusion, these data indicate that the pool of PIP3 produced by PI3Kγ occupies a central stage in the molecular pathophysiology of heart failure, which is dominated by abnormal β-adrenergic stimulation. In this context, PI3Kγ escapes physiological feedback control mechanisms and orchestrates key aspects of myocardial damage and remodelling, such as β-adrenergic desensitization and downregulation, myocardial inflammation and fibrosis. Overall, these works suggest that concomitant inhibition of PI3Kγ-PIP3 signaling in cardiomyocytes and leukocytes is required to promote cardiac protection.

IP3

Although IP3 receptors (IP3-Rs) are mainly located on the sarcoplasmic reticulum where they induce Ca^{2+}mobilization regulating contractility, specific pools of IP3-Rs can be located to the nucleus and exert a role in cardiac remodeling (Wu et al. 2006). The involvement of IP3-mediated Ca^{2+} signaling in cardiac hypertrophy was first described by the work of Barac et al. showing that the IP3 pathway is crucial for the development of Fas-mediated hypertrophy in a model of cultured rat neonatal ventricular myocytes (Barac et al. 2005). More recently, it has been demonstrated that a pool of IP3-Rs located within, or close to, the nucleus can specifically regulate DNA modifying enzymes involved in controlling hypertrophic growth of cardiac myocytes. In isolated rabbit myocytes, endothelin-1 stimulation, which activates plasmalemmal G-protein coupled receptors and IP3 production, elicits local nuclear envelope Ca^{2+} release via IP3-Rs. In turn, such local Ca^{2+} release activates nuclear CamKII, which triggers HDAC5 phosphorylation and nuclear export, eventually derepressing transcription. Interestingly, this Ca^{2+} pathway cannot be activated by the global Ca^{2+} transients that cause contraction at each heartbeat, thus demonstrating that IP3-mediated signaling is highly compartmentalized in cardiomyocytes (Wu et al. 2006). The generation of heart-specific transgenic mice with both gain- and loss-of-function for IP3-R signaling has then allowed to investigate the importance of the IP3-mediated hypertrophic response in vivo, both in physiological and pathological conditions (Nakayama et al. 2010). Mice overexpressing the IP3-R2 subtype in the heart display mild baseline hypertrophy at three months of age, which is not further enhanced by two weeks of pressure-overload stimulation. By contrast, IP3-R2 overexpression significantly enhances basal hypertrophy following two weeks of isoproterenol infusion, in response to Gαq overexpression and/or to exercise stimulation. Accordingly, overexpression of an IP3 chelating protein totally abolishes cardiac hypertrophy in response to isoproterenol and angiotensin II infusion, but not pressure-overload stimulation (Nakayama et al. 2010). Altogether these studies point to a central role for IP3-mediated Ca^{2+} signaling not only in the modulation of excitation–contraction coupling (ECC), but also in the regulation of cardiac hypertrophy in response to selected stimuli.

PIP2

Among major regulators of cardiac hypertrophy is the Gq signaling axis. In cardiomyocytes, heterotrimeric G proteins of the Gq family transduce signals from a variety of receptors that bind α_1-adrenergic agonists, endothelin, purine nucleotides, or angiotensin and activate PLC, thus promoting PIP2 hydrolysis into IP3 and DAG (Dorn and Force 2005). Overexpression of the wild-type α subunit of Gq (GαqWT) induces hypertrophy in vitro, in isolated cardiomyocytes (Adams et al. 1998). Similarly, mice with cardiac-specific overexpression of Gαq show hypertrophic cardiomyocytes (D'Angelo et al. 1997). Furthermore, in these animals, cardiac hypertrophy rapidly progresses toward heart failure when Gαq-dependent signaling is further enhanced or prolonged by concomitant expression of PLC, Gαq-coupled receptors, and stress stimuli (D'Angelo et al. 1997). Accordingly, expression of a constitutively active form of Gαq ($G\alpha q^{Q209L}$) does not cause hypertrophy, but induces cardiomyocyte apoptosis which correlates with enhanced PLC-mediated hydrolysis of PIP2 (D'Angelo et al. 1997). The simplest explanation of these findings might be that depletion of PIP2 levels leads to reduced amounts of PIP3 and, in turn, to attenuated Akt/PKB survival signaling, eventually driving cardiomyocyte apoptosis. However, $G\alpha q^{Q209L}$ expressing cardiomyocytes contain elevated concentration of PIP3, suggesting that the apoptotic process is mainly linked to the loss of PIP2. Indeed, it has been suggested that PIP2 depletion alone might be sufficient to trigger apoptosis as PIP2 is able to reduce caspase activation either by a direct mechanism or by keeping procaspases complexed with gelsolin (Azuma et al. 2000; Mejillano et al. 2001). Consistently, $G\alpha q^{Q209L}$-induced apoptosis is caspase dependent (D'Angelo et al. 1997). It is thus interesting to speculate that PIP2 depletion regulates cardiomyocyte apoptosis and subsequent heart failure.

Arrhythmias

A prominent mechanism of death in patients with heart failure is arrhythmia, where electrical activation of the heart occurs so rapidly that effective filling and pumping of blood cannot occur (Wang and Hill 2010). Cardiac rhythm disturbances or arrhythmias refer to a large and heterogeneous group of conditions characterized by abnormal electrical activity of the heart. Despite their incidence, the molecular bases of arrhythmogenesis are still incompletely understood. In the last decade, PIs have emerged as potential arrhythmogenic factors (Woodcock et al. 2009), with both membrane PIs and soluble signaling molecules potentially involved. In particular, membrane PIP2 and the soluble IP3 have been shown to impact on cardiomyocyte electrical activity, thus contributing to arrhythmogenesis, with PIP3 playing only a minor role in this pathological context.

PIP2

The arrhythmogenic activity of membrane PIP2 is linked to its ability to regulate a wide array of cardiac ion channels and exchangers, including inward rectifying K^+ channels (K_{IR}), repolarizing K^+ channels (K_V), and pacemaker channels.

Inwardly rectifying K^+ (K_{IR}) channels are important regulators of resting membrane potential and cell excitability. The activity of K_{IR}, including $K_{IR}2$, $K_{IR}3$, and $K_{IR}6$ isoforms is critically dependent on the integrity of channel interactions with PIP2. Opening of K_{IR} requires PIP2 binding to basic and polar amino acids in cytoplasmic domains, whereas depletion of PIP2 acts to close the channel (Huang et al. 1998). $K_{IR}2$ family members are responsible for the current that maintains resting membrane potential, in both atrial and ventricular myocytes. Several mutations in $K_{IR}2.1$ channel proteins have been shown to cause the Andersen-Tawil syndrome, a condition characterized by periodic paralysis, dysmorphic features, and cardiac arrhythmias (Plaster et al. 2001; Lopes et al. 2002). Some of these mutations map to the region of $K_{IR}2.1$ that interacts with PIP2 and are shown to allosterically decrease channel-PIP2 interactions. The weakening of channel-PIP2 interactions, in turn, leads to inhibition of $K_{IR}2.1$ currents (Lopes et al. 2002) and mutant $K_{IR}2.1$ has been shown to be more susceptible to modulation by stimuli that decrease membrane concentrations of PIP2 (Lopes et al. 2002; Ma et al. 2007). Also, $K_{IR}3$ channels are strictly dependent on PIP2 for their activity. $K_{IR}3$ isoenzymes are muscarinic potassium channels (K_{ACh}), functioning in atrial and pacemaker myocytes. K_{ACh} belongs to the G protein regulated inward rectifying K^+ channel family (GIRK), which are modulated via activation of the heterotrimeric G protein, G_i, causing release of $G\beta\gamma$ subunits (Hommers et al. 2003). Blockade of PIP2 binding to channels impairs the stimulatory effects of $G\beta\gamma$ on channel activity and such effects can be reversed by restoring PIP2 (Sui et al. 1998). $K_{IR}3$ currents play a key role in atrial tachycardia-induced electric remodeling and in the pathogenesis of atrial fibrillation in dogs (Cha et al. 2006). However, there is currently no evidence that PIP2 plays a role in this pathogenetic mechanism. Similarly, whether PIP2-mediated regulation of $K_{IR}6$ channels can lead to cardiac dysfunction or arrhythmogenesis is currently unknown (Haider et al. 2007).

Besides its impact on K_{IR} channel function, PIP2 affects the activity of repolarizing K^+ channels (K_V), thus contributing to the termination of the cardiac action potential. Among K_V channels, $K_V11.1$ (also known as HERG) is responsible for the rapid phase of repolarization, while $K_V7.1$ (KCNQ1/KCNE1) mainly regulates the slow repolarization. Both $K_V11.1$ and $K_V7.1$ have six trans-membrane spanning regions that form an ion pore, together with a long C-terminal tail and a relatively short cytosolic N-terminal tail and are regulated by PIP2, although its interaction has not been studied as extensively as for K_{IR} channels (Li et al. 2005b; Bian and McDonald 2007). Changes in the concentration of intracellular PIP2 have been found to alter both the current amplitude and the voltage-dependent gating of heterologously expressed HERG channels (Bian et al. 2001): PIP2 significantly increases the current amplitude, accelerates the voltage-dependent activation, and slows the voltage-dependent inactivation of the channel. As a result of these biophysical regulations, elevation of membrane PIP2 concentration triggers more effective currents, while PIP2 depletion has an opposite effect (Bian et al. 2001). Similarly, in rabbit ventricular myocytes, native HERG currents are reduced by 15–20 % upon application of epinephrine, a stimulus known to trigger $G\alpha q$-mediated activation of PLC, and subsequent depletion of membrane PIP2 (Bian et al. 2004).

HERG function is also modulated by β-adrenergic receptor activation and the cAMP/Protein kinase A (PKA) axis. PKA phosphorylates the channel in its polybasic region responsible for PIP2 binding and causes its opening (Bian and McDonald 2007). Weakening of PIP2-HERG channel interaction is potentially involved in the induction of an inherited form of arrhythmia, the long-QT (LQT) syndrome, a genetic disease characterized by prolonged cardiac repolarization, cardiac arrhythmias, and a high risk of sudden death (Li et al. 1998). Several of the known LQT mutations are characterized by either alterations or deletions of the polybasic PIP2 interacting site (Schwartz et al. 2001). However, the detailed mechanism linking aberrant PIP2-HERG interaction to cardiac arrhythmias or acquired heart diseases is still incomplete and needs further investigation.

On the contrary, the involvement of PIP2-mediated regulation of KCNQ1/KCNE1 in arrhythmogenesis has been extensively studied. In the heart, KCNQ1 assembles with KCNE1 to form a channel complex constituting the slow component of the delayed rectifier current (Sanguinetti et al. 1996). Intracellular PIP2 regulates KCNQ1/KCNE1 channel activity via stabilization of the open state of the channel, leading to increased current amplitude, slowed deactivation kinetics, and a shift in the activation curve toward negative potentials (Loussouarn et al. 2003; Li et al. 2005b). The PIP2-binding sequence is part of an endogenous inhibitory region on KCNQ1, and PIP2 association prevents this inhibition (Oliver et al. 2004). Mutations in this channel have been linked to diverse forms of arrhythmia, including atrial fibrillation and inherited arrhythmias, such as long QT and short QT syndromes (Chen et al. 2003; Bellocq et al. 2004). Some of these mutants are in residues likely to be important for PIP2 interactions. For instance, R555C and R539W mutations, which have been associated to the LQT syndrome, have the same consequences on the channel biophysical properties as a decrease in PIP2 concentrations (Park et al. 2005). In particular, the mutant channels display reduced affinity for PIP2 compared to the wild-type protein and an alteration in PIP2 binding in mutants R555C and R539W fully explains the channel dysfunction. In patch clamp studies, addition of excess PIP2 reverses the lowered activity of the mutant channels and returns channel activity to normal. This confirms the importance of PIP2 binding for the optimal functioning of the channel and raises the possibility that changes in PIP2 availability could initiate arrhythmia (Park et al. 2005). A recent work suggests that PIP2 sensitivity of KCNQ1 can be controlled by the auxiliary subunit, KCNE1 (Li et al. 2011). KCNE1 increases the PIP2 sensitivity of the channel by several orders of magnitude. Mutations of the key residues in KCNE1 that are determinants of PIP2 sensitivity, R67C, R67H, K70M, and K70N, are associated with LQT syndrome. These mutations reduce the channel current and PIP2 sensitivity. Interestingly, application of supernormal levels of exogenous PIP2 is able to rescue wild-type channel function. Thus, decreased sensitivity to PIP2 is at the base of this inherited rhythm dysfunction.

PIP2 also regulates the pacemaker current by controlling the hyperpolarization-activated cyclic nucleotide-gated channels (HCN). PIP2 shifts the voltage of the pacemaker channels toward depolarized potentials and thus increases the spontaneous firing rate. Although these channels have been suggested to be important in the development of atrial fibrillation, there is currently no evidence that this involves PIP2 (Zolles et al. 2006).

Altogether these results unveil a central role for PIP2 in the control of cardiomyocyte electrophysiology and suggest that PIP2-mediated arrhythmogenesis is mainly due to defective regulation of the $K_{IR}2$, HERG, and KCNQ1/KCNE1 channels.

IP3

In the heart, IP3 binding to IP3-Rs on the sarcoplasmic reticulum (SR) generates Ca^{2+} fluxes that amplify the Ca^{2+} signal occurring during the excitation–contraction coupling (ECC) (Marks 2000). During ECC, action potential-induced membrane depolarization leads to the opening of voltage-gated Ca^{2+} channels on the plasma membrane, resulting in Ca^{2+} influx. In turn, entering Ca^{2+} activates the ryanodine receptors (RyRs) which lead to massive Ca^{2+} release from the SR, via a mechanism initiating contraction, known as Ca^{2+}-induced Ca^{2+}-release (CICR) (Bers 2002). Given their low expression levels in cardiomyocytes compared to the more abundant RyRs, IP3-Rs do not contribute to the modulation of global intracellular Ca^{2+} levels, rather they control spatially restricted Ca^{2+} pools within specific subcellular microdomains. The predominant IP3-R isoform in cardiomyocytes, IP3-R2, co-localizes with RyR and locally increases Ca^{2+} levels sensitizing RyR to CICR and enhancing ECC efficiency (Mackenzie et al. 2002). In agreement, direct stimulation of IP3-Rs has positive inotropic effects in both atrial and ventricular myocytes (Proven et al. 2006).

Atrial myocytes express IP3-R2 at higher levels than ventricular myocytes and several reports have underlined the importance of IP3/IP3-R2 signaling in atrial ECC (Mackenzie et al. 2002). For instance, in mouse atrial myocytes, selective activation of IP3-R2 causes an increase in basal levels of Ca^{2+}, an enhancement of action potential-induced Ca^{2+} transients and of fractional SR Ca^{2+} release (Li et al. 2005a). IP3-Rs-mediated Ca^{2+} signaling can interfere with the highly orchestrated Ca^{2+} responses mediated by the RyR, thereby predisposing to arrhythmia. IP3-R2 stimulation of atrial cells induces spontaneous arrhythmogenic Ca^{2+} release events, a potential source of ectopic beats (Li et al. 2005a). Similarly, in cat atrial myocytes IP3 causes spontaneous Ca^{2+} transients, Ca^{2+} waves as well as Ca^{2+} alternans, all disturbances in Ca^{2+} signaling related to cardiac arrhythmias (Zima and Blatter 2004). In rat, IP3-R activation increases the amplitude of electrically induced Ca^{2+} transients and triggers premature extra Ca^{2+} transients (Mackenzie et al. 2002). Spontaneous Ca^{2+} release events are fully abrogated in IP3-R2-deficient atrial myocytes (Li et al. 2005a) thus suggesting a crucial role for IP3/IP3-R signaling in the initiation and perpetuation of atrial fibrillation (AF), the most common form of cardiac arrhythmia (Li et al. 2005a).

IP3-R2 is also present in ventricular myocytes, although its expression is 3.5-fold lower than in atrial myocytes (Domeier et al. 2008). However, the contribution of IP3 signaling to ECC and arrhythmogenesis in these cells is still controversial and appears to be species specific. For example, IP3-R-dependent Ca^{2+} signaling produces positive inotropic effects and evokes spontaneous pro-arrhythmic Ca^{2+} signals in rat (Proven et al. 2006). In contrast, IP3-R-mediated arrhythmogenic events are detected neither in rabbit (Domeier et al. 2008) nor in cat ventricular myocytes (Zima and Blatter 2004). Therefore, it appears that atrial rather than ventricular arrhythmia can be associated with perturbations in IP3 signaling. In this respect, it has been suggested that ventricular arrhythmias, apparently associated with IP3, derive primarily from the conductive tissue where highest levels of IP3-Rs are found (Woodcock et al. 2009).

PIP3

Different from the case of IP3 and PIP2, evidence linking PIP3 signaling to cardiac arrhythmias are few. Two reports demonstrate that enhanced PI3Kα-PIP3 signaling can lead to atrial fibrillation and heart failure-associated ventricular arrhythmia (Pretorius et al. 2009; Yang et al. 2012). Cardiac-specific expression of a dominant-negative mutant of PI3Kα in the DCM-Tg model of dilated cardiomyopathy leads to atrial fibrillation. Interestingly, in atrial appendages from patients with atrial fibrillation, PI3K activity is lower than in tissues from patients in sinus rhythm. However, no evidence is reported that arrhythmia is mediated by low PIP3 levels (Pretorius et al. 2009). Consistent with a positive role for the PI3Kα-PIP3 axis in the protection against arrhythmia, enhanced PIP3 signaling due to PI3Kα overexpression has been shown to mitigate the arrhythmogenic electrical remodeling associated to pathological hypertrophy and heart failure. Increased activation of PI3Kα leads to a transcriptional upregulation of the repolarizing K^+ channel (Yang et al. 2012). Whether other PIP3 pools produced by distinct PI3K isoforms are necessary for cardiac rhythm control is not known and further studies are required to better define whether PIP3 directly controls channel activity.

Conclusion

In the cardiovascular system PIs do not represent merely structural components of cell membranes; rather they function as substrates for enzymes that generate second messengers modulating fundamental processes, such as cardiac hypertrophy and contractility. Besides, PIs can function as direct regulators of signaling proteins, such as cardiac ion channels, eventually contributing to cell excitability and heart rhythm modulation. Aberrant PI metabolism is known to contribute to diverse human cardiovascular pathologies, including atherosclerosis, heart failure, and arrhythmia and molecules involved in PI signaling pathways are thus considered attractive drug targets for therapy. However, in view of the pleiotropic cellular processes controlled by each PI, it can be predicted that pharmacological modulation of PI signaling is likely to incur into undesired, important side effects. For clinical interventions to be effective, there is urgent need to elucidate the mechanisms by which PI effectors that bind to the same lipid, but that mediate different processes, are independently regulated. The understanding of the spatio-temporal compartmentalization of each PI and of their related signaling in specific pathophysiological settings is crucial to achieve this goal. Such knowledge could potentially pave the way of successful PI manipulation for human therapy.

References

Adams JW, Sakata Y, Davis MG, Sah VP, Wang Y, Liggett SB, Chien KR, Brown JH, Dorn GW 2nd (1998) Enhanced Galphaq signaling: a common pathway mediates cardiac hypertrophy and apoptotic heart failure. Proc Natl Acad Sci U S A 95:10140–10145

Azuma T, Koths K, Flanagan L, Kwiatkowski D (2000) Gelsolin in complex with phosphatidylinositol 4,5-bisphosphate inhibits caspase-3 and-9 to retard apoptotic progression. J Biol Chem 275:3761–3766

Barac YD, Zeevi-Levin N, Yaniv G, Reiter I, Milman F, Shilkrut M, Coleman R, Abassi Z, Binah O (2005) The 1,4,5-inositol trisphosphate pathway is a key component in Fas-mediated hypertrophy in neonatal rat ventricular myocytes. Cardiovasc Res 68:75–86

Bellocq C, van Ginneken AC, Bezzina CR, Alders M, Escande D, Mannens MM, Baro I, Wilde AA (2004) Mutation in the KCNQ1 gene leading to the short QT-interval syndrome. Circulation 109:2394–2397

Bers DM (2002) Cardiac excitation-contraction coupling. Nature 415:198–205

Bian JS, McDonald TV (2007) Phosphatidylinositol 4,5-bisphosphate interactions with the HERG K(+) channel. Pflugers Arch 455:105–113

Bian J, Cui J, McDonald TV (2001) HERG K(+) channel activity is regulated by changes in phosphatidyl inositol 4,5-bisphosphate. Circ Res 89:1168–1176

Bian JS, Kagan A, McDonald TV (2004) Molecular analysis of PIP2 regulation of HERG and IKr. Am J Physiol Heart Circ Physiol 287:H2154–H2163

Bristow MR (1998) Why does the myocardium fail? insights from basic science. Lancet 352(Suppl 1):SI8–SI14

Bristow MR, Ginsburg R, Minobe W, Cubicciotti RS, Sageman WS, Lurie K, Billingham ME, Harrison DC, Stinson EB (1982) Decreased catecholamine sensitivity and beta-adrenergic-receptor density in failing human hearts. N Engl J Med 307:205–211

Canobbio I, Stefanini L, Cipolla L, Ciraolo E, Gruppi C, Balduini C, Hirsch E, Torti M (2009) Genetic evidence for a predominant role of PI3Kbeta catalytic activity in ITAM- and integrin-mediated signaling in platelets. Blood 114:2193–2196

Cantley LC (2002) The phosphoinositide 3-kinase pathway. Science 296:1655–1657

Cha TJ, Ehrlich JR, Chartier D, Qi XY, Xiao L, Nattel S (2006) Kir3-based inward rectifier potassium current: potential role in atrial tachycardia remodeling effects on atrial repolarization and arrhythmias. Circulation 113:1730–1737

Chang JD, Sukhova GK, Libby P, Schvartz E, Lichtenstein AH, Field SJ, Kennedy C, Madhavarapu S, Luo J, Wu D, Cantley LC (2007) Deletion of the phosphoinositide 3-kinase p110gamma gene attenuates murine atherosclerosis. Proc Natl Acad Sci U S A 104:8077–8082

Chen YH, Xu SJ, Bendahhou S, Wang XL, Wang Y, Xu WY, Jin HW, Sun H, Su XY, Zhuang QN, Yang YQ, Li YB, Liu Y, Xu HJ, Li XF, Ma N, Mou CP, Chen Z, Barhanin J, Huang W (2003) KCNQ1 gain-of-function mutation in familial atrial fibrillation. Science 299:251–254

Consonni A, Cipolla L, Guidetti G, Canobbio I, Ciraolo E, Hirsch E, Falasca M, Okigaki M, Balduini C, Torti M (2012) Role and regulation of phosphatidylinositol 3-kinase beta in platelet integrin alpha2beta1 signaling. Blood 119:847–856

Costa C, Barberis L, Ambrogio C, Manazza AD, Patrucco E, Azzolino O, Neilsen PO, Ciraolo E, Altruda F, Prestwich GD, Chiarle R, Wymann M, Ridley A, Hirsch E (2007) Negative feedback regulation of Rac in leukocytes from mice expressing a constitutively active phosphatidylinositol 3-kinase gamma. Proc Natl Acad Sci U S A 104:14354–14359

Crackower MA, Oudit GY, Kozieradzki I, Sarao R, Sun H, Sasaki T, Hirsch E, Suzuki A, Shioi T, Irie-Sasaki J, Sah R, Cheng HY, Rybin VO, Lembo G, Fratta L, Oliveira-dos-Santos AJ, Benovic JL, Kahn CR, Izumo S, Steinberg SF, Wymann MP, Backx PH, Penninger JM (2002) Regulation of myocardial contractility and cell size by distinct PI3K-PTEN signaling pathways. Cell 110:737–749

Czech MP (2000) PIP2 and PIP3: complex roles at the cell surface. Cell 100:603–606

Damilano F, Perino A, Hirsch E (2010) PI3K kinase and scaffold functions in heart. Ann N Y Acad Sci 1188:39–45

Damilano F, Franco I, Perrino C, Schaefer K, Azzolino O, Carnevale D, Cifelli G, Carullo P, Ragona R, Ghigo A, Perino A, Lembo G, Hirsch E (2011) Distinct effects of leukocyte and cardiac phosphoinositide 3-kinase gamma activity in pressure overload-induced cardiac failure. Circulation 123:391–399

D'Angelo DD, Sakata Y, Lorenz JN, Boivin GP, Walsh RA, Liggett SB, Dorn GW 2nd (1997) Transgenic Galphaq overexpression induces cardiac contractile failure in mice. Proc Natl Acad Sci U S A 94:8121–8126

Domeier TL, Zima AV, Maxwell JT, Huke S, Mignery GA, Blatter LA (2008) IP3 receptor-dependent Ca2+ release modulates excitation-contraction coupling in rabbit ventricular myocytes. Am J Physiol Heart Circ Physiol 294:H596–H604

Dorn GW 2nd, Force T (2005) Protein kinase cascades in the regulation of cardiac hypertrophy. J Clin Invest 115:527–537

Fougerat A, Gayral S, Gourdy P, Schambourg A, Ruckle T, Schwarz MK, Rommel C, Hirsch E, Arnal JF, Salles JP, Perret B, Breton-Douillon M, Wymann MP, Laffargue M (2008) Genetic and pharmacological targeting of phosphoinositide 3-kinase-gamma reduces atherosclerosis and favors plaque stability by modulating inflammatory processes. Circulation 117:1310–1317

Fougerat A, Gayral S, Malet N, Briand-Mesange F, Breton-Douillon M, Laffargue M (2009) Phosphoinositide 3-kinases and their role in inflammation: potential clinical targets in atherosclerosis? Clin Sci (Lond) 116:791–804

Frangogiannis NG (2008) The immune system and cardiac repair. Pharmacol Res 58:88–111

Ghigo A, Morello F, Perino A, Damilano F, Hirsch E (2011) Specific PI3K isoform modulation in heart failure: lessons from transgenic mice. Curr Heart Fail Rep 8:168–175

Haider S, Tarasov AI, Craig TJ, Sansom MS, Ashcroft FM (2007) Identification of the PIP2-binding site on Kir6.2 by molecular modelling and functional analysis. EMBO J 26:3749–3759

Hamilton CA, Thorin E, McCulloch J, Dominiczak MH, Reid JL (1994) Chronic exposure of bovine aortic endothelial cells to native and oxidized LDL modifies phosphatidylinositol metabolism. Atherosclerosis 107:55–63

Hirsch E, Katanaev VL, Garlanda C, Azzolino O, Pirola L, Silengo L, Sozzani S, Mantovani A, Altruda F, Wymann MP (2000) Central role for G protein-coupled phosphoinositide 3-kinase gamma in inflammation. Science 287:1049–1053

Hirsch E, Lembo G, Montrucchio G, Rommel C, Costa C, Barberis L (2006) Signaling through PI3Kgamma: a common platform for leukocyte, platelet and cardiovascular stress sensing. Thromb Haemost 95:29–35

Hishikawa K, Nakaki T, Marumo T, Hayashi M, Suzuki H, Kato R, Saruta T (1994) Pressure promotes DNA synthesis in rat cultured vascular smooth muscle cells. J Clin Invest 93:1975–1980

Hommers LG, Lohse MJ, Bunemann M (2003) Regulation of the inward rectifying properties of G-protein-activated inwardly rectifying K+ (GIRK) channels by Gbeta gamma subunits. J Biol Chem 278:1037–1043

Huang CL, Feng S, Hilgemann DW (1998) Direct activation of inward rectifier potassium channels by PIP2 and its stabilization by Gbetagamma. Nature 391:803–806

Ishikawa Y, Asaoka Y, Taniguchi T, Tsunemitsu M, Fukuzaki H (1989) Phosphatidylinositol turnover in human monocyte-derived macrophages by native and acetyl LDL. FEBS Lett 246:35–38

Jackson SP, Schoenwaelder SM, Goncalves I, Nesbitt WS, Yap CL, Wright CE, Kenche V, Anderson KE, Dopheide SM, Yuan Y, Sturgeon SA, Prabaharan H, Thompson PE, Smith GD, Shepherd PR, Daniele N, Kulkarni S, Abbott B, Saylik D, Jones C, Lu L, Giuliano S, Hughan SC, Angus JA, Robertson AD, Salem HH (2005) PI 3-kinase p110beta: a new target for antithrombotic therapy. Nat Med 11:507–514

Kasirer-Friede A, Cozzi MR, Mazzucato M, De Marco L, Ruggeri ZM, Shattil SJ (2004) Signaling through GP Ib-IX-V activates alpha IIb beta 3 independently of other receptors. Blood 103:3403–3411

Kim S, Mangin P, Dangelmaier C, Lillian R, Jackson SP, Daniel JL, Kunapuli SP (2009) Role of phosphoinositide 3-kinase beta in glycoprotein VI-mediated Akt activation in platelets. J Biol Chem 284:33763–33772

Laporte SA, Oakley RH, Holt JA, Barak LS, Caron MG (2000) The interaction of beta-arrestin with the AP-2 adaptor is required for the clustering of beta 2-adrenergic receptor into clathrin-coated pits. J Biol Chem 275:23120–23126

Lee SB, Rhee SG (1995) Significance of PIP2 hydrolysis and regulation of phospholipase C isozymes. Curr Opin Cell Biol 7:183–189

Li H, Chen Q, Moss AJ, Robinson J, Goytia V, Perry JC, Vincent GM, Priori SG, Lehmann MH, Denfield SW, Duff D, Kaine S, Shimizu W, Schwartz PJ, Wang Q, Towbin JA (1998) New mutations in the KVLQT1 potassium channel that cause long-QT syndrome. Circulation 97:1264–1269

Li X, Zima AV, Sheikh F, Blatter LA, Chen J (2005a) Endothelin-1-induced arrhythmogenic Ca2+ signaling is abolished in atrial myocytes of inositol-1,4,5-trisphosphate(IP3)-receptor type 2-deficient mice. Circ Res 96:1274–1281

Li Y, Gamper N, Hilgemann DW, Shapiro MS (2005b) Regulation of Kv7 (KCNQ) K+ channel open probability by phosphatidylinositol 4,5-bisphosphate. J Neurosci 25:9825–9835

Li Z, Zhang G, Feil R, Han J, Du X (2006) Sequential activation of p38 and ERK pathways by cGMP-dependent protein kinase leading to activation of the platelet integrin alphaIIb beta3. Blood 107:965–972

Li Y, Zaydman MA, Wu D, Shi J, Guan M, Virgin-Downey B, Cui J (2011) KCNE1 enhances phosphatidylinositol 4,5-bisphosphate (PIP2) sensitivity of IKs to modulate channel activity. Proc Natl Acad Sci U S A 108:9095–9100

Libby P (2002) Inflammation in atherosclerosis. Nature 420:868–874

Lin RC, Weeks KL, Gao XM, Williams RB, Bernardo BC, Kiriazis H, Matthews VB, Woodcock EA, Bouwman RD, Mollica JP, Speirs HJ, Dawes IW, Daly RJ, Shioi T, Izumo S, Febbraio MA, Du XJ, McMullen JR (2010) PI3K(p110 alpha) protects against myocardial infarction-induced heart failure: identification of PI3K-regulated miRNA and mRNA. Arterioscler Thromb Vasc Biol 30:724–732

Lopes CM, Zhang H, Rohacs T, Jin T, Yang J, Logothetis DE (2002) Alterations in conserved Kir channel-PIP2 interactions underlie channelopathies. Neuron 34:933–944

Loussouarn G, Park KH, Bellocq C, Baro I, Charpentier F, Escande D (2003) Phosphatidyl-inositol-4,5-bisphosphate, PIP2, controls KCNQ1/KCNE1 voltage-gated potassium channels: a functional homology between voltage-gated and inward rectifier K+ channels. EMBO J 22:5412–5421

Ma D, Tang XD, Rogers TB, Welling PA (2007) An andersen-Tawil syndrome mutation in Kir2.1 (V302M) alters the G-loop cytoplasmic K+ conduction pathway. J Biol Chem 282:5781–5789

Mackenzie L, Bootman MD, Laine M, Berridge MJ, Thuring J, Holmes A, Li WH, Lipp P (2002) The role of inositol 1,4,5-trisphosphate receptors in Ca(2+) signalling and the generation of arrhythmias in rat atrial myocytes. J Physiol 541:395–409

Marks AR (2000) Cardiac intracellular calcium release channels: role in heart failure. Circ Res 87:8–11

Martin V, Guillermet-Guibert J, Chicanne G, Cabou C, Jandrot-Perrus M, Plantavid M, Vanhaesebroeck B, Payrastre B, Gratacap MP (2010) Deletion of the p110beta isoform of phosphoinositide 3-kinase in platelets reveals its central role in Akt activation and thrombus formation in vitro and in vivo. Blood 115:2008–2013

McMullen JR, Shioi T, Zhang L, Tarnavski O, Sherwood MC, Kang PM, Izumo S (2003) Phosphoinositide 3-kinase(p110alpha) plays a critical role for the induction of physiological, but not pathological, cardiac hypertrophy. Proc Natl Acad Sci U S A 100:12355–12360

McMullen JR, Amirahmadi F, Woodcock EA, Schinke-Braun M, Bouwman RD, Hewitt KA, Mollica JP, Zhang L, Zhang Y, Shioi T, Buerger A, Izumo S, Jay PY, Jennings GL (2007) Protective effects of exercise and phosphoinositide 3-kinase(p110alpha) signaling in dilated and hypertrophic cardiomyopathy. Proc Natl Acad Sci U S A 104:612–617

Mejillano M, Yamamoto M, Rozelle AL, Sun HQ, Wang X, Yin HL (2001) Regulation of apoptosis by phosphatidylinositol 4,5-bisphosphate inhibition of caspases, and caspase inactivation of phosphatidylinositol phosphate 5-kinases. J Biol Chem 276:1865–1872

Myers DE, Fidge NH, Stanton H, Larkins RG (1992) The effects of low density lipoprotein and high density lipoprotein on phosphoinositide hydrolysis in bovine aortic endothelial cells. Atherosclerosis 92:9–16

Naga Prasad SV, Esposito G, Mao L, Koch WJ, Rockman HA (2000) Gbetagamma-dependent phosphoinositide 3-kinase activation in hearts with in vivo pressure overload hypertrophy. J Biol Chem 275:4693–4698

Naga Prasad SV, Barak LS, Rapacciuolo A, Caron MG, Rockman HA (2001) Agonist-dependent recruitment of phosphoinositide 3-kinase to the membrane by beta-adrenergic receptor sequestration kinase 1. A role in receptor. J Biol Chem 276:18953–18959

Naga Prasad SV, Laporte SA, Chamberlain D, Caron MG, Barak L, Rockman HA (2002) Phosphoinositide 3-kinase regulates beta2-adrenergic receptor endocytosis by AP-2 recruitment to the receptor/beta-arrestin complex. J Cell Biol 158:563–575

Nakayama H, Bodi I, Maillet M, DeSantiago J, Domeier TL, Mikoshiba K, Lorenz JN, Blatter LA, Bers DM, Molkentin JD (2010) The IP3 receptor regulates cardiac hypertrophy in response to select stimuli. Circ Res 107:659–666

Nienaber JJ, Tachibana H, Naga Prasad SV, Esposito G, Wu D, Mao L, Rockman HA (2003) Inhibition of receptor-localized PI3K preserves cardiac beta-adrenergic receptor function and ameliorates pressure overload heart failure. J Clin Invest 112:1067–1079

Oliver D, Lien CC, Soom M, Baukrowitz T, Jonas P, Fakler B (2004) Functional conversion between A-type and delayed rectifier K+ channels by membrane lipids. Science 304:265–270

Park KH, Piron J, Dahimene S, Merot J, Baro I, Escande D, Loussouarn G (2005) Impaired KCNQ1-KCNE1 and phosphatidylinositol-4,5-bisphosphate interaction underlies the long QT syndrome. Circ Res 96:730–739

Pasquet JM, Bobe R, Gross B, Gratacap MP, Tomlinson MG, Payrastre B, Watson SP (1999) A collagen-related peptide regulates phospholipase Cgamma2 via phosphatidylinositol 3-kinase in human platelets. Biochem J 342(Pt 1):171–177

Patrucco E, Notte A, Barberis L, Selvetella G, Maffei A, Brancaccio M, Marengo S, Russo G, Azzolino O, Rybalkin SD, Silengo L, Altruda F, Wetzker R, Wymann MP, Lembo G, Hirsch E (2004) PI3Kgamma modulates the cardiac response to chronic pressure overload by distinct kinase-dependent and -independent effects. Cell 118:375–387

Perino A, Ghigo A, Ferrero E, Morello F, Santulli G, Baillie GS, Damilano F, Dunlop AJ, Pawson C, Walser R, Levi R, Altruda F, Silengo L, Langeberg LK, Neubauer G, Heymans S, Lembo G, Wymann MP, Wetzker R, Houslay MD, Iaccarino G, Scott JD, Hirsch E (2011) Integrating cardiac PIP3 and cAMP signaling through a PKA anchoring function of p110gamma. Mol Cell 42:84–95

Perrino C, Naga Prasad SV, Patel M, Wolf MJ, Rockman HA (2005) Targeted inhibition of beta-adrenergic receptor kinase-1-associated phosphoinositide-3 kinase activity preserves beta-adrenergic receptor signaling and prolongs survival in heart failure induced by calsequestrin overexpression. J Am Coll Cardiol 45:1862–1870

Perrino C, Naga Prasad SV, Noma T, Mao L, Yan Z, Kim HS, Smithies O, Rockman HA (2006) Intermittent pressure overload triggers hypertrophy-independent cardiac dysfunction and vascular rarefaction. J Clin Invest 116:1547–1560

Perrino C, Schroder JN, Lima B, Villamizar N, Nienaber JJ, Milano CA, Naga Prasad SV (2007) Dynamic regulation of phosphoinositide 3-kinase-gamma activity and beta-adrenergic receptor trafficking in end-stage human heart failure. Circulation 116:2571–2579

Plaster NM, Tawil R, Tristani-Firouzi M, Canun S, Bendahhou S, Tsunoda A, Donaldson MR, Iannaccone ST, Brunt E, Barohn R, Clark J, Deymeer F, George AL Jr, Fish FA, Hahn A, Nitu A, Ozdemir C, Serdaroglu P, Subramony SH, Wolfe G, Fu YH, Ptacek LJ (2001) Mutations in Kir2.1 cause the developmental and episodic electrical phenotypes of Andersen's syndrome. Cell 105:511–519

Pretorius L, Du XJ, Woodcock EA, Kiriazis H, Lin RC, Marasco S, Medcalf RL, Ming Z, Head GA, Tan JW, Cemerlang N, Sadoshima J, Shioi T, Izumo S, Lukoshkova EV, Dart AM, Jennings GL, McMullen JR (2009) Reduced phosphoinositide 3-kinase (p110alpha) activation increases the susceptibility to atrial fibrillation. Am J Pathol 175:998–1009

Proven A, Roderick HL, Conway SJ, Berridge MJ, Horton JK, Capper SJ, Bootman MD (2006) Inositol 1,4,5-trisphosphate supports the arrhythmogenic action of endothelin-1 on ventricular cardiac myocytes. J Cell Sci 119:3363–3375

Ragab A, Severin S, Gratacap MP, Aguado E, Malissen M, Jandrot-Perrus M, Malissen B, Ragab-Thomas J, Payrastre B (2007) Roles of the C-terminal tyrosine residues of LAT in GPVI-induced platelet activation: insights into the mechanism of PLC gamma 2 activation. Blood 110:2466–2474

Redfield MM (2002) Heart failure–an epidemic of uncertain proportions. N Engl J Med 347:1442–1444

Resink TJ, Tkachuk VA, Bernhardt J, Buhler FR (1992) Oxidized low density lipoproteins stimulate phosphoinositide turnover in cultured vascular smooth muscle cells. Arterioscler Thromb 12:278–285

Ridley AJ, Schwartz MA, Burridge K, Firtel RA, Ginsberg MH, Borisy G, Parsons JT, Horwitz AR (2003) Cell migration: integrating signals from front to back. Science 302:1704–1709

Rockman HA, Koch WJ, Lefkowitz RJ (2002) Seven-transmembrane-spanning receptors and heart function. Nature 415:206–212

Ross R (1986) The pathogenesis of atherosclerosis–an update. N Engl J Med 314:488–500

Ross R (1995) Cell biology of atherosclerosis. Annu Rev Physiol 57:791–804

Ross R (1999) Atherosclerosis–an inflammatory disease. N Engl J Med 340:115–126

Sanguinetti MC, Curran ME, Zou A, Shen J, Spector PS, Atkinson DL, Keating MT (1996) Coassembly of K(V)LQT1 and minK (IsK) proteins to form cardiac I(Ks) potassium channel. Nature 384:80–83

Sasaki T, Sasaki J, Sakai T, Takasuga S, Suzuki A (2007) The physiology of phosphoinositides. Biol Pharm Bull 30:1599–1604

Schwartz PJ, Priori SG, Spazzolini C, Moss AJ, Vincent GM, Napolitano C, Denjoy I, Guicheney P, Breithardt G, Keating MT, Towbin JA, Beggs AH, Brink P, Wilde AA, Toivonen L, Zareba W, Robinson JL, Timothy KW, Corfield V, Wattanasirichaigoon D, Corbett C, Haverkamp W, Schulze-Bahr E, Lehmann MH, Schwartz K, Coumel P, Bloise R (2001) Genotype-phenotype correlation in the long-QT syndrome: gene-specific triggers for life-threatening arrhythmias. Circulation 103:89–95

Shioi T, Kang PM, Douglas PS, Hampe J, Yballe CM, Lawitts J, Cantley LC, Izumo S (2000) The conserved phosphoinositide 3-kinase pathway determines heart size in mice. EMBO J 19:2537–2548

Stojanovic A, Marjanovic JA, Brovkovych VM, Peng X, Hay N, Skidgel RA, Du X (2006) A phosphoinositide 3-kinase-AKT-nitric oxide-cGMP signaling pathway in stimulating platelet secretion and aggregation. J Biol Chem 281:16333–16339

Sui JL, Petit-Jacques J, Logothetis DE (1998) Activation of the atrial KACh channel by the betagamma subunits of G proteins or intracellular Na+ ions depends on the presence of phosphatidylinositol phosphates. Proc Natl Acad Sci U S A 95:1307–1312

Wang Y, Hill JA (2010) Electrophysiological remodeling in heart failure. J Mol Cell Cardiol 48:619–632

Woodcock EA, Kistler PM, Ju YK (2009) Phosphoinositide signalling and cardiac arrhythmias. Cardiovasc Res 82:286–295

Wu X, Zhang T, Bossuyt J, Li X, McKinsey TA, Dedman JR, Olson EN, Chen J, Brown JH, Bers DM (2006) Local InsP3-dependent perinuclear Ca2+ signaling in cardiac myocyte excitation-transcription coupling. J Clin Invest 116:675–682

Yang KC, Jay PY, McMullen JR, Nerbonne JM (2012) Enhanced cardiac PI3Kalpha signalling mitigates arrhythmogenic electrical remodelling in pathological hypertrophy and heart failure. Cardiovasc Res 93:252–262

Yap CL, Anderson KE, Hughan SC, Dopheide SM, Salem HH, Jackson SP (2002) Essential role for phosphoinositide 3-kinase in shear-dependent signaling between platelet glycoprotein Ib/V/IX and integrin alpha(IIb)beta(3). Blood 99:151–158

Yin H, Stojanovic A, Hay N, Du X (2008) The role of Akt in the signaling pathway of the glycoprotein Ib-IX induced platelet activation. Blood 111:658–665

Zima AV, Blatter LA (2004) Inositol-1,4,5-trisphosphate-dependent Ca(2+) signalling in cat atrial excitation-contraction coupling and arrhythmias. J Physiol 555:607–615

Zolles G, Klocker N, Wenzel D, Weisser-Thomas J, Fleischmann BK, Roeper J, Fakler B (2006) Pacemaking by HCN channels requires interaction with phosphoinositides. Neuron 52:1027–1036

Chapter 3
Phosphoinositides in Insulin Action and Diabetes

Dave Bridges and Alan R. Saltiel

Abstract Phosphoinositides play an essential role in insulin signaling, serving as a localization signal for a variety of proteins that participate in the regulation of cellular growth and metabolism. This chapter will examine the regulation and localization of phosphoinositide species, and will explore the roles of these lipids in insulin action. We will also discuss the changes in phosphoinositide metabolism that occur in various pathophysiological states such as insulin resistance and diabetes.

Introduction

Phosphoinositides play an essential role in insulin signaling, serving as a localization signal for a variety of proteins that participate in the regulation of cellular growth and metabolism. This chapter will examine the regulation and localization of phosphoinositide species, and will explore the roles of these lipids in insulin action. We will also discuss the changes in phosphoinositide metabolism that occur in various pathophysiological states such as insulin resistance and diabetes.

The Synthesis and Degradation of Phosphoinositides

Inositol phospholipids can exist in one of eight molecular species, the monophosphorylated phosphatidylinositols, PI(3)P, PI(4)P and PI(5)P, and the polyphoshorylated

D. Bridges · A. R. Saltiel (✉)
Life Sciences Institute, University of Michigan,
210 Washtenaw Ave, Ann Arbor, MI 48109, USA
e-mail: saltiel@lsi.umich.edu

M. Falasca (ed.), *Phosphoinositides and Disease*, Current Topics in Microbiology and Immunology 362, DOI: 10.1007/978-94-007-5025-8_3,

phoshatidylinositols (PI(4,5)P_2 and PI(3,4,5)P_3. Polyphosphoinositides are enzymatically generated by phosphorylation or dephosphorylation. Each phosphoinositide-metabolizing enzyme is able to phosphorylate or dephosphorylate only one specific position on the inositol ring. This reaction is dependent not only on activating signals, but also on the location where the catalysis occurs.

PI(3,4,5)P_3 and PI(3,4)P_2

In most cells, PI(3,4)P_2 and PI(3,4,5)P_3 are the lowest abundance phosphoinositides and are generated mainly by phosphorylation of PI(4)P and PI(4,5)P_2 via different classes of PI3-Kinases. Four Class I PI3-Kinase enzymes (α, β, δ, and γ) have been identified that interact with regulatory subunits. The Class IB PI3K, p110γ, is activated by GPCR signaling and plays an important role in inflammation (reviewed in Wymann et al. 2003).

The other three Class I PI3K isoforms (class IA) are present in complexes with the adaptor proteins p55 and p85. These regulatory subunits contain phosphotyrosine-binding SH2 domains, and recruit the catalytic p110 enzymes to receptors or receptor-associated adapter proteins upon tyrosine phosphorylation (reviewed in Mellor et al. 2012; Vanhaesebroeck et al. 2012). In addition to binding the catalytic subunits, the regulatory subunits may also be activated by other protein–protein interactions. p85 binds the small GTPases Ras (Khwaja et al. 1997; Rodriguez-Viciana et al. 1994) and Rab5 (Chamberlain et al. 2004), as well as the lipid phosphatase PTEN (Barber et al. 2006; Chagpar et al. 2010; Rabinovsky et al. 2009). Once recruited to these sites, the PI3K phosphorylates PI(4,5)P_2 to produce PI(3,4,5)P_3 and initiate downstream signaling events.

PI(3,4,5)P_3 is typically transient in nature and is rapidly degraded following its synthesis. There are three mechanisms by which the levels of PI(3,4,5)P_3 can be reduced, each of which depends on the activity of a lipid phosphatase acting at the $3'$, $4'$, or $5'$ position of the inositol ring. The best studied negative regulator of PI(3,4,5)P_3 is PTEN, a 3-phosphatase that catalyzes the conversion of PI(3,4,5)P_3 to PI(4,5)P_2. PTEN was first identified as a tumor suppressor in 1997 (Li et al. 1997; Steck et al. 1997), then as a PI(3,4,5)P_3 3-phosphatase in 1998 (Maehama and Dixon 1998).

Another mechanism through which PI(3,4,5)P_3 levels are reduced involves the 5-phosphatases SHIP-2 and INPP5K (reviewed in Ooms et al. 2009). In contrast to PTEN, the $5'$-phosphatases generate PI(3,4)P_2 rather than PI(4,5)P_2 from PI(3,4,5)P_3. SHIP-2 contains several important regulatory motifs in addition to its 5-phosphatase domain including a phosphotyrosine-binding SH2 domain, a PTB domain-binding NPXY motif and several polyproline motifs that could bind SH3 domains and other regulatory motifs (reviewed in Erneux et al. 2011). In addition to these protein–protein interaction domains, SHIP-2 is also tyrosine phosphorylated in response to growth factors, increasing its catalytic activity (Habib et al. 1998; Prasad et al. 2009). Based on these data, SHIP-2 appears to be recruited to its substrate

$PI(3,4,5)P_3$ in a phosphotyrosine-dependent manner. In contrast, much less is known about the cell biology of the other 5-phosphatases such as SKIP. SKIP has an amino-terminal SKITCH domain that is essential for its recruitment to the plasma membrane in response to growth factors, but the molecular basis by which this domain functions is unclear (Gurung et al. 2003). Depletion of either SHIP or INPP5K results in increased $PI(3,4,5)_3$ levels, while enhancing Akt signaling in both cultured cells and animals (Clément et al. 2001; Ijuin and Takenawa 2003, 2012; Ijuin et al. 2008; Rommel et al. 2001; Sleeman et al. 2005; Wada et al. 2001).

Although $PI(3,4)P_2$ is a degradation product of $PI(3,4,5)P_3$, some data indicate that this lipid can also support Akt activation via interactions with the PH domain of the kinase (James et al. 1996). Furthermore, $PI(3,4)P_2$ can itself be degraded by the 4 phosphatases INPP4A and INPP4B (Norris et al. 1995, 1997). Work by several groups has shown that like PTEN, INPP4B functions as a tumor suppressor, and that depletion of this enzyme leads to increased Akt signaling (Fedele et al. 2010; Gewinner et al. 2009). These data suggest that sustained levels of $PI(3,4)P_2$ are able to maintain Akt signaling, though whether this is different from $PI(3,4,5)P_3$-supported Akt activity has not yet been established. Notably, in the case of sustained Akt activation in the absence of INPP4B, the end product of this phosphoinositide pathway is PI(3)P rather than $PI(4,5)P_2$. It will be interesting to see if negative regulation of Akt by 4-phosphatases differs from negative regulation by 5-phosphatases, as this will be important for insulin-sensitizing interventions that aim to inhibit these catalytic activities.

PI(3)P

PI3Ks can also phosphorylate PI to produce PI(3)P. Two PI3K classes, the class II PI3K (PI3KC2A, PI3KC2B, and PI3KC2G) and the class III PI3K (Vps34, also known as PI3KC3) have been shown to generate PI(3)P in vivo and in vitro. Vps34 is responsible for all detectable PI3K activity in yeast (Schu et al. 1993). However, it is not known what proportion of PI(3)P is produced by Vps34 in mammals. In yeast, Vps34 exists in two functionally distinct complexes, one of which regulates autophagy while the other regulates vesicle trafficking (Kihara et al. 2001). The Vps34 complex is recruited to endosomes by Rab5 (Christoforidis et al. 1999), where it synthesizes early endosomal PI(3)P. Like the Class I PI3K's, Vps34 is regulated by an adaptor protein (Vps15 also known as p150), which is also a serine/threonine protein kinase (reviewed in Backer 2008).

Less is known about the three Class II PI3-Kinases (see Mazza and Maffucci 2011 for a more detailed review). In addition to their kinase domains, class II PI3Ks include phosphoinositide-binding PX and C2 domains. Although most PX domains are able to bind PI(3)P (Yu and Lemmon 2001), lipid-binding studies with the class II PI3-Kinases have shown an affinity of this domain for $PI(4,5)P_2$ (Stahelin et al. 2004). The significance of this interaction has not been established, especially in light of subcellular localization studies demonstrating that PI3K-C2α

is located in the trans-golgi network, via a PX and C2 domain-independent process (Domin et al. 2000). In contrast to class I and class III PI3-Kinases, there have been no regulatory subunits described for the class II enzymes. These kinases have been reported to be activated by calcium (Wen et al. 2008), clathrin (Domin et al. 2000; Gaidarov et al. 2001), and the small GTPase TC10 (Maffucci et al. 2003), although the precise mechanism of activation is currently uncertain. Also unknown is the relative redundancy of the three class II PI3Ks in the regulation of mammalian PI(3)P levels.

A third potential source of PI(3)P is through a cascade involving class I PI3K enzymes that generate $PI(3,4,5)P_3$, a lipid which is then dephosphorylated by 4- and 5-phosphatases to produce PI(3)P. The identification of the class I PI3K p110β as well as 4- and 5-phosphatases as Rab5 effectors suggests that Rab5 might also control this pathway (Shin et al. 2005).

The primary PI(3)P-degrading enzymes in eukaryotic cells are the myotubularins. Although there are no described myotubularins in budding yeast, there are multiple myotubularin isoforms in worms, flies, and mammalian cells (reviewed in Begley and Dixon 2005; Nicot and Laporte 2008). Due to this redundancy, it is difficult to determine precisely which myotubularins have non-redundant roles in a particular signaling process.

Regulation of $PI(3,5)P_2$ and PI(5)P Levels

In contrast to the multi-gene regulation of most other phosphoinositides, in both yeast (Duex et al. 2006a; Gary et al. 2002) and mammals (Ikonomov et al. 2011), the PI5-Kinase PIKfyve (also known as Fab1) is solely required for the synthesis of $PI(3,5)P_2$. PIKfyve exists in a complex with at least 4 other proteins in yeast, the scaffolding proteins Vac7 and Vac14, a 5-phosphatase named Fig4 (Sac3 in mammals), and a $PI(3,5)P_2$ sensor protein Atg18 (Botelho et al. 2008; Gary et al. 2002; Jin et al. 2008).

Both Vac14 and Sac3 have been identified in mammals and are also part of the PIKfyve complex (Ikonomov et al. 2007, 2009, 2010; Sbrissa et al. 2007, 2008; Sbrissa and Shisheva 2005). The functional role of Fig4/Sac3 is complicated by the fact that while it acts as a 5-phosphatase, and is able to dephosphorylate $PI(3,5)P_2$ in vitro (Ikonomov et al. 2009; Rudge et al. 2004; Yuan et al. 2007); its deletion in yeast or mammalian cells paradoxically results in a decrease in $PI(3,5)P_2$ levels (Botelho et al. 2008; Chow et al. 2007; Duex et al. 2006b). The current model suggests that Fig4 is required for assembly of the PIKfyve complex, and it is therefore a necessary component, while also potentially functioning as a $PI(3,5)P_2$ phosphatase in vivo. Atg18, which by homology has four mammalian isoforms (WIPI1-4), is a negative regulator of $PI(3,5)P_2$ synthesis (Efe et al. 2007). The current hypothesis is that Atg18 directly binds $PI(3,5)P_2$ (Dove et al. 2004) upon which it inhibits the kinase activity of PIKfyve. Vac7, which is required for

PIKfyve function in yeast (Bonangelino et al. 1997; Duex et al. 2006b; Gary et al. 2002) and has not yet been identified in mammals.

Heterozygosity of PIKfyve or loss of its regulator Vac14 (also known as ArPIKfyve) correlates with near equivalent reductions in PI(5)P (Ikonomov et al. 2011; Zhang et al. 2007). These data suggest that PIKfyve is also required for PI(5)P synthesis. It is currently uncertain as to whether PIKfyve is required for synthesis of all of the PI(5)P in the cell, whether it is generated directly via phosphorylation of PI, or indirectly in concert with one or more myotubularin enzymes that catalyze the conversion of PI(3,5)P_2 to PI(5)P.

Generation of PI(3,5)P_2 requires the substrate PI(3)P, and therefore defects in PI(3)P synthesis may also result in reductions in both PI(3,5)P_2 and PI(5)P. In addition to using PI(3)P as a substrate, PIKfyve is recruited to sites of PI(3)P via a presumed PI(3)P-binding FYVE domain at its amino terminus. Therefore, the levels of PI(3,5)P_2 may be regulated by both PI(3)P availability due to substrate flux and direct interaction with the enzyme to change its cellular localization. Whether there are other mechanisms involved in the regulation of this enzyme remains uncertain, although there have been reports that it is directly phosphorylated by Akt (Berwick et al. 2004).

In addition to the dephosphorylation of PI(3,5)P_2 to produce PI(5)P by myotubularins, some PI(3,5)P_2 may also be dephosphorylated into PI(3)P by the essential PIKfyve complex component factor- induced gene; (also known as Sac3). While this protein is required for activity of PIKfyve in yeast and mammalian systems, it can also support PI(3,5)P_2 dephosphorylation at the 5′ position in vitro (Sbrissa et al. 2007; Yuan et al. 2007). The extent to which this phosphatase activity is relevant in regenerating PI(3)P from PI(3,5)P_2 in vivo is currently unknown.

The flux of PI(3)P into PI(3,5)P_2 is substantial, as cell culture models in which PIKfyve or its regulators are disrupted also result in substantially increased levels of PI(3)P (Chow et al. 2007; Ikonomov et al. 2011; Jin et al. 2008; Zhang et al. 2007). Caution should therefore be taken in the analysis of PIKfyve-deficient systems as the phenotype could be the result of reduced PI(3,5)P_2, reduced PI(5)P, or increased PI(3)P.

Metabolism of PI(4)P

Similar to PI(3)P, PI4-Kinases are essential not only in generating the pool of PI(4)P, but also in generating the precursor to PI(4,5)P_2. There are four PI4-Kinases in mammalian cells, PI4KIIα, PI4KIIβ, PI4KIIIα, and PI4KIIIβ as grouped by wortmannin sensitivity (Balla and Balla 2006). The major pool of PI(4)P is in the Golgi apparatus, and PI4-Kinase enzymes appear in this organelle, although there is also some PI(4)P detectable at the plasma membrane. In addition to PI4-Kinase activity, PI(4)P levels are regulated by phosphorylation to PI(4,5)P_2 as described below, or by dephosphorylation by the Sac1 phosphatase (Cleves et al. 1989; Foti et al. 2001).

Synthesis and Degradation of PI(4,5)P_2

There are two primary routes through which the synthesis of PI(4,5)P_2 occurs: the phosphorylation of PI(4)P by one of three PIP5-Kinases, PI5Kα, PI5Kβ, PI5Kγ, (Boronenkov and Anderson 1995; Divecha et al. 1995; Zhang et al. 1997), or the degradation of PI(3,4,5)P_3 by the 3′ phosphatase PTEN (as described above). There is also an alternative route, in which PI(5)P can be phosphorylated by one of the two PIP4K isoforms to generate PI(4,5)P_2 (Bultsma et al. 2010; Rameh et al. 1997). This pathway is thought to generate only a small amount of total PI(4,5)P_2, and may be essential for a subset of trafficking or nuclear processes (Bultsma et al. 2010; Jones et al. 2006; Sarkes and Rameh 2010).

In most conditions, the degradation of PI(4,5)P_2 is largely accomplished through the activity of 5-phosphatases, reconverting the lipid back to PI(4)P. There are 14 5-phosphatases in the mouse genome, with varying and sometimes poorly characterized substrate specificity toward PI(4,5)P_2, PI(3,4,5)P_3, PI(4,5)P_2, PI(5)P, and even soluble inositol phosphates. Of these, two enzymes (INPP5B and OCRL) have been noted as essential mediators of endocytosis, likely via the removal of PI(4,5)P_2 from maturing early endosomes (Bohdanowicz et al. 2011; Mao et al. 2009; Shin et al. 2005; Vicinanza et al. 2011; Williams et al. 2007; Zoncu et al. 2009). The in vivo role of the other 5-phosphatase enzymes will require careful study of PI(4,5)P_2 and PI(3,4,5)P_3 dynamics, as it is difficult to separate effects of these two lipids.

Relative Levels and Locations of Phosphoinositides

Phosphatidylinositol is a minor membrane constituent comprising approximately only 5 % of membrane lipids. Moreover, phosphatidylinositol itself comprises greater than 90 % of the total cellular phosphoinositides (Auger et al. 1989; Lodhi et al. 2008; Patton et al. 1982). Therefore, the overall abundance of any phosphorylated phosphatidylinositol species is quite low. In most eukaryotic cells, PI(4)P and PI(4,5)P_2 are the most abundant polyphosphorylated phosphoinositides, followed by PI(5)P and PI(3)P and with PI(3,5)P_2, PI(3,4)P_2, and PI(3,4,5)P_3 all present at lower levels (Auger et al. 1989). While the total abundance of phosphoinositides varies widely, their local concentration is largely determined by their restricted synthesis. Fluorescent lipid-binding probes that bind endogenous phosphoinositide species in live cells have advanced our understanding of phosphatidylinositol biology by allowing us to monitor the cellular dynamics of phosphatidylinositol synthesis, degradation, and location (reviewed in Downes et al. 2005). While some phosphorylated phosphoinositides such as PI(4,5)P_2, PI(3,4)P_2, and PI(3,4,5)P_3 are largely found at the plasma membrane, PI(3)P and potentially PI(4,5)P_2 reside in endosomal structures, while PI(4)P is largely present in the Golgi apparatus (reviewed in Di Paolo and de Camilli 2006). There is also substantial overlap in these regions, with PI(3)P present at the plasma membrane of cultured adipocytes and myocytes (see below). As of the writing of this review, there are no widely used probes for visualizing either PI(3,5)P_2 or PI(5)P, so their subcellular distribution is still unclear.

Acute Changes in Phosphoinositides

Phosphorylated phosphoinositides are interconverted during normal trafficking processes or acutely in response to extracellular stimuli. As a spatially restricted and rapidly interchangeable module, phosphoinositides are used in multiple signaling paradigms. Work by Cantley and coworkers in the early 1980s showed that tyrosine kinase-dependent growth factor signaling increased the levels of all four D3 phosphorylated inositol lipids, especially PI(3,4,5)P_3 (Auger et al. 1989). While PI(3,4)P_2- and PI(3,4,5)P_3-dependent signaling processes have been widely studied, the roles of growth factor-stimulated changes in PI(3)P and PI(3,5)P_2 have only recently been the focus of investigation.

Because lipids are altered both by trafficking and hormonal signaling events, and because processes such as insulin-stimulated glucose uptake and glucose-stimulated insulin release are also trafficking events, it is at times difficult to tease apart which lipid alterations are due to primary hormonal signals, and which are due to trafficking responses to those signals.

Phosphoinositides are Key Second Messengers in Insulin Action

Phosphoinositides are essential second messengers for insulin signal transduction. Once generated, these lipids serve to localize the proteins with which they bind in specific compartments in cells, leading to activation of protein kinases, G proteins and other pathways.

Regulation of Akt Signaling Pathways

Akt was first identified as the homolog of a viral oncogene v-Akt (Staal and Hartley 1988), and has three isoforms in mammalian genomes. In addition to a carboxy-terminal AGC kinase domain, each isoform of Akt has an amino-terminal Pleckstrin homology (PH) domain which is able to bind PI(3,4)P_2 or PI(3,4,5)P_3 (James et al. 1996). Correspondingly, the activation and phosphorylation of Akt by growth factors is a wortmannin-sensitive process (Alessi et al. 1996; Burgering and Coffer 1995; Franke et al. 1995).

The activation of Akt is a multi-step process involving the recruitment of Akt to the plasma membrane via direct binding to PI(3,4,5)P_3 or PI(3,4)P_2, bringing the enzyme within proximity of two activating protein kinases, PDK1 and mTORC2 (mTOR Complex 2), which phosphorylate Akt on Threonine 308 and Serine 473, respectively. Phosphorylation of both of these residues is required for complete activation of Akt (Alessi et al. 1997; Sarbassov et al. 2005). The role of phosphatidylinositol-3 kinases in mTORC2 activity is unclear, and the degree to which this kinase activity is increased by insulin remains uncertain. However, PDK1 has a PH domain which is itself able to bind plasma membrane lipids (Currie et al. 1999). Ablation of the PDK1 lipid-binding activity results in a complete blockade in Akt phosphorylation in embryonic fibroblasts (McManus et al. 2004).

Another route for Akt activation is through the innate immune kinases IKKε and TBK1. Several recent reports suggest that overexpression of these kinases results in Akt phosphorylation on both Threonine 308 and Serine 473 (Guo et al. 2011; Ou et al. 2011; Xie et al. 2011b). These groups report differing results with respect to the sensitivity of this activation to wortmannin, so it is currently unclear if PI3Ks are essential for this alternate route of Akt activation.

Regulation of mTORC1 Signaling

The nutrient- and growth factor-responsive protein kinase complex TOR complex I (TORC1) is regulated by phosphoinositides via at least two separate mechanisms. Early studies revealed that the PI3K inhibitor wortmannin blocks activation of mTORC1 by insulin (Petritsch et al. 1995; Weng et al. 1995), and it was later shown that this inhibition was at least in part due to the Akt-mediated inactivation of the Rheb GAP complex TSC1/2 (Dan et al. 2002; Inoki et al. 2002, 2003; Manning et al. 2002; Tee et al. 2003).

In addition to this PI(3,4,5)P_3-dependent effect, recent work has also suggested that mTORC1 activity may be influenced by generation of PI(3)P. Genetic screens in yeast have shown synthetic lethality between deletion mutants of the Vps34 complex and Tor1 (Zurita-Martinez et al. 2007). Furthermore, there have been several reports of Vps34 dependence of TORC1 activity in mammalian cells (Byfield et al. 2005; Nobukuni et al. 2005; Yoon et al. 2011). This work has also implicated the PI(3)P binding protein phospholipase D as another activator of TORC1 (Fang et al. 2001, 2003; Stahelin et al. 2004; Xu et al. 2011; Yoon et al. 2011), although the direct mechanism by which phospholipase D affects TORC1 function is still unclear.

Work primarily by David Sabatini's group has also implicated lysosomal localization as an important component of mTORC1 activation. A second set of small GTPases, the Rag GTPases, along with the vacuolar ATPase are also important regulators of TORC1 activation and lysosomal localization (Binda et al. 2009; Kim et al. 2008; Sancak et al. 2008, 2010; Zoncu et al. 2011). Since phosphoinositides such as PI(3)P and PI(3,5)P_2 are important components of vacuolar function and the vacuolar proton pump (Gary et al. 1998), it will be interesting to see if these lipids also play a role in lysosomal localization and activation of the TORC1 complex.

Phosphoinositides Play Key Roles Downstream of Early Signaling in Insulin Action

Insulin is released from the pancreas postprandially and stimulates the uptake and storage of glucose and other nutrients into muscle and fat tissue. Phosphoinositides regulate each of these processes at multiple steps.

GLUT4 Trafficking

Insulin stimulates glucose uptake from blood into peripheral tissues. In adipocytes and muscle cells, this is largely dependent on the facilitative glucose transporter GLUT4 (Abel et al. 2001; Zisman et al. 2000). This transporter is normally sequestered in internal storage vesicles that exocytose and fuse with the plasma membrane upon insulin stimulation, allowing glucose to travel down a concentration gradient into cells (recently reviewed in Foley et al. 2011; Rowland et al. 2011). The recycling, translocation, and fusion of GLUT4-containing vesicles in response to insulin can be broken down into several general steps, each of which is regulated by phosphoinositides.

The movement of GLUT4-containing vesicles toward and then fusion with the plasma membrane comprises multiple steps. Early inhibitor studies showed that the PI3-Kinase inhibitor wortmannin blocks insulin-stimulated glucose uptake and GLUT4 translocation in multiple cell types (Clarke et al. 1994; Kanai et al. 1993; Okada et al. 1994). Further work employing Akt inhibitors and knockdowns have suggested that the effects of wortmannin may be mediated through a PI(3,4,5)P_3 - Akt signaling pathway (Gonzalez and McGraw 2009; Green et al. 2008; Jiang et al. 2003; Ng et al. 2008; Tan et al. 2011). Once activated by PI(3,4,5)P_3, Akt has several important targets in GLUT4 trafficking. The Rab10 family GTPase activating protein (GAP) AS160 is phosphorylated and inactivated by insulin, allowing for translocation of GLUT4 storage vesicles (reviewed in Eguez et al. 2005; Foley et al. 2011; Sano et al. 2003). Inactivation of AS160 may lead to the increased activity of its targets, the Rab10 family GTPases. However, activation of these G proteins by insulin has not been demonstrated, and their downstream effectors have not been defined (Sano et al. 2007; Sun et al. 2010).

Another recently described Akt substrate in GLUT4 translocation is the RGC1/2 complex (Chen et al. 2011; Gridley et al. 2006). Akt inhibits this Ral GAP allowing for activation of that GTPase on Glut4-containing vesicles.

Once activated, RalA targets these vesicles to the plasma membrane by interacting with the exocyst complex, which is assembled in response to insulin and is required for efficient fusion of GLUT4 vesicles (Chen et al. 2007; Inoue et al. 2003, 2006).

The final fusion step of GLUT4 trafficking has been proposed to be regulated by two other Akt substrates. Synip, a masking protein for the targeting SNARE syntaxin-4 has been proposed to be phosphorylated and disinhibited by Akt as an important positive regulator of GLUT4 fusion (Okada et al. 2007; Yamada et al. 2005), but this finding has been controversial (Sano et al. 2005). A more recent report has implicated CDP138 in the regulation of GLUT4 fusion. Loss of this Akt substrate appears to have no effect on translocation, but blocks fusion events (Xie et al. 2011a). Interestingly, the C2 domain of this protein has been shown to interact with mixed intracellular lipids, but whether it interacts with specific phosphoinositides is currently unknown.

Apart from PI(3,4,5)P_3-dependent Akt substrates, PI(4,5)P_2 has also been proposed to be necessary for vesicle exocytosis. This phospholipid is essential for clustering and activation of t-SNAREs in a variety of exocytic contexts (James et al. 2008; Mima and Wickner 2009; Vicogne et al. 2006), and also interacts directly with a number of exocyst components. Both Exo70 (He et al. 2007; Liu et al. 2007)and Sec3 (Yamashita et al. 2010) interact with PI(4,5)P_2, but the relevance of these interactions to insulin-stimulated GLUT4 trafficking has not yet been reported.

In addition to its role in GLUT4 exocytosis, PI(4,5)P_2 is also an essential mediator of endocytosis (reviewed in Di Paolo and de Camilli 2006). Since GLUT4 must be retrieved from the plasma membrane and sorted, endocytosis is an essential compartment of GLUT4 biology. Cargo is internalized first into compartments that are PI(3,4,5)P_3 and PI(4,5)P_2 positive. These very early endosomal structures are also positive for the small GTPase Rab5 and the adaptor protein APPL1. As these vesicles mature, the levels of PI(3,4,5)P_3 and PI(4,5)P_2 decrease, whereas the levels of PI(3)P increase (Bohdanowicz et al. 2011; Zoncu et al. 2009). In most cells, PI(3)P exists primarily on early endosomal structures. Along with Rab5, PI(3)P serves as a coincidence detector for EEA1 and plays an essential role in homotypic endosomal fusion. This aggregation process feeds into the recycling endosome and is important for the internalization and sorting of membrane components, including GLUT4.

PI(3)P is likely generated by the recruitment of the two Rab5 PI3 kinase effectors, p110β and Vps34 (Christoforidis et al. 1999; Kurosu and Katada 2001). These very early APPL1 positive endosomal structures also serve as signaling platforms for Akt and EGF signaling (Bohdanowicz et al. 2011; Saito et al. 2007; Schenck et al. 2008; Tan et al. 2010). The effect of these transient structures appears to be context dependent. In some instances, the loss of APPL1 leads to decreased Akt signaling (Saito et al. 2007; Schenck et al. 2008), whereas in other instances, such as phagocytosis and EGF signaling, APPL1 serves as a negative regulator of signaling (Bohdanowicz et al. 2011; Miaczynska et al. 2004). This contradiction may be resolved by the temporal nature of these signaling events. If a particular signaling context requires sustained Akt activation, then arresting endosomes at the PI(3,4,5)P_3 or PI(3,4)P_2-positive stage may allow for sustained activation of Akt. On the other hand, rapid agonist-mediated activation could require transient generation of APPL1 positive structures, and disruption of those may inhibit the extent of Akt activation.

These studies have been largely performed in undifferentiated dividing cells, and it will be interesting to test the role of APPL1- and PI(3)P-positive endosomal structures in GLUT4 internalization. Ablation of APPL1 is inhibitory toward insulin-stimulated glucose uptake in cultured adipocytes, but it is difficult to decouple the effects of this knockdown on Akt signaling from other potential trafficking effects (Saito et al. 2007). GLUT4 continuously traffics between the plasma membrane and various pools of internalized, recycled, and pre-exocytic vesicles. Therefore, internal pools of PI(3)P are expected to be essential for the efficient movement of GLUT4 through these stages.

In addition to the roles of PI(3)P in endosomal trafficking, cultured muscle and adipocytes also have a second major pool of this lipid at or near the plasma membrane (Falasca et al. 2007; Kong et al. 2006; Lodhi et al. 2008; Maffucci et al. 2003). The precise role of this PI(3)P pool is currently unclear, but reductions in this pool are co-incident with reductions in insulin-stimulated GLUT4 trafficking and glucose transport (Chaussade et al. 2003; Falasca et al. 2007; Kong et al. 2006; Lodhi et al. 2008; Maffucci et al. 2003). Furthermore, exogenous transfection of this lipid into cultured cells results in increased translocation, but not fusion of GLUT4 with the plasma membrane (Ishiki et al. 2005). These data suggest that in insulin-responsive tissues, PI(3)P may play a role in the recruitment of GLUT4 to a location proximal to sites of fusion.

In addition to the well-documented roles of PI(4,5)P_2 and PI(3,4,5)P_3 in GLUT4 trafficking, there have also been reports of positive roles of PI(3,5)P_2 or PI(5)P in the regulation of insulin-stimulated glucose transport. Overexpression of dominant interfering mutants, and knockdown experiments suggests a correlation between reduced PI(3,5)P_2 or PI(5)P levels with reduced insulin-stimulated GLUT4 translocation and glucose uptake (Ikonomov et al. 2002; Sbrissa et al. 1999, 2004). The stages of GLUT4 trafficking and the potential direct targets of PI(3,5)P_2 or PI(5)P are unknown.

Insulin Secretion

Phosphoinositides may also play a role in the glucose-dependent secretion of insulin from pancreatic beta cells. In line with a general role in SNARE assembly, PI(4,5)P_2 has also been shown to be essential for insulin release from pancreatic beta cells. Direct application of PI(4)P or PI(4,5)P_2 to cultured beta cells promotes the priming of insulin granules (Olsen et al. 2003; Waselle et al. 2005). Furthermore, either sequestration of PI(4,5)P_2 by overexpression of the PI(4,5)P_2-binding GFP-PLCδ1 (Lawrence and Birnbaum 2003; Waselle et al. 2005), or by anti-

$PI(4,5)P_2$ antibodies (Olsen et al. 2003) reduces insulin secretion from beta cells. Finally, siRNA-mediated depletion of either PI4Kβ (which generates PI(4)P) or PIP5Kγ (which generates $PI(4,5)P_2$ from PI(4)P) also reduces insulin secretion in cultured beta cells (Waselle et al. 2005).

PI(3)P may play a role in regulated exocytosis of insulin vesicles. Work by Tania Maffucci's group has uncovered a positive role for PI3K-C2α, which generates PI(3)P, in insulin exocytosis (Dominguez et al. 2010), perhaps reflecting a general pro-exocytic role of this phospholipid, such as has been described in neurosecretory cells (Meunier et al. 2005; Wen et al. 2008).

Glycogen and Lipid Synthesis

Both glycogen and lipid storage in metabolically responsive tissues are largely dependent on Akt signaling pathways. In muscle and adipose tissue, both substrate availability (through glucose uptake as described above) and lipid and glycogen synthesis are sensitive to wortmannin (Le Marchand-Brustel et al. 1995; Shepherd et al. 1995). The direct pathways by which $PI(3,4,5)P_3$ regulates lipid synthesis involves both Akt- and mTORC1- dependent effects and transcriptional and post-translational changes (Huffman et al. 2002; Laplante and Sabatini 2010; Peterson et al. 2011; Yecies et al. 2011). Whether there are other Akt-independent roles of phosphoinositides in lipid and glycogen storage is not currently known.

Genetics and Pharmacology of Phosphoinositide Disruption

Due to the essential nature of phosphoinositides in several biological processes, there have been intensive studies directed toward understanding their roles in whole-animal physiology.

$PI(3,4,5)P_3$ and $PI(3,4)P_2$

The majority of animal models describing phosphoinositide disruption involve enzymes involved in the synthesis and degradation of $PI(3,4,5)P_3$ and $PI(4,5)P_2$. As described above, $PI(3,4,5)P_3$ is generated by class I PI3-Kinases. Ablation of the most widely expressed catalytic isoforms (α and β) both result in embryonic lethality (Bi et al. 1999, 2002). Somewhat surprisingly, a kinase-dead knock-in for p110β is not embryonic lethal, suggesting that the absence of the protein rather than the loss of catalytic activity is what leads to the lethality (Ciraolo et al. 2008). Transgenic mice expressing kinase-dead p110β have mild peripheral insulin resistance. Expression of an inactivating kinase domain mutation in p110α is also embryonic lethal when homozygous, and peripherally insulin resistant as a heterozygote (Foukas et al. 2006). While heterozygotes of p110α and p110β have not been shown to have a metabolic phenotype, compound double p110α/β heterozygotes have mild glucose and insulin intolerance (Brachmann et al. 2005). Together, these data suggest that whole-animal reductions of p110α, and to a lesser extent p110β, lead to peripheral insulin resistance.

To test tissue-specific roles of these isoforms, floxed alleles have been generated for both p110α and p110β. Liver-specific knockout of p110α, either through use of albumin-Cre or transfection with adenoviral Cre ablates insulin-stimulated PI3K activity, with associated reductions in Akt phosphorylation. Based on both insulin-tolerance tests, and hyperinsulinemic/euglycemic-clamp studies, these animals have reduced insulin-induced hypoglycemia, associated with a lack of suppression of endogenous glucose production (Sopasakis et al. 2010). This is consistent with the phenotype of hepatic insulin resistance for both acute and chronic depletion of p110α in the liver. Interestingly, chronic reductions of liver p110α on a normal chow diet does not cause significant reductions in the serum lipid profile (Chattopadhyay et al. 2011; Sopasakis et al. 2010), as it does for the acute reduction of liver p110α or the liver insulin receptor knockout mice (Michael et al. 2000; Sopasakis et al. 2010). There were, however, reductions in high-fat diet-induced hypercholesterolemia and hepatosteatosis (Chattopadhyay et al. 2011).

Although p110β is reported to provide only a small fraction of insulin-induced PI(3,4,5)P_3, liver-specific knockout of this enzyme also causes substantial hepatic insulin resistance as measured by insulin-tolerance tests, but surprisingly has only very limited effects on Akt phosphorylation (Chattopadhyay et al. 2011; Jia et al. 2008). These mice had no detectable differences in serum or hepatic lipid profiles, even when challenged by a high-fat diet (Chattopadhyay et al. 2011).

Negative Regulators of PI(3,4,5)P_3

Since decreased PI(3,4,5)P_3 results in decreased insulin signaling, insulin resistance could be potentially ameliorated by increasing PI(3,4,5)P_3 levels. Therefore, negative regulators of PI(3,4,5)P_3 have been explored as model systems by which the deleterious effects of high-fat diet-induced insulin resistance may be corrected. Converse to the PI(3,4,5)P_3 reductions present in Class I PI3K knockout models, PTEN knockout mice have increased levels of this phosphoinositide due to an inability to dephosphorylate PI(3,4,5)P_3 into PI(4,5)P_2.

Homozygous deletion of PTEN causes embryonic lethality (Podsypanina et al. 1999; Suzuki et al. 1998), but heterozygotes have reduced fasting glucose levels and enhanced insulin sensitivity (Wong et al. 2007). Antisense oligonucleotides have also been used to probe PTEN function in whole animals. These oligonucleotides studies, which reduced PTEN levels 70–90 % in liver and adipose tissue, caused normalization of blood glucose and insulin in ob/ob mice (Butler et al. 2002). PTEN has been extensively studied in tissue-specific knockout models (reviewed in

Knobbe et al. 2008). Mice with a targeted knockout of PTEN in skeletal and cardiac muscle using Mck-Cre were resistant to high-fat diet-induced insulin resistance (Wijesekara et al. 2005), but surprisingly there were no detectable changes in body weight either under normal or high-fat diet fed conditions. These mice were also characterized by enhanced Akt phosphorylation and glycogen accumulation in their soleus muscles. It is unclear why enhanced insulin sensitivity of these mice is only present after a high-fat diet challenge and not under normal diet conditions.

Two groups have independently generated Albumin-Cre driven liver-specific PTEN knockout mice (Horie et al. 2004; Stiles et al. 2004). Both groups report substantial hepatomegaly, suppressed fasting glucose levels, and increased insulin sensitivity. These data are consistent with a positive role for PI(3,4,5)P_3 in cell division and the suppression of gluconeogenesis. One group also reported substantial hepatosteatosis and accumulation of glycogen in the liver-specific PTEN knockout mice, likely due to accelerated insulin-induced lipogenesis and glycogenesis (Horie et al. 2004). Some of these mice also developed hepatocellular carcinomas, underlying the important role of PTEN both in normal physiology and as a tumor suppressor.

Insulin leads to increased glucose uptake and lipid/glycogen storage in adipose tissue. Targeted deletion of PTEN from white adipose tissue using the AP2-Cre promoter revealed increased Akt signaling, an accumulation of lipids in white adipose tissue depots, and an increase in whole-body insulin sensitivity (Kurlawalla-Martinez et al. 2005).

PI(3,4,5)P_3 and PTEN have also been implicated in the regulation of pancreatic function. Two groups have explored the phenotype of beta-cell-specific knockouts of PTEN, driven by RIP (Rat Insulin Promoter)-Cre (Nguyen et al. 2006; Stiles et al. 2006). Both groups report hyperplasia of islets, increased glucose induced insulin secretion, decreased fasting glucose, and reduced insulin sensitivity. These data highlight a role for PTEN in suppressing islet growth and dampening glucose-induced insulin secretion from islet cells.

PI(3,4,5)P_3 levels can also be reduced through a 5-phosphatase such as SHIP2 or SKIP. Whole-body SHIP2 knockout mice are viable although animals were slightly smaller than littermate controls (Sleeman et al. 2005). On normal chow diet, these mice had no apparent defects in glucose homeostasis, although they did have increased insulin-stimulated Akt phosphorylation. These animals were also resistant to high-fat diet-induced hyperglycemia and hyperinsulinemia, consistent with a role of SHIP2 in the development of insulin resistance (Sleeman et al. 2005).

Another 5-phosphatase that has been suggested to negatively regulate PI(3,4,5)P_3 levels is INPP5K (also known as SKIP). While homozygous deletion of this enzyme is embryonic lethal, heterozygotes are insulin sensitive with modest resistance to diet-induced obesity. These data are consistent with a negative role of INPP5K in Akt signaling (Ijuin and Takenawa 2003, 2012; Ijuin et al. 2008; Xiong et al. 2009).

Phosphoinositide-Metabolizing Enzymes as Drug Targets in Diabetes

Due to their major roles in cancer, diabetes, and inflammation, phosphoinositides have been attractive therapeutic targets for a wide variety of pharmaceuticals. While overgrowth of cells in a PI3-Kinase dependent manner is an important target for anti-cancer therapy (Wong et al. 2010), the opposite is generally desired to improve insulin sensitivity where increases in PI(3,4,5)P_3 can restore insulin action. The primary proposed pharmacological targets are therefore the 3-, 4-, and 5-phosphatases that negatively regulate PI(3,4,5)P_3 levels. Of these both 3- and 5-phosphatase targeting drugs have been described.

The increased insulin sensitivity in PTEN loss of function models led to the hypothesis that pharmacological inhibition of PTEN may improve insulin resistance. Bisperoxovanadium compounds inhibit PTEN in vitro and result in increased Akt activation in cells (Rosivatz et al. 2006; Schmid et al. 2004). These vanadate derivatives have dramatic positive effects on the stimulation of glucose transport in cells (Clark et al. 1985; Duckworth et al. 1988; Fisher et al. 2002; Nolte et al. 2003; Yu et al. 1997). In animals, these chemicals cause reductions in circulating glucose levels consistent stimulation of glucose uptake and suppression of gluconeogenesis (Brand and Hamel 1999; Westergaard et al. 1999). Complicating these results however is the insight that bisperoxovanadium compounds increase tyrosine phosphorylation of the insulin receptor and IRS proteins (Nolte et al. 2003; Yu et al. 1997). These findings suggest that these drugs also inhibit the Protein Tyrosine Phosphatase family tyrosine phosphatases, another important set of negative regulators of insulin signaling that work upstream of PI3K activity.

Another lipid phosphatase that has been suggested as a drug target is 5-phosphatase SHIP-2. An inhibitor to this enzyme developed by Astellas Pharmaceuticals has shown some promise, as it also increases Akt phosphorylation and glucose uptake in cells, as well as hypoglycemia in mice (Suwa et al. 2009).

Summary

Since their discovery as major modulators of hormonal action, phosphoinositides have been under intensive investigation. The study of these lipids has led to major advances in our understanding of intracellular trafficking, signal transduction and organelle biogenesis.

Acknowledgments The authors would like to thank Drs. Dara Leto, David Buchner, Binbin Lu, Irit Hochberg and Diane Fingar (University of Michigan), and Dr. Alan Cheng (University of Louisville) for critical discussions and suggestions during the writing of this chapter.

References

Abel ED, Peroni O, Kim JK et al (2001) Adipose-selective targeting of the GLUT4 gene impairs insulin action in muscle and liver. Nature 409:729–733. doi:10.1038/35055575

Alessi DR, Andjelkovic M, Caudwell B et al (1996) Mechanism of activation of protein kinase B by insulin and IGF-1. EMBO J 15:6541–6551

Alessi DR, James SR, Downes CP et al (1997) Characterization of a 3-phosphoinositide-dependent protein kinase which phosphorylates and activates protein kinase Balpha. Curr Biol 7:261–269. doi:S0960-9822(06)00122-9 [pii]

Auger KR, Serunian LA, Soltoff SP et al (1989) PDGF-dependent tyrosine phosphorylation stimulates production of novel polyphosphoinositides in intact cells. Cell 57:167–175

Backer JM (2008) The regulation and function of class III PI3Ks: novel roles for Vps34. Biochem J 410:1–17. doi:10.1042/BJ20071427

Balla A, Balla T (2006) Phosphatidylinositol 4-kinases: old enzymes with emerging functions. Trends Cell Biol 16:351–361. doi:10.1016/j.tcb.2006.05.003

Barber DF, Alvarado-Kristensson M, González-García A et al (2006) PTEN regulation, a novel function for the p85 subunit of phosphoinositide 3-kinase. Science's STKE Signal Transduct Knowl Environ 2006:pe49. doi:10.1126/stke.3622006pe49

Begley M, Dixon J (2005) The structure and regulation of myotubularin phosphatases. Curr Opin Struct Biol 15:614–620

Berwick DC, Dell G, Welsh G et al (2004) Protein kinase B phosphorylation of PIKfyve regulates the trafficking of GLUT4 vesicles. J Cell Sci 117:5985–5993

Bi L, Okabe I, Bernard DJ et al (1999) Proliferative defect and embryonic lethality in mice homozygous for a deletion in the p110 subunit of phosphoinositide 3-kinase. J Biol Chem 274:10963–10968

Bi L, Okabe I, Bernard DJ, Nussbaum RL (2002) Early embryonic lethality in mice deficient in the p110 catalytic subunit of PI 3-kinase. New York 172:169–172. doi:10.1007/s00335-001-2123-x

Binda M, Péli-Gulli M-P, Bonfils G et al (2009) The Vam6 GEF controls TORC1 by activating the EGO complex. Mol Cell 35:563–573. doi:10.1016/j.molcel.2009.06.033

Bohdanowicz M, Balkin DM, De Camilli P, Grinstein S (2011) Recruitment of OCRL and Inpp 5B to phagosomes by Rab5 and APPL1 depletes phosphoinositides and attenuates Akt signaling. Mol Biol Cell 23:176–187 doi:10.1091/mbc.E11-06-0489

Bonangelino C, Catlett NL, Weisman LS (1997) Vac7p, a novel vacuolar protein, is required for normal vacuole inheritance and morphology. Mol Cell Biol 17:6847–6858

Boronenkov IV, Anderson RA (1995) The sequence of phosphatidylinositol-4-phosphate 5-kinase defines a novel family of lipid kinases. J Biol Chem 270:2881–2884

Botelho RJ, Efe JA, Teis D, Emr SD (2008) Assembly of a Fab1 phosphoinositide kinase signaling complex requires the Fig4 phosphoinositide phosphatase. Mol Biol Cell 19:4273–4286

Brachmann SM, Ueki K, Engelman JA et al (2005) Phosphoinositide 3-kinase catalytic subunit deletion and regulatory subunit deletion have opposite effects on insulin sensitivity in mice. Society 25:1596–1607. doi:10.1128/MCB.25.5.1596

Brand RM, Hamel FG (1999) Transdermally delivered peroxovanadium can lower blood glucose levels in diabetic rats. Int J Pharm 183:117–123

Bultsma Y, Keune W-J, Divecha N (2010) PIP4Kbeta interacts with and modulates nuclear localization of the high-activity PtdIns5P-4-kinase isoform PIP4Kalpha. Biochem J 430: 223–235. doi:10.1042/BJ20100341

Burgering BM, Coffer PJ (1995) Protein kinase B (c-Akt) in phosphatidylinositol-3-OH kinase signal transduction. Nature 376:599–602. doi:10.1038/376599a0

Butler M, McKay RA, Popoff IJ et al (2002) Specific inhibition of PTEN expression reverses hyperglycemia in diabetic mice. Diabetes 51:1028–1034

Byfield MP, Murray JT, Backer JM (2005) hVps34 is a nutrient-regulated lipid kinase required for activation of p70 S6 kinase. J Biol Chem 280:33076–33082. doi:10.1074/jbc.M507201200 M507201200 [pii]

Chagpar RB, Links PH, Pastor MC et al (2010) Direct positive regulation of PTEN by the p85 subunit of phosphatidylinositol 3-kinase. Proc Natl Acad Sci U S A 107:5471–5476. doi:10.1073/pnas.0908899107

Chamberlain M, Berry T, Pastor M, Anderson DH (2004) The p85{alpha} subunit of phosphatidylinositol 3′-kinase binds to and stimulates the GTPase activity of Rab proteins. J Biol Chem 279:48607–48614

Chattopadhyay M, Selinger ES, Ballou LM, Lin RZ (2011) Ablation of PI3K p110-a prevents high-fat diet-induced liver steatosis. Liver. doi:10.2337/db10-0869

Chaussade C, Pirola L, Bonnafous S et al (2003) Expression of myotubularin by an adenoviral vector demonstrates its function as a phosphatidylinositol 3-phosphate [PtdIns(3)P] phosphatase in muscle cell lines: involvement of PtdIns(3)P in insulin-stimulated glucose transport. Mol Endocrinol (Baltimore, Md) 17:2448–2460. doi:10.1210/me.2003-0261

Chen XW, Leto D, Chiang SH et al (2007) Activation of RalA is required for insulin-stimulated Glut4 trafficking to the plasma membrane via the exocyst and the motor protein Myo1c. Dev Cell 13:391–404. doi:10.1016/j.devcel.2007.07.007 S1534-5807(07)00268-7 [pii]

Chen X-W, Leto D, Xiong T et al (2011) A Ral GAP complex links PI 3-kinase/Akt signaling to RalA activation in insulin action. Mol Biol Cell 22:141–152. doi:10.1091/mbc.E10-08-0665

Chow C, Zhang Y, Dowling JJ et al (2007) Mutation of FIG4 causes neurodegeneration in the pale tremor mouse and patients with CMT4J. Nature 448:68–72

Christoforidis S, Miaczynska M, Ashman K et al (1999) Phosphatidylinositol-3-OH kinases are Rab5 effectors. Nat Cell Biol 1:249–252

Ciraolo E, Iezzi M, Marone R et al (2008) Phosphoinositide 3-kinase p110beta activity: key role in metabolism and mammary gland cancer but not development. Science Signal 1:ra3. doi:10.1126/scisignal.1161577

Clark AS, Fagan JM, Mitch WE (1985) Selectivity of the insulin-like actions of vanadate on glucose and protein metabolism in skeletal muscle. Biochem J 232:273–276

Clarke JF, Young PW, Yonezawa K et al (1994) Inhibition of the translocation of GLUT1 and GLUT4 in 3T3-L1 cells by the phosphatidylinositol 3-kinase inhibitor, wortmannin. Biochem J 300(Pt 3):631–635

Clément S, Krause U, Desmedt F et al (2001) The lipid phosphatase SHIP2 controls insulin sensitivity. Nature 409:92–97. doi:10.1038/35051094

Cleves AE, Novick PJ, Bankaitis VA (1989) Mutations in the SAC1 gene suppress defects in yeast Golgi and yeast actin function. J Cell Biol 109:2939–2950. doi:10.1016/0168-9525(90)90063-C

Currie RA, Walker KS, Gray A et al (1999) Role of phosphatidylinositol 3,4,5-trisphosphate in regulating the activity and localization of 3-phosphoinositide-dependent protein kinase-1. Biochem J 337(Pt 3):575–583

Dan HC, Sun M, Yang L et al (2002) Phosphatidylinositol 3-kinase/Akt pathway regulates tuberous sclerosis tumor suppressor complex by phosphorylation of tuberin. J Biol Chem 277:35364–35370. doi:10.1074/jbc.M205838200

Di Paolo G, De Camilli P (2006) Phosphoinositides in cell regulation and membrane dynamics. Nature 443:651–657. doi:10.1038/nature05185

Divecha N, Truong O, Hsuan JJ et al (1995) The cloning and sequence of the C isoform of PtdIns4P 5-kinase. Biochem J 309(Pt 3):715–719

Domin J, Gaidarov I, Smith ME et al (2000) The class II phosphoinositide 3-kinase PI3K-C2alpha is concentrated in the trans-Golgi network and present in clathrin-coated vesicles. J Biol Chem 275:11943–11950

Dominguez V, Raimondi C, Somanath S et al (2010) Class II phosphoinositide 3-kinase regulates exocytosis of insulin granules in pancreatic beta cells. J Biol Chem 1–22: . doi:10.1074/jbc.M110.200295

Dove SK, Piper RC, McEwen RK et al (2004) Svp1p defines a family of phosphatidylinositol 3,5-bisphosphate effectors. EMBO J 23:1922–1933. doi:10.1038/sj.emboj.7600203 7600203 [pii]

Downes CP, Gray A, Lucocq JM (2005) Probing phosphoinositide functions in signaling and membrane trafficking. Trends Cell Biol 15:259–268. doi:10.1016/j.tcb.2005.03.008

Duckworth WC, Solomon SS, Liepnieks J et al (1988) Insulin-like effects of vanadate in isolated rat adipocytes. Endocrinology 122:2285–2289

Duex JE, Nau J, Kauffman E, Weisman LS (2006a) Phosphoinositide 5-phosphatase Fig4p Is required for both acute rise and subsequent fall in stress-induced phosphatidylinositol 3,5-bisphosphate levels. Eukaryot Cell 5:723–731

Duex JE, Tang F, Weisman LS (2006b) The Vac14p-Fig4p complex acts independently of Vac7p and couples PI3,5P2 synthesis and turnover. J Cell Biol 172:693–704

Efe J, Botelho RJ, Emr SD (2007) Atg18 regulates organelle morphology and Fab1 kinase activity independent of its membrane recruitment by phosphatidylinositol 3,5-bisphosphate. Mol Biol Cell 18:4232–4244

Eguez L, Lee A, Chavez JA et al (2005) Full intracellular retention of GLUT4 requires AS160 Rab GTPase activating protein. Cell Metab 2:263–272. doi:10.1016/j.cmet.2005.09.005

Erneux C, Edimo WE, Deneubourg L, Pirson I (2011) SHIP2 multiple functions: a balance between a negative control of PtdIns(3,4,5)P(3) level, a positive control of PtdIns(3,4)P(2) production, and intrinsic docking properties. J Cell Biochem 112:2203–2209. doi:10.1002/jcb.23146

Falasca M, Hughes WE, Dominguez V et al (2007) The role of phosphoinositide 3-kinase C2alpha in insulin signaling. J Biol Chem 282:28226–28236. doi:10.1074/jbc.M704357200

Fang Y, Vilella-Bach M, Bachmann R et al (2001) Phosphatidic acid-mediated mitogenic activation of mTOR signaling. Science (New York, NY) 294:1942–1945. doi:10.1126/science.1066015

Fang Y, Park I-H, Wu A-L et al (2003) PLD1 regulates mTOR signaling and mediates Cdc42 activation of S6K1. Curr Biol 13:2037–2044. doi:10.1016/j.cub.2003.11.021

Fedele CG, Ooms LM, Ho M et al (2010) Inositol polyphosphate 4-phosphatase II regulates PI3K/Akt signaling and is lost in human basal-like breast cancers. Proc Natl Acad Sci U S A 107:22231–22236. doi:10.1073/pnas.1015245107

Fisher JS, Nolte LA, Kawanaka K et al (2002) Glucose transport rate and glycogen synthase activity both limit skeletal muscle glycogen accumulation. Am J Physiol Endocrinol Metab 282:E1214–E1221. doi:10.1152/ajpendo.00254.2001

Foley K, Boguslavsky S, Klip A (2011) Endocytosis, recycling and regulated exocytosis of glucose transporter 4 (GLUT4). Biochemistry. doi:10.1021/bi2000356

Foti M, Audhya A, Emr SD (2001) Sac1 lipid phosphatase and Stt4 phosphatidylinositol 4-kinase regulate a pool of phosphatidylinositol 4-phosphate that functions in the control of the actin cytoskeleton and vacuole morphology. Mol Biol Cell 12:2396–2411

Foukas LC, Claret M, Pearce W et al (2006) Critical role for the p110alpha phosphoinositide-3-OH kinase in growth and metabolic regulation. Nature 441:366–370. doi:10.1038/nature04694

Franke TF, Yang SI, Chan TO et al (1995) The protein kinase encoded by the Akt proto-oncogene is a target of the PDGF-activated phosphatidylinositol 3-kinase. Cell 81:727–736

Gaidarov I, Smith ME, Domin J, Keen JH (2001) The class II phosphoinositide 3-kinase C2alpha is activated by clathrin and regulates clathrin-mediated membrane trafficking. Mol Cell 7:443–449

Gary JD, Wurmser AE, Bonangelino C et al (1998) Fab1p is essential for PtdIns(3)P 5-kinase activity and the maintenance of vacuolar size and membrane homeostasis. J Cell Biol 143:65–79

Gary J, Sato T, Stefan C et al (2002) Regulation of Fab1 phosphatidylinositol 3-phosphate 5-kinase pathway by Vac7 protein and Fig4, a polyphosphoinositide phosphatase family member. Mol Biol Cell 13:1238–1251

Gewinner C, Wang ZC, Richardson A et al (2009) Evidence that inositol polyphosphate 4-phosphatase type II is a tumor suppressor that inhibits PI3K signaling. Cancer Cell 16:115–125. doi:10.1016/j.ccr.2009.06.006

Gonzalez E, McGraw TE (2009) Insulin-modulated Akt subcellular localization determines Akt isoform-specific signaling. Proc Natl Acad Sci U S A 106:7004–7009. doi:10.1073/pnas.0901933106

Green CJ, Göransson O, Kular GS et al (2008) Use of Akt inhibitor and a drug-resistant mutant validates a critical role for protein kinase B/Akt in the insulin-dependent regulation of glucose and system A amino acid uptake. J Biol Chem 283:27653–27667. doi:10.1074/jbc.M802623200

Gridley S, Chavez JA, Lane WS, Lienhard GE (2006) Adipocytes contain a novel complex similar to the tuberous sclerosis complex. Cell Signal 18:1626–1632. doi:10.1016/j.cellsig.2006.01.002
Guo J-P, Coppola D, Cheng JQ (2011) IKBKE activates Akt independent of phosphatidylinositol 3-kinase/PDK1/mTORC2 and PH domain to sustain malignant transformation. J Biol Chem. doi:10.1074/jbc.M111.287433
Gurung R, Tan A, Ooms LM et al (2003) Identification of a novel domain in two mammalian inositol-polyphosphate 5-phosphatases that mediates membrane ruffle localization. The inositol 5-phosphatase skip localizes to the endoplasmic reticulum and translocates to membrane ruffles following epide. J Biol Chem 278:11376–11385. doi:10.1074/jbc.M209991200
Habib T, Hejna JA, Moses RE, Decker SJ (1998) Growth factors and insulin stimulate tyrosine phosphorylation of the 51C/SHIP2 protein. J Biol Chem 273:18605–18609
He B, Xi F, Zhang X et al (2007) Exo70 interacts with phospholipids and mediates the targeting of the exocyst to the plasma membrane. EMBO J 26:4053–4065. doi:10.1038/sj.emboj.7601834
Horie Y, Suzuki A, Kataoka E et al (2004) Hepatocyte-specific Pten deficiency results in steatohepatitis and hepatocellular carcinomas. J Clin Investig. doi:10.1172/JCI200420513.1774
Huffman TA, Mothe-Satney I, Lawrence JJC (2002) Insulin-stimulated phosphorylation of lipin mediated by the mammalian target of rapamycin. Proc Natl Acad Sci U S A 99:1047–1052
Ijuin T, Takenawa T (2003) SKIP negatively regulates insulin-induced GLUT4 translocation and membrane ruffle formation. Mol Cell Biol 23:1209–1220. doi:10.1128/MCB.23.4.1209
Ijuin T, Takenawa T (2012) Regulation of insulin signalling and glucose transporter 4 (GLUT4) exocytosis by the phosphatidylinositol 3,4,5-trisphosphate (PIP3) phosphatase, SKIP. J Biol Chem. doi:10.1074/jbc.M111.335539
Ijuin T, Yu YE, Mizutani K et al (2008) Increased insulin action in SKIP heterozygous knockout mice. Mol Cell Biol 28:5184–5195. doi:10.1128/MCB.01990-06
Ikonomov OC, Sbrissa D, Mlak K, Shisheva A (2002) Requirement for PIKfyve enzymatic activity in acute and long-term insulin cellular effects. Endocrinology 143:4742–4754
Ikonomov OC, Sbrissa D, Dondapati R, Shisheva A (2007) ArPIKfyve-PIKfyve interaction and role in insulin-regulated GLUT4 translocation and glucose transport in 3T3-L1 adipocytes. Exp Cell Res 313:2404–2416. doi:10.1016/j.yexcr.2007.03.024 S0014-4827(07)00113-9 [pii]
Ikonomov OC, Sbrissa D, Ijuin T et al (2009) Sac3 is an Insulin-regulated PtdIns(3,5)P2 phosphatase: gain in insulin responsiveness through Sac3 downregulation in adipocytes. J Biol Chem. doi:10.1074/jbc.M109.025361. M109.025361 [pii]
Ikonomov OC, Sbrissa D, Fligger J et al (2010) ArPIKfyve regulates Sac3 protein abundance and turnover: disruption of the mechanism by Sac3I41T mutation causing Charcot-Marie-Tooth 4J disorder. J Biol Chem. doi:10.1074/jbc.C110.154658
Ikonomov OC, Sbrissa D, Delvecchio K et al (2011) The phosphoinositide kinase PIKfyve is vital in early embryonic development: preimplantation lethality of PIKfyve−/− embryos but normality of PIKfyve+/− mice. J Biol Chem 286:13404–13413. doi:10.1074/jbc.M111.222364
Inoki K, Li Y, Zhu T et al (2002) TSC2 is phosphorylated and inhibited by Akt and suppresses mTOR signalling. Nat Cell Biol 4:648–657. doi:10.1038/ncb839 ncb839 [pii]
Inoki K, Li Y, Xu T, Guan K-L (2003) Rheb GTPase is a direct target of TSC2 GAP activity and regulates mTOR signaling. Genes Dev 17:1829–1834. doi:10.1101/gad.1110003
Inoue M, Chang L, Hwang J et al (2003) The exocyst complex is required for targeting of Glut4 to the plasma membrane by insulin. Nature 422:629–633. doi:10.1038/nature01533 nature01533 [pii]
Inoue M, Chiang SH, Chang L et al (2006) Compartmentalization of the exocyst complex in lipid rafts controls Glut4 vesicle tethering. Mol Biol Cell 17:2303–2311. doi:10.1091/mbc.E06-01-0030 E06-01-0030 [pii]
Ishiki M, Randhawa VK, Poon V et al (2005) Insulin regulates the membrane arrival, fusion, and C-terminal unmasking of glucose transporter-4 via distinct phosphoinositides. J Biol Chem 280:28792–28802. doi:10.1074/jbc.M500501200
James SR, Downes CP, Gigg R et al (1996) Specific binding of the Akt-1 protein kinase to phosphatidylinositol 3,4,5-trisphosphate without subsequent activation. Biochem J 315 (Pt 3):709–713

James DJ, Khodthong C, Kowalchyk JA, Martin TFJ (2008) Phosphatidylinositol 4,5-bisphosphate regulates SNARE-dependent membrane fusion. J Cell Biol 182:355–366. doi:10.1083/jcb.200801056

Jia S, Liu Z, Zhang S et al (2008) Essential roles of PI (3) K—p110b in cell growth, metabolism and tumorigenesis. Nature. doi:10.1038/nature07091

Jiang ZY, Zhou QL, Coleman KA et al (2003) Insulin signaling through Akt/protein kinase B analyzed by small interfering RNA-mediated gene silencing. Proc Natl Acad Sci U S A 100:7569–7574. doi:10.1073/pnas.1332633100

Jin N, Chow CY, Liu L et al (2008) VAC14 nucleates a protein complex essential for the acute interconversion of PI3P and PI(3,5)P(2) in yeast and mouse. EMBO J 27:3221–3234. doi:10.1038/emboj.2008.248 emboj2008248 [pii]

Jones DR, Bultsma Y, Keune W-J et al (2006) Nuclear PtdIns5P as a transducer of stress signaling: an in vivo role for PIP4Kbeta. Mol Cell 23:685–695. doi:10.1016/j.molcel.2006.07.014

Kanai F, Ito K, Todaka M et al (1993) Insulin-stimulated GLUT4 translocation is relevant to the phosphorylation of IRS-1 and the activity of PI3-kinase. Biochem Biophys Res Commun 195:762–768. doi:10.1006/bbrc.1993.2111

Khwaja A, Rodriguez-Viciana P, Wennström S et al (1997) Matrix adhesion and Ras transformation both activate a phosphoinositide 3-OH kinase and protein kinase B/Akt cellular survival pathway. EMBO J 16:2783–2793. doi:10.1093/emboj/16.10.2783

Kihara A, Noda T, Ishihara N, Ohsumi Y (2001) Two distinct Vps34 phosphatidylinositol 3-kinase complexes function in autophagy and carboxypeptidase Y sorting in Saccharomyces cerevisiae. J Cell Biol 152:519–530

Kim E, Goraksha-Hicks P, Li L et al (2008) Regulation of TORC1 by Rag GTPases in nutrient response. Nat Cell Biol 10:935–945. doi:10.1038/ncb1753 ncb1753 [pii]

Knobbe CB, Lapin V, Suzuki A, Mak TW (2008) The roles of PTEN in development, physiology and tumorigenesis in mouse models: a tissue-by-tissue survey. Oncogene 27:5398–5415. doi:10.1038/onc.2008.238

Kong AM, Horan KA, Sriratana A et al (2006) Phosphatidylinositol 3-phosphate [PtdIns3P] is generated at the plasma membrane by an inositol polyphosphate 5-phosphatase: endogenous PtdIns3P can promote GLUT4 translocation to the plasma membrane. Mol Cell Biol 26:6065–6081. doi:10.1128/MCB.00203-06

Kurlawalla-Martinez C, Stiles B, Wang Y et al (2005) Insulin hypersensitivity and resistance to streptozotocin-induced diabetes in mice lacking PTEN in adipose tissue. Mol Cell Biol 25:2498–2510. doi:10.1128/MCB.25.6.2498-2510.2005

Kurosu H, Katada T (2001) Association of phosphatidylinositol 3-kinase composed of p110beta-catalytic and p85-regulatory subunits with the small GTPase Rab5. J Biochem 130:73–78

Laplante M, Sabatini DM (2010) mTORC1 activates SREBP-1c and uncouples lipogenesis from gluconeogenesis. Proc Natl Acad Sci U S A 107:3281–3282. doi:10.1073/pnas.1000323107

Lawrence JTR, Birnbaum MJ (2003) ADP-ribosylation factor 6 regulates insulin secretion through plasma membrane phosphatidylinositol 4,5-bisphosphate. Proc Natl Acad Sci U S A 100:13320–13325. doi:10.1073/pnas.2232129100

Le Marchand-Brustel Y, Gautier N, Cormont M, Van Obberghen E (1995) Wortmannin inhibits the action of insulin but not that of okadaic acid in skeletal muscle: comparison with fat cells. Endocrinology 136:3564–3570

Li J, Yen C, Liaw D et al (1997) PTEN, a putative protein tyrosine phosphatase gene mutated in human brain, breast, and prostate cancer. Science (New York, NY) 275:1943–1947

Liu J, Zuo X, Yue P, Guo W (2007) Phosphatidylinositol 4, 5-bisphosphate mediates the targeting of the Exocyst to the plasma membrane for exocytosis in mammalian cells. Mol Biol Cell 18:4483–4492. doi:10.1091/mbc.E07

Lodhi I, Bridges D, Chiang S-H et al (2008) Insulin stimulates phosphatidylinositol 3-phosphate production via the activation of Rab5. Mol Biol Cell. doi:10.1091/mbc.E08-01-0105

Maehama T, Dixon J (1998) The tumor suppressor, PTEN/MMAC1, dephosphorylates the lipid second messenger, phosphatidylinositol 3,4,5-trisphosphate. J Biol Chem 273:13375–13378

Maffucci T, Brancaccio A, Piccolo E et al (2003) Insulin induces phosphatidylinositol-3-phosphate formation through TC10 activation. EMBO J 22:4178–4189. doi:10.1093/emboj/cdg402

Manning BD, Tee AR, Logsdon MN et al (2002) Identification of the tuberous sclerosis complex-2 tumor suppressor gene product tuberin as a target of the phosphoinositide 3-kinase/akt pathway. Mol Cell 10:151–162

Mao Y, Balkin DM, Zoncu R et al (2009) A PH domain within OCRL bridges clathrin-mediated membrane trafficking to phosphoinositide metabolism. EMBO J 28:1831–1842. doi:10.1038/emboj.2009.155

Mazza S, Maffucci T (2011) Class II phosphoinositide 3-kinase C2alpha: what we learned so far. J Biochem Mol Biol 2:168–182

McManus EJ, Collins BJ, Ashby PR et al (2004) The in vivo role of PtdIns(3,4,5)P3 binding to PDK1 PH domain defined by knockin mutation. EMBO J 23:2071–2082. doi:10.1038/sj.emboj.7600218

Mellor P, Furber LA, Nyarko JNK, Anderson DH (2012) Multiple roles for the p85α isoform in the regulation and function of PI3K signalling and receptor trafficking. Biochem J 441:23–37. doi:10.1042/BJ20111164

Meunier FA, Osborne SL, Hammond GRV et al (2005) Phosphatidylinositol 3-kinase C2alpha is essential for ATP-dependent priming of neurosecretory granule exocytosis. Mol Biol Cell 16:4841–4851. doi:10.1091/mbc.E05-02-0171

Miaczynska M, Christoforidis S, Giner A et al (2004) APPL proteins link Rab5 to nuclear signal transduction via an endosomal compartment. Cell 116:445–456

Michael MD, Kulkarni RN, Postic C et al (2000) Loss of insulin signaling in hepatocytes leads to severe insulin resistance and progressive hepatic dysfunction. Mol Cell 6:87–97

Mima J, Wickner W (2009) Phosphoinositides and SNARE chaperones synergistically assemble and remodel SNARE complexes for membrane fusion. Proc Natl Acad Sci U S A 106:16191–16196

Ng Y, Ramm G, Lopez JA, James DE (2008) Rapid activation of Akt2 is sufficient to stimulate GLUT4 translocation in 3T3-L1 adipocytes. Cell Metab 7:348–356. doi:10.1016/j.cmet.2008.02.008

Nguyen K-TT, Tajmir P, Lin CH et al (2006) Essential role of Pten in body size determination and pancreatic beta-cell homeostasis in vivo. Mol Cell Biol 26:4511–4518. doi:10.1128/MCB.00238-06

Nicot A-S, Laporte J (2008) Endosomal phosphoinositides and human diseases. Traffic (Copenhagen, Denmark) 9:1240–1249

Nobukuni T, Joaquin M, Roccio M et al (2005) Amino acids mediate mTOR/raptor signaling through activation of class 3 phosphatidylinositol 3OH-kinase. Proc Natl Acad Sci U S A 102:14238–14243. doi:10.1073/pnas.0506925102

Nolte LA, Han D-H, Hansen PA et al (2003) A peroxovanadium compound stimulates muscle glucose transport as powerfully as insulin and contractions combined. Diabetes 52:1918–1925

Norris FA, Auethavekiat V, Majerus PW (1995) The isolation and characterization of cDNA encoding human and rat brain inositol polyphosphate 4-phosphatase. J Biol Chem 270:16128–16133

Norris FA, Atkins RC, Majerus PW (1997) The cDNA cloning and characterization of inositol polyphosphate 4-phosphatase type II. Evidence for conserved alternative splicing in the 4-phosphatase family. J Biol Chem 272:23859–23864

Okada T, Kawano Y, Sakakibara T et al (1994) Essential role of phosphatidylinositol 3-kinase in insulin-induced glucose transport and antilipolysis in rat adipocytes. Studies with a selective inhibitor wortmannin. J Biol Chem 269:3568–3573

Okada S, Ohshima K, Uehara Y et al (2007) Synip phosphorylation is required for insulin-stimulated Glut4 translocation. Biochem Biophys Res Commun 356:102–106. doi:10.1016/j.bbrc.2007.02.095

Olsen HL, Hoy M, Zhang W et al (2003) Phosphatidylinositol 4-kinase serves as a metabolic sensor and regulates priming of secretory granules in pancreatic beta cells. Proc Natl Acad Sci U S A 100:5187–5192. doi:10.1073/pnas.0931282100

Ooms LM, Horan KA, Rahman P et al (2009) The role of the inositol polyphosphate 5-phosphatases in cellular function and human disease. Biochem J 419:29–49. doi:10.1042/BJ20081673

Ou Y-H, Torres M, Ram R et al (2011) TBK1 directly engages Akt/PKB survival signaling to support oncogenic transformation. Mol Cell 41:458–470. doi:10.1016/j.molcel.2011.01.019
Patton GM, Fasulo JM, Robins SJ (1982) Separation of phospholipids and individual molecular species of phospholipids by high-performance liquid chromatography. J Lipid Res 23:190–196
Peterson TRR, Sengupta SSS, Harris TEE et al (2011) mTOR complex 1 regulates Lipin 1 localization to control the SREBP pathway. Cell 146:408–420. doi:10.1016/j.cell.2011.06.034
Petritsch C, Woscholski R, Edelmann HM et al (1995) Selective inhibition of p70 S6 kinase activation by phosphatidylinositol 3-kinase inhibitors. Eur J Biochem/FEBS 230:431–438
Podsypanina K, Ellenson LH, Nemes A et al (1999) Mutation of Pten/Mmac1 in mice causes neoplasia in multiple organ systems. Proc Natl Acad Sci U S A 96:1563–1568
Prasad NK, Werner ME, Decker SJ (2009) Specific tyrosine phosphorylations mediate signal-dependent stimulation of SHIP2 inositol phosphatase activity, while the SH2 domain confers an inhibitory effect to maintain the basal activity. Biochemistry 48:6285–6287. doi:10.1021/bi900492d
Rabinovsky R, Pochanard P, McNear C et al (2009) p85 Associates with unphosphorylated PTEN and the PTEN-associated complex. Mol Cell Biol 29:5377–5388. doi:10.1128/MCB.01649-08
Rameh LE, Tolias KF, Duckworth BC, Cantley LC (1997) A new pathway for synthesis of phosphatidylinositol-4,5-bisphosphate. Nature 390:192–196. doi:10.1038/36621
Rodriguez-Viciana P, Warne PH, Dhand R et al (1994) Phosphatidylinositol-3-OH kinase as a direct target of Ras. Nature 370:527–532. doi:10.1038/370527a0
Rommel C, Bodine SC, Clarke BA et al (2001) Mediation of IGF-1-induced skeletal myotube hypertrophy by PI(3)K/Akt/mTOR and PI(3)K/Akt/GSK3 pathways. Nat Cell Biol 3:1009–1013. doi:10.1038/ncb1101-1009
Rosivatz E, Matthews JG, McDonald NQ et al (2006) A small molecule inhibitor for phosphatase and tensin homologue deleted on chromosome 10 (PTEN). ACS Chem Biol 1:780–790. doi:10.1021/cb600352f
Rowland AF, Fazakerley DJ, James DE (2011) Mapping insulin/GLUT4 circuitry. traffic (Copenhagen, Denmark). doi:10.1111/j.1600-0854.2011.01178.x
Rudge S, Anderson DH, Emr SD (2004) Vacuole size control: regulation of PtdIns(3,5)P2 levels by the vacuole-associated Vac14-Fig4 complex, a PtdIns(3,5)P2-specific phosphatase. Mol Biol Cell 15:24–36
Saito T, Jones CC, Huang S et al (2007) The interaction of Akt with APPL1 is required for insulin-stimulated Glut4 translocation. J Biol Chem 282:32280–32287. doi:10.1074/jbc.M704150200
Sancak Y, Peterson TR, Shaul YD et al (2008) The Rag GTPases bind raptor and mediate amino acid signaling to mTORC1. Science (New York, NY) 320:1496–1501
Sancak Y, Bar-Peled L, Zoncu R et al (2010) Ragulator-rag complex targets mTORC1 to the lysosomal surface and is necessary for its activation by amino acids. Cell 141:290–303
Sano H, Kane S, Sano E et al (2003) Insulin-stimulated phosphorylation of a Rab GTPase-activating protein regulates GLUT4 translocation. J Biol Chem 278:14599–14602. doi:10.1074/jbc.C300063200
Sano H, Kane S, Sano E, Lienhard GE (2005) Synip phosphorylation does not regulate insulin-stimulated GLUT4 translocation. Biochem Biophys Res Commun 332:880–884. doi:10.1016/j.bbrc.2005.05.027
Sano H, Eguez L, Teruel MN et al (2007) Rab10, a target of the AS160 Rab GAP, is required for insulin-stimulated translocation of GLUT4 to the adipocyte plasma membrane. Cell Metab 5:293–303. doi:10.1016/j.cmet.2007.03.001
Sarbassov DD, Guertin DA, Ali SM, Sabatini DM (2005) Phosphorylation and regulation of Akt/PKB by the rictor-mTOR complex. Science (New York, NY) 307:1098–1101. doi:10.1126/science.1106148
Sarkes D, Rameh LE (2010) A novel HPLC-based approach makes possible the spacial characterization of cellular PtdIns5P and other phosphoinositides. Biochem J 384:375–384. doi:10.1042/BJ20100129

Sbrissa D, Shisheva A (2005) Acquisition of unprecedented phosphatidylinositol 3,5-bisphosphate rise in hyperosmotically stressed 3T3-L1 adipocytes, mediated by ArPIKfyve-PIKfyve pathway. J Biol Chem 280:7883–7889. doi:10.1074/jbc.M412729200 M412729200 [pii]

Sbrissa D, Ikonomov OC, Shisheva A (1999) PIKfyve, a mammalian ortholog of yeast Fab1p lipid kinase, synthesizes 5-phosphoinositides. Effect of insulin. J Biol Chem 274:21589–21597

Sbrissa D, Ikonomov OC, Strakova J, Shisheva A (2004) Role for a novel signaling intermediate, phosphatidylinositol 5-phosphate, in insulin-regulated F-actin stress fiber breakdown and GLUT4 translocation. Endocrinology 145:4853–4865. doi:10.1210/en.2004-0489

Sbrissa D, Ikonomov OC, Fu Z et al (2007) Core protein machinery for mammalian phosphatidylinositol 3,5-bisphosphate synthesis and turnover that regulates the progression of endosomal transport. Novel Sac phosphatase joins the ArPIKfyve-PIKfyve complex. J Biol Chem 282:23878–23891. doi:10.1074/jbc.M611678200 M611678200 [pii]

Sbrissa D, Ikonomov OC, Fenner H, Shisheva A (2008) ArPIKfyve homomeric and heteromeric interactions scaffold PIKfyve and Sac3 in a complex to promote PIKfyve activity and functionality. J Mol Biol 384:766–779

Schenck A, Goto-Silva L, Collinet C et al (2008) The endosomal protein App l1 mediates Akt substrate specificity and cell survival in vertebrate development. Cell 133:486–497. doi:10.1016/j.cell.2008.02.044

Schmid AC, Byrne RD, Vilar R, Woscholski R (2004) Bisperoxovanadium compounds are potent PTEN inhibitors. FEBS Lett 566:35–38. doi:10.1016/j.febslet.2004.03.102

Schu PV, Takegawa K, Fry MJ et al (1993) Phosphatidylinositol 3-kinase encoded by yeast VPS34 gene essential for protein sorting. Science (New York, NY) 260:88–91

Shepherd PR, Navé BT, Siddle K (1995) Insulin stimulation of glycogen synthesis and glycogen synthase activity is blocked by wortmannin and rapamycin in 3T3-L1 adipocytes: evidence for the involvement of phosphoinositide 3-kinase and p70 ribosomal protein-S6 kinase. Biochem J 305(Pt 1):25–28

Shin HW, Hayashi M, Christoforidis S et al (2005) An enzymatic cascade of Rab5 effectors regulates phosphoinositide turnover in the endocytic pathway. J Cell Biol 170:607–618. doi:10.1083/jcb.200505128 jcb.200505128 [pii]

Sleeman MW, Wortley KE, Lai K-MV et al (2005) Absence of the lipid phosphatase SHIP2 confers resistance to dietary obesity. Nat Med 11:199–205. doi:10.1038/nm1178

Sopasakis VR, Liu P, Suzuki R et al (2010) Specific roles of the p110a isoform of phosphatidylinsositol 3-kinase in hepatic insulin signaling and metabolic regulation. Cell Metab 11:220–230. doi:10.1016/j.cmet.2010.02.002

Staal SP, Hartley JW (1988) Thymic lymphoma induction by the AKT8 murine retrovirus. J Exp Med 167:1259–1264

Stahelin R, Ananthanarayanan B, Blatner N et al (2004) Mechanism of membrane binding of the phospholipase D1 PX domain. J Biol Chem 279:54918–54926

Steck PA, Pershouse MA, Jasser SA et al (1997) Identification of a candidate tumour suppressor gene, MMAC1, at chromosome 10q23.3 that is mutated in multiple advanced cancers. Nat Genet 15:356–362. doi:10.1038/ng0497-356

Stiles B, Wang Y, Stahl A et al (2004) Liver-specific deletion of negative regulator Pten results in fatty liver and insulin hypersensitivity [corrected]. Proc Natl Acad Sci U S A 101:2082–2087. doi:10.1073/pnas.0308617100

Stiles BL, Kuralwalla-Martinez C, Guo W et al (2006) Selective deletion of Pten in pancreatic beta cells leads to increased islet mass and resistance to STZ-induced diabetes. Mol Cell Biol 26:2772. doi:10.1128/MCB.26.7.2772

Sun Y, Bilan PJ, Liu Z, Klip A (2010) Rab8A and Rab13 are activated by insulin and regulate GLUT4 translocation in muscle cells. Proc Natl Acad Sci U S A 2010:6–11. doi:10.1073/pnas.1009523107

Suwa A, Yamamoto T, Sawada A et al (2009) Discovery and functional characterization of a novel small molecule inhibitor of the intracellular phosphatase, SHIP2. Br J Pharmacol 158:879–887. doi:10.1111/j.1476-5381.2009.00358.x

Suzuki A, De la Pompa JL, Stambolic V et al (1998) High cancer susceptibility and embryonic lethality associated with mutation of the PTEN tumor suppressor gene in mice. Curr Biol CB 8:1169–1178

Tan Y, You H, Wu C et al (2010) App l1 is dispensable for mouse development, and loss of Appl1 has growth factor-selective effects on Akt signaling in murine embryonic fibroblasts. J Biol Chem 285:6377–6389. doi:10.1074/jbc.M109.068452

Tan S-X, Ng Y, James DE (2011) Next generation Akt inhibitors provide greater specificity-effects on glucose metabolism in adipocytes. Biochem J. doi:10.1042/BJ20110040

Tee AR, Manning BD, Roux PP et al (2003) Tuberous sclerosis complex gene products, Tuberin and Hamartin, control mTOR signaling by acting as a GTPase-activating protein complex toward Rheb. Curr Biol 13:1259–1268

Vanhaesebroeck B, Stephens L, Hawkins P (2012) PI3K signalling: the path to discovery and understanding. Nat Rev Mol Cell Biol 13:195–203. doi:10.1038/nrm3290

Vicinanza M, Campli AD, Polishchuk E et al (2011) OCRL controls trafficking through early regulation of endosomal actin. EMBO J 1–16: . doi:10.1038/emboj.2011.354

Vicogne J, Vollenweider D, Smith JR et al (2006) Asymmetric phospholipid distribution drives in vitro reconstituted SNARE-dependent membrane fusion. Proc Natl Acad Sci U S A 103:14761–14766. doi:10.1073/pnas.0606881103

Wada T, Sasaoka T, Funaki M et al (2001) Overexpression of SH2-containing inositol phosphatase 2 results in negative regulation of insulin-induced metabolic actions in 3T3-L1 adipocytes via its 5′-phosphatase catalytic activity. Mol Cell Biol 21:1633–1646. doi:10.1128/MCB.21.5.1633-1646.2001

Waselle L, Gerona RRL, Vitale N et al (2005) Role of phosphoinositide signaling in the control of insulin exocytosis. Mol Endocrinol (Baltimore, Md) 19:3097–3106. doi:10.1210/me.2004-0530

Wen PJ, Osborne SL, Morrow IC et al (2008) Ca^{2+}-regulated pool of phosphatidylinositol-3-phosphate produced by phosphatidylinositol 3-kinase C2alpha on neurosecretory vesicles. Mol Biol Cell 19:5593–5603. doi:10.1091/mbc.E08-06-0595

Weng QP, Andrabi K, Kozlowski MT et al (1995) Multiple independent inputs are required for activation of the p70 S6 kinase. Mol Cell Biol 15:2333–2340

Westergaard N, Brand CL, Lewinsky RH et al (1999) Peroxyvanadium compounds inhibit glucose-6-phosphatase activity and glucagon-stimulated hepatic glucose output in the rat in vivo. Arch Biochem Biophys 366:55–60. doi:10.1006/abbi.1999.1181

Wijesekara N, Konrad D, Eweida M et al (2005) Muscle-specific Pten deletion protects against insulin resistance and diabetes. Society 25:1135–1145. doi:10.1128/MCB.25.3.1135

Williams C, Choudhury R, McKenzie E, Lowe M (2007) Targeting of the type II inositol polyphosphate 5-phosphatase INPP5B to the early secretory pathway. J Cell Sci 120:3941–3951. doi:10.1242/jcs.014423

Wong JT, Kim PTW, Peacock JW et al (2007) Pten (phosphatase and tensin homologue gene) haploinsufficiency promotes insulin hypersensitivity. Diabetologia 50:395–403. doi:10.1007/s00125-006-0531-x

Wong K-K, Engelman JA, Cantley LC (2010) Targeting the PI3K signaling pathway in cancer. Curr Opin Genet Dev 20:87–90. doi:10.1016/j.gde.2009.11.002

Wymann MP, Björklöf K, Calvez R et al (2003) Phosphoinositide 3-kinase gamma: a key modulator in inflammation and allergy. Biochem Soc Trans 31:275–280. doi:10.1042/

Xie X, Gong Z, Mansuy-Aubert V et al (2011a) C2 domain-containing phosphoprotein CDP138 regulates GLUT4 insertion into the plasma membrane. Cell Metab 14:378–389. doi:10.1016/j.cmet.2011.06.015

Xie X, Zhang D, Zhao B et al (2011b) I{kappa}B kinase varepsilon and TANK-binding kinase 1 activate AKT by direct phosphorylation. Proc Natl Acad Sci U S A. doi:10.1073/pnas.1016132108

Xiong Q, Deng C-Y, Chai J et al (2009) Knockdown of endogenous SKIP gene enhanced insulin-induced glycogen synthesis signaling in differentiating C2C12 myoblasts. BMB Rep 42:119–124

Xu L, Salloum D, Medlin PS et al (2011) Phospholipase D mediates nutrient input to mTORC1. J Biol Chem 286:25477–25486. doi:10.1074/jbc.M111.249631

Yamada E, Okada S, Saito T et al (2005) Akt2 phosphorylates Synip to regulate docking and fusion of GLUT4-containing vesicles. J Cell Biol 168:921–928. doi:10.1083/jcb.200408182

Yamashita M, Kurokawa K, Sato Y et al (2010) Structural basis for the Rho- and phosphoinositide-dependent localization of the exocyst subunit Sec3. Nat Struct Mol Biol 17:180–186. doi:10.1038/nsmb.1722

Yecies JL, Zhang HH, Menon S et al (2011) Akt stimulates hepatic SREBP1c and Lipogenesis through Parallel mTORC1-Dependent and Independent Pathways. Cell Metab 14:21–32. doi:10.1016/j.cmet.2011.06.002

Yoon M-S, Du G, Backer JM et al (2011) Class III PI-3-kinase activates phospholipase D in an amino acid-sensing mTORC1 pathway. J Cell Biol. doi:10.1083/jcb.201107033

Yu JW, Lemmon MA (2001) All phox homology (PX) domains from Saccharomyces cerevisiae specifically recognize phosphatidylinositol 3-phosphate. J Biol Chem 276:44179–44184. doi:10.1074/jbc.M108811200

Yu ZW, Jansson PA, Posner BI et al (1997) Peroxovanadate and insulin action in adipocytes from NIDDM patients. Evidence against a primary defect in tyrosine phosphorylation. Diabetologia 40:1197–1203. doi:10.1007/s001250050807

Yuan Y, Gao X, Guo N et al (2007) rSac3, a novel Sac domain phosphoinositide phosphatase, promotes neurite outgrowth in PC12 cells. Cell Res 17:919–932. doi:10.1038/cr.2007.82

Zhang X, Loijens JC, Boronenkov IV et al (1997) Phosphatidylinositol-4-phosphate 5-kinase isozymes catalyze the synthesis of 3-phosphate-containing phosphatidylinositol signaling molecules. J Biol Chem 272:17756–17761

Zhang Y, Zolov SN, Chow C et al (2007) Loss of Vac14, a regulator of the signaling lipid phosphatidylinositol 3,5-bisphosphate, results in neurodegeneration in mice. Proc Natl Acad Sci 104:17518–17523

Zisman A, Peroni OD, Abel ED et al (2000) Targeted disruption of the glucose transporter 4 selectively in muscle causes insulin resistance and glucose intolerance. Nat Med 6:924–928. doi:10.1038/78693

Zoncu R, Perera RM, Balkin DM et al (2009) A phosphoinositide switch controls the maturation and signaling properties of APPL endosomes. Cell 136:1110–1121. doi:10.1016/j.cell.2009.01.032 S0092-8674(09)00080-4 [pii]

Zoncu R, Bar-Peled L, Efeyan A et al (2011) mTORC1 senses lysosomal amino acids through an inside-out mechanism that requires the vacuolar H^{+}-ATPase. Science (New York, NY) 334:678–683. doi:10.1126/science.1207056

Zurita-Martinez SA, Puria R, Pan X et al (2007) Efficient Tor signaling requires a functional class C Vps protein complex in Saccharomyces cerevisiae. Genetics 176:2139–2150

Chapter 4
Phosphoinositides in Neuroexocytosis and Neuronal Diseases

Peter J. Wen, Shona L. Osborne and Frederic A. Meunier

Abstract Phosphoinositides (PIs) are a family of phospholipids derived from phosphatidylinositol (PtdIns), whose location, synthesis, and degradation depend on specific PI kinases and phosphatases. PIs have emerged as fundamental regulators of secretory processes, such as neurotransmitter release, hormone secretion, and histamine release in allergic responses. In neurons and neuroendocrine cells, regulated secretion requires the calcium-dependent fusion of transmitter-containing vesicles with the plasma membrane. The role played by PIs in exocytosis is best exemplified by the Ca^{2+}-dependent binding of vesicular Synaptotagmin1 to the plasma membrane PtdIns(4,5)P_2, and the recently demonstrated role of PtdIns(4,5)P_2 in the mobilization of secretory vesicles to the plasma membrane. New evidence has also recently emerged of an alternative PI pathway that can control exocytosis positively (via PtdIn3P) or negatively (via PtdIns(3,5)P_2). However, the positive or negative effectors for these pathways remain to be established. Reducing PtdIns(3,5)P_2 potentiates neuroexocytosis but leads to neuronal degeneration and has been linked to certain forms of Charcot-Marie-Tooth disease and amyotrophic lateral sclerosis. The goal of this review is to describe the role of PIs in neuroexocytosis and explore the current hypotheses linking these effects to human diseases.

Abbreviations

$A\beta-1$ 42	Amyloid beta peptide
Arf-1	ADP-ribosylation factor 1
CAPS	Ca^{2+}-dependent activator protein for secretion
Cdc42	Cell division control protein 42 homolog
CMT	Charcot-Marie-Tooth

P. J. Wen · S. L. Osborne · F. A. Meunier (✉)
Queensland Brain Institute, The University of Queensland,
Brisbane, QLD 4072, Australia
e-mail: f.meunier@uq.edu.au

M. Falasca (ed.), *Phosphoinositides and Disease*, Current Topics in Microbiology and Immunology 362, DOI: 10.1007/978-94-007-5025-8_4,

ER	Endoplasmic reticulum
FAD	Familial Alzheimer's disease
FYVE	Fab1, YGL023,VPS27, and EEA1
GLUT4	Glucose transporter type 4
LDCVs	Large dense core vesicles
Mtmr	Myotubularin-related protein
NCS-1	Neuronal calcium sensor 1
PIKfyve	Phosphoinositide kinase for five positions containing a Fyve finger
TGN	Trans-Golgi network
PI	Phosphoinositide
SNARE	Soluble NSF attachment protein receptor
Syt	Synaptotagmin

The Discovery of the Phosphoinositide Pathway and its Role in Exocytosis

Cellular membranes are mostly made of phospholipids. Phosphoinositides (PIs) are an important branch of the phospholipid family, consisting of a fatty-acyl chain embedded in the membrane bilayer, linked to a myo-inositol head group via a glycerol backbone. This head group can be phosphorylated or de-phosphorylated by a series of phosphatidylinositol (PtdIns) kinases and phosphatases on the 3', 4', or 5' -OH position to generate seven different combinations. The roles of PI were first discovered by Hokin and Hokin in 1950s, who found unexpected high turnover of a phosphate-based moiety in preformed phospholipids in response to the stimulation of pancreatic slices, sympathetic ganglia or brain cortex with acetylcholine (Hokin and Hokin 1954, 1955a, b, 1958, 1960; Hokin et al. 1960). These studies not only triggered a great amount of interest in the nature of these lipids and their role in cell signaling but also initiated a race to establish the nature of the link between PI turnover, Ca^{2+} levels, and exocytosis. This led to the discovery of $Ins(1,4,5)P_3$ that is released following cleavage of $PtdIns(4,5)P_2$ by phospholipase C, and is responsible for Ca^{2+} mobilization from intracellular stores.

Since then, multiple lines of evidence have accumulated linking PIs and the process of exocytosis, whereby secretory vesicles fuse with the plasma membrane in a regulated fashion, allowing cells to secrete neurotransmitters in the brain, hormones, such as insulin, digestive enzymes, and many other critical factors. Dysregulation of exocytosis have important pathological consequences linked to conditions such as Alzheimer's disease, diabetes, epilepsy, and asthma. We will focus in this chapter on the role of PIs in exocytosis from neurons to endocrine cells, referred to as neuroexocytosis, and how this may be linked to neuronal diseases.

Role of Phosphoinositides and Neuroexocytosis

Each of the seven different PIs varies in its abundance and subcellular localization in mammalian cells. Among all PIs, PtdIns4P, and PtdIns(4,5)P_2 account for the majority in mammalian cells. PtdIns4P is a highly abundant PI in the trans-Golgi network (TGN) and a key regulator of vesicular trafficking emanating from the TGN. It mediates this function by acting as a signpost for the recruitment of clathrin-adaptor proteins, such as AP-1 (Wang et al. 2003), Arf-binding proteins (GGAs) (Wang et al. 2007), and Gga2p (Demmel et al. 2008) to the TGN. The binding of these adaptor proteins to the TGN also requires a small GTPase ADP ribosylation factor 1 (Arf-1), which in turn stimulates the production of PtdIns4P by activating type III PI4Kinase β (PI4KIIIβ) on the Golgi complex (Godi et al. 1999), suggesting that Arf-1 is both a co-operative and synergistic factor for protein recruitment to the TGN. Neuronal calcium sensor 1 (NCS-1) can also activate PI4KIIIβ and overexpression of NCS-1 enhances exocytosis in Ca^{2+}-dependent (Koizumi et al. 2002; Rajebhosale et al. 2003) and -independent (Haynes et al. 2005) manner in PC12 cells. An open question remains as to whether this potentiation is a consequence of through enhanced PtdIns4P production and/or the subsequent production of PtdIns(4,5)P_2, a positive regulator for neuroexocytosis (Milosevic et al. 2005; Wen et al. 2011b).

PtdIns4P is not only localized at the TGN, since as a pool of PtdIns4P can be generated at the plasma membrane by the type III PI4kinase α (PI4KIIIα) (Balla et al. 2005), and disruption of PI4KIIIα blocks the re-synthesis of PtdIns(4,5)P_2 on the plasma membrane after its breakdown by angiotensin II (Balla et al. 2008). Interestingly, PI4KIIIα is localized to the endoplasmic reticulum (ER) (Wong et al. 1997) and its contribution to PtdIns4P pools on the plasma membrane suggests that the ER can make direct or indirect contacts with the plasma membrane to generate PtdIns4P. Evidence has now shown that a highly dynamic ER compartment containing PtdIns synthesizing enzymes can make multiple contacts with other membranes, suggesting that the ER also serves as a major lipid distribution platform alongside the TGN (Kim et al. 2011).

PtdIns(4,5)P_2 is the most abundant PI on the plasma membrane and this pool of PtdIns(4,5)P_2 has been linked to exocytosis from studies carried out in neurons (Di Paolo et al. 2004) and secretory cells including mast cells (Hammond et al. 2006; Kapp-Barnea et al. 2003; Way et al. 2000), pancreatic β cells (Olsen et al. 2003; Tomas et al. 2010), pituitary cells (Sedej et al. 2009), and bovine chromaffin cells (Aikawa and Martin 2003; Aoyagi et al. 2005; Eberhard et al. 1990; Grishanin et al. 2004; Hay et al. 1995; Hay and Martin 1992, 1993; James et al. 2008, 2010; Milosevic et al. 2005; Wen et al. 2011b). PtdIns(4,5)P_2 coordinates exocytosis by acting in *cis* through binding to the plasma membrane-localized proteins such as calcium-dependent activator protein for secretion (CAPS) (Loyet et al. 1998) and in *trans* by altering the binding of vesicular proteins such as synaptotagmin I (Syt I) to the plasma membrane (Kuo et al. 2011). Notably, CAPS potentiates secretion through its direct binding to both PtdIns(4,5)P_2 and syntaxin-1, indicating CAPS acts in *cis* with target-soluble NSF attachment protein receptors (t-SNARE) to facilitate SNARE complex assembly (James et al. 2008, 2009, 2010).

As most PtdIns(4,5)P_2 effectors including CAPS and Syt I are known to be calcium sensitive proteins, it is therefore difficult to uncouple the contribution of PtdIns(4,5)P_2 from that of calcium. Through pharmacological inhibition of Class I PI3-kinase δ (PI3Kδ) in bovine chromaffin cells, a recent study has uncovered a way to elicit an increase in PtdIns(4,5)P_2 pools on the plasma membrane without causing any change in calcium level (Wen et al. 2011b). This strategy was used to demonstrate that an increase in PtdIns(4,5)P_2 is sufficient to promote the mobilization of secretory vesicles to the plasma membrane via cell division control protein 42 homolog (Cdc42)-mediated actin polymerization, highlighting a novel link between PtdIns(4,5)P_2 signaling and secretory vesicle dynamics (Wen et al. 2011b).

Although considerable efforts have focused on elucidating the effects of PtdIns4P and PtdIns(4,5)P_2 in neuroexocytosis, less is known about the role of 3'phosphorylated PIs in this process. PtdIns3P is the most abundant form of 3-phosphorylated PIs and is mainly produced by Class III PI3-kinase (hvps34p) but also Class II PI3-kinase (α, β, γ) in mammalian cells. PtdIns3P produced by hvps34p is localized to early/late endosome and is essential for endosomal membrane trafficking (Johnson et al. 2006; Murray et al. 2002) and autophagy (Levine and Klionsky 2004). The alpha isoform of Class II PI3-kinase (PI3K-C2α) is essential for ATP-dependent priming step of secretory vesicle exocytosis in neurosecretory cells (Meunier et al. 2005), raising the possibility that PtdIns3P could be synthesized by PI3K-C2α on secretory vesicles during exocytosis. This was later confirmed in a study that highlighted a functionally distinct PtdIns3P pool on a subpopulation of large dense core vesicles (LDCVs) produced by PI3K-C2α and tightly regulated by Ca^{2+} (Wen et al. 2008). This suggests that PtdIns3P is a dynamic lipid second messenger that can relay the Ca^{2+} signal elicited during stimulation of exocytosis to prime LDCVs. It is also possible that those vesicles bearing PtdIns3P could represent primed vesicles that preferentially undergo fusion upon stimulation of exocytosis. Such an increase in PtdIns3P production on LDCVs was proposed to serve as a recruitment factor for molecule(s) acting downstream of PI3K-C2α activity to promote ATP-dependent priming. Further work will be required to determine the nature of the priming effector(s).

The PtdIns3P signal can be turned off by the action of phosphoinositide phosphatases (myotubularin family) or by the addition of a further phosphate by the phosphoinositide kinase for five position containing a Fyve finger (PIKfyve) to generate PtdIns(3,5)P_2. PIKfyve is localized to early/late endosome in mammalian cells (Cabezas et al. 2006; de Lartigue et al. 2009; Ikonomov et al. 2006; Jefferies et al. 2008; Rutherford et al. 2006). In neurosecretory cells, PIKfyve is localized on both the early endosome and LDCVs (Osborne et al. 2008). Treatment with PIKfyve selective inhibitor YM201636 or siRNA knockdown of PIKfyve potentiates neuroexocytosis (Osborne et al. 2008), suggesting PIKfyve normally exerts a negative regulatory role on this process. Interestingly, a cytosolic fraction of PIKfyve is recruited onto secretory vesicles in an activity-dependent manner (Osborne et al. 2008) through PIKfyve binding to PtdIns3P, generating PtdIns(3,5)P_2 which downregulates exocytosis via an unknown mechanism. Whether PIKfyve is acting downstream of PI3K-C2α to fine-tuning priming remains to be determined (Wen et al. 2011a).

The positive and negative regulatory roles of PtdIns3P and PtdIns(3,5)P_2, respectively, highlight a mechanism whereby neurosecretory cells could regulate vesicle priming in a timely and spatially regulated manner. However, their essential roles in exocytosis are not confined to neurosecretory cells. In adipocytes, exogenous addition of PtdIns3P promotes translocation of both overexpressed and endogenous glucose transporter type 4 (GLUT4) to the cell surface (Maffucci et al. 2003) while depleting PtdIns3P by overexpression of myotubularins (Chaussade et al. 2003) or PI3K-C2α knockdown (Falasca et al. 2007) reduces insulin-induced GLUT4 translocation and glucose uptake. Intriguingly, insulin promotes PtdIns3P production at the level of the plasma membrane through the translocation of PI3K-C2α from an endosomal compartment to the plasma membrane in a manner dependent on a small GTPase, TC10 (Falasca et al. 2007; Maffucci et al. 2003). However, the molecular basis of the TC10-mediated PI3K-C2α activation remains unclear and whether a similar mechanism also exists in neurosecretory cells is still unknown.

In adipocytes, a pathway linking the insulin-induced PtdIns(3,5)P_2 production to PIKfyve and its associated regulator ArPIKfyve has recently been characterized (Ikonomov et al. 2007). This suggests that the regulation of PIKfyve activity is not solely dependent on the availability of its substrate PtdIns3P but also relies on ArPIKfyve. Loss of PIKfyve or ArPIKfyve activity was found to severely attenuate insulin-induced PtdIns(3,5)P_2 production and GLUT4 translocation (Ikonomov et al. 2007), suggesting that PIKfyve has a positive regulatory role in insulin-induced GLUT4 trafficking. This positive regulatory role is in contrast to the negative role found in neurosecretory cells (Osborne et al. 2008). It would therefore be interesting to test the nature of these differences, which are likely to be attributable to divergent modes of membrane trafficking pathways adopted by these two cell types.

Identification of Phosphoinositide Effectors

PIs exert their cellular effects through binding and recruitment of effector proteins or through the allosteric modulation of protein function. The identification of PI binding proteins is not only critical for understanding how PIs can regulate cellular events, but has also provided the research world with valuable tools for studying the sub-cellular localization and production of PIs.

Many PI binding proteins have been identified through the presence of known PI binding domains, such as FYVE, PH, and C2 domains. However, many other proteins bind through atypical regions that are often polybasic in nature. Therefore, a number of different laboratories have combined PI binding assays with mass spectrometry to identify novel effectors. While a number of studies have employed resin-conjugated PIs and liposome-based methods to identify PI effectors, few of the proteins identified in these screens have been validated as PI binding proteins, nor have the specificity and affinity of their binding to different PIs been determined (Catimel et al. 2008, 2009). One of the inherent difficulties in identifying specific effectors is that many proteins bind to more than one PI with varying affinities. Thus, it is essential that studies are performed using liposome binding or surface plasmon resonance assays to determine the binding specificities and relative affinities of the candidate proteins (Narayan and Lemmon 2006). Some of the more promising studies in this regard have employed sequential steps to enrich for proteins that bind to the PI of interest, coupled with sensitive mass spectrometry, and validation of potential targets (Dixon et al. 2011; Lewis et al. 2011). The importance of validation was highlighted in the study by Dixon et al. (2011) where a protein identified in a screen for PtdIns(3,4)P_2 binding proteins was found to bind with a higher affinity to PtdIns(3,4,5)P_3 (Dixon et al. 2011).

While much is known about the PtdIns(4,5)P_2 effectors for neuroexocytosis, such as Synaptotagmin1 and CAPS, little is known about the 3-phosphorylated phosphoinositide effectors. At present, the yeast svp1p/Atg18 is the best characterized PtdIns(3,5)P_2 binding protein where it regulates retrograde trafficking from the endosome to the vacuole (Dove et al. 2004). The mammalian equivalent of Atg 18/svp1p is the WIPI family of proteins (WIPI1-4), of which WIPI1α is the best characterized. However, WIPIα binds to PtdIns3P in addition to PtdIns(3,5)P_2 (Jeffries et al. 2004), and it is believed that it is PtdIns3P, rather than PtdIns(3,5)P_2 binding, that is relevant for its role in autophagy (Noda et al. 2010). WIPI2 is also involved in the regulation of autophagy (Polson et al. 2010) and it remains to be seen whether any of the WIPI family members are cellular effectors of PtdIns(3,5)P_2.

A recent study combining sub-cellular fractionation, lipid pull down, and tandem mass spectrometry has highlighted the need for studies to identify PI effectors specific for cellular processes. Using biotinylated PtdIns(4,5)P_2 as bait, a number of new effectors for exocytosis, including Synaptotagmin7, were identified (Osborne et al. 2007) and later confirmed by genetic evidence (Schonn et al. 2008). Similar techniques are likely to identify effectors for other types of PIs and lipids, including PtdIns(3,5)P_2 (Anderson et al. 2010). More work is clearly needed to identify and characterize other effectors and assign them to the plethora of functions played by PIs.

Links to Neurodegenerative Disease

Dyshomeostasis of PtdIns(4,5)P_2, PtdIns3P, and PtdIn(3,5)P_2 levels in the brain have been linked to different forms of neurodegenerative disease (Wen et al. 2011a). One of the important regulators of PtdIns(4,5)P_2 levels in the brain is synaptojanin 1 (synj1). Synj1 is an inositol 5-phosphatase mapped to human chromosome 21 and its mutation has been linked to Down's syndrome (Herrera et al. 2009). Overexpression of a single gene copy of synj1 in a mouse model of Down's syndrome (Ts65Dn) leads to reduction in PtdIns(4,5)P_2 levels in the brain and a deficit in learning as evaluated by the Morris water maze (Voronov et al. 2008). Familial Alzheimer's disease (FAD) encoding presenilin mutations also results in aberrant PtdIns(4,5)P_2 metabolism, perturbed calcium currents and elevated amyloid-beta peptide $A\beta_{1-42}$ levels (Landman et al. 2006). Expression of the synj1 5-phosphatase domain can mimic the effect of elevated $A\beta_{1-42}$ levels in FAD, suggesting that PtdIns(4,5)P_2 itself is an essential signalling lipid in the pathogenesis of FAD (Landman et al. 2006). Synj1 heterozygous mice, which have a low rate of PtdIns(4,5)P_2 breakdown, are protected from the detrimental effects of $A\beta_{1-42}$ on synaptic and cognitive functions (Berman et al. 2008). These studies suggest that imbalance in PtdIns(4,5)P_2 level is sufficient to cause learning and memory deficits.

Myotubularin related protein 2 (Mtmr2) is a member of the myotubularin family of PI 3-phosphatases. Mtmr2 is highly expressed in neurons and can dephosphorylate both PtdIns3P and PtdIns(3,5)P_2 (Berger et al. 2002). Mutation of Mtmr2 resulting in the loss of its phosphatase activity has been linked to Charcot-Marie-Tooth disease 4B1 (CMT4B1), a motor and sensory neuropathy resulting from demyelination (Berger et al. 2002). Although the mechanism underlying this disease is still unclear, the reduced levels of PtdIns3P and PtdIns(3,5)P_2 in CMT4B1 are likely to cause severe defects in membrane homeostasis and vesicle trafficking pathways. Interestingly, knockout of myotubularin-related protein 13 (Mtmr13), a pseudophosphatase bearing no phosphatase activity, also leads to CMT-like neuropathy in mice (Robinson et al. 2008). This, however, can be partially explained by the reduced expression of Mtmr2 ($\sim$50 %) in Mtmr13 knockdown mice and the knock on effect on Mtmr2 activity (Berger et al. 2006). The essential role of PtdIns(3,5)P_2 in neuronal function has been further highlighted in mice lacking Vac14 and FIG4, which are important regulators of PtdIns(3,5)P_2 synthesis. In mice deficient in Vac14, there is massive degeneration in the midbrain and peripheral sensory neurons (Zhang et al. 2007). Closer examination revealed enlarged vacuoles and apparent defects in endosome to TGN trafficking (Zhang et al. 2007). Similarly, mutations in FIG4 also cause massive neurodegeneration in mice, with neurons containing enlarged vacuoles (Chow et al. 2007), an effect consistent with the essential role of PIKfyve in membrane trafficking and homeostasis. The neurodegenerative phenotypes induced by reduction in PtdIns(3,5)P_2 levels could also stem from elevated Ca^{2+} entry and increased susceptibility to excitotoxicity (Tsuruta et al. 2009) and/or elevated levels of autophagy (Ferguson et al. 2009, 2010). Clearly, defects in PtdIns3P or PtdIns(3,5)P_2 synthesis underlie a variety of neurodegenerative disorders and more work is needed to understand the precise causality.

Acknowledgments This work was supported by two National Health and Medical Research Council (NHMRC) project grants (to F.A.M. and S.L.O.). F.A.M. is a Senior Research Fellow of the NHMRC.

References

Aikawa Y, Martin TF (2003) ARF6 regulates a plasma membrane pool of phosphatidylinositol(4,5)bisphosphate required for regulated exocytosis. J Cell Biol 162(4):647–659

Anderson RJ, Osborne SL, Meunier FA, Painter GF (2010) Regioselective approach to phosphatidylinositol 3,5-bisphosphates: syntheses of the native phospholipid and biotinylated short-chain derivative. J Org Chem 75(11):3541–3551. doi:10.1021/jo100393c

Aoyagi K, Sugaya T, Umeda M, Yamamoto S, Terakawa S, Takahashi M (2005) The activation of exocytotic sites by the formation of phosphatidylinositol 4,5-bisphosphate microdomains at syntaxin clusters. J Biol Chem 280(17):17346–17352. doi:10.1074/jbc.M413307200 M413307200 [pii]

Balla A, Tuymetova G, Tsiomenko A, Varnai P, Balla T (2005) A plasma membrane pool of phosphatidylinositol 4-phosphate is generated by phosphatidylinositol 4-kinase type-III alpha: studies with the PH domains of the oxysterol binding protein and FAPP1. Mol Biol Cell 16(3):1282–1295. doi:10.1091/mbc.E04-07-0578

Balla A, Kim YJ, Varnai P, Szentpetery Z, Knight Z, Shokat KM, Balla T (2008) Maintenance of hormone-sensitive phosphoinositide pools in the plasma membrane requires phosphatidylinositol 4-kinase IIIalpha. Mol Biol Cell 19(2):711–721. doi:10.1091/mbc.E07-07-0713

Berger P, Bonneick S, Willi S, Wymann M, Suter U (2002) Loss of phosphatase activity in myotubularin-related protein 2 is associated with Charcot-Marie-Tooth disease type 4B1. Hum Mol Genet 11(13):1569–1579

Berger P, Berger I, Schaffitzel C, Tersar K, Volkmer B, Suter U (2006) Multi-level regulation of myotubularin-related protein-2 phosphatase activity by myotubularin-related protein-13/set-binding factor-2. Hum Mol Genet 15(4):569–579. doi:10.1093/hmg/ddi473 ddi473 [pii]

Berman DE, Dall'Armi C, Voronov SV, McIntire LB, Zhang H, Moore AZ, Staniszewski A, Arancio O, Kim TW, Di Paolo G (2008) Oligomeric amyloid-beta peptide disrupts phosphatidylinositol-4,5-bisphosphate metabolism. Nat Neurosci 11(5):547–554. doi:10.1038/nn.2100 nn.2100 [pii]

Cabezas A, Pattni K, Stenmark H (2006) Cloning and subcellular localization of a human phosphatidylinositol 3-phosphate 5-kinase, PIKfyve/Fab1. Gene 371(1):34–41. doi:10.1016/j.gene.2005.11.009 S0378-1119(05)00702-X [pii]

Catimel B, Schieber C, Condron M, Patsiouras H, Connolly L, Catimel J, Nice EC, Burgess AW, Holmes AB (2008) The PI(3,5)P2 and PI(4,5)P2 interactomes. J Proteome Res 7(12):5295–5313

Catimel B, Yin MX, Schieber C, Condron M, Patsiouras H, Catimel J, Robinson DE, Wong LS, Nice EC, Holmes AB, Burgess AW (2009) PI(3,4,5)P3 Interactome. J Proteome Res 8(7):3712–3726. doi:10.1021/pr900320a

Chaussade C, Pirola L, Bonnafous S, Blondeau F, Brenz-Verca S, Tronchere H, Portis F, Rusconi S, Payrastre B, Laporte J, Van Obberghen E (2003) Expression of myotubularin by an adenoviral vector demonstrates its function as a phosphatidylinositol 3-phosphate [PtdIns(3)P] phosphatase in muscle cell lines: involvement of PtdIns(3)P in insulin-stimulated glucose transport. Mol Endocrinol 17(12):2448–2460. doi:10.1210/me.2003-0261

Chow CY, Zhang Y, Dowling JJ, Jin N, Adamska M, Shiga K, Szigeti K, Shy ME, Li J, Zhang X, Lupski JR, Weisman LS, Meisler MH (2007) Mutation of FIG4 causes neurodegeneration in the pale tremor mouse and patients with CMT4J. Nature 448(7149):68–72. doi:10.1038/nature05876 nature05876 [pii]

De Lartigue J, Polson H, Feldman M, Shokat K, Tooze SA, Urbe S, Clague MJ (2009) PIKfyve regulation of endosome-linked pathways. Traffic 10(7):883–893

Demmel L, Gravert M, Ercan E, Habermann B, Muller-Reichert T, Kukhtina V, Haucke V, Baust T, Sohrmann M, Kalaidzidis Y, Klose C, Beck M, Peter M, Walch-Solimena C (2008) The clathrin adaptor Gga2p is a phosphatidylinositol 4-phosphate effector at the Golgi exit. Mol Biol Cell 19(5):1991–2002. doi:10.1091/mbc.E06-10-0937

Di Paolo G, Moskowitz HS, Gipson K, Wenk MR, Voronov S, Obayashi M, Flavell R, Fitzsimonds RM, Ryan TA, De Camilli P (2004) Impaired PtdIns(4,5)P2 synthesis in nerve terminals produces defects in synaptic vesicle trafficking. Nature 431(7007):415–422

Dixon MJ, Gray A, Boisvert FM, Agacan M, Morrice NA, Gourlay R, Leslie NR, Downes CP, Batty IH (2011) A screen for novel phosphoinositide 3-kinase effector proteins. Mol Cell Proteomics 10 (4):M110 003178. doi:10.1074/mcp.M110.003178

Dove SK, Piper RC, McEwen RK, Yu JW, King MC, Hughes DC, Thuring J, Holmes AB, Cooke FT, Michell RH, Parker PJ, Lemmon MA (2004) Svp1p defines a family of phosphatidylinositol 3,5-bisphosphate effectors. EMBO j 23(9):1922–1933. doi:10.1038/sj.emboj.7600203

Eberhard DA, Cooper CL, Low MG, Holz RW (1990) Evidence that the inositol phospholipids are necessary for exocytosis. Loss of inositol phospholipids and inhibition of secretion in permeabilized cells caused by a bacterial phospholipase C and removal of ATP. Biochem J 268(1):15–25

Falasca M, Hughes WE, Dominguez V, Sala G, Fostira F, Fang MQ, Cazzolli R, Shepherd PR, James DE, Maffucci T (2007) The role of phosphoinositide 3-kinase C2alpha in insulin signalling. J Biol Chem 282(38):28226–28236. doi:10.1074/jbc.M704357200 M704357200 [pii]

Ferguson CJ, Lenk GM, Meisler MH (2009) Defective autophagy in neurons and astrocytes from mice deficient in PI(3,5)P2. Hum Mol Genet 18(24):4868–4878. doi:10.1093/hmg/ddp460 ddp460 [pii]

Ferguson CJ, Lenk GM, Meisler MH (2010) PtdIns(3,5)P2 and autophagy in mouse models of neurodegeneration. Autophagy 6(1):170–171 10626 [pii]

Godi A, Pertile P, Meyers R, Marra P, Di Tullio G, Iurisci C, Luini A, Corda D, De Matteis MA (1999) ARF mediates recruitment of PtdIns-4-OH kinase-beta and stimulates synthesis of PtdIns(4,5)P2 on the Golgi complex. Nat Cell Biol 1(5):280–287

Grishanin RN, Kowalchyk JA, Klenchin VA, Ann K, Earles CA, Chapman ER, Gerona RR, Martin TF (2004) CAPS acts at a prefusion step in dense-core vesicle exocytosis as a PIP2 binding protein. Neuron 43(4):551–562

Hammond GR, Dove SK, Nicol A, Pinxteren JA, Zicha D, Schiavo G (2006) Elimination of plasma membrane phosphatidylinositol (4,5)-bisphosphate is required for exocytosis from mast cells. J Cell Sci 119(Pt 10):2084–2094. doi:10.1242/jcs.02912 119/10/2084 [pii]

Hay JC, Martin TF (1992) Resolution of regulated secretion into sequential Mg ATP-dependent and calcium-dependent stages mediated by distinct cytosolic proteins. J Cell Biol 119(1):139–151

Hay JC, Martin TF (1993) Phosphatidylinositol transfer protein required for ATP-dependent priming of Ca(2+)-activated secretion. Nature 366(6455):572–575

Hay JC, Fisette PL, Jenkins GH, Fukami K, Takenawa T, Anderson RA, Martin TF (1995) ATP-dependent inositide phosphorylation required for Ca(2+)-activated secretion. Nature 374(6518):173–177

Haynes LP, Thomas GM, Burgoyne RD (2005) Interaction of neuronal calcium sensor-1 and ADP-ribosylation factor 1 allows bidirectional control of phosphatidylinositol 4-kinase beta and trans-Golgi network-plasma membrane traffic. J Biol Chem 280(7):6047–6054

Herrera F, Chen Q, Fischer WH, Maher P, Schubert DR (2009) Synaptojanin-1 plays a key role in astrogliogenesis: possible relevance for Down's syndrome. Cell Death Differ 16(6):910–920. doi:10.1038/cdd.2009.24 cdd200924 [pii]

Hokin LE, Hokin MR (1955a) Effects of acetylcholine on phosphate turnover in phospholipids of brain cortex in vitro. Biochim Biophys Acta 16(2):229–237

Hokin LE, Hokin MR (1955b) Effects of acetylcholine on the turnover of phosphoryl units in individual phospholipids of pancreas slices and brain cortex slices. Biochim Biophys Acta 18(1):102–110

Hokin LE, Hokin MR (1958) Acetylcholine and the exchange of inositol and phosphate in brain phosphoinositide. J Biol Chem 233(4):818–821

Hokin MR, Hokin LE (1954) Effects of acetylcholine on phospholipids in the pancreas. J Biol Chem 209(2):549–558

Hokin LE, Hokin MR (1960) The role of phosphatidic acid and phosphoinositide in transmembrane transport elicited by acetylcholine and other humoral agents. Int Rev Neurobiol 2:99–136

Hokin MR, Hokin LE, Shelp WD (1960) The effects of acetylcholine on the turnover of phosphatidic acid and phosphoinositide in sympathetic ganglia, and in various parts of the central nervous system in vitro. J Gen Physiol 44:217–226

Ikonomov OC, Sbrissa D, Shisheva A (2006) Localized PtdIns 3,5–P2 synthesis to regulate early endosome dynamics and fusion. Am J Physiol Cell Physiol 291(2):C393–C404. doi:10.1152/ajpcell.00019.2006 00019.2006 [pii]

Ikonomov OC, Sbrissa D, Dondapati R, Shisheva A (2007) ArPIKfyve-PIKfyve interaction and role in insulin-regulated GLUT4 translocation and glucose transport in 3T3-L1 adipocytes. Exp Cell Res 313(11):2404–2416. doi:10.1016/j.yexcr.2007.03.024 S0014-4827(07)00113-9 [pii]

James DJ, Khodthong C, Kowalchyk JA, Martin TF (2008) Phosphatidylinositol 4,5-bisphosphate regulates SNARE-dependent membrane fusion. J Cell Biol 182(2):355–366. doi:10.1083/jcb.200801056 jcb.200801056 [pii]

James DJ, Kowalchyk J, Daily N, Petrie M, Martin TF (2009) CAPS drives trans-SNARE complex formation and membrane fusion through syntaxin interactions. Proc Nat Acad Sci USA 106(41):17308–17313. doi:10.1073/pnas.0900755106

James DJ, Khodthong C, Kowalchyk JA, Martin TF (2010) Phosphatidylinositol 4,5-bisphosphate regulation of SNARE function in membrane fusion mediated by CAPS. Adv Enzyme Regul 50(1):62–70. doi:10.1016/j.advenzreg.2009.10.012

Jefferies HB, Cooke FT, Jat P, Boucheron C, Koizumi T, Hayakawa M, Kaizawa H, Ohishi T, Workman P, Waterfield MD, Parker PJ (2008) A selective PIKfyve inhibitor blocks PtdIns(3,5)P(2) production and disrupts endomembrane transport and retroviral budding. EMBO Rep 9(2):164–170. doi:10.1038/sj.embor.7401155 7401155 [pii]

Jeffries TR, Dove SK, Michell RH, Parker PJ (2004) PtdIns-specific MPR pathway association of a novel WD40 repeat protein, WIPI49. Mol Biol Cell 15(6):2652–2663. doi:10.1091/mbc.E03-10-0732

Johnson EE, Overmeyer JH, Gunning WT, Maltese WA (2006) Gene silencing reveals a specific function of hVps34 phosphatidylinositol 3-kinase in late versus early endosome. J Cell Sci 119(Pt 7):1219–1232

Kapp-Barnea Y, Melnikov S, Shefler I, Jeromin A, Sagi-Eisenberg R (2003) Neuronal calcium sensor-1 and phosphatidylinositol 4-kinase beta regulate IgE receptor-triggered exocytosis in cultured mast cells. J immunol 171(10):5320–5327

Kim YJ, Guzman-Hernandez ML, Balla T (2011) A highly dynamic ER-derived phosphatidylinositol-synthesizing organelle supplies phosphoinositides to cellular membranes. Dev Cell 21(5):813–824. doi:10.1016/j.devcel.2011.09.005

Koizumi S, Rosa P, Willars GB, Challiss RA, Taverna E, Francolini M, Bootman MD, Lipp P, Inoue K, Roder J, Jeromin A (2002) Mechanisms underlying the neuronal calcium sensor-1-evoked enhancement of exocytosis in PC12 cells. J Biol Chem 277(33):30315–30324

Kuo W, Herrick DZ, Cafiso DS (2011) Phosphatidylinositol 4,5-bisphosphate alters synaptotagmin 1 membrane docking and drives opposing bilayers closer together. Biochemistry 50(13):2633–2641. doi:10.1021/bi200049c

Landman N, Jeong SY, Shin SY, Voronov SV, Serban G, Kang MS, Park MK, Di Paolo G, Chung S, Kim TW (2006) Presenilin mutations linked to familial Alzheimer's disease cause an imbalance in phosphatidylinositol 4,5-bisphosphate metabolism. Proc Nat Acad Sci U S A 103(51):19524–19529. doi:10.1073/pnas.0604954103

Levine B, Klionsky DJ (2004) Development by self-digestion: molecular mechanisms and biological functions of autophagy. Dev Cell 6(4):463–477 S1534580704000991 [pii]

Lewis AE, Sommer L, Arntzen MO, Strahm Y, Morrice NA, Divecha N, D'Santos CS (2011) Identification of nuclear phosphatidylinositol 4,5-bisphosphate-interacting proteins by neomycin extraction. Mol Cell Proteomics 10 (2):M110 003376. doi:10.1074/mcp.M110.003376

Loyet KM, Kowalchyk JA, Chaudhary A, Chen J, Prestwich GD, Martin TF (1998) Specific binding of phosphatidylinositol 4,5-bisphosphate to calcium-dependent activator protein for secretion (CAPS), a potential phosphoinositide effector protein for regulated exocytosis. J Biol Chem 273(14):8337–8343

Maffucci T, Brancaccio A, Piccolo E, Stein RC, Falasca M (2003) Insulin induces phosphatidylinositol-3-phosphate formation through TC10 activation. EMBO J 22(16):4178–4189

Meunier FA, Osborne SL, Hammond GR, Cooke FT, Parker PJ, Domin J, Schiavo G (2005) Phosphatidylinositol 3-kinase C2alpha is essential for ATP-dependent priming of neurosecretory granule exocytosis. Mol Biol Cell 16(10):4841–4851

Milosevic I, Sorensen JB, Lang T, Krauss M, Nagy G, Haucke V, Jahn R, Neher E (2005) Plasmalemmal phosphatidylinositol-4,5-bisphosphate level regulates the releasable vesicle pool size in chromaffin cells. J Neurosci 25(10):2557–2565

Murray JT, Panaretou C, Stenmark H, Miaczynska M, Backer JM (2002) Role of Rab5 in the recruitment of hVps34/p150 to the early endosome. Traffic 3(6):416–427 tra030605 [pii]

Narayan K, Lemmon MA (2006) Determining selectivity of phosphoinositide-binding domains. Methods 39(2):122–133. doi:10.1016/j.ymeth.2006.05.006

Noda T, Matsunaga K, Taguchi-Atarashi N, Yoshimori T (2010) Regulation of membrane biogenesis in autophagy via PI3P dynamics. Semin Cell Dev Biol 21(7):671–676. doi:10.1016/j.semcdb.2010.04.002

Olsen HL, Hoy M, Zhang W, Bertorello AM, Bokvist K, Capito K, Efanov AM, Meister B, Thams P, Yang SN, Rorsman P, Berggren PO, Gromada J (2003) Phosphatidylinositol 4-kinase serves as a metabolic sensor and regulates priming of secretory granules in pancreatic beta cells. Proc Natl Acad Sci USA 100(9):5187–5192

Osborne SL, Wallis TP, Jimenez JL, Gorman JJ, Meunier FA (2007) Identification of secretory granule phosphatidylinositol 4,5-bisphosphate-interacting proteins using an affinity pull down strategy. Mol Cell Proteomics 6(7):1158–1169. doi:10.1074/mcp.M600430-MCP200 M600430-MCP200 [pii]

Osborne SL, Wen PJ, Boucheron C, Nguyen HN, Hayakawa M, Kaizawa H, Parker PJ, Vitale N, Meunier FA (2008) PIKfyve negatively regulates exocytosis in neurosecretory cells. J Biol Chem 283(5):2804–2813. doi:10.1074/jbc.M704856200 M704856200 [pii]

Polson HE, de Lartigue J, Rigden DJ, Reedijk M, Urbe S, Clague MJ, Tooze SA (2010) Mammalian Atg18 (WIPI2) localizes to omegasome-anchored phagophores and positively regulates LC3 lipidation. Autophagy 6(4)

Rajebhosale M, Greenwood S, Vidugiriene J, Jeromin A, Hilfiker S (2003) Phosphatidylinositol 4-OH kinase is a downstream target of neuronal calcium sensor-1 in enhancing exocytosis in neuroendocrine cells. J Biol Chem 278(8):6075–6084

Robinson FL, Niesman IR, Beiswenger KK, Dixon JE (2008) Loss of the inactive myotubularin-related phosphatase Mtmr13 leads to a Charcot-Marie-Tooth 4B2-like peripheral neuropathy in mice. Proc Natl Acad Sci U S A 105(12):4916–4921. doi:10.1073/pnas.0800742105 0800742105 [pii]

Rutherford AC, Traer C, Wassmer T, Pattni K, Bujny MV, Carlton JG, Stenmark H, Cullen PJ (2006) The mammalian phosphatidylinositol 3-phosphate 5-kinase (PIKfyve) regulates endosome-to-TGN retrograde transport. J Cell Sci 119(Pt 19):3944–3957. doi:10.1242/jcs.03153 jcs.03153 [pii]

Schonn JS, Maximov A, Lao Y, Sudhof TC, Sorensen JB (2008) Synaptotagmin-1 and -7 are functionally overlapping Ca2+ sensors for exocytosis in adrenal chromaffin cells. Proc Nat Acad Sci USA 105(10):3998–4003. doi:10.1073/pnas.0712373105

Sedej S, Gurung IS, Binz T, Rupnik M (2009) Phosphatidylinositol-4,5-bisphosphate-dependent facilitation of the ATP-dependent secretory activity in mouse pituitary cells. Ann N Y Acad Sci 1152:165–173. doi:10.1111/j.1749-6632.2008.04002.x

Tomas A, Yermen B, Regazzi R, Pessin JE, Halban PA (2010) Regulation of insulin secretion by phosphatidylinositol-4,5-bisphosphate. Traffic 11(1):123–137. doi:10.1111/j.1600-0854.2009.00996.x TRA996 [pii]

Tsuruta F, Green EM, Rousset M, Dolmetsch RE (2009) PIKfyve regulates CaV1.2 degradation and prevents excitotoxic cell death. J Cell Biol 187 (2):279-294. doi:jcb.200903028 [pii] 10.1083/jcb.200903028

Voronov SV, Frere SG, Giovedi S, Pollina EA, Borel C, Zhang H, Schmidt C, Akeson EC, Wenk MR, Cimasoni L, Arancio O, Davisson MT, Antonarakis SE, Gardiner K, De Camilli P, Di Paolo G (2008) Synaptojanin 1-linked phosphoinositide dyshomeostasis and cognitive deficits in mouse models of Down's syndrome. Proc Natl Acad Sci U S A 105(27):9415–9420. doi:10.1073/pnas.0803756105 0803756105 [pii]

Wang YJ, Wang J, Sun HQ, Martinez M, Sun YX, Macia E, Kirchhausen T, Albanesi JP, Roth MG, Yin HL (2003) Phosphatidylinositol 4 phosphate regulates targeting of clathrin adaptor AP-1 complexes to the Golgi. Cell 114(3):299–310

Wang J, Sun HQ, Macia E, Kirchhausen T, Watson H, Bonifacino JS, Yin HL (2007) PI4P promotes the recruitment of the GGA adaptor proteins to the trans-Golgi network and regulates their recognition of the ubiquitin sorting signal. Mol Biol Cell 18(7):2646–2655. doi:10.1091/mbc.E06-10-0897

Way G, O'Luanaigh N, Cockcroft S (2000) Activation of exocytosis by cross-linking of the IgE receptor is dependent on ADP-ribosylation factor 1-regulated phospholipase D in RBL- 2H3 mast cells: evidence that the mechanism of activation is via regulation of phosphatidylinositol 4,5-bisphosphate synthesis. Biochem J 346(1):63–70

Wen PJ, Osborne SL, Morrow IC, Parton RG, Domin J, Meunier FA (2008) Ca2+ -regulated pool of phosphatidylinositol-3-phosphate produced by phosphatidylinositol 3-kinase C2alpha on neurosecretory vesicles. Mol Biol Cell 19(12):5593–5603. doi:10.1091/mbc.E08-06-0595 E08-06-0595 [pii]

Wen PJ, Osborne SL, Meunier FA (2011a) Dynamic control of neuroexocytosis by phosphoinositides in health and disease. Prog Lipid Res 50(1):52–61. doi:10.1016/j.plipres.2010.08.001

Wen PJ, Osborne SL, Zanin M, Low PC, Wang HT, Schoenwaelder SM, Jackson SP, Wedlich-Soldner R, Vanhaesebroeck B, Keating DJ, Meunier FA (2011b) Phosphatidylinositol(4,5)bisphosphate coordinates actin-mediated mobilization and translocation of secretory vesicles to the plasma membrane of chromaffin cells. Nat Commun 2:491. doi:10.1038/ncomms1500

Wong K, Meyers dd R, Cantley LC (1997) Subcellular locations of phosphatidylinositol 4-kinase isoforms. J Biol Chem 272(20):13236–13241

Zhang Y, Zolov SN, Chow CY, Slutsky SG, Richardson SC, Piper RC, Yang B, Nau JJ, Westrick RJ, Morrison SJ, Meisler MH, Weisman LS (2007) Loss of Vac14, a regulator of the signalling lipid phosphatidylinositol 3,5-bisphosphate, results in neurodegeneration in mice. Proc Natl Acad Sci U S A 104(44):17518–17523. doi:10.1073/pnas.0702275104 0702275104 [pii]

Chapter 5
Effects of Phosphoinositides and Their Derivatives on Membrane Morphology and Function

Banafshé Larijani and Dominic L. Poccia

Abstract Currently, one of the fundamental problems in the study of membrane function and morphology is that the roles of proteins and lipids are usually investigated separately. In most cases proteins are predominant, with lipids taking a subsidiary role. This polarised view is in part due to the more straightforward and familiar techniques used to investigate proteins. Here, we summarise how phospholipids can be studied in cells with new tools that can acutely (rapidly and specifically) modify phospholipid composition of membranes in subcellular compartments. We point out some of the important physical effects that phosphoinositides in particular can have in altering membrane bilayer morphology, and provide specific examples to illustrate the roles that these phospholipids may play in maintaining the geometry of endomembranes.

Physical Properties of Phospholipids Affecting Membrane Morphology and Fusion

It is only in the past few years that researchers have revised traditional models of membrane dynamics by studying the involvement of phosphoinositides and their derivatives such as diacylglycerol (DAG) (Mima et al. 2008; Byrne et al. 2009; Wang et al. 2002; Chernomordik and Kozlov 2008; Dumas et al. 2010; Goni and

Present Address:
B. Larijani (✉)
Cell Biophysics Laboratory, Cancer Research UK, London Research Institute, 44 Lincoln's Inn Fields, London, WC2A 3LY, UK
e-mail: banafshe.larijani@cancer.org.uk

D. L. Poccia
Department of Biology, Amherst College, Amherst, MA 01002, USA
e-mail: dlpoccia@amherst.edu

M. Falasca (ed.), *Phosphoinositides and Disease*, Current Topics in Microbiology and Immunology 362, DOI: 10.1007/978-94-007-5025-8_5,

Alonso 1999). These phospholipids have been recognised as important second messengers but their effects on membrane dynamics and morphology have not been integrated with their roles as signalling molecules. Both in vitro and in vivo studies on the participation of phosphoinositides in the regulation of membrane dynamics implicate structural roles and have resulted in a re-evaluation of, for example, protein-regulated fusion models to include the involvement of higher phosphorylated phosphoinositides (Mima et al. 2008, James et al. 2008; Lynch et al. 2008; Mima and Wickner 2009a, b) and their derivatives.

The roles of various lipids in membrane dynamics have been studied extensively (Chernomordik and Kozlov 2008; Dumas et al. 2010; Larijani and Poccia 2009; Martin et al. 1997). There are two main ways by which lipids affect the physical properties of membranes. The first is that they contribute to variations in membrane fluidity, which is assessed by the order parameter of the localised membranous regions. The order parameter determines the dynamics of the membrane. Changes in membrane fluidity (molecular order parameter) may influence lipid–lipid interactions, which leads to lipid mixing and lipid–protein interactions, which can lead to "productive" associations of proteins with lipids. The second is that lipids affect localised membrane morphology, which is dependent on the chemical structure of the lipids composing a monolayer. The localised morphology is determined by a parameter known as spontaneous curvature. Lipids whose hydrophobicity and head group size are similar have a spontaneous curvature close to zero. These lipids assemble spontaneously into lamellar structures. Lipids with positive spontaneous curvature (e.g. lysophosphatidylcholine and polyphosphoinositides) favour bending away from the head groups whereas those with negative spontaneous curvature (DAG, cholesterol and phosphatidylethanolamine) favour bending towards the head groups. Asymmetry of lipid composition of the bilayer can thus contribute to bending of the bilayer, depending on whether elevated amounts of lipids favouring positive or negative spontaneous curvature are located in the monolayer. Moreover, some lipids, such as DAG, render membranes more disordered and destabilise the membrane by forming intermediate non-lamellar structures. Other lipids, such as cholesterol, render the membrane more ordered and also form non-lamellar structures. Therefore, a lipid such as DAG would be a perfect candidate for the induction of dynamic processes, such as fission and fusion, where both destabilisation of the lamellar structure, negative curvature and lipid mixing are required to form one continuous bilayer from two separate bilayers.

Much experimental evidence shows that model membranes enriched in negatively curved lipids spontaneously fuse whereas incorporation of positively curved lipids hinders fusion (Chernomordik and Kozlov 2005; Das and Rand 1984; Dufourc et al. 1992; Kozlovsky et al. 2002; Kozlovsky and Kozlov 2002). In natural membranes, proteins inserted into the bilayer may also contribute in overcoming the energy barrier required for membrane fusion, although one of the main functions of proteins in membrane trafficking events that involve fusion is to ensure that membrane vesicles are targeted specifically to the appropriate subcellular compartments.

There is evidence to indicate that phosphoinositides may "prepare" the membrane environment for localised membrane fusion. Solid-state NMR spectroscopy studies of model membranes that have high levels of 18:0/20:4 DAG and polyunsaturated phosphatidylinositol document that they are very fluid even at low temperatures (Byrne et al. 2007; Larijani and Dufourc 2006). Charged phosphoinositides may also reorganise membranes and cause phase separation (Gambhir et al. 2004). Recently deuterium and ^{31}P solid-state NMR studies have shown that model membranes with a complex mixture of polyphosphoinositides at elevated levels, resembling in composition naturally occurring fusogenic membrane vesicles, are more fluid than classical liquid-ordered membranes. The elevated fluidity of such membranes results in increasing molecular motion, which may promote the initial interactions of fusogenic membrane vesicles to more rigid membrane domains (Garnier-Lhomme et al. 2009), and hence facilitate lipid mixing during membrane fusion.

Another effect of enhanced membrane fluidity would be on protein–lipid interactions. An increase of membrane surface kinetics, due to variations in molecular motion, would increase the probability of significant interactions between proteins and the membrane. This would augment the probability of weak protein–lipid interactions, prior to their more specific associations via, for example, phosphoinositide or DAG binding domains. One instance of this may be the initial recruitment of PDK1 and PKB to regions of membranes with negative charged phospholipids prior to their specific interactions with PtdIns(3,4,5)P_3 (Knight and Falke 2009; Lucas and Cho 2011). The same phenomenon may also play a role in the recruitment of bacterial PI-PLC or eukaryotic PLCγ to regions of membranes where respectively phosphatidylinositol or PtdIns(4,5)P_2 are localised and subsequently hydrolysed to DAG (Dumas et al. 2010; Byrne et al. 2007; Larijani et al. 2001). As a result, in such regions localised production of DAG could induce fusion.

^{31}P NMR studies of membranes with elevated levels of phosphoinositides demonstrate that, in addition to effects on fluidity, polyphosphoinositides impose their orientation dynamics on other lipids. Under these circumstances, the polyphosphoinositides could induce a clustering effect, which may enhance the "bulging" of a region of negative potential and hence facilitate the interaction with phosphoinositide-interacting domains and polybasic-rich residues of various proteins (Gambhir et al. 2004; Zhendre et al. 2011). These types of interactions may counterbalance membranes with high amounts of charged phosphoinositides and render them stable.

Furthermore, another function of the spontaneous curvature of lipids may be determination of organelle morphology where organelle geometry would be dependent on the balance of negatively and/or positively spontaneously curved lipids in each monolayer. Organelle morphologies, therefore, may not be solely dependent on the localisation of specific scaffolding or other proteins (Mattaj 2004; McMahon and Gallop 2005; Shibata et al. 2010; Zimmerberg and Kozlov 2006) but also on the asymmetry of non-lamellar bilayer lipids, discussed further below.

Acute and Inducible Modification of Lipids in Endomembranes

A tool that can specifically recruit lipid-modifying enzymes to endomembranes was created to destabilise the symmetry of endomembranes (Fili et al. 2006) so that the bilayer would no longer be at equilibrium and the flip-flop of lipids, which usually occurs to restore symmetry, would not be sufficient (Ruiz-Arguello et al. 1998). This specific chemical biology tool is a regulated dimerisation device, which can induce an acute lipid asymmetry in various subcellular compartments. This approach has unique advantages over classical usage of general inhibitors or RNAi depletion of mRNAs, since the membrane compartment of choice is targeted and lipid modifications occur rapidly within this compartment.

An example of acute induction is the in vivo application of a rapalogue-induced depletion that locally modifies PtdIns(3)P and PtdIns(3,5)P_2 to PtdIns by the recruitment of human myotubularin 1 to Rab5 positive endosomes. Induction results in extensive endosomal tubularisation due to the local and acute depletion of PtdIns(3)P and PtdIns(3,5)P_2. Also the normal maturation of the endosomal compartment and the flux of receptors through it are significantly affected by acute removal of these two phosphoinositides. The effect of locally depleting phospholipids of higher spontaneous positive curvature (PtdIns(3)P and PtdIns(3,5)P_2 to PtdIns) on the cytosolic face of the endosomes would alter the morphology of the endosomal compartment and render it more tubularised (curved). As shown by molecular dynamics experiments and deuterium solid-state NMR spectroscopy, unsaturated phosphatidylinositol induces a higher molecular order than its unsaturated phosphorylated forms. Therefore, enhanced formation of phosphatidylinositol would also induce a more fluid membrane, which could facilitate the elongation of the endosomal membranes.

The inducible dimerisation device has also been used to acutely modify PtdIns(4)P in the Golgi complex (Szentpetery et al. 2010). However, the outcome of this study was mostly focused on regulatory aspect of PtdIns(4)P on the recruitment of clathrin and clathrin adaptor proteins to the Golgi membrane. Modifications of Golgi morphology upon the depletion of PtdIns(4)P were not discussed.

In summary, a major advantage of such new methods is that they result in rapid and specifically localised alterations of membrane lipid composition, and therefore unlike with knockdowns of enzymes or mutations of proteins, where the cell membranes have time to adjust to a new steady state, the effects on the targeted subcellular compartments can be observed in short period of time. Most importantly, other subcellular compartments are not directly affected and morphological studies can be carried on the compartment of interest. Furthermore, with these methods the effect of lipid composition on each subcellular compartment can be studied without requiring the purification of each compartment and quantification of its lipid composition.

Role of Phosphoinositides and DAG on Endomembrane Function and Morphology

Localisation and quantification of specific phosphoinositides to endomembranes present many challenges. Isolation of highly purified components for direct quantification is usually difficult. Other approaches potentially involving less perturbation of the cells use lipid-specific probes on either live or dead cells, but these are also subject to various pitfalls and artefacts (Szabadkai et al. 2006; Balla et al. 2000; Hammond et al. 2009). PtdIns-specific antibodies are often used on fixed tissues accompanied, however, by the variability introduced with fixation protocols and different sources of antibodies (Hammond et al. 2009). Protein domains such as PH, PX, ENTH, C1 and C2, and FYVE, which recognise various phosphoinositides have varying binding affinities (Lindsay et al. 2006; Watt et al. 2002). Linked to GFP these can be used as probes in live cells, but even when displaying preference for single compartments, it is not always clear whether they report relative intracellular levels of their target phosphoinositides, especially given that they have to compete for lipid binding with endogenous proteins or in some cases can bind to pools of IP_3. Domain binding may also depend upon auxiliary proteins for its affinity or specificity. Finally as with all probes in live cells, high levels of the probe have the potential to alter cellular physiology and can result in relocation of the target. Perhaps, the most specific localisation and best quantification of the phosphoinositides have been by transmission electron microscopy methods. Such studies illustrate that PtdIns(4,5)P_2 is mainly localised to the plasma membrane, although the endoplasmic reticulum (ER) and the Golgi network are the subcellular compartments that contain the second highest levels of this phosphoinositide. Spatially restricted PtdIns(4,5)P_2 could be important in maintaining Golgi organisation and facilitating protein transport. A high proportion of PtdIns(4,5)P_2 is localised in the Golgi stack where it may play a role in maintenance of Golgi morphology (Lindsay et al. 2006; Watt et al. 2002; van Meer 1998).

Although most literature suggests that the main function of phosphoinositides in these subcellular compartments is to serve as protein binding partners or allosteric regulators of phosphoinositide binding proteins, it is also important to realise that the localisation of these lipids and their derivatives can affect the geometry of the membrane. Recent reviews (Mattaj 2004; Shibata et al. 2010) on ER morphology have stressed that membrane resident proteins such as the reticulons and Ypo1 stabilise curvature by "scaffolding" or "hydrophobic wedging". However, these models ignore the effects of lipids in determining local geometry. First, it is difficult to imagine that proteins "wedge" in a bilayer without interaction with specific lipids. Second, there is little known about the potential asymmetry of lipids in endomembranes, in particular those leading to spontaneous negative or positive curvature. Therefore, it will be essential to take into account how both lipids and proteins modify local geometry of membranes to have a more detailed understanding of the mechanisms involved in this complex process.

Phosphoinositides Associated with the Nucleus

In this section, we take up issues of the phosphoinositide composition of the nuclear envelope (NE), the amounts present in the envelope versus the nuclear interior and potential roles for PtdIns derivatives in nuclear membranes. We assess possible problems with current techniques and indicate some future directions that might prove fruitful.

NE Structure and Definitions. The NE constitutes a functionally important endomembrane system, which is both part of, and distinct from, the ER. It mediates nucleo-cytoplasmic transport, gene regulation, and various signalling pathways and its malfunctions have been linked to several dystrophies (Akhtar and Gasser 2007; Webster et al. 2009). The NE consists of an outer membrane continuous with the ER and an inner membrane continuous with the outer membrane only at the regions of the nuclear pores, which span the cisternal or perinuclear space between the two membranes. The cisternal space is continuous with the lumen of the ER.

The protein composition of the inner membrane is distinct from the ER during interphase of the cell cycle (Prunuske and Ullman 2006) but proteins of the inner membrane can be absorbed into the ER at mitosis (Ellenberg et al. 1997). The inner membrane is underlain with a proteinaceous nuclear lamina, which is one of many constituents represented in the complex and dynamic nuclear interior (Austin and Bellini 2010). We will use the term *nuclear interior* here to distinguish nuclear membrane constituents from the remainder of the nucleus including the lamina, although the term NE is typically used to include both the membranes and the associated lamina. The nuclear interior includes structures like the chromosomes, lamina, nucleoli, various nuclear bodies, actin filaments, possibly a nuclear matrix and the nucleoplasm with many RNAs, proteins, and other soluble components (Austin and Bellini 2010; Mao et al. 2011).

The fundamental structure of both nuclear membranes is expected to derive from amphipathic phospholipids arranged predominantly as a bilayer. Thus, there are four monolayers to consider in the envelope. One monolayer of each bilayer faces the common perinuclear cisternum coextensive with the ER lumen. The outer membrane surface faces the cytosol like the cytoplasmic face of the ER and the inner membrane faces the nucleoplasm, lamina, or other interior structures. The pores could provide a barrier to lipid diffusion within the membranes allowing different lipid compositions to be maintained in the two bilayers during interphase.

NE Lipid Composition. The typical phospholipid composition of each monolayer or bilayer of the interphase NE is not known. However, precursor membrane fractions that contribute a small proportion of the sea urchin NE formed in vitro have been assayed by 2D NMR or HPLC-ESI tandem mass spectrometry and these contain surprisingly large amounts of PtdIns species. One of these fractions is a population from eggs required for nuclear membrane fusion called MV1 (60–70 mol % PtdIns of total phospholipids, two-thirds of those as phosphoinositides of which 25 % are $PtdInsP_2$) (Byrne et al. 2007; Larijani et al. 2000).

Another is the detergent-resistant remnants of the sperm NE that are also incorporated into the male pronuclear envelope (50 mol % PtdIns species, 12 mol % of which is $PtdInsP_2$ consisting predominantly of alkyl acyl species; 1.4/1 molar ratio of phospholipid/sterols; the PtdIns molecules largely highly unsaturated arachidonyl (20:4) species) (Garnier-Lhomme et al. 2009). By contrast in the same species, other membranes, mostly ER-derived, which contribute the bulk of the forming NE, as quantified by the same methods, have substantially lower levels of PtdIns species (25 mol %, almost all unphosphorylated) (Byrne et al. 2007).

In MV1, a majority of PtdInsP and $PtdInsP_2$ are diacyl forms. MV1 is enriched in the PtdIns(4,5)P_2 hydrolysing enzyme PLCγ and its activator Src Family Kinase 1 (SFK1) (Byrne et al. 2009; Byrne et al. 2007). Hydrolysis of MV1 $PtdInsP_2$ to DAG with human recombinant PLCγ resulted in preferential utilisation of the highly unsaturated diacyl species 18:0/20:4 and 18:0/22:5. Normalised to total phospholipids, the ER membranes do not have high levels of PLCγ (Byrne et al. 2007) or SFK1 (Byrne et al. 2009).

Phosphoinositides in the Nuclear Interior. Further complicating the architecture of the NE are "invaginations". Although the NE is often thought of as approximating a sphere around the nuclear contents, it is clear that in many cell types the envelope can penetrate deeply within the nucleus to form a "nuclear reticulum" (Malhas et al. 2011). In Type I invaginations, only the inner nuclear membrane penetrates; in Type II, both bilayers penetrate. Either type will place a portion of the inner nuclear membrane (and its phospholipids) deep within the nucleus, increasing the surface area of the nucleus, and decreasing the distances between the envelope and the chromosomes or other interior bodies.

The fact that phospholipids of the NE can penetrate the interior of the nucleus raises the issue of whether all lipids that are apparently located in the interior remain associated with membranes or reside in the nuclear interior itself. Evidence for interior membranes unconnected to the NE is scant (Keune et al. 2011) but they may occur (Malhas et al. 2011) which further complicates assignment of interior phospholipids to non-membranous origins. However, a growing literature has documented the existence of PtdIns derivatives within the nucleus apparently not associated with the membranes, although the subject is not without controversy.

Unambiguous localisation of phospholipids to the nuclear interior is a challenge. While there are several methods to demonstrate the non-membrane location of interior lipids, none are perfect. Many biochemical investigations of this compartment depend on isolation of nuclei with intact nuclear envelopes subsequently solubilised by non-ionic detergents (and evaluated by electron microscopy). The use of detergents is problematic because of possible rearrangements of lipids, especially the more hydrophilic ones like the phosphoinositides, occurring during treatment raising thorny issues previously faced by investigators of detergent treated plasma membrane lipid rafts or domains (Chamberlain 2004; Shaw 2006) or even earlier of the nuclear matrix (Pederson 2000). This is a potential complication for study of the NE remnants described above (Garnier-Lhomme et al. 2009), although a fluorescent MARCKs peptide probe for $PtdInsP_2$ (Garnier-Lhomme et al. 2009) recognised similar structures in both live and detergent-treated cells.

The PtdIns species of isolated mammalian nuclei are not the same before and after detergent extraction and these species appear to change with the cell cycle or state of cell differentiation (Keune et al. 2011; Ogiso et al. 2010; Martelli et al. 2011). Differential pools of DAGs have also been reported for nucleus and cytoplasm (D'Santos et al. 1999). It is also important that the amounts of intranuclear lipids may be quite small compared to those in the membranes. A careful quantification has shown that the majority of the DAG, PtdIns and PtdCho in the NE are removed by low levels of non-ionic detergents, but under the same conditions 40 % of the $PtdInsP_2$ remains (Vann et al. 1997). If not due to rearrangements due to the detergent, this $PtdInsP_2$ would constitute a legitimate nuclear interior pool. Such membrane-depleted nuclei are capable of a range of metabolic conversions of the phosphoinositide pathways (Keune et al. 2011; Cocco et al. 2009).

Despite these difficulties, and cytological problems of phosphoinositide localisation discussed in the previous section, many studies point to roles of interior nuclear phosphoinositides in differentiation and cell division, regulation of gene expression and mRNA processing, and stress signalling responses, which we cannot address here (Keune et al. 2011; Mellman and Anderson 2009; Okada and Ye 2009; Ramazzotti et al. 2011; Barlow et al. 2010). The locations of and roles for PtdIns species and their modifying enzymes in intranuclear signalling are discussed in detail in many excellent recent reviews (Keune et al. 2011; Martelli et al. 2011; Cocco et al. 2009; Ramazzotti et al. 2011; Barlow et al. 2010; Cocco et al. 2010; Irvine 2006).

Putative Roles for Phosphoinositides in the NE. A possible role for phosphoinositides residing in the NE is in signalling mechanisms extensively discussed in reviews of interior nuclear PtdIns (Keune et al. 2011; Martelli et al. 2011; Barlow et al. 2010). Other potential roles for PtdIns derivatives in the NE are structural. For example, alterations of DAG levels can have profound effects on NE morphology. Lipins are lipid phosphatases that convert phosphatidic acid to DAG and function in lipid biosynthetic pathways that produce DAG to be subsequently utilised in triacylglycerol formation which can then be stored in lipid droplets as triglycerides or used in the Kennedy pathway for synthesis of PtdEth and PtdCho (Siniossoglou 2009). Deletion of the lipin gene PAH1 in yeast results in irregular nuclei with expanded NEs. Whether the altered morphology is due to increase lipid synthesis or physical effects on the membrane is not known. In *Caenorhabditis elegans*, down-regulation of lipin gene expression regulation by RNAi also results in NE disruption (Golden et al. 2009; Gorjanacz and Mattaj 2009).

As noted above, PtdIns species and DAG have very different physical consequences in membranes, affecting bilayer stability, curvature, and fluidity. The most investigated role of these molecules in NE dynamics is in membrane fusion of precursor vesicles during reconstruction of the sperm NE of the male pronucleus following fertilisation. Isolation of a membrane vesicle fraction (MV1) containing 50–70 % PtdIns species, well above the levels necessary as a signalling molecule, suggested structural roles for these lipids (Larijani et al. 2001; Larijani et al. 2000).

Demonstration of DAG function in the fusion of MV1 with ER membranes led to a model of MV1 as a potentially fusogenic fraction capable of rapid, local generation of DAG. As mentioned, generation of DAG could alter bilayer structure by favouring non-lamellar structures (Larijani and Dufourc 2006) and also leads to negative intrinsic membrane curvature, which favours fusion stalk formation (Chernomordik and Kozlov 2008).

Evidence for this model includes the in vitro requirement for MV1 in GTP-initiated fusion of NE precursor membrane vesicles (mostly derived from ER), the functional substitution of liposomes containing 75 % DAG–25 % PtdCho or 25 % PtdCho–75 % PtdIns prehydrolysed to DAG by bacterial PI-PLC which does not produce the Ca^{++} mediator inositol triphosphate (Barona et al. 2005), and direct FRET-FLIM measurements of fusion by mixing of lipid dyes incorporated into the precursor vesicles showing that fusion proceeds from the two regions of the nuclear surface to which MV1 binds (Dumas et al. 2010). Bacterial PI-PLC preferentially acted on highly unsaturated PtdIns, especially the 18:0/22:4 species (Barona et al. 2005) and NE formation by the endogenous enzyme resulted in depletion of 18:0/22:4 $PtdInsP_2$ (Byrne et al. 2005). Again the presence of polyunsaturated lipids, whether of PtdIns species or DAG disorders membranes and increases membrane fluidity (Larijani and Dufourc 2006).

In summary, many physical characteristics of the molecules involved in the PtdIns pathway leading to NE formation can contribute to alterations of membrane morphology that can facilitate membrane fusion. However, since these data are largely derived from a cell-free system, which mimics but may not totally replicate in vivo dynamics, it will be important to establish whether NE formation in vivo, during pronuclear or mitotic reconstitution, occurs similarly.

Nonetheless, the system points to a novel role, in addition to signalling, for $PtdInsP_2$ and DAG in controlling natural membrane structure and fusion. Molecules like DAG when asymmetrically distributed across the bilayer provide negative spontaneous curvature necessary for formation of fusion pores (Chernomordik and Kozlov 2005). $PtdInsP_2$ on the other hand leads to positive curvature and would inhibit fusion. However, $PtdInsP_2$ also favours membrane fluidity, which could facilitate initial stages of membrane apposition or lipid mixing (Larijani and Dufourc 2006; Zhendre et al. 2011).

In membranes with low levels of these phosphoinositides or DAG, such as the plasma membrane of mammalian cells, the physical effects would not be prominent unless the lipids were clustered. In membranes with high levels like MV1 the structural effects could be considerable. Clearly, in vitro studies of these natural membranes are consistent with theoretical studies and investigations of protein-free synthetic membrane fusion (Basanez et al. 1996). However, it will be of great importance in the future to link these approaches to assessments of natural membrane composition, structure, and function in vivo. Moreover, rapid depletion of individual lipids restricted to only the ER or NE compartments will be necessary to show structural effects of the lipids on membranes in living cells.

Acknowledgments BL was supported by core funding to London Research Institute from Cancer Research UK. DLP was supported by an Amherst College Faculty Research Award from the H. Axel Schupf'57 Fund for Intellectual Life.

References

Mima J, Hickey CM, Xu H, Jun Y, Wickner W (2008) Reconstituted membrane fusion requires regulatory lipids, SNAREs and synergistic SNARE chaperones. EMBO J 27:2031–2042

Byrne RD, Larijani B, Poccia DL (2009) Tyrosine kinase regulation of nuclear envelope assembly. Adv Enzyme Regul 49:148–156

Wang L, Seeley ES, Wickner W, Merz AJ (2002) Vacuole fusion at a ring of vertex docking sites leaves membrane fragments within the organelle. Cell 108:357–369

Chernomordik LV, Kozlov MM (2008) Mechanics of membrane fusion. Nat Struct Mol Biol 15:675–683

Dumas F, Byrne RD, Vincent B, Hobday TM, Poccia DL et al (2010) Spatial regulation of membrane fusion controlled by modification of phosphoinositides. PLoS ONE 5:e12208

Goni FM, Alonso A (1999) Structure and functional properties of diacylglycerols in membranes. Prog Lipid Res 38:1–48

James DJ, Khodthong C, Kowalchyk JA, Martin TF (2008) Phosphatidylinositol 4,5-bisphosphate regulates SNARE-dependent membrane fusion. J Cell Biol 182:355–366

Lynch KL, Gerona RR, Kielar DM, Martens S, McMahon HT et al (2008) Synaptotagmin-1 utilizes membrane bending and SNARE binding to drive fusion pore expansion. Mol Biol Cell 19:5093–5103

Mima J, Wickner W (2009a) Phosphoinositides and SNARE chaperones synergistically assemble and remodel SNARE complexes for membrane fusion. Proc Natl Acad Sci U S A 106:16191–16196

Mima J, Wickner W (2009b) Complex lipid requirements for SNARE- and SNARE chaperone-dependent membrane fusion. J Biol Chem 284:27114–27122

Larijani B, Poccia DL (2009) Nuclear envelope formation: mind the gaps. Annu Rev Biophys 38:107–124

Martin TF, Loyet KM, Barry VA, Kowalchyk JA (1997) The role of PtdIns(4,5)P2 in exocytotic membrane fusion. Biochem Soc Trans 25:1137–1141

Chernomordik LV, Kozlov MM (2005) Membrane hemifusion: crossing a chasm in two leaps. Cell 123:375–382

Das S, Rand RP (1984) Diacylglycerol causes major structural transitions in phospholipid bilayer membranes. Biochem Biophys Res Commun 124:491–496

Dufourc EJ, Mayer C, Stohrer J, Althoff G, Kothe G (1992) Dynamics of phosphate head groups in biomembranes-comprehensive analysis using P31 nuclear magnetic resonance lineshape and relaxation-time measurements. Biophys J 61:42–57

Kozlovsky Y, Chernomordik LV, Kozlov MM (2002) Lipid intermediates in membrane fusion: formation, structure, and decay of hemifusion diaphragm. Biophys J 83:2634–2651

Kozlovsky Y, Kozlov MM (2002) Stalk model of membrane fusion: solution of energy crisis. Biophys J 82:882–895

Byrne RD, Garnier-Lhomme M, Han K, Dowicki M, Michael N et al (2007) PLCgamma is enriched on poly-phosphoinositide-rich vesicles to control nuclear envelope assembly. Cell Signal 19:913–922

Larijani B, Dufourc EJ (2006) Polyunsaturated phosphatidylinositol and diacylglycerol substantially modify the fluidity and polymorphism of biomembranes: a solid-state deuterium NMR study. Lipids 41:925–932

Gambhir A, Hangyas-Mihalyne G, Zaitseva I, Cafiso DS, Wang J et al (2004) Electrostatic sequestration of PIP2 on phospholipid membranes by basic/aromatic regions of proteins. Biophys J 86:2188–2207

Garnier-Lhomme M, Byrne RD, Hobday TM, Gschmeissner S, Woscholski R et al (2009) Nuclear envelope remnants: fluid membranes enriched in sterols and polyphosphoinositides. PLoS ONE 4:e4255

Knight JD, Falke JJ (2009) Single-molecule fluorescence studies of a PH domain: new insights into the membrane docking reaction. Biophys J 96:566–582

Lucas N, Cho W (2011) Phosphatidylserine binding is essential for plasma membrane recruitment and signalling function of 3-phosphoinositide-dependent kinase-1. J Biol Chem 286:41265–41272

Larijani B, Barona TM, Poccia DL (2001) Role for phosphatidylinositol in nuclear envelope formation. Biochem J 356:495–501

Zhendre V, Grelard A, Garnier-Lhomme M, Buchoux S, Larijani B et al (2011) Key role of polyphosphoinositides in dynamics of fusogenic nuclear membrane vesicles. PLoS ONE 6:e23859

Mattaj IW (2004) Sorting out the nuclear envelope from the endoplasmic reticulum. Nat Rev Mol Cell Biol 5:65–69

McMahon HT, Gallop JL (2005) Membrane curvature and mechanisms of dynamic cell membrane remodelling. Nature 438:590–596

Shibata Y, Shemesh T, Prinz WA, Palazzo AF, Kozlov MM et al (2010) Mechanisms determining the morphology of the peripheral ER. Cell 143:774–788

Zimmerberg J, Kozlov MM (2006) How proteins produce cellular membrane curvature. Nat Rev Mol Cell Biol 7:9–19

Fili N, Calleja V, Woscholski R, Parker PJ, Larijani B (2006) Compartmental signal modulation: endosomal phosphatidylinositol 3-phosphate controls endosome morphology and selective cargo sorting. Proc Natl Acad Sci U S A 103:15473–15478

Ruiz-Arguello MB, Goni FM, Alonso A (1998) Vesicle membrane fusion induced by the concerted activities of sphingomyelinase and phospholipase C. J Biol Chem 273:22977–22982

Szentpetery Z, Varnai P, Balla T (2010) Acute manipulation of Golgi phosphoinositides to assess their importance in cellular trafficking and signaling. Proc Natl Acad Sci U S A 107:8225–8230

Szabadkai G, Bianchi K, Varnai P, De Stefani D, Wieckowski MR et al (2006) Chaperone-mediated coupling of endoplasmic reticulum and mitochondrial Ca2 + channels. J Cell Biol 175:901–911

Balla T, Bondeva T, Varnai P (2000) How accurately can we image inositol lipids in living cells? Trends Pharmacol Sci 21:238–241

Hammond GR, Schiavo G, Irvine RF (2009) Immunocytochemical techniques reveal multiple, distinct cellular pools of PtdIns4P and PtdIns(4,5)P(2). Biochem J 422:23–35

Lindsay Y, McCoull D, Davidson L, Leslie NR, Fairservice A et al (2006) Localization of agonist-sensitive PtdIns(3,4,5)P3 reveals a nuclear pool that is insensitive to PTEN expression. J Cell Sci 119:5160–5168

Watt SA, Kular G, Fleming IN, Downes CP, Lucocq JM (2002) Subcellular localization of phosphatidylinositol 4,5-bisphosphate using the pleckstrin homology domain of phospholipase C delta1. Biochem J 363:657–666

van Meer G (1998) Lipids of the Golgi membrane. Trends Cell Biol 8:29–33

Akhtar A, Gasser SM (2007) The nuclear envelope and transcriptional control. Nat Rev Genet 8:507–517

Webster M, Witkin KL, Cohen-Fix O (2009) Sizing up the nucleus: nuclear shape, size and nuclear-envelope assembly. J Cell Sci 122:1477–1486

Prunuske AJ, Ullman KS (2006) The nuclear envelope: form and reformation. Curr Opin Cell Biol 18:108–116

Ellenberg J, Siggia ED, Moreira JE, Smith CL, Presley JF et al (1997) Nuclear membrane dynamics and reassembly in living cells: targeting of an inner nuclear membrane protein in interphase and mitosis. J Cell Biol 138:1193–1206

Austin CM, Bellini M (2010) The dynamic landscape of the cell nucleus. Mol Reprod Dev 77:19–28
Mao YS, Zhang B, Spector DL (2011) Biogenesis and function of nuclear bodies. Trends Genet 27:295–306
Larijani B, Poccia DL, Dickinson LC (2000) Phospholipid identification and quantification of membrane vesicle subfractions by 31P–1H two-dimensional nuclear magnetic resonance. Lipids 35:1289–1297
Malhas A, Goulbourne C, Vaux DJ (2011) The nucleoplasmic reticulum: form and function. Trends Cell Biol 21:362–373
Keune W, Bultsma Y, Sommer L, Jones D, Divecha N (2011) Phosphoinositide signalling in the nucleus. Adv Enzyme Regul 51:91–99
Chamberlain LH (2004) Detergents as tools for the purification and classification of lipid rafts. FEBS Lett 559:1–5
Shaw AS (2006) Lipid rafts: now you see them, now you don't. Nat Immunol 7:1139–1142
Pederson T (2000) Half a century of "the nuclear matrix". Mol Biol Cell 11:799–805
Ogiso H, Nakamura K, Yatomi Y, Shimizu T, Taguchi R (2010) Liquid chromatography/mass spectrometry analysis revealing preferential occurrence of non-arachidonate-containing phosphatidylinositol bisphosphate species in nuclei and changes in their levels during cell cycle. Rapid Commun Mass Spectrom 24:436–442
Martelli AM, Ognibene A, Buontempo F, Fini M, Bressanin D et al (2011) Nuclear phosphoinositides and their roles in cell biology and disease. Crit Rev Biochem Mol Biol 46:436–457
D'Santos CS, Clarke JH, Irvine RF, Divecha N (1999) Nuclei contain two differentially regulated pools of diacylglycerol. Curr Biol 9:437–440
Vann LR, Wooding FB, Irvine RF, Divecha N (1997) Metabolism and possible compartmentalization of inositol lipids in isolated rat-liver nuclei. Biochem J 327(Pt 2):569–576
Cocco L, Faenza I, Follo MY, Billi AM, Ramazzotti G et al (2009) Nuclear inositides: PI-PLC signalling in cell growth, differentiation and pathology. Adv Enzyme Regul 49:2–10
Mellman DL, Anderson RA (2009) A novel gene expression pathway regulated by nuclear phosphoinositides. Adv Enzyme Regul 49:11–28
Okada M, Ye K (2009) Nuclear phosphoinositide signalling regulates messenger RNA export. RNA Biol 6:12–16
Ramazzotti G, Faenza I, Fiume R, Matteucci A, Piazzi M et al (2011) The physiology and pathology of inositide signalling in the nucleus. J Cell Physiol 226:14–20
Barlow CA, Laishram RS, Anderson RA (2010) Nuclear phosphoinositides: a signalling enigma wrapped in a compartmental conundrum. Trends Cell Biol 20:25–35
Cocco L, Follo MY, Faenza I, Billi AM, Ramazzotti G et al (2010) Inositide signaling in the nucleus: from physiology to pathology. Adv Enzyme Regul 50:2–11
Irvine RF (2006) Nuclear inositide signalling—expansion, structures and clarification. Biochim Biophys Acta 1761:505–508
Siniossoglou S (2009) Lipins, lipids and nuclear envelope structure. Traffic 10:1181–1187
Golden A, Liu J, Cohen-Fix O (2009) Inactivation of the C. elegans lipin homolog leads to ER disorganization and to defects in the breakdown and reassembly of the nuclear envelope. J Cell Sci 122:1970–1978
Gorjanacz M, Mattaj IW (2009) Lipin is required for efficient breakdown of the nuclear envelope in *C. elegans*. J Cell Sci 122:1963–1969
Barona T, Byrne RD, Pettitt TR, Wakelam MJ, Larijani B et al (2005) Diacylglycerol induces fusion of nuclear envelope membrane precursor vesicles. J Biol Chem 280:41171–41177
Byrne RD, Barona TM, Garnier M, Koster G, Katan M et al (2005) Nuclear envelope assembly is promoted by phosphoinositide-specific phospholipase C with selective recruitment of phosphatidylinositol-enriched membranes. Biochem J 387:393–400
Basanez G, Nieva JL, Rivas E, Alonso A, Goni FM (1996) Diacylglycerol and the promotion of lamellar-hexagonal and lamellar-isotropic phase transitions in lipids: implications for membrane fusion. Biophys J 70:2299–2306

Chapter 6
Molecular Analysis of Protein–Phosphoinositide Interactions

Tatiana G. Kutateladze

Abstract Diverse biological processes including cell growth and survival require transient association of proteins with cellular membranes. A large number of these proteins are drawn to a bilayer through binding of their modular domains to phosphoinositide (PI) lipids. Seven PI isoforms are found to concentrate in distinct pools of intracellular membranes, and this lipid compartmentalization provides an efficient way for recruiting PI-binding proteins to specific cellular organelles. The atomic-resolution structures and membrane docking mechanisms of a dozen PI effectors have been elucidated in the last decade, offering insight into the molecular basis for regulation of the PI-dependent signaling pathways. In this chapter, I summarize the mechanistic aspects of deciphering the 'PI code' by the most common PI-recognizing domains and discuss similarities and differences in the membrane anchoring mechanisms.

Introduction

Phosphoinositides (PIs), phosphorylated derivatives of phosphatidylinositol (PtdIns), are essential components of eukaryotic cell membranes.[1] They are involved in regulation of various fundamental biological processes, including cell growth and survival, membrane trafficking and cytoskeletal dynamics (reviewed in Di Paolo and De Camilli 2006; Hurley 2006; Lemmon 2008; Roth 2004). Although PIs comprise

[1] This chapter is adapted from Kutateladze (2010).

T. G. Kutateladze (✉)
Department of Pharmacology, University of Colorado Denver School of Medicine, 12801 East 17th Avenue, Aurora, CO 80045, USA
e-mail: Tatiana.Kutateladze@ucdenver.edu

M. Falasca (ed.), *Phosphoinositides and Disease*, Current Topics in Microbiology and Immunology 362, DOI: 10.1007/978-94-007-5025-8_6,

approximately 1 % of cellular lipids, they play pivotal roles in major signaling pathways, serving as docking sites for protein effectors and as precursors of secondary messengers. The inositol headgroup of PIs can be reversibly phosphorylated at three positions, D3, D4, and D5, and all seven PI isoforms, including three mono-phosphorylated [PtdIns(3)P, PtdIns(4)P, and PtdIns(5)P], three bis-phosphorylated [PtdIns(3,4)P_2, PtdIns(3,5)P_2, and PtdIns(4,5)P_2], and one tris-phosphorylated PtdIns(3,4,5)P_3 species have been identified in eukaryotic cells.

The amount and spatial and temporal distribution of PIs in the cell vary substantially (Vanhaesebroeck et al. 2001). The largest pool of these lipids comprises PtdIns(4)P and PtdIns(4,5)P_2, whereas PtdIns(3,4,5)P_3 is undetectable in unstimulated cells. Some PIs, such as PtdIns(4)P and PtdIns(4,5)P_2, are constitutively present in membranes, others are transiently produced in response to the activation of cell surface receptors and other stimuli. The level and turnover of PIs are tightly controlled by a set of PI-specific enzymes. Found exclusively in the cytosolic leaflet of membrane bilayers, PIs are readily accessible to PI kinases and phosphatases capable of attaching and removing phosphate groups, respectively, and to phospholipases that cleave the lipids. Because PI-modifying enzymes are heterogeneously localized in the cell, PIs are clustered in distinct intracellular membranes and thus each PI essentially serves as a marker of an organelle. For example, the plasma membrane is enriched in PtdIns(4,5)P_2, whereas PtdIns(4)P and PtdIns(3)P are detected primarily in the Golgi and early endosomes, respectively. The unique distribution of PIs may provide the mechanism for fine-tuning the membrane trafficking flow and controlling the proper sequence of signaling events.

A major breakthrough in understanding the significance of PI signaling was the identification of protein effectors that recognize individual PIs (Fig. 6.1). The pleckstrin homology (PH) domain was the first effector found to associate with PIs (Harlan et al. 1994). The list of PI-binding domains has since grown rapidly and at present contains 15 modules that display a wide range of affinities and selectivities toward lipid membranes. It includes the AP180 *N*-terminal homology (ANTH), Barkor/Atg14(L) autophagosome targeting sequence (BATS), conserved region-2 of protein kinase C (C2), Dock homology region-1 (DHR-1), epsin *N*-terminal homology (ENTH), 4.1, ezrin, radixin, moiesin (FERM), Fab1, YOTB, Vac1 and EEA1 (FYVE), Golgi phosphoprotein 3 (GOLPH3), PtdIns(4)P binding of SidM/DrrA (P4M), postsynaptic density 95, disk large, zonula occludens (PDZ), β-propellers that bind PIs (PROPPINs), phosphotyrosine binding (PTB), Phox homology (PX), SH3YL1, Ysc84p/Lsb4p, Lsb3p and plant FYVE proteins (SYLF), and Tubby modules (reviewed in Kutateladze 2010; Moravcevic et al. 2012). The recognition of a unique arrangement of phosphate groups around the inositol ring, which can be referred to as a 'PI code', by these domains results in the recruitment of the host proteins to specific intracellular compartments (Kutateladze 2010). Many of these proteins are modular in architecture and contain other lipid- and protein-binding domains or possess catalytic activities. These in turn trigger phosphorylation/dephosphorylation of other membrane-associated complexes and adaptors and promote the interconversion of PIs, subsequently

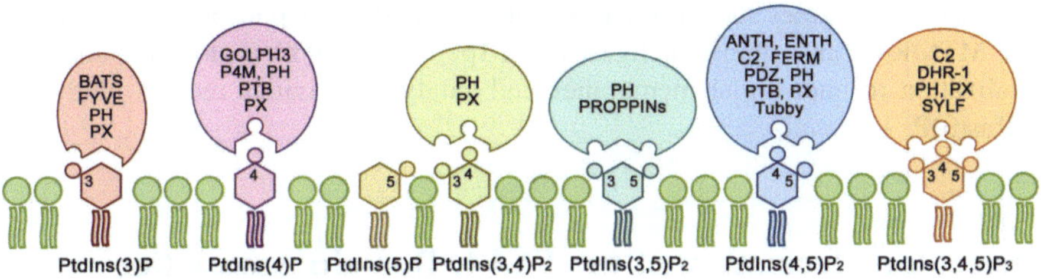

Fig. 6.1 PI-recognizing effectors. Signaling domains and their target PIs are shown

leading to the activation or termination of signaling cascades. A crosstalk between PIs, PI-modifying enzymes, and effectors capable of 'reading' the PI code constitutes one of the most intriguing and complex signaling networks in the cell, which we have only begun to understand. This chapter focuses on the structural and mechanistic aspects of deciphering the PI code by the most common PI-binding protein effectors.

The FYVE Domain Targets PtdIns(3)P

The FYVE domain is a ~70-residue zinc-binding finger which is found in 53 human proteins (SMART database) (reviewed in Kutateladze 2006). It binds PtdIns(3)P with high specificity and affinity and bridges a number of cytosolic proteins with PtdIns(3)P-enriched early endosomes, multivesicular bodies (MVB), and phagosomes (Burd and Emr 1998; Gaullier et al. 1998; Patki et al. 1998). A small fraction of PtdIns(3)P has been identified in the nucleus and the Golgi apparatus, and FYVE domain-containing proteins DFCP1 and Alfy preferentially localize to these sites (Ridley et al. 2001; Simonsen et al. 2004). The FYVE domain is defined by the three conserved sequences: the WxxD, RR/KHHCR and RVC motifs that form a highly positively charged binding site for PtdIns(3)P. Whereas topologically the FYVE domain belongs to a larger family of zinc-coordinating RING fingers, it can be distinguished from other DNA- and protein-binding members of the RING superfamily by the presence of these three signature motifs.

FYVE domain-containing proteins have diverse biological functions. One of the largest subsets of FYVE proteins is involved in the regulation of endocytic trafficking and fusion of endosomal membranes with transport vesicles and other organelles (Gillooly et al. 2001). This subset includes well-characterized mammalian proteins EEA1, Endofin, FENS-1, FYCO1, Hrs, Rabenosin-5, Rabip4 and WDFY2, and yeast proteins Vac1p and Vps27p. Another fast growing subset plays a critical role in signal transduction and TGFβ/Smad activation (Hrs and SARA), adipocyte differentiation (ProF), leukocyte signaling (FGD2), cytoskeletal reorganization (EhFP), autophagosome formation (DFCP1), and apoptosis (Phafin1/2).

A number of enzymes such as kinases (Fab1 and PIKfyve), phosphatases (MTMR3 and MTMR4), and ubiquitin ligases (Pib1p) contain FYVE domain, and their localization to endosomal membranes and catalytic activities require binding to PtdIns(3)P.

Molecular Mechanism of the FYVE Domain Association with Membranes

While specific recognition of PtdIns(3)P is a major characteristic of the FYVE finger, it localizes to membranes through a multivalent mechanism that also involves nonspecific electrostatic contacts with acidic lipids other than PtdIns(3)P (Diraviyam et al. 2003; Kutateladze et al. 2004; Stahelin et al. 2002), activation of a histidine switch (He et al. 2009; Lee et al. 2005; Mertens et al. 2007), hydrophobic insertion into the bilayer (Blatner et al. 2004; Brunecky et al. 2005; Diraviyam et al. 2003; Kutateladze et al. 2004; Kutateladze and Overduin 2001; Misra and Hurley 1999; Sankaran et al. 2001; Stahelin et al. 2002), and in some cases dimerization of the host protein (Callaghan et al. 1999; Dumas et al. 2001; Hayakawa et al. 2004; Lawe et al. 2000). Each of these components uniquely contributes to the FYVE domain specificity and increases the binding affinity for PtdIns(3)P embedded in membranes to the low nM level (Blatner et al. 2004; Ridley et al. 2001; Stahelin et al. 2002).

The FYVE domain contains a variable-length loop next to the PtdIns(3)P binding pocket. Upon binding to PtdIns(3)P, the hydrophobic residues at the tip of this loop [termed the turret loop or membrane insertion/interaction loop (MIL)] insert into the bilayer. The MIL is flanked by a set of basic and polar residues that are positioned at the level of the lipid headgroups when the protein penetrates the membrane. These residues make nonspecific electrostatic contacts with acidic phospholipids, such as phosphatidylserine (PS) and phosphatidic acid (PA). The strong positive potential around the MIL can also drive the initial membrane docking and facilitate association with PtdIns(3)P. It has recently been shown that interaction of the FYVE domain with PtdIns(3)P is pH-dependent and can be regulated by a histidine switch comprising a pair of adjacent His residues in the RR/KHHCR motif. The FYVE domain binds PtdIns(3)P when both histidine residues are positively charged and releases the lipid upon their deprotonation. Membrane association of FYVE domain-containing proteins can be further enhanced by bivalent or multivalent interactions with PtdIns(3)P. For example, a central region of EEA1 forms a parallel coiled coil homodimer that juxtaposes two *C*-terminal FYVE domains, allowing for the simultaneous interaction with two PtdIns(3)P headgroups (Dumas et al. 2001).

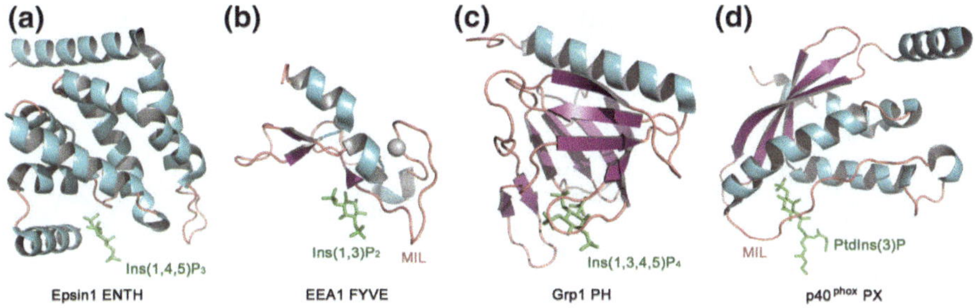

Fig. 6.2 The crystal structures of **a** the ENTH domain of Epsin1 in complex with inositol 1,4,5-trisphosphate, a headgroup of PtdIns(4,5)P_2 (1H0A) (Ford et al. 2002), **b** the EEA1 FYVE domain in complex with inositol 1,3-bisphosphate, a headgroup of PtdIns(3)P (1JOC) (Dumas et al. 2001), **c** the Grp1 PH domain in complex with inositol 1,3,4,5-tetrakisphosphate, a headgroup of PtdIns(3,4,5)P_3 (1FGY) (Lietzke et al. 2000) and **d** the p40^{phox} PX domain in complex with dibutanoyl PtdIns(3)P (1H6H) (Bravo et al. 2001)

Structural Basis of PtdIns(3)P Recognition by the FYVE Domain

The three-dimensional structures of the FYVE domain of human EEA1 bound to inositol 1,3-bisphosphate (Dumas et al. 2001) and dibutanoyl PtdIns(3)P (Kutateladze and Overduin 2001) and the ligand-free FYVE domains of EEA1 (Kutateladze and Overduin 2001), human Endofin, fly Hrs (Mao et al. 2000), Leishmania Major Lm5-1 (Mertens et al. 2007), human RUFY and yeast Vps27p (Misra and Hurley 1999) have been determined by X-ray crystallography and NMR spectroscopy. The structures reveal a similar overall fold that consists of two double-stranded antiparallel β sheets and a *C*-terminal α helix (Fig. 6.2). An additional *N*-terminal α-helical turn is seen in the structures of the FYVE domain of EEA1 and Endofin, and a short α helix connecting $\beta2$ and $\beta3$ is present in EEA1, Endofin, Lm5-1, and RUFY. The functionally important $\beta1$ strand spans three residues of the RR/KHHCR motif and pairs with the $\beta2$ strand, which links two zinc-binding clusters. The zinc ions are bound by four CxxC motifs in a cross-braced topology. One zinc ion is coordinated by the first and third cysteine motifs, whereas another zinc ion is bound by the second and fourth motifs in all human proteins. In yeast Vps27p, the fourth Cys residue is replaced by a His.

Structural insight into PtdIns(3)P recognition by the FYVE domain is provided by the crystal structure of the EEA1 FYVE domain in complex with inositol 1,3-bisphosphate (Dumas et al. 2001) (Fig. 6.3a). The structure reveals that the WxxD, RR/KHHCR, and RVC motifs are centrally involved in coordination of the inositol headgroup. Critical hydrogen bonds are formed between the 3-phospate group of PtdIns(3)P and the RR/KHHCR motif, particularly the guanidino moiety of the last arginine of this motif (R1375), the imidazole ring of the first histidine (H1372), and the backbone amide of the second histidine (H1373). The 1-phosphate group of PtdIns(3)P is bound by the first arginine of the motif (R1370) and

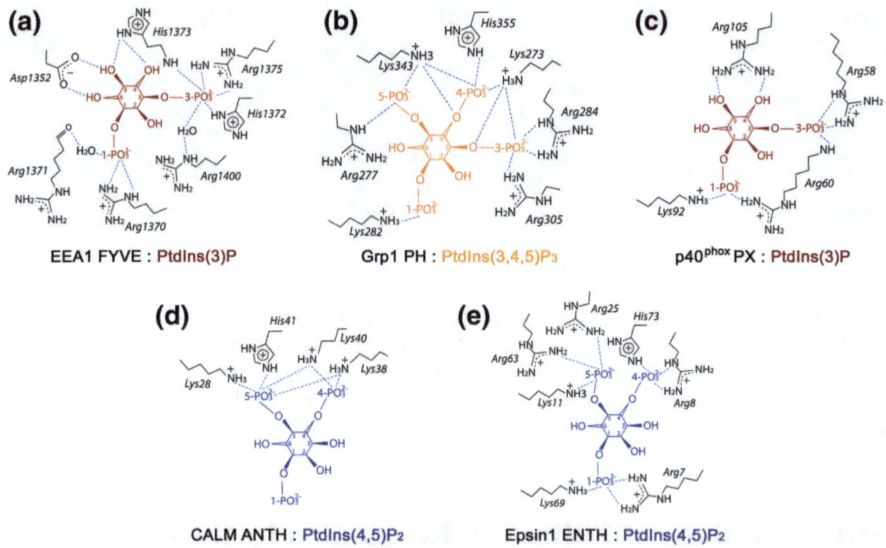

Fig. 6.3 Schematic diagrams showing PtdIns(3)P headgroup coordination by **a** EEA1 FYVE (1JOC) and **c** p40^{phox} PX (1H6H) domains; PtdIns(3,4,5)P_3 headgroup coordination by **b** Grp1 PH domain (1FGY); and PtdIns(4,5)P_2 headgroup coordination by **d** CALM ANTH (1HFA) and **e** Epsin1 ENTH (1H0A) domains. Only charged residues of the proteins are depicted for clarity

through a water-mediated contact with the backbone carboxyl group of the second arginine (R1371). R1400 of the RVC motif, positioned between H1372 and R1375, forms another water-mediated hydrogen bond to the 3-phosphate group. Coordination of the 4-, 5-, and 6-hydroxyl groups of the inositol ring is crucial for stereospecificity and the exclusion of alternatively phosphorylated PIs. The 4- and 5-hydroxyl groups are hydrogen bonded to the imidazole ring of H1373, whereas the carboxylate of D1352 in the *N*-terminal WxxD motif makes contacts with the hydroxyl groups at the 5 and 6 positions.

The PH Domain Binds Various PIs

The PH domains comprise one of the largest families of signaling modules and are the most thoroughly characterized among the PI binding domains (reviewed in DiNitto and Lambright 2006; Lemmon and Ferguson 2000). The PH domain was identified within a set of human proteins in 1993 and derives its name from the two homologous regions of pleckstrin, the major protein kinase C substrate in platelets (Haslam et al. 1993; Mayer et al. 1993). Since then, it has been found in 561 human proteins involved in intracellular signaling, membrane trafficking, cytoskeletal structure, and lipid modifications (SMART). The PH domain contains ~120 residues that are folded in a highly conserved three-dimensional structure despite little sequence similarity between the family members. As a result of high sequence variability, the PH domains have diverse functions and interact with numerous ligands, including proteins, acidic phospholipids, inositol polyphosphates, and PIs. Many of those able to recognize PIs do so weakly and promiscuously, however, a subset of PH domains (about 10–20 % of all PH modules) binds individual PIs specifically and strongly, most commonly PtdIns(3,4,5)P_3, and PtdIns(4,5)P_2, as well as PtdIns(3,4)P_2. The binding affinity of the PH domains for PIs varies significantly, ranging from low nM to low μM (DiNitto and Lambright 2006). Among the best characterized PtdIns(3,4,5)P_3 effectors are the PH domains of ARNO, Btk, Gap1, Grp1, and cytohesin-1 (Cronin et al. 2004; Fukuda et al. 1996; Klarlund et al. 1997). The PLCδ1 PH domain is specific for PtdIns(4,5)P_2, whereas PH domains of TAPP1, Centaurin β2, PEPP1, and FAPP1 prefer PtdIns(3,4)P_2, PtdIns(3,5)P_2, PtdIns(3)P, and PtdIns(4)P, respectively, and Akt/PKB, PDK1, and DAPP1 bind both PtdIns(3,4)P_2 and PtdIns(3,4,5)P_3 (DiNitto and Lambright 2006; Lemmon 2004; Lemmon and Ferguson 2000).

PH domains are present in GTP/GDP exchange factors (ARNO, cytohesin, FGD2, Grp1), GTPase activating proteins (Centaurin β2, Gap1), lipid-metabolizing (PLCδ1, PLD1/2), lipid-transport (FAPP2) and cytoskeletal (dynamin, β-spectrin) proteins, kinases (Akt/PKB, Btk/Itk, CERK), phosphatases (PHLPP1), and other macromolecules that are implicated in vital biological processes including growth, proliferation, metabolism, cell polarization and migration, receptor endo- and exocytosis, membrane budding and trafficking, actin rearrangement, immune responses, and apoptosis.

Molecular Mechanism of Membrane Docking by the PH Domain

The PI-binding site of the PH domain is formed by three variable loops flanking the open end of a β barrel (described below and shown in Fig. 6.2c). Like the binding sites of all other PI-binding modules, it contains a cluster of basic lysine and arginine residues that make direct contacts with the phosphate groups of the lipid. The PH domain is electrostatically polarized and displays a strong positive electrostatic potential around the binding site that contributes to both specific PI binding and nonspecific electrostatic interactions with other anionic lipids in membranes (DiNitto and Lambright 2006; He et al. 2008; Lemmon and Ferguson 2000; Manna et al. 2007; Singh and Murray 2003). Single-molecule fluorescent studies reveal that the Grp1 PH domain can interact with PS via one or more secondary binding sites and that the electrostatic search mechanism speeds its association with PtdIns(3,4,5)P_3 (Knight and Falke 2009).

PtdIns(3,4,5)P_3 binding of the Grp1 PH domain can be further enhanced by acidification of the media and subsequent protonation of the histidine residue (He et al. 2008) that forms a critical hydrogen bond to the 4-phosphate group of the

PI (Ferguson et al. 2000; Lietzke et al. 2000). The pH-dependence is less pronounced for the PH domain than for the FYVE domain—unlike FYVE domains which all contain two invariable histidine residues in the binding pocket, a single histidine is present in only a small set of PH domains (He et al. 2008). Additionally, PH domains of ARNO, DAPP1, Fapp1, Grp1, PLCδ1, and TAPP1 have been demonstrated to penetrate PI-containing monolayers to various degrees (Flesch et al. 2005; He et al. 2008; Lumb et al. 2011; Manna et al. 2007; Stahelin et al. 2007), and, at least in the case of Grp1, the membrane insertion is triggered by specific recognition of the inositol headgroup and is increased in an acidic environment (He et al. 2008).

Several PH domains have been found to interact with two ligands. The PH domains of Grp1, Fapp1, and oxysterol binding protein recognize not only PIs but also membrane-attached Arf GTPases (Balla et al. 2005; Cohen et al. 2007; Godi et al. 2004; Levine and Munro 2002). Recent biochemical and structural analysis of the Fapp1 PH domain reveals that the PtdIns(4)P and Arf1 binding sites are separate, which allow for the association with both ligands simultaneously and independently (He et al. 2011a, b). This mode of 'coincidence detection' increases affinities and specificities of the PH domains toward membranes enriched in unique PIs and Arfs. The membrane recruitment is further augmented at the sites where Arfs are active. A phospholipid or sphingolipid molecule can also serve as a second ligand in coincidence detection as reported for the PH domains of Akt1 and Slm1 (Gallego et al. 2010; Huang et al. 2011).

Structural Basis of the PH Domain-PtdIns(3,4,5)P_3 Interaction

Over 100 three-dimensional crystal and NMR structures of various canonical PH domains have been deposited in the PDB. Of them, two are complexes with PtdIns(3,4,5)P_3 [Btk and PDK1 (Komander et al. 2004)], 13 are complexes of eight PH domains with inositol 1,3,4,5-tetrakisphosphate (IP4) [ARNO (Cronin et al. 2004), Btk (Baraldi et al. 1999), DAPP1/PHISH (Ferguson et al. 2000), Grp1 (DiNitto et al. 2007; Ferguson et al. 2000; Lietzke et al. 2000), Kindlin (Liu et al. 2011), PDK1 (Komander et al. 2004), PEPP1, and PKB/Akt (Carpten et al. 2007; Milburn et al. 2003; Thomas et al. 2002)], two are complexes with inositol 1,4,5-trisphosphate (IP3) [ARNO (Cronin et al. 2004) and PLCδ1 (Ferguson et al. 1995)], two are complexes with inositol 1,2,3,5,6-pentakisphosphate [pleckstrin (Jackson et al. 2007)], and one is a complex with inositol 1,3,4,5,6-pentakisphosphate [Grp1 (Ferguson et al. 2000)]. Additionally, PH domains of ArhGAP9 (Ceccarelli et al. 2007) and β-spectrin (Hyvonen et al. 1995) have been shown to non-canonically interact with IPs. The canonical PH domain folds into a seven-stranded β-barrel, capped by an amphipathic α helix at one open end, while the opposite end is framed by three variable loops (Fig. 6.2c). The variable loops form a large PI-binding pocket, and their length and primary sequence define the specificity of the PH domain.

Details of how the PH domain recognizes the pattern of sequential 3-, 4-, and 5-phosphate groups in PtdIns(3,4,5)P_3 are revealed by the crystal structure of the Grp1 PH domain in complex with IP4, an isolated headgroup of PtdIns(3,4,5)P_3 (Ferguson et al. 2000; Lietzke et al. 2000) (Fig. 6.3b). The inositol ring lies in the center of a deep positively charged pocket formed by the $\beta 1$–$\beta 2$, $\beta 3$–$\beta 4$, and $\beta 6$–$\beta 7$ loops and the strands they connect. The distal phosphates are buried the furthest in the pocket, whereas the 1-phosphate group is positioned near the tips of the loops. A network of hydrogen bonds, formed between conserved lysine and arginine residues in $\beta 2$, $\beta 3$, $\beta 4$, and $\beta 7$ and all four phosphate groups of IP4, fully restrains the inositol molecule. A unique β-hairpin in the long $\beta 6$–$\beta 7$ loop of Grp1 is involved in additional hydrogen bonding contacts with the 5-phosphate group of IP4. These additional hydrogen bonds with the 5-phosphate account for the high specificity of the Grp1 PH domain toward PtdIns(3,4,5)P_3. In general, the $\beta 1$–$\beta 2$ loop of PH domains functions as a platform for the interaction with PIs (Lemmon 2008). It contains the sequence motif Kx_n(K/R)xR, which makes the most critical contacts with the phosphate groups of the lipid (Lemmon 2008).

The PX Domain Prefers PtdIns(3)P

The PX domain consists of ~130 residues and is found in 83 human signaling and regulatory proteins (SMART) (reviewed in Seet and Hong 2006). It is named after the two phagocyte NADPH oxidase (phox) subunits $p40^{phox}$ and $p47^{phox}$, in which it was first identified in 1996 (Ponting et al. 1996). Of all PIs, PtdIns(3)P appears to be the primary target of PX domain-containing proteins, as the majority of them associate with PtdIns(3)P-enriched endosomes and vacuoles [KIF16B, $p40^{phox}$, PXK, sorting nexins (SNXs), Vam7p], although binding to other PIs has also been reported for Bem1, CISK, CPK, FISH, NOXO1, $p47^{phox}$, PI3K-C2α, PLD1, and SNXs. All yeast PX domains bind PtdIns(3)P, however, only four with relatively high affinity (K_d ~ 2–3 μM) (Yu and Lemmon 2001).

PX domain-containing proteins play essential roles in endocytosis, protein sorting, membrane trafficking, transcription, cell polarity, and signaling (Seet and Hong 2006). SNXs, found in both yeast and mammalian cells, comprise the largest subset of proteins harboring this domain. Human SNXs are involved in endosomal sorting and recycling, in internalization, transport and lysosomal degradation of epidermal growth factor receptor, and in membrane tubulation. Yeast SNXs are required for regulation of protein retrieval and recycling traffic from prevacuolar/late endosomes to late Golgi. The t-SNARE Vam7p mediates fusion of multiple transport intermediates with the vacuole. The subunits of neutrophilic NADPH oxidase complex, $p40^{phox}$ and $p47^{phox}$ are implicated in phagocyte-mediated destruction of ingested microbes. The cytokine-independent survival kinase (CISK), PI 3-kinases, and the adaptor protein FISH play roles in cell signaling.

Molecular Mechanism of the PX Domain Association with Membranes

Similar to the FYVE and PH modules, the PX domain associates with PtdIns(3)P- or other PI-containing membranes via multiple interactions. Specific recognition of the inositol headgroup is often facilitated by nonspecific electrostatic contacts with acidic membrane surfaces and is accompanied by a hydrophobic insertion into the bilayer. The PX domain was identified as a PI binding module independently by several groups in 2001 (Cheever et al. 2001; Ellson et al. 2001; Kanai et al. 2001; Song et al. 2001; Xu et al. 2001; Yu and Lemmon 2001). The $p40^{phox}$, SNX3, and Vam7p PX domains were found to recognize PtdIns(3)P, whereas the $p47^{phox}$ PX domain was shown to prefer PtdIns(3,4)P_2.

Several mechanistic studies have demonstrated that the PI binding induces membrane penetration of the hydrophobic residues in the variable loop $\alpha1$–$\alpha2$ (MIL). X-ray reflectivity experiments show that the $p40^{phox}$ PX domain penetrates 9Å into the lipid layer, with the side chains of a tyrosine and a valine inserted most deeply (Malkova et al. 2006). The corresponding hydrophobic residues in the $p47^{phox}$ and Vam7p PX domains also penetrate membrane-mimetics (Cheever et al. 2001; Lee et al. 2006; Stahelin et al. 2003). Alignment of the PX domain sequences reveals some conservation of the hydrophobic residues despite the fact that overall the $\alpha1$–$\alpha2$ loop is highly variable. Bem1, CISK, CPK, FISH, Grd19p, $p40^{phox}$, $p47^{phox}$ and SNX3 contain VPYV, IFG, MVLG, VYVGV, ILF, ILL, WFDG, and LPF sequences, respectively, in place of the hydrophobic residues in $p40^{phox}$, $p47^{phox}$ and Vam7p, and the MIL occupies analogous conformations in the PX domain structures (Bravo et al. 2001; Hiroaki et al. 2001; Karathanassis et al. 2002; Kutateladze 2007; Lu et al. 2002; Xing et al. 2004; Zhou et al. 2003).

Basic residues located in and around the PI binding pocket and the MIL of the $p40^{phox}$, $p47^{phox}$ and Vam7p PX domains are involved in nonspecific electrostatic contacts with the negatively charged lipids (Karathanassis et al. 2002; Lee et al. 2006; Stahelin et al. 2003). Electrostatic interactions have been shown to alleviate recognition of the PI lipid, enhance affinity, and induce hydrophobic insertion of the PX domains (Malkova et al. 2006; Stahelin et al. 2004, 2003, 2006). Coincidence detection of PtdIns(3,4)P_2, and another lipid (PS or PA) in a separate well-defined binding site is essential for membrane targeting of the $p47^{phox}$ PX domain (Karathanassis et al. 2002; Stahelin et al. 2003).

Structural Basis of PtdIns(3)P Recognition by the PX Domain

The atomic-resolution crystal and solution structures of three PX domains bound to PtdIns(3)P [Grd19p (Zhou et al. 2003), $p40^{phox}$ (Bravo et al. 2001), and SNX9 (Pylypenko et al. 2007)] and 17 PX domains in the ligand-free form [Bem1p

(Stahelin et al. 2007), CISK (Xing et al. 2004), Grd19p (Zhou et al. 2003), KIF16B (Blatner et al. 2007), Nischarin, NOXO1b, p40^{phox} (Honbou et al. 2007), p47^{phox} (Hiroaki et al. 2001; Karathanassis et al. 2002), PI3K-C2α (Parkinson et al. 2008; Stahelin et al. 2006), PI3K-C2γ, SNX1 (Zhong et al. 2005), SNX7, SNX9 (Pylypenko et al. 2007), SNX12, SNX17, SNX22 (Song et al. 2007) and Vam7p (Lu et al. 2002)] have been determined. The structures show a similar fold that consists of a three-stranded β sheet, packed against a helical subdomain composed of three to four α-helices (Fig. 6.2d). An additional 3_{10} helix is present in the structures of the Bem1p, CISK, p40^{phox} and PI3K-C2α PX domains and another helix $\alpha0$ is formed by the residues *N*-terminal to the β sheet in p40^{phox} (Bravo et al. 2001; Stahelin et al. 2006, 2007; Xing et al. 2004). The $\alpha1$ and $\alpha2$ helices are connected by a long variable loop (MIL), which in Bem1p, CISK, p40^{phox}, p47^{phox} and PI3K-C2α contains a type II polyproline helix (Bravo et al. 2001; Karathanassis et al. 2002; Stahelin et al. 2006, 2007; Xing et al. 2004). The $\beta1$ strand has a β-bulge that twists the β sheet, forming one wall of the lipid binding pocket.

In the p40^{phox} PX complex PtdIns(3)P is bound in a relatively narrow and deep (7Å) groove formed by the three elements: the loop connecting $\beta3$ and $\alpha1$, a part of MIL closest to $\alpha2$, and the *N*-terminal halves of $\beta2$ and $\alpha2$ (Bravo et al. 2001) (Figs. 6.2d and 6.3c). The 3-phosphate group of the lipid is restrained through the formation of hydrogen bonds with the guanidino moiety of Arg58 in the $\beta3/\alpha1$ loop and with backbone amides of Tyr59 and Arg60. The side chains of Lys92 and Arg60 are involved in the hydrogen bonding contacts with the 1-phosphate, whereas the 4- and 5-hydroxyl groups of PtdIns(3)P are hydrogen bonded to Arg105. The three motifs essential for PI binding, $RRYx_2Fx_2Lx_3L$ of $\beta3/\alpha1$, Px_2PxK of the MIL, and RR/Kx_2L of $\alpha2$ are present in the majority of PX domain sequences.

Other PI Effectors

The number of PI effectors is increasing rapidly and, in addition to the FYVE, PH and PX domains described above, includes the ANTH, BATS, C2, DHR-1, ENTH, FERM, GOLPH3, P4M, PDZ, PROPPINs, PTB, SYLF, and Tubby modules. The ANTH domain and its structural relative ENTH bind strongly and specifically to PtdIns(4,5)P_2 in the plasma membrane (Itoh and De Camilli 2006). The BATS domain localizes to highly curved membranes enriched in PtdIns(3)P (Fan et al. 2011). Although the majority of C2 domains associate with the most common anionic and zwitterionic lipids such as PS and phosphocholine (PC), some show preference for PtdIns(3,4,5)P_3 and PtdIns(4,5)P_2 (Cho and Stahelin 2006). A recently characterized atypical C2 module, the DHR-1 domain, binds PtdIns(3,4,5)P_3 and is required for targeting of the Rho family guanine exchange factor, Dock1, to the plasma membrane and for triggering cell polarization (Premkumar et al. 2010). Another effector of PtdIns(4,5)P_2 is the FERM domain (Hamada et al. 2000).

GOLPH3 and P4M target specifically PtdIns(4)P in the Golgi apparatus (Dippold et al. 2009; Schoebel et al. 2010; Wood et al. 2009), whereas PDZ recognizes PtdIns(4,5)P_2 in the plasma membrane (Zimmermann 2006). Human and yeast PROPPINs bind PtdIns(3,5)P_2 in membranes of endosomes, lysosomes and vacuoles (Dove et al. 2009), and the fly PROPPIN Dm3 associates with both PtdIns(3,5)P_2 and PtdIns(3)P (Lemmon 2008). Interaction with PtdIns(4,5)P_2 and PtdIns(4)P has been reported for the PTB domain that normally binds phosphotyrosine peptides (DiNitto and Lambright 2006). The SYLF domain interacts preferentially with PtdIns(3,4,5)P_3 and is involved in the regulation of dorsal ruffle formation (Hasegawa et al. 2011). The Tubby domain localizes to the plasma membrane through binding to PtdIns(4,5)P_2 and to a lesser degree to PtdIns(3,4,5)P_3 and PtdIns(3,4)P_2 (Santagata et al. 2001; Szentpetery et al. 2009). Several distinct proteins, for example, AKAP79, CAP23, GAP43, MARKS, NHE3, PAR-3, profilin and WASP, recognize PIs via clusters of basic residues, however, the molecular details of these interactions have not been characterized (Caroni 2001; Dell'Acqua et al. 1998; Goldschmidt-Clermont et al. 1990; Horikoshi et al. 2011; Mohan et al. 2010; Rohatgi et al. 2000; Wang et al. 2001).

Although PI effectors have diverse and unrelated structures, their membrane docking and PI binding mechanisms share many similarities. First, all effectors possess a highly basic binding site composed of at least three positively charged residues [three basic residues and a histidine in CALM ANTH (Ford et al. 2001)], and as many as six positively charged residues [six basic residues and a histidine in Epsin1 ENTH (Ford et al. 2002) and Grp1 PH (Ferguson et al. 2000; Lietzke et al. 2000)] (Fig. 6.3). Second, membrane binding involves some or all components of multiple anchoring. Association with PI-containing membranes can be augmented by nonspecific electrostatic interactions, hydrophobic insertion, protonation of a histidine switch, coincidence detection and increased avidity, with each component contributing to binding energetics. Cooperation of multiple interactions is particularly essential for the recruitment of less selective PI-binding modules, including many PH domains. The affinity and specificity can be further increased due to cooperative binding of multiple PI-recognizing domains present in the same protein and association of the adjacent regions with various membrane elements and membrane-attached proteins.

Concluding Remarks

Phosphoinositide-binding domains have emerged as a family of 'PI code' readers, and considerable effort has been put forth by many groups to determine their role in mediating acute and constitutive membrane signaling. This chapter focuses on the mechanistic aspects of single PI effectors; however, these domains are often found next to other PI-binding modules and PI-modifying catalytic domains. Of those discussed here, pleckstrin and Centaurin contain two and five PH domains, respectively; FGD1 contains two PH domains separated by a FYVE finger; PLCγ1 has two PH domains, a catalytic phospholipase module and a C2 domain; PLD1C contains PX, PH and phospholipase domains; DFCP1 contains tandem FYVE fingers; and PDZ and PX domains are present in SNX27. Depending on the specificities, these modules may act either in concert or compete for targeting their host proteins to particular subcellular membranes and regions. The location and duration of membrane association by proteins containing multiple distinct PI effectors can be mediated by activities of organelle-specific PI kinases, phosphatases and lipases. The crosstalk between PIs can provide a mechanism for the regulation of temporal and spatial membrane localization of these proteins and may be essential for controlling signaling cascades. The link between dysregulation of the PI signaling network and numerous diseases suggests a strong therapeutic potential (Engelman et al. 2006; McCrea and De Camilli 2009; Prestwich 2004; Xu et al. 2006; Zhang et al. 2010), and further mechanistic studies will be essential to fully understand and exploit this potential.

Acknowledgments The research in TGK laboratory is supported by the NIH grants GM096863 and CA 113472.

References

Balla A, Tuymetova G, Tsiomenko A, Varnai P, Balla T (2005) Mol Biol Cell 16:1282–1295

Baraldi E, Carugo KD, Hyvonen M, Surdo PL, Riley AM, Potter BV, O'Brien R, Ladbury JE, Saraste M (1999) Structure 7:449–460

Blatner NR, Stahelin RV, Diraviyam K, Hawkins PT, Hong W, Murray D, Cho W (2004) J Biol Chem 279:53818–53827

Blatner NR, Wilson MI, Lei C, Hong W, Murray D, Williams RL, Cho W (2007) EMBO J 26:3709–3719

Bravo J, Karathanassis D, Pacold CM, Ellson CD, Anderson KE, Butler PJ, Lavenir I, Perisic O, Hawkins PT, Stephens L, Williams RL (2001) Mol Cell 8:829–839

Brunecky R, Lee S, Rzepecki PW, Overduin M, Prestwich GD, Kutateladze AG, Kutateladze TG (2005) Biochemistry 44:16064–16071

Burd CG, Emr SD (1998) Mol Cell 2:157–162

Callaghan J, Simonsen A, Gaullier JM, Toh BH, Stenmark H (1999) Biochem J 338:539–543

Caroni P (2001) EMBO J 20:4332–4336

Carpten JD, Faber AL, Horn C, Donoho GP, Briggs SL, Robbins CM, Hostetter G, Boguslawski S, Moses TY, Savage S, Uhlik M, Lin A, Du J, Qian YW, Zeckner DJ, Tucker-Kellogg G, Touchman J, Patel K, Mousses S, Bittner M, Schevitz R, Lai MH, Blanchard KL, Thomas JE (2007) Nature 448:439–444

Ceccarelli DF, Blasutig IM, Goudreault M, Li Z, Ruston J, Pawson T, Sicheri F (2007) J Biol Chem 282:13864–13874

Cheever ML, Sato TK, de Beer T, Kutateladze TG, Emr SD, Overduin M (2001) Nat Cell Biol 3:613–618

Cho W, Stahelin RV (2006) Biochim Biophys Acta 1761:838–849

Cohen LA, Honda A, Varnai P, Brown FD, Balla T, Donaldson JG (2007) Mol Biol Cell 18:2244–2253

Cronin TC, DiNitto JP, Czech MP, Lambright DG (2004) EMBO J 23:3711–3720

Dell'Acqua ML, Faux MC, Thorburn J, Thorburn A, Scott JD (1998) EMBO J 17:2246–2260

Di Paolo G, De Camilli P (2006) Nature 443:651–657

DiNitto JP, Lambright DG (2006) Biochim Biophys Acta 1761:850–867
DiNitto JP, Delprato A, Gabe Lee MT, Cronin TC, Huang S, Guilherme A, Czech MP, Lambright DG (2007) Mol Cell 28:569–583
Dippold HC, Ng MM, Farber-Katz SE, Lee SK, Kerr ML, Peterman MC, Sim R, Wiharto PA, Galbraith KA, Madhavarapu S, Fuchs GJ, Meerloo T, Farquhar MG, Zhou H, Field SJ (2009) Cell 139:337–351
Diraviyam K, Stahelin RV, Cho W, Murray D (2003) J Mol Biol 328:721–736
Dove SK, Dong K, Kobayashi T, Williams FK, Michell RH (2009) Biochem J 419:1–13
Dumas JJ, Merithew E, Sudharshan E, Rajamani D, Hayes S, Lawe D, Corvera S, Lambright D (2001) Mol Cell 8:947–958
Ellson CD, Gobert-Gosse S, Anderson KE, Davidson K, Erdjument-Bromage H, Tempst P, Thuring JW, Cooper MA, Lim ZY, Holmes AB, Gaffney PR, Coadwell J, Chilvers ER, Hawkins PT, Stephens LR (2001) Nat Cell Biol 3:679–682
Engelman JA, Luo J, Cantley LC (2006) Nat Rev Genet 7:606–619
Fan W, Nassiri A, Zhong Q (2011) Proc Natl Acad Sci U S A 108:7769–7774
Ferguson KM, Lemmon MA, Schlessinger J, Sigler PB (1995) Cell 83:1037–1046
Ferguson KM, Kavran JM, Sankaran VG, Fournier E, Isakoff SJ, Skolnik EY, Lemmon MA (2000) Mol Cell 6:373–384
Flesch FM, Yu JW, Lemmon MA, Burger KN (2005) Biochem J 389:435–441
Ford MG, Pearse BM, Higgins MK, Vallis Y, Owen DJ, Gibson A, Hopkins CR, Evans PR, McMahon HT (2001) Science 291:1051–1055
Ford MG, Mills IG, Peter BJ, Vallis Y, Praefcke GJ, Evans PR, McMahon HT (2002) Nature 419:361–366
Fukuda M, Kojima T, Kabayama H, Mikoshiba K (1996) J Biol Chem 271:30303–30306
Gallego O, Betts MJ, Gvozdenovic-Jeremic J, Maeda K, Matetzki C, Aguilar-Gurrieri C, Beltran-Alvarez P, Bonn S, Fernandez-Tornero C, Jensen LJ, Kuhn M, Trott J, Rybin V, Muller CW, Bork P, Kaksonen M, Russell RB, Gavin AC (2010) Mol Syst Biology 6:430
Gaullier JM, Simonsen A, D'Arrigo A, Bremnes B, Stenmark H, Aasland R (1998) Nature 394:432–433
Gillooly DJ, Simonsen A, Stenmark H (2001) Biochem J 355:249–258
Godi A, Di Campli A, Konstantakopoulos A, Di Tullio G, Alessi DR, Kular GS, Daniele T, Marra P, Lucocq JM, De Matteis MA (2004) Nat Cell Biol 6:393–404
Goldschmidt-Clermont PJ, Machesky LM, Baldassare JJ, Pollard TD (1990) Science 247: 1575–1578
Hamada K, Shimizu T, Matsui T, Tsukita S, Hakoshima T (2000) EMBO J 19:4449–4462
Harlan JE, Hajduk PJ, Yoon HS, Fesik SW (1994) Nature 371:168–170
Hasegawa J, Tokuda E, Tenno T, Tsujita K, Sawai H, Hiroaki H, Takenawa T, Itoh T (2011) J Cell Biology 193:901–916
Haslam RJ, Koide HB, Hemmings BA (1993) Nature 363:309–310
Hayakawa A, Hayes SJ, Lawe DC, Sudharshan E, Tuft R, Fogarty K, Lambright D, Corvera S (2004) J Biol Chem 279:5958–5966
He J, Haney RM, Vora M, Verkhusha VV, Stahelin RV, Kutateladze TG (2008) J Lipid Res 49:1807–1815
He J, Vora M, Haney RM, Filonov GS, Musselman CA, Burd CG, Kutateladze AG, Verkhusha VV, Stahelin RV, Kutateladze TG (2009) Proteins 76:852–860
He J, Gajewiak J, Scott JL, Gong D, Ali M, Best MD, Prestwich GD, Stahelin RV, Kutateladze TG (2011a) Chem Biol 18:1312–1319
He J, Scott JL, Heroux A, Roy S, Lenoir M, Overduin M, Stahelin RV, Kutateladze TG (2011b) J Biol Chem 286:18650–18657
Hiroaki H, Ago T, Ito T, Sumimoto H, Kohda D (2001) Nat Struct Biol 8:526–530
Honbou K, Minakami R, Yuzawa S, Takeya R, Suzuki NN, Kamakura S, Sumimoto H, Inagaki F (2007) EMBO J 26:1176–1186
Horikoshi Y, Hamada S, Ohno S, Suetsugu S (2011) Cell Struct Funct 36:97–102
Huang BX, Akbar M, Kevala K, Kim HY (2011) J Cell Biology 192:979–992

Hurley JH (2006) Biochim Biophys Acta 1761:805–811
Hyvonen M, Macias MJ, Nilges M, Oschkinat H, Saraste M, Wilmanns M (1995) EMBO J 14:4676–4685
Itoh T, De Camilli P (2006) Biochim Biophys Acta 1761:897–912
Jackson SG, Zhang Y, Haslam RJ, Junop MS (2007) BMC Struct Biol 7:80
Kanai F, Liu H, Field SJ, Akbary H, Matsuo T, Brown GE, Cantley LC, Yaffe MB (2001) Nat Cell Biol 3:675–678
Karathanassis D, Stahelin RV, Bravo J, Perisic O, Pacold CM, Cho W, Williams RL (2002) EMBO J 21:5057–5068
Klarlund JK, Guilherme A, Holik JJ, Virbasius JV, Chawla A, Czech MP (1997) Science 275:1927–1930
Knight JD, Falke JJ (2009) Biophys J 96:566–582
Komander D, Fairservice A, Deak M, Kular GS, Prescott AR, Peter Downes C, Safrany ST, Alessi DR, van Aalten DM (2004) Embo J 23:3918–3928
Kutateladze TG (2006) Biochim Biophys Acta 1761:868–877
Kutateladze TG (2007) Prog Lipid Res 46:315–327
Kutateladze TG (2010) Nat Chem Biol 6:507–513
Kutateladze TG, Overduin M (2001) Science 291:1793–1796
Kutateladze TG, Capelluto DGS, Ferguson CG, Cheever ML, Kutateladze AG, Prestwich GD, Overduin M (2004) J Biol Chem 279:3050–3057
Lawe DC, Patki V, Heller-Harrison R, Lambright D, Corvera S (2000) J Biol Chem 275:3699–3705
Lee SA, Eyeson R, Cheever ML, Geng J, Verkhusha VV, Burd C, Overduin M, Kutateladze TG (2005) Proc Natl Acad Sci U S A 102:13052–13057
Lee SA, Kovacs J, Stahelin RV, Cheever ML, Overduin M, Setty TG, Burd CG, Cho W, Kutateladze TG (2006) J Biol Chem 281:37091–37101
Lemmon MA (2004) Biochem Soc Trans 32:707–711
Lemmon MA (2008) Nat Rev Mol Cell Biol 9:99–111
Lemmon MA, Ferguson KM (2000) Biochem J 350(Pt 1):1–18
Levine TP, Munro S (2002) Curr Biol 12:695–704
Lietzke SE, Bose S, Cronin T, Klarlund J, Chawla A, Czech MP, Lambright DG (2000) Mol Cell 6:385–394
Liu J, Fukuda K, Xu Z, Ma YQ, Hirbawi J, Mao X, Wu C, Plow EF, Qin J (2011) J Biol Chem 286:43334–43342
Lu J, Garcia J, Dulubova I, Sudhof TC, Rizo J (2002) Biochemistry 41:5956–5962
Lumb CN, He J, Xue Y, Stansfeld PJ, Stahelin RV, Kutateladze TG, Sansom MS (2011) Structure 19:1338–1346
Malkova S, Stahelin RV, Pingali SV, Cho W, Schlossman ML (2006) Biochemistry 45: 13566–13575
Manna D, Albanese A, Park WS, Cho W (2007) J Biol Chem 282:32093–32105
Mao Y, Nickitenko A, Duan X, Lloyd TE, Wu MN, Bellen H, Quiocho FA (2000) Cell 100: 447–456
Mayer BJ, Ren R, Clark KL, Baltimore D (1993) Cell 73:629–630
McCrea HJ, De Camilli P (2009) Physiology (Bethesda) 24:8–16
Mertens HD, Callaghan JM, Swarbrick JD, McConville MJ, Gooley PR (2007) Protein Sci 16:2552–2559
Milburn CC, Deak M, Kelly SM, Price NC, Alessi DR, Van Aalten DM (2003) Biochem J 375:531–538
Misra S, Hurley JH (1999) Cell 97:657–666
Mohan S, Tse CM, Gabelli SB, Sarker R, Cha B, Fahie K, Nadella M, Zachos NC, Tu-Sekine B, Raben D, Amzel LM, Donowitz M (2010) J Biol Chem 285:34566–34578
Moravcevic K, Oxley CL, Lemmon MA (2012) Structure 20:15–27
Parkinson GN, Vines D, Driscoll PC, Djordjevic S (2008) BMC Struct Biol 8:13
Patki V, Lawe DC, Corvera S, Virbasius JV, Chawla A (1998) Nature 394:433–434

Ponting CP, Onda T, Kitagawa M, Takeda O, Sago H, Kubonoya K, Iinuma K, Bradley LA, Canick JA, Krasikov NE, Ponting NR, Grier RE (1996) Protein Sci 5:2353–2357
Premkumar L, Bobkov AA, Patel M, Jaroszewski L, Bankston LA, Stec B, Vuori K, Cote JF, Liddington RC (2010) J Biol Chem 285:13211–13222
Prestwich GD (2004) Chem Biol 11:619–637
Pylypenko O, Lundmark R, Rasmuson E, Carlsson SR, Rak A (2007) EMBO J 26:4788–4800
Ridley SH, Ktistakis N, Davidson K, Anderson KE, Manifava M, Ellson CD, Lipp P, Bootman M, Coadwell J, Nazarian A, Erdjument-Bromage H, Tempst P, Cooper MA, Thuring JW, Lim ZY, Holmes AB, Stephens LR, Hawkins PT (2001) J Cell Sci 114:3991–4000
Rohatgi R, Ho HY, Kirschner MW (2000) J Cell Biol 150:1299–1310
Roth MG (2004) Physiol Rev 84:699–730
Sankaran VG, Klein DE, Sachdeva MM, Lemmon MA (2001) Biochemistry 40:8581–8587
Santagata S, Boggon TJ, Baird CL, Gomez CA, Zhao J, Shan WS, Myszka DG, Shapiro L (2001) Science 292:2041–2050
Schoebel S, Blankenfeldt W, Goody RS, Itzen A (2010) EMBO Rep 11:598–604
Seet LF, Hong W (2006) Biochim Biophys Acta 1761:878–896
Simonsen A, Birkeland HC, Gillooly DJ, Mizushima N, Kuma A, Yoshimori T, Slagsvold T, Brech A, Stenmark H (2004) J Cell Sci 117:4239–4251
Singh SM, Murray D (2003) Protein Sci 12:1934–1953
Song X, Xu W, Zhang A, Huang G, Liang X, Virbasius JV, Czech MP, Zhou GW (2001) Biochemistry 40:8940–8944
Song J, Zhao KQ, Newman CL, Vinarov DA, Markley JL (2007) Protein Sci 16:807–814
Stahelin RV, Long F, Diraviyam K, Bruzik KS, Murray D, Cho W (2002) J Biol Chem 277:26379–26388
Stahelin RV, Burian A, Bruzik KS, Murray D, Cho W (2003) J Biol Chem 278:14469–14479
Stahelin RV, Ananthanarayanan B, Blatner NR, Singh S, Bruzik KS, Murray D, Cho W (2004) J Biol Chem 279:54918–54926
Stahelin RV, Karathanassis D, Bruzik KS, Waterfield MD, Bravo J, Williams RL, Cho W (2006) J Biol Chem 281:39396–39406
Stahelin RV, Karathanassis D, Murray D, Williams RL, Cho W (2007) J Biol Chem 282:25737–25747
Szentpetery Z, Balla A, Kim YJ, Lemmon MA, Balla T (2009) BMC Cell Biol 10:67
Thomas CC, Deak M, Alessi DR, van Aalten DM (2002) Curr Biol 12:1256–1262
Vanhaesebroeck B, Leevers SJ, Ahmadi K, Timms J, Katso R, Driscoll PC, Woscholski R, Parker PJ, Waterfield MD (2001) Annu Rev Biochem 70:535–602
Wang J, Arbuzova A, Hangyas-Mihalyne G, McLaughlin S (2001) J Biol Chem 276:5012–5019
Wood CS, Schmitz KR, Bessman NJ, Setty TG, Ferguson KM, Burd CG (2009) J Cell Biol 187:967–975
Xing Y, Liu D, Zhang R, Joachimiak A, Songyang Z, Xu W (2004) J Biol Chem 279:30662–30669
Xu Y, Hortsman H, Seet L, Wong SH, Hong W (2001) Nat Cell Biol 3:658–666
Xu Y, Lee SA, Kutateladze TG, Sbrissa D, Shisheva A, Prestwich GD (2006) J Am Chem Soc 128:885–897
Yu JW, Lemmon MA (2001) J Biol Chem 276:44179–44184
Zhang H, He J, Kutateladze TG, Sakai T, Sasaki T, Markadieu N, Erneux C, Prestwich GD (2010) ChemBioChem 11:388–395
Zhong Q, Watson MJ, Lazar CS, Hounslow AM, Waltho JP, Gill GN (2005) Mol Biol Cell 16:2049–2057
Zhou CZ, de La Sierra-Gallay IL, Quevillon-Cheruel S, Collinet B, Minard P, Blondeau K, Henckes G, Aufrere R, Leulliot N, Graille M, Sorel I, Savarin P, de la Torre F, Poupon A, Janin J, van Tilbeurgh H (2003) J Biol Chem 278:50371–50376
Zimmermann P (2006) Biochim Biophys Acta 1761:947–956

Chapter 7
PIKfyve and its Lipid Products in Health and in Sickness

Assia Shisheva

Abstract PIKfyve, a phosphoinositide 5-kinase synthesizing PtdIns(3,5)P_2 and PtdIns5P in a cellular context, belongs to an evolutionarily ancient gene family of PtdIns(3,5)P_2-synthesizing enzymes that, except for plants, are products of a single-copy gene across species. In the dozen years after its discovery, enormous progress has been made in characterizing the numerous PIKfyve cellular functions and the regulatory mechanisms that govern these functions. It became clear that PIKfyve does not act alone but, rather, it engages the scaffolding regulator ArPIKfyve and the phosphatase Sac3 to make a multiprotein "PAS" complex, so called for the first letters of the protein names. This complex relays antagonistic signals, one for synthesis, another for turnover of PtdIns(3,5)P_2, whose dysregulated coordination is linked to several human diseases. The physiological significance for each protein in the PAS complex is underscored by the early lethality of the mouse models with disruption in any of the three genes. This chapter summarizes our current knowledge of the diverse and complex functionality of PIKfyve and PtdIns(3,5)P_2/PtdIns5P products with particular highlights on recent discoveries of inherited or somatic mutations in PIKfyve and Sac3 linked to human disorders.

Introduction

The phosphorylated derivatives of phosphatidylinositol (PtdIns), called phosphoinositides (PIs), are eukaryotic cell membrane-anchored phospholipids, whose cytosol-exposed hydrophilic inositol headgroup could be reversibly phosphorylated

A. Shisheva (✉)
Department of Physiology, Wayne State University School of Medicine,
540 E. Canfield, Detroit, Michigan 48201, USA
e-mail: ashishev@med.wayne.edu

M. Falasca (ed.), *Phosphoinositides and Disease*, Current Topics in Microbiology and Immunology 362, DOI: 10.1007/978-94-007-5025-8_7,

at positions D3, D4, and/or D5, singly or in all possible combinations to yield seven PIs: PtdIns3P, PtdIns4P, PtdIns5P, PtdIns(3,4)P_2, PtdIns(3,5)P_2, PtdIns(4,5)P_2 and PtdIns(3,4,5)P_3. PIs are heterogeneously localized within cells, where in conjunction with a membrane protein, they serve as co-receptors for specific membrane targeting of cytosolic effector proteins through coincidence detection (Carlton and Cullen 2005; Di Paolo and De Camilli 2006; Lemmon 2008). PI metabolism is tightly controlled in time and space by the action of various kinases, phosphatases, and phospholipases, each exhibiting specific intracellular localization (Vicinanza et al. 2008; Sasaki et al. 2009). This ensures unique PI distribution, despite the constant membrane flow through the intracellular compartments. PIs play important and indispensable roles in regulating diverse and essential cellular processes, underscored by the growing number of human diseases associated with mutations in the genes encoding the PI-metabolizing enzymes.

The mammalian enzyme that makes PtdIns(3,5)P_2 from PtdIns3P, and PtdIns5P from PtdIns is PIKfyve (Shisheva 2001; Shisheva 2008b). It also possesses an intrinsic protein Ser kinase activity, a feature shared by other lipid kinases. PIKfyve belongs to an evolutionarily ancient gene family of PtdIns(3,5)P_2-synthesizing enzymes that, except for plants, are products of a single-copy gene across species. Intriguingly, PIKfyve has been discovered in a functional context rather than by sequence homology to the yeast ortholog Fab1, through screens for transcripts that, like GLUT4, are expressed in a fat/muscle enriched manner (Shisheva et al. 1999). Subsequent studies have revealed that PIKfyve is either undetectable or expressed to various degrees in other cells and tissues but its highest content is in brain and fat. Notably, PIKfyve is a remarkably long-lived protein ($t1/2 = 33$ h) as determined in metabolically labeled adipose cells (Ikonomov et al. 2010). In the dozen years after its discovery it became clear that PIKfyve and the products of its enzymatic activity exhibit pleiotropic functions, regulating various aspects of key cellular processes such as membrane trafficking, stress- or hormone-induced signaling, ion channel activity, cytoskeletal dynamics, nuclear transport, gene transcription and cell cycle progression. In many instances it is still unclear whether the functional outcome is associated with selective changes in the PIKfyve-generated PtdIns(3,5)P_2, PtdIns5P, and/or phosphoprotein. Adding to the complexity of PIKfyve functionality and regulation is the recently established stable complex between PIKfyve and the antagonistic PtdIns(3,5)P_2-specific phosphatase Sac3, which orchestrates the coordinated regulation of PtdIns(3,5)P_2 synthesis and turnover (Sbrissa et al. 2007, 2008). Whereas our mechanistic understanding of how PIKfyve governs these numerous biological processes is rather limited, the vital importance of PIKfyve for life in mammals is evidenced by the established preimplantation lethality of mouse embryos null for *pikfyve* (Ikonomov et al. 2011). Likewise, loss of PIKfyve function in *Caenorhabditis elegans* and *Drosophila* is associated with developmental defects and early lethality (Nicot et al. 2006; Rusten et al. 2006). It is therefore not surprising that dysfunction of PIKfyve and PtdIns(3,5)P_2/PtdIns5P underlie the pathogenic mechanisms of several hereditary human diseases. Somatic mutations in PIKfyve are recently found in ovarian and other cancers. This chapter provides a comprehensive and critical view of our current understanding of PIKfyve functionality in mammalian cells with mechanistic emphasis on genetically modified mouse models and human disorders. Where appropriate, comparison with orthologous systems in lower eukaryotes is provided; for details on those see Dove et al. 2009.

PIKfyve Splice Forms, Domains, Localization, and Enzymology

Protein Domain Organization, Splice Forms, and Localization

PIKfyve (gene symbol, PIKFYVE) belongs to a family of evolutionarily conserved large PI 5-kinases that exhibit similar domain organization across a great evolutionary distance (Shisheva 2001; Michell et al. 2006). A mouse and a human cDNA clone are currently available (Shisheva et al. 1999; Cabezas et al. 2006). However, most of the studies are performed with the cDNA clone encoding the mouse 2052-residue PIKfyveL splice form, first isolated in our laboratory (Shisheva et al. 1999) and distributed within the community (Berwick et al. 2004; Rutherford et al. 2006; Seebohm et al. 2007; Shojaiefard et al. 2007; Gehring et al. 2009a; b; Klaus et al. 2009a; 2009b; Tsuruta et al. 2009; Alesutan et al. 2010; Sopjani et al. 2010). Therefore, the numbering provided below reflects the mouse 2052-residue protein form. It incorporates exon 4 (encoding residues 108–118) that is missing in the human gene but lacks exon 12 (encoding residues 490–545), present in both the human and mouse genes (Shisheva et al. 1999; Cabezas et al. 2006) (Fig. 1a). The specific functions of these insertions are not yet known.

The N-terminal region contains the PtdIns3P-binding FYVE finger domain (residues 164–230) that targets the protein to PtdIns3P-enriched endosome membranes (Shisheva 2008b). Next is the DEP domain (residues 376–449) with still uncharacterized function, which in addition to mammalian PIKfyve proteins, is also seen in *Drosophila*. The middle part of the molecule (residues 560–1438) is occupied by two domains: Cpn60_TCP1, with sequence similarity to molecular chaperonins, and a CHK homology region, with conserved C, H, and K residues uniquely displayed by the PIKfyve orthologs. The region of conserved Lys is homologous to spectrin repeats found in many cytoskeletal proteins. Through this middle part, PIKfyve interacts with a number of protein partners, including the ArPIKfyve-Sac3 heterodimer, kinesin adapter JLP, Rab9 effector p40 and $Ca_V1.2$ Ca^{2+} channel (see PIKfyve Molecular Interactions and Their Functional Roles). The catalytic domain (residues 1801–2038) shares sequence similarities with the catalytic domains in PI4Ks and PIP5Ks but not with those in PI3Ks and PtdIns4Ks. It is responsible for all three kinase activities of PIKfyve, i.e., PtdIns(3,5)P_2, PtdIns5P, and phosphoprotein biosynthesis, with Lys at position 1831 functioning as a single ATP binding site (Sbrissa et al. 2000). Lys-1999 supports preferentially

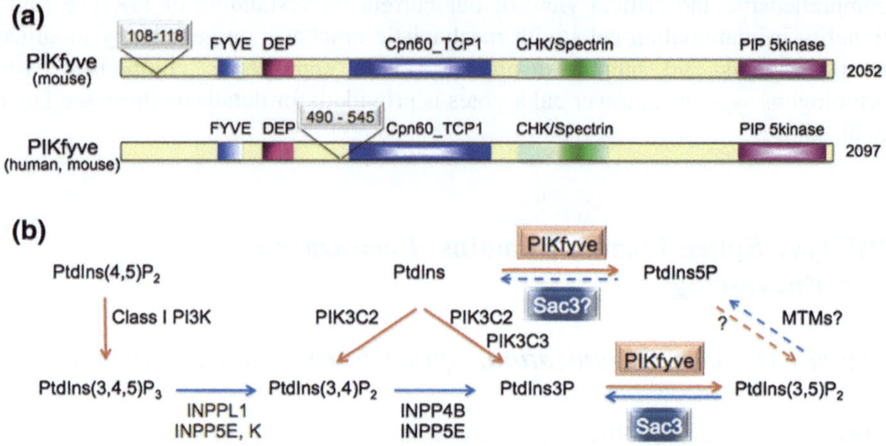

Fig. 7.1 *PIKfyve domain structure, splice forms, and metabolic reactions generating PtdIns(3,5)P₂ and PtdIns5P*. **a,** Schematic diagram of the domain organization and positions of inserts in the mouse and human PIKfyve splice variants. **b,** Metabolic reactions leading to generation of Ptdns3P, PtdIns5P and PtdIns(3,5)P_2 and the enzymes PIKfyve and Sac3 supporting these reactions. Reactions indicated with dotted arrows are either speculative or require further clarification. Whereas PIK3C3 is the main supplier of PtdIns3P, PIK3C2 and class IA PI3 K could directly or indirectly produce PtdIns3P. Depicted are known pathways for PtdIns3P production from PtdIns(3,4,5)P_3 and PtdIns(3,4)P_2 through turnover by INPPs. Whether PIKfyve uses these PIK3C3-independent PtdIn3P pools for PtdIns(3,5)P_2 synthesis is uncertain. PIK, phosphoinositide kinase; INPP, inositol polyphosphate phosphatase; synthesis, red arrows; turnover, blue arrows. See text for details

PtdIns(3,5)P_2 over PtdIns5P catalysis, whereas Lys-2000 shows preferences for PtdIns5P over PtdIns(3,5)P_2 phosphorylation (Ikonomov et al. 2002a). Specific residues supporting PIKfyve protein kinase activity are unknown. The powerful ability of the PIKfyve^{K1831} mutant to dominantly interfere with functionality of endogenous PIKfyve (Ikonomov et al. 2001) has been largely explored in the field (Kim et al. 2007; Osborne et al. 2008; Tsuruta et al. 2009; Hill et al. 2010; Kerr et al. 2010). It should be emphasized that the K1831 mutation abrogates equally well all three PIKfyve enzymatic activities and therefore could not be solely used for conclusions about the role for PtdIns(3,5)P_2.

Biochemical fractionation data in native 3T3L1 adipocytes, fibroblasts, PC12, and chromaffin cells indicate both particulate and soluble pools of endogenous PIKfyve (Shisheva et al. 2001, 2002; Osborne et al. 2008). Immunofluorescence microscopy analyses carried out predominantly with heterologously expressed PIKfyve reveal punctate/vesicular distribution (Shisheva et al. 1999; Ikonomov et al. 2001; Berwick et al. 2004; Cabezas et al. 2006; Rutherford et al. 2006). This localization pattern is lost in the presence of dominant negative GDP-bound Rab5S34N or upon cell treatment with wortmannin, indicative of a PtdIns3P-dependent PIKfyve targeting mechanism downstream of active Rab5 and wortmannin-sensitive PI3Ks (Sbrissa et al. 2002b; Ikonomov et al. 2006). PIKfyve^{WT}—

containing puncta appear to be positive for various endocytic markers of early endosomes (EEA1, SNX1, Hrs, and Rab5), late endosomes (Rab7, CI-MPR, LAMP1, lgp100), intralumenal vesicles of the multivesicular endosomes (LBPA) and TGN (CI-MPR), with relative proportions varying among studies (Ikonomov et al. 2001; Cabezas et al. 2006; Ikonomov et al. 2006; Rutherford et al. 2006). Several factors, including the cell type, levels of intracellular PtdIns3P, or extent and duration of PIKfyve^{WT} overexpression, could account for the variation. Endogenous PIKfyve is successfully visualized only in 3T3L1 adipocytes and chromaffin cells by two groups using independently generated antibodies for immunofluorescence microscopy (Shisheva et al. 1999; Osborne et al. 2008). Although endogenous PIKfyve displays a similar characteristic punctate distribution in both cell types, details about the vesicle identity have not been investigated. Thus, the exact endosome compartments populated by endogenous PIKfyve remain unclear.

Enzymatic Products: Occurrence, Detection, Regulation, and Localization

PtdIns(3,5)P_2

PtdIns(3,5)P_2 is of very low abundance comprising up to 0.8 % of the total PIs in mammalian cells (Ikonomov et al. 2001; Sbrissa and Shisheva 2005). Its synthesis from PtdIns3P was experimentally verified in 1997 in both mammalian and yeast cells (Dove et al. 1997; Whiteford et al. 1997). That PIKfyve was the main source for this reaction in mammals became clear in 1999 subsequent to its identification (Shisheva et al. 1999) and characterization of the lipid kinase activity both in vitro and in yeast strains (McEwen et al. 1999; Sbrissa et al. 1999). HPLC-based data for increased and decreased [^{32}P]PtdIns(3,5)P_2 and [^{32}P]PtdIns3P levels, respectively, in ^{32}Pi-labeled mammalian cells ectopically expressing PIKfyve^{WT}, have further implicated PIKfyve in PtdIns(3,5)P_2 biosynthesis from PtdIns3P (Ikonomov et al. 2001). Consistent with previous observations, more recent HPLC analyses following PIKfyve protein depletion or genetic ablation have demonstrated decreased PtdIns(3,5)P_2 and increased PtdIns3P in [^{3}H]inositol-labeled 3T3L1 adipocytes and mouse embryonic fibroblasts (MEF) (Sbrissa et al. 2005; Ikonomov et al. 2011).

Although there is no doubt that PIKfyve makes PtdIns(3,5)P_2 from PtdIns3P, the enzyme source that supplies PtdIns3P has not been completely resolved. The conservation of the PtdIns(3,5)P_2 pathway in yeast, where the wortmannin-sensitive Vps34 is the sole PtdIns3P-generating kinase, together with observations for wortmannin-sensitivity of intracellular PtdIns(3,5)P_2 production (Whiteford et al. 1997; Backer 2008), suggests that PIKfyve uses the PtdIns3P pool generated by the endosome-localized mammalian Vps34 (gene symbol, PIK3C3). In mammals, however, both class I and class II PI3Ks (gene symbol, PIK3C2) could directly or indirectly produce PtdIns3P (Munday et al. 1999; Ivetac et al. 2005; Shin et al. 2005; Kong et al. 2006; Ooms et al. 2006) and, thus, a Vps34-unrelated PtdIns3P pool may also serve as a precursor in PIKfyve-catalyzed PtdIns(3,5)P_2 synthesis (Fig. 7.1b). This notion is further supported by data demonstrating a widespread distribution of PtdIns3P among various mammalian organelles, including early/late endosomes, intralumenal vesicles of multivesicular endosomes (MVEs), plasma membrane and the nucleus, yet the Vps34-dependent PtdIns3P production is confined to MVEs (Gillooly et al. 2000; Johnson et al. 2006; Falasca et al. 2007). Moreover, both class I and class II PI3K activities are regulated by extracellular stimuli (Falasca and Maffucci 2009; Vanhaesebroeck et al. 2010), which may indirectly affect PIKfyve-catalyzed PtdIns(3,5)P_2 production by increased availability of PtdIns3P. Consistent with this idea, the IL2-stimulated elevation of PtdIns(3,5)P_2 in a mouse T-lymphocyte cell line is linked to class IA PI3K activation (Jones et al. 1999). Similarly, PIKfyve associates with class IA PI3K in 3T3L1 adipocytes, a cell line responding to acute insulin challenge by elevating [^{3}H]PtdIns(3,5)P_2 production (Sbrissa et al. 2001; Ikonomov et al. 2009c). Intriguingly, insulin action in myocytes and adipocytes activates PIK3C2α, thereby elevating the plasma membrane PtdIns3P pool (Falasca et al. 2007). Whether PIKfyve utilizes this PIK3C2α-derived PtdIns3P is questionable, with a caveat mainly related to the wortmannin-resistance of the PIK3C2 activity (Falasca and Maffucci 2009), whereas both PtdIns(3,5)P_2 synthesis and PIKfyve targeting to PtdIns3P-enriched membranes occur in a wortmannin-sensitive manner (Whiteford et al. 1997; Sbrissa et al. 2002b; Ikonomov et al. 2006; Rutherford et al. 2006). More studies are required to determine the contribution of the PIK3C3-independent PtdIns3P pools to the PIKfyve-catalyzed PtdIns(3,5)P_2 synthesis.

In addition to IL2 in T-cells, and insulin in 3T3L1 adipocytes, other mitogenic signals, such as EGF in COS cells and PMA in platelets, have also been shown to elevate steady-state levels of PtdIns(3,5)P_2 (Banfic et al. 1998; Tsujita et al. 2004). Likewise, cellular stresses such as UV light and hyperosmolarity up-regulate PtdIns(3,5)P_2 production (Jones et al. 1999; Sbrissa and Shisheva 2005). Although in yeast, hyperosmotic stress profoundly increases PtdIns(3,5)P_2 ($\sim$20-fold), of the numerous mammalian cell types tested, only 3T3L1 adipocytes hyperosmotically elevate PtdIns(3,5)P_2 (Dove et al. 1997; Sbrissa and Shisheva 2005). Because even in this case, the elevation is manyfold less than in yeast, it is conceivable that the ancient osmo-regulatory mechanism evolutionarily degenerated as a result of relatively constant mammalian extracellular osmotic potential and reduced risk of dehydration (Sbrissa and Shisheva 2005). As fat cells have exceptionally low cytoplasm, constituting a negligible proportion of the fat-laden cell volume, they might have retained the hyperosmotic stress response as one means to combat cytosolic water loss.

The data outlined above evaluate changes in PtdIns(3,5)P_2 levels by HPLC inositol-head group analyses, defined as "*sine qua non*" in PI quantification (Backer 2010). Immunocytochemistry with the previously available monoclonal anti-PtdIns(3,5)P_2 antibody (discontinued by the manufacturer) has also been explored to quantify intracellular PtdIns(3,5)P_2 (Tsuruta et al. 2009; Alesutan et al. 2010). Regrettably, this antibody lacks the required specificity and is unsuitable for quantitative imaging (Shisheva 2008a, and personal communications by P. Cullen, T. Takenawa and S. Dove). Therefore, the reported up-regulation of PtdIns(3,5)P_2 by glutamate stimulation of cortical neurons (Tsuruta et al. 2009) requires further validation.

PtdIns(3,5)P_2 is thought to be localized on late endosomes. Whereas this might be correct, it is based on circumstantial evidence for PtdIns3P and PIKfyve distribution, rather than direct experimental data with PtdIns(3,5)P_2 specific reagents. Indeed, both PtdIns(3,5)P_2-binding molecules and antibodies have been described, but as discussed above, they both appear to be inadequate as bioreporters for morphological definition of PtdIns(3,5)P_2 distribution in the intracellular context, where PtdIns(3,5)P_2 is $\sim$2 orders of magnitude less than PtdIns4P and PtdIns(4,5)P_2. Accordingly, recent observations for PtdIns(3,5)P_2 localization to sarcoplasmic reticulum and the smooth muscle layer in mouse myotubes and aorta, respectively (Shen et al. 2009; Silswal et al. 2011), which are solely based on the antibody require alternative verification.

PtdIns5P

PtdIns5P was first identified in 1997 in mouse fibroblasts (Rameh et al. 1997). It is undetectable in many cell types by conventional HPLC analyses due to its poor chromatographic separation from the descending arm of the abundant PtdIns4P. Usage of alternative or optimized approaches such as the PtdIns5P mass assay or PtdIns5P resolution by HPLC on a double column have led to the conclusion that PtdIns5P is present in all mammalian and plant cells examined thus far (Morris et al. 2000; Meijer et al. 2001; Sbrissa et al. 2002a; Sarkes and Rameh 2010). Steady-state levels of PtdIns5P vary among cell types but they likely exceed those of PtdIns(3,5)P_2 and PtdIns3P by 6–8- and 2–4-fold, respectively, as evidenced from studies where these PIs are simultaneously quantified by HPLC (Sbrissa et al. 2002a; Zhang et al. 2007; Ikonomov et al. 2011). PtdIns5P levels are quite substantial in differentiated 3T3L1 adipocytes, constituting $\sim$11 % of those for the abundant PtdIns4P (Sbrissa et al. 2002a), as well as in transformed lymphoma and insulinoma cell lines (Coronas et al. 2008; Sarkes and Rameh 2010).

PIKfyve was implicated in PtdIns5P biosynthesis based on data from our laboratory documenting that purified PIKfyve preparations synthesize in vitro PtdIns5P along with PtdIns(3,5)P_2 and that infection with PIKfyve^{WT} baculovirus and adenovirus increases levels of PtdIns5P in insect and mammalian cells, respectively (Sbrissa et al. 1999, 2002a). Data for higher or lower PtdIns5P mass levels in HEK293 cell lines stably expressing PIKfyve^{WT} or dominant-negative kinase-deficient PIKfyve^{K1831E}, respectively, further corroborated this conclusion (Sbrissa et al. 2002a). Concordantly, hypo-osmotic shock in 3T3L1 fibroblasts down-regulates both PtdIns5P and PtdIns(3,5)P_2, consistent with their concomitant production by the same pathway. Despite this evidence, the field appeared to be somewhat reluctant in linking PtdIns5P with PIKfyve catalysis, mainly due to studies in yeast where PtdIns5P is not detected (McEwen et al. 1999). However, more recent data based on PIKfyve pharmacological inhibition or protein knockdown seem to generate a new wave of experimental support for the claim that PIKfyve is a major source of intracellular PtdIns5P biosynthesis in mammals. For instance, PIKfyve protein depletion or activity inhibition by curcumin results in profound ablation of the PtdIns5P pool as determined by PtdIns5P mass assays (Coronas et al. 2008). The notion for direct PIKfyve-catalyzed PtdIns5P biosynthesis is further underscored by recent HPLC-based quantitation of PIs in fibroblasts derived from genetically modified mouse models. Thus, steady-state levels of PtdIns5P appear to be ~7-fold higher than those of PtdIns(3,5)P_2 yet they both decrease similarly (by ~40 %) in the heterozygous PIKfyve$^{WT/KO}$ mice (Ikonomov et al. 2011). In the same vein, fibroblasts from knockout mice for ArPIKfyve, an associated regulator that activates PIKfyve, display an equal reduction in PtdIns(3,5)P_2 and PtdIns5P levels (by 50 %) yet the latter is ~7-fold higher than the former (Zhang et al. 2007). Finally, recent studies in several mammalian cell types reveal that the PIKfyve inhibitor YM201638 arrests both PtdIns(3,5)P_2 and PtdIns5P biosynthesis, being even more effective towards the latter (Sbrissa et al. 2012). Clearly, these data and considerations are consistent with both lipids being produced by PIKfyve-dependent biosynthesis.

Experimental evidence also points to several members of the myotubularin family of lipid phosphatases (gene symbols, MTM and MTMR) in PtdIns5P production by removal of 3' phosphate from PtdIns(3,5)P_2 (Walker et al. 2001; Berger et al. 2002; Tronchere et al. 2004). However, these data are in disagreement with an earlier study documenting robust PtdIns(3,5)P_2 conversion to PtdIns3P in pulse-labeled mouse fibroblasts (Whiteford et al. 1997). PtdIns5P production from PtdIns(3,5)P_2 is also incompatible with the findings that PtdIns5P levels exceed by manyfold those of PtdIns(3,5)P_2 yet both lipids decrease equally in PIKfyve$^{WT/KO}$ or ArPIKfyve$^{-/-}$ mouse models, as outlined above. It should also be emphasized that levels of PIs under myotubularin knockdown and/or knockout conditions have never been quantified by HPLC despite the availability of several KO mouse lines (Al-Qusairi et al. 2009; Bolis et al. 2009; Shen et al. 2009; Hnia et al. 2011). Such analysis will certainly shed light on the contribution of MTM/MTMR to the intracellular PtdIns5P pool. The reported reduction in PtdIns(3,5)P_2 in MTMR14 KO mice by immunostaining with the PtdIns(3,5)P_2 antibody (Shen et al. 2009) does not suffice for a quantitative measure of PtdIns(3,5)P_2 levels, as we discussed above. Thus, the available evidence indicates that whereas myotubularin-dependent production of PtdIns5P from PtdIns(3,5)P_2 might take place under yet-to-be identified circumstances, the principal pathway for PtdIns5P is through direct PIKfyve-catalyzed biosynthesis. Revisiting the PtdIns5P absence by the optimized HPLC analysis in yeast seems warranted, particularly in light of data for the presence of a putative downstream effector of Ptdns5P in yeast (Casamayor and Snyder 2003).

Whether the PtdIns5P pool is hydrolyzed by a specific phosphatase is unknown; the previously implicated PTPMT1 phosphatase is now found to have a different physiological substrate (Zhang et al. 2011). A putative function of the Sac3 phosphatase in PtdIns5P turnover is yet to be determined (see below). Interestingly, phospholipase C-δ1 could acquire preferential specificity for PtdIns5P over PtdIns(4,5)P_2 hydrolysis under high Ca^{2+}, which could potentially have physiological significance (Liu et al. 2003; Roberts et al. 2005).

A putative PIKfyve-independent PtdIns5P pool has been detected in the nucleus where it binds ING2, a protein regulating transcription and p53-dependent apoptosis (Gozani et al. 2003). This pool is controlled by the nuclear type I PtdIns(4,5)P_2 4-phosphatase that, in conjunction with the PIPKIIβ kinase, plays a role in UV stress, apoptosis, and cell cycle progression (Clarke et al. 2001; Jones et al. 2006; Zou et al. 2007). In addition to nuclear localization, substantial PtdIns5P amounts are detected in Golgi and at the plasma membrane in a mouse insulinoma cell line as revealed by biochemical fractionation and HPLC quantitation (Sarkes and Rameh 2010). Based on quantitative considerations Sarkes and Rameh conclude that these PtdIns5P pools occur from PtdIns synthesis rather than PtdIns(3,5)P_2 or PtdIns(4,5)P_2 hydrolysis and, thus, implicating PIKfyve. It should also be pointed out that as with PtdIns(3,5)P_2, there are no specific PtdIns5P probes or antibodies; the tandem PHD domain from ING2 appears to be unsuitable as a bioreporter for PtdIns5P distribution in cells (personal communications by N. Divecha and T. Takenawa). Thus, apart from its demonstrated localization to the nucleus the precise intracellular sites of PtdIns5P remain highly conjectural.

In addition, intracellular PtdIns5P can be up-regulated by various extracellular cues such as thrombin, in platelets (Morris et al. 2000), by insulin, in 3T3L1 adipocytes or CHO-T cells (Sbrissa et al. 2004b; Sarkes and Rameh 2010) and upon bacterial invasion (Niebuhr et al. 2002). In contrast, decreases in PtdIns5P levels are shown to occur in 3T3L1 cells upon hypo-osmotic stimulation (Sbrissa et al. 2002a). PtdIns5P appears to mediate both insulin- and bacteria-induced F-actin remodeling, but whether this is coupled with its postulated role in exocytosis is currently unclear (see To the Plasma Membrane).

Phosphoproteins

That PIKfyve displays also a protein kinase activity has been first recognized by the ability of recombinant PIKfyve to self-phosphorylate (Sbrissa et al. 2000). Mass spectrometry of such preparations found several Ser residues, residing between the FYVE and DEP domains, to be autophosphorylated (Ser310, Ser318, Ser329 and Ser340; Shisheva and Hunt, unpublished) but their specific function is yet to be addressed. PIKfyve could also phosphorylate several protein substrates in vitro, including its associated partner, the Rab9 effector p40 (Ikonomov et al. 2003b) (see Rab9 Effector p40 and JLP Adapter). It is proposed that in the cellular context, this phosphorylation facilitates Rab9p40 membrane tethering. Because the PIKfyve-associated Sac3 and ArPIKfyve (see ArPIKfyve and Sac3: The PAS Regulatory Trio) are phosphorylated in quiescent cells (Ikonomov et al. 2009c) they both have been considered as candidate substrates of the PIKfyve protein kinase activity. Indeed, PIKfyve-catalyzed protein phosphorylation could be potentially an efficient mechanism for regulating the association and/or activity of the Sac3 phosphatase or ArPIKfyve scaffold within the PAS complex. However, data from protein kinase assays conducted both in vitro and in vivo with heterologously expressed or endogenous proteins are inconsistent with ArPIKfyve or Sac3 phosphorylation operating in a manner dependent on the PIKfyve protein kinase (Ikonomov et al. 2009c, Sbrissa and Shisheva, unpublished). Thus, the physiological meaning of the protein phosphorylation signals relayed by the PIKfyve protein kinase remains elusive.

PIKfyve Molecular Interactions and Their Functional Roles

PtdIns3P

Like most FYVE finger-containing proteins, the PIKfyve FYVE domain binds PtdIns3P with high affinity (Kd $\sim$550 nM) and specificity, as evidenced by in vitro liposome binding assays (Sbrissa et al. 2002b). Because wortmannin renders the PIKfyve^{WT} vesicular distribution diffuse in several cell types (Sbrissa et al. 2002b; Ikonomov et al. 2006; Rutherford et al. 2006), this interaction is necessary and, likely, sufficient for PIKfyve intracellular targeting to PtdIns3P-enriched endosomal microdomains. Concordantly, PIKfyve mutants with a defective FYVE domain display diffuse distribution (Sbrissa et al. 2002b; Berwick et al. 2004). As PtdIns3P is of low abundance (1–4 % of total PIs), this substrate-binding mechanism likely assures robust PIKfyve recruitment at sites where the enzyme action is required.

ArPIKfyve and Sac3: The PAS Regulatory Trio

ArPIKfyve (gene symbol, VAC14) is an evolutionarily conserved protein that in mammals comprises 782 residues (Sbrissa et al. 2004a). ArPIKfyve harbors several HEAT repeats and coiled-coil motifs (Fig. 7.2) both known for their role in protein–protein interactions (Andrade et al. 2001; Strauss and Keller 2008). It was first characterized in our laboratory as a physically associated regulator of PIKfyve (hence the name) based on its ability to coimmunoprecipitate with PIKfyve and up-regulate both PtdIns(3,5)P_2 and PtdIns5P synthesis in vitro or in intact cells (Sbrissa et al. 2004a; Sbrissa and Shisheva 2005). HPLC inositol-head group profiles in ArPIKfyve KO mice, demonstrating selective yet equal reduction in steady-state levels of both lipids (Zhang et al. 2007), corroborate the idea that ArPIKfyve controls PIKfyve-catalyzed production of PtdIns(3,5)P_2 and PtdIns5P. The situation became quite complex following our identification of the

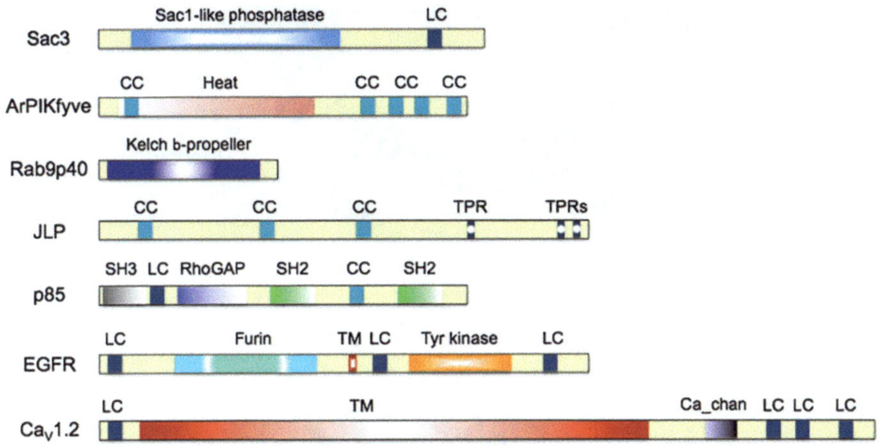

Fig. 7.2 *PIKfyve physically interacting partners*. The domain architecture of ArPIKfyve, Sac3, Rab9p40, JLP, p85, EGFR, and Ca$_V$1.2, known to physically associate with PIKfyve, are illustrated. Relevant functional domains and motifs are indicated. LC, low complexity; CC, coiled-coil motif; Heat, heat repeats; TPR, tetratricopeptide repeats; Furin, furin repeats; TM, transmembrane domain

mammalian Sac3 phosphatase (gene symbol, Fig. 7.4) and the demonstration that endogenous Sac3 also coimmunoprecipitates with PIKfyve as well as with ArPIKfyve (Sbrissa et al. 2007). Subsequent data for a PIKfyve efficient binding only to preformed ArPIKfyve-Sac3 heterodimers, but not to the individual proteins, led to the idea that the three proteins form a ternary complex, named the PAS complex (for PIKfyve-ArPIKfyve-Sac3) to orchestrate the coordinated regulation of PtdIns(3,5)P_2 synthesis and turnover (Sbrissa et al. 2008). The interactions between the three proteins within this complex appear to be quite stable (sustain washes with strong detergents), related to the ability of ArPIKfyve to homooligomerize that scaffolds the complex. ArPIKfyve also fulfills stabilizing functions in the PIKfyve-independent binary ArPIKfyve-Sac3 assembly. Thus, under cellular depletion of ArPIKfyve, the Sac3 protein levels are profoundly reduced as a result of rapid Sac3 degradation by the proteasome pathway (Sbrissa et al. 2008; Ikonomov et al. 2010). Based on these observations we concluded that Sac3 is unlikely to exist in cells without ArPIKfyve (Ikonomov et al. 2010). Recent data for the absence of immunoreactive Sac3 in ArPIKfyve$^{-/-}$ fibroblasts (Lenk et al. 2011) have confirmed this conclusion.

Insights into the protein interactions within the PAS complex are provided based on triple co-immunoprecipitation analyses with a battery of truncated or point mutants of the three proteins (Ikonomov et al. 2009b). The results are consistent with a model whereby the ArPIKfyve C-terminal coiled-coil motifs play a prominent role in organizing and maintaining the integrity of the PAS complex (Fig. 7.3). The observations that a C-terminal peptide of the ArPIKfyve homodimerization sites not only disassembles the PAS complex but also reduces

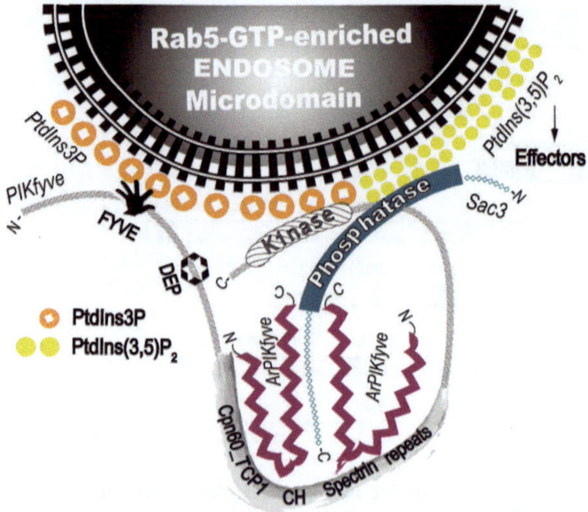

Fig. 7.3 *A model of the interacting domains in the PAS complex and membrane recruitment*. The PtdIns(3)P–PtdIns(3,5)P_2 interconversion on mammalian endosomal membranes is achieved through a multiprotein complex constituted by the kinase PIKfyve, the PtdIns(3,5)P_2-specific phosphatase Sac3, and the scaffolding regulator ArPIKfyve. The complex assembles on PtdIns3P-enriched early endosome microdomains formed by recruited PI3Ks on active GTP-bound Rab5. The PtdIns3P-binding FYVE domain in PIKfyve promotes the interaction. The conserved domains in the PIKfyve middle region (Cpn60_TCP1 and the CHK/Spectrin homology domains) interact with the preformed ArPIKfyve-Sac3 subcomplex, organized via the C-terminal-coiled-coil motifs of an ArPIKfyve homodimer (or higher-order homooligomer) and the C-terminus of a Sac3 monomer. The enzymes act in concert to coordinate dynamic endosome remodeling, presumably through regulated conformation changes within the complex. Modified from Ikonomov et al. 2009b

PtdIns(3,5)P)$_2$/PtdIns5P production and attenuates PIKfyve functionality indicate that PAS complex formation is required to activate PIKfyve-catalyzed PtdIns(3,5)P_2 and PtdIns5P production (Sbrissa et al. 2008). Thus, in this complex the Sac3 phosphatase fulfills an unusual role of activating the PIKfyve kinase by allowing the PIKfyve–ArPIKfyve interaction to occur. The findings for severely reduced PtdIns(3,5)P_2 in pale-tremor mice that are null for Sac3 (Chow et al. 2007) are thus consistent with disrupted ability of PIKfyve to produce proper PtdIns(3,5)P_2 amounts in the absence of Sac3 and impeded formation of the regulatory PAS core complex.

Given the profound decline (by ~70 %) in steady-state levels of PtdIns(3,5)P_2 in pale-tremor mice (Chow et al. 2007), one wonders whether Sac3 functions as a phosphatase or it is only engaged in the PIKfyve activation. Data demonstrating a subtle rise in the HPLC-measurable PtdIns(3,5)P_2 in several mammalian cell types upon Sac3 protein knockdown indicate that Sac3 is indeed an active phosphatase and suggest that at lower levels, it promotes PtdIns(3,5)P_2 synthesis over turnover (Sbrissa et al. 2007; Ikonomov et al. 2009c). Remarkably, it appears that Sac3 retains its phosphatase activity within the PAS complex. Thus, the cell vacuolation

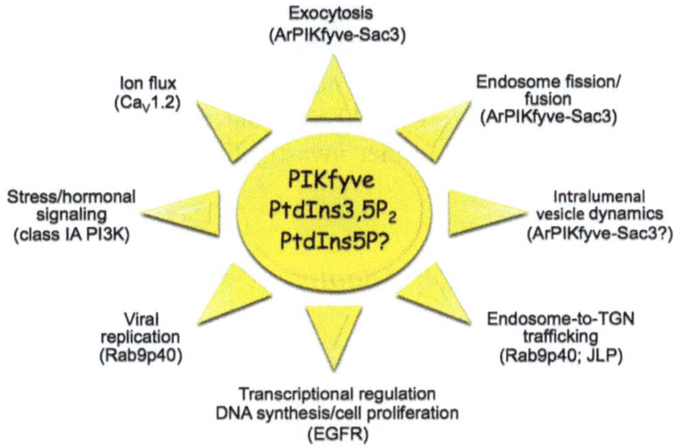

Fig. 7.4 *Functions of PIKfyve and its products PtdIns(3,5)P$_2$ and PtdIns5P*. Depicted are the main functions of PIKfyve discussed in this chapter. Indicated are the protein partners that have been shown to physically interact with PIKfyve and contribute to function (in parenthesis). It is still unknown whether all listed PIKfyve functions are mediated by PtdIns(3,5)P$_2$ or whether some may require PtdIns5P

capacity of dominant-negative kinase-deficient PIKfyve^{K1831E} is dramatically attenuated under coexpression of phosphatase-deficient Sac3^{D488A} and ArPIKfyve but profoundly exacerbated if PIKfyve^{K1831E} is coexpressed with the active Sac3^{WT} phosphatase and ArPIKfyve (Ikonomov et al. 2009b). These findings suggest that the PAS core complex relays two antagonistic inputs, one for PtdIns(3,5)P$_2$ synthesis and another for turnover, consistent with the requirement of a tight PtdIns(3,5)P$_2$ homeostatic control (Sbrissa et al. 2008; Ikonomov et al. 2009b). It is conceivable that such a phosphorylation-dephosphorylation cycle elevates the PtdIns(3,5)P$_2$ flux, resulting in reduced diffusion and enhanced local availability (Vicinanza et al. 2008). How the two opposing activities are coordinated and integrated in time within a single complex is currently a mystery. In addition, experimental evidence is provided to indicate that the Sac3 phosphatase is also active within the ArPIKfyve-Sac3 subcomplex (Ikonomov et al. 2010). The relative proportion of Sac3 hydrolyzing activity in the binary versus the ternary PAS complex is yet to be resolved.

Whether Sac3 hydrolyzes PtdIns5P along with PtdIns(3,5)P$_2$ in mammalian cell contexts is currently unclear. Indeed, Sac3 belongs to the group of relatively non-specific evolutionarily conserved Sac1 homology-containing inositol polyphosphate phosphatases (Fig. 7.2) able to hydrolyze phosphate from any of the D-3, -4 or -5 positions in inositol in in vitro assays (Hughes et al. 2000; Blagoveshchenskaya and Mayinger 2009; Majerus and York 2009). Although recombinant Fig4, the yeast Sac3 ortholog, hydrolyzes exclusively the D-5 phosphate in PtdIns(3,5)P$_2$ (Rudge et al. 2004), such specificity is not reproduced with human and rat Sac3 proteins (Sbrissa et al. 2007; Yuan et al. 2007). Rather, both are more promiscuous in in vitro assays, hydrolyzing PtdIns4P, Ptdns3P,

PtdIns5P, and each of the higher 5′-phosphorylated PIs (Sbrissa et al. 2007; Yuan et al. 2007). The PtdIns5P levels in the HPLC-resolved PI profiles under Sac3-null or protein knockdown conditions have not been determined (Chow et al. 2007; Sbrissa et al. 2007; Ikonomov et al. 2009c), and thus, possible Sac3-dependent hydrolysis of PtdIns5P requires further investigation.

Rab9 Effector p40 and JLP Adapter

Yeast two-hybrid screens with the conserved Cpn60_TCP1 domain of PIKfyve have identified several potential partners, two of which are already characterized in detail. These are the Rab9 effector p40 and the motor protein adapter JLP, both implicated in the endosome-to-TGN transport (Ikonomov et al. 2003b; Ikonomov et al. 2009a).

Rab9 effector p40 (gene symbol, RABEPK), a 372-residue protein, comprises six *kelch* repeats (Fig. 7.2), with the C-terminal four found relevant in the interaction with PIKfyve as determined by pull-down assays with the recombinant proteins (Ikonomov et al. 2003b). Because the PIKfyve-Rab9p40 interaction is not detected with the endogenous proteins, it is either of low affinity or requires induced post-translational modifications. The Rab9p40 structure, thought to fold into a single blade of a β-propeller, is highly reminiscent of the WD40-repeat-based seven-bladed β-propeller in yeast Atg18 and mammalian orthologous WIPI proteins that bind to and function as downstream effectors of PtdIns(3,5)P_2 (Hudson and Cooley 2008; Dove et al. 2009). Despite the structural similarity and presence of basic-residue clusters to support PI-binding, Rab9p40 appears to not interact with PtdIns(3,5)P_2 (Sbrissa et al. 2005). Rather, it might function as an effector of the PIKfyve protein kinase as concluded by data for Rab9p40 in vitro phosphorylation and membrane localization occurring in a manner dependent on active PIKfyve enzyme (Ikonomov et al. 2003b). Interestingly, in contrast to inhibited PIKfyve lipid kinase, the PIKfyve protein kinase activity remains unaffected by ArPIKfyve depletion from the PAS complex, achieved through detergent stripping or siRNA (Ikonomov et al. 2007; Ikonomov et al. 2009b). Thus, if the PIKfyve protein kinase controls a physiologically relevant phosphorylation of Rab9p40 in the context of endosome-to-TGN transport, this would occur independently of the PAS complex.

JLP (gene symbol, SPAG9) belongs to the group of scaffolds of the JNK/p38 MAP kinase signaling molecules, which also function as adapters of kinesin motor proteins, promoting microtubule-dependent membrane trafficking (Schnapp 2003; Dhanasekaran et al. 2007). JLP is a 1307 residue protein that harbors several coiled-coil domains and tetratricopeptide repeats (TPRs) (Fig. 7.2), both known to facilitate assembly of multiprotein complexes (D'Andrea and Regan 2003; Strauss and Keller 2008). The interaction with the PIKfyve Cpn60_TCP1 domain engages the JLP-C-terminus that harbors three out of the six TPRs as determined by pull-down assays with recombinant proteins (Ikonomov et al. 2009a). Endogenous PIKfyve and JLP associate only under milder conditions, indicative of low affinity interactions. Neither ArPIKfyve nor Sac3 coimmunoprecipitates with JLP indicating that PIKfyve interacts with JLP independently of the PAS complex. Inhibitory peptides from PIKfyve-JLP contact sites delay the endosome-to-TGN transport of furin (microtubule-dependent) but not that of TGN38 (microtubule-independent), suggesting that JLP links PIKfyve with motor-driven endosome processing/TGN trafficking on microtubules (Ikonomov et al. 2009a).

EGFR and Class IA PI3K

The epidermal growth factor receptor tyrosine kinase (gene symbol, EGFR) plays a prominent role in malignant transformation of many organs. An interaction between EGFR (Fig. 7.2) and PIKfyve has been identified by mass spectrometry of EGFR immunoprecipitates from human bladder cancer cells (Kim et al. 2007). The association is enhanced ~10-fold upon EGFR stimulation by heparin-binding EGF-like growth factor (HB-EGF), an autocrine urothelial cell mitogen, up-regulated in bladder cancers and accumulated in the nucleus. EGFR appears to be routed to the nucleus in a manner dependent on PIKfyve activity as evidenced by inhibited EGFR nuclear translocation in the presence of PIKfyve protein knock-down or kinase-deficient PIKfyve^{K1831E} whose EGFR association is intact (Kim et al. 2007). Further molecular details and plausible PIKfyve nuclear localization upon HB-EGF stimulation remain to be seen.

Interaction of PIKfyve with class IA PI3K (gene symbol, PIK3C/PIK3R) has also been suggested based on p85/p110 co-immunoprecipitation with PIKfyve (Sbrissa et al. 2001). Concordantly, PIKfyve immunoprecipitates from insulin-stimulated 3T3L1 adipocytes make more PtdIns3P and PtdIns(3,5)P_2 in vitro in a wortmannin-sensitive manner, consistent with the idea that the class IA PI3K function upstream of PIKfyve (Sbrissa et al. 1999, 2001). It is conceivable that subpopulations of the two kinases form a physiologically relevant complex as one mechanism to channel the PtdIns3P substrate for PIKfyve-catalyzed PtdIns(3,5)P_2 production. Alternatively or additionally, class IA PI3K might use PIKfyve-synthesized PtdIns5P to produce PtdIns(3,5)P_2 (Fig. 7.1b), a potential pathway that might be relevant in certain cell types or stimulations and requires further research. The interacting regions in PIKfyve and p85 (or p110) are yet to be determined.

$Ca_V1.2$

$Ca_V1.2$ (gene symbol, CACNA1C), an L-type voltage-gated Ca^{2+}-channel, is a main port for Ca^{2+} entry into the brain (Tsien et al. 1991). It is identified as a PIKfyve partner by mass spectrometry of neuroblastoma cell proteins pulled down with GST-fusions of the intracellular $Ca_V1.2$ domain (Tsuruta et al. 2009). The interaction engages the PIKfyve C-terminal half and a conserved peptide stretch spanning residues 1709–1808 of the $Ca_V1.2$ cytosolic region (Fig. 7.2). Reportedly, the two proteins associate only after sustained neuronal activation of the glutamate receptors, which promotes PIKfyve-assisted lysosome targeting and degradation of $Ca_V1.2$ to prevent excitotoxic neuronal death. Intriguingly, transport activity/trafficking of several other cell-surface or intracellular Ca^{2+} channels as well as that of other ion channels are reportedly regulated by PIKfyve (see Ion Flux in the Control of Inter- and Intra-Endosomal PIKfyve Functions) but physical interactions with PIKfyve have not yet been established.

Cell Biological Processes Controlled by PIKfyve and Lipid Products

Membrane Homeostasis in the Endosomal System

The first evidence that PIKfyve activity controls endomembrane homeostasis came from our observations in 2001 for massive endomembrane vacuolation induced by expression of the kinase-deficient $PIKfyve^{K1831E}$ mutant in several mammalian cell types, which could be rescued by cell delivery of $PIKfyve^{WT}$ adenovirus (Ikonomov et al. 2001). That the aberrant morphology is associated with abrogated PtdIns(3,5)P_2 rather than PtdIns5P production is verified by PIKfyve point mutants selectively deficient in either PtdIns(3,5)P_2 or PtdIns5P synthesis as well as by rescue experiments with the individual lipids (Ikonomov et al. 2002a). The membrane enlargement first arises at the EEA1 and Rab5-positive early endosomes and then gradually develops to culminate with the appearance of heterogeneous populations of large translucent vacuoles of hybrid origin throughout the cytoplasm (Ikonomov et al. 2006). The aberrant vacuolation phenotype was subsequently reproduced by other maneuvers of PIKfyve suppression, such as siRNA-mediated protein depletion and pharmacological inhibition of the lipid kinase activity (Rutherford et al. 2006; Jefferies et al. 2008; Osborne et al. 2008; de Lartigue et al. 2009). Morphological data in different cell types under PIKfyve dysfunction triggered by any of these three means concur that membrane swelling affects heterogeneous populations of vesicles of early endosomes, late endosomes or both. The nature of the swollen endocytic compartments varies among different studies, with the variability likely related to the progressive nature of the morphological aberrations, the cell type, and the treatments. An additional source of discrepancy worth considering is levels of active Sac3. As outlined above, PtdIns(3,5)P_2 production is controlled by the PAS complex, where Sac3 relays opposing activities for PtdIns(3,5)P_2 synthesis and turnover (Sbrissa et al. 2008; Ikonomov et al. 2009b). High levels of active Sac3 in this complex exacerbate, whereas low levels ease the aberrant morphology (Ikonomov et al. 2009b). These observations might explain the more striking aberrant vacuolation phenotype and robust PtdIns(3,5)P_2 reduction under PIKfyve pharmacological inhibition vs.

siRNA-mediated PIKfyve depletion (Sbrissa and Shisheva 2005; Rutherford et al. 2006; Jefferies et al. 2008). Thus, under the former treatment, active Sac3 is retained in the PAS complex, whereas under the latter, it is concomitantly eliminated.

Inter-Endosomal Dynamics

The cellular mechanisms that trigger early endosome enlargement and subsequent endomembrane vacuolation under PtdIns(3,5)P_2 deficiency are poorly defined. Some clues, however, have started to emerge. The excess membrane could be provided by a number of mechanisms, one being the deregulated balance between fission and fusion in the endosomal system, viewed as key events in the formation of a dynamic network of endosomal structures in the course of endocytic transport progression (Murray and Wolkoff 2003; Gruenberg and Stenmark 2004; Rink et al. 2005; Lindmo and Stenmark 2006). Two studies have directly addressed the role of PtdIns(3,5)P_2 in early endosome plasticity by exploring in vitro assays reconstituting fission or fusion events in the endosomal system in the presence of cytosols with depleted or enhanced levels of PIKfyve, ArPIKfyve, and/or Sac3 (Ikonomov et al. 2006; Sbrissa et al. 2007). These studies allow two important conclusions: first, the biogenesis/fission (or maturation) of transport intermediates from early endosomes (referred to as endosomal carrier vesicles/multivesicular bodies, ECV/MVBs), thought to underlie the endocytic traffic progression, is enhanced by membrane PtdIns(3,5)P_2 production (Sbrissa et al. 2007) and, second, the homotypic and/or heterotypic fusion efficiency within the early/late endosome compartments is negatively regulated by PtdIns(3,5)P_2 (Ikonomov et al. 2006). These observations are consistent with dynamic PtdIns(3,5)P_2 synthesis and turnover controlled by PIKfyve and Sac3 enzymes being associated with fission and fusion events, respectively, in the endosomal system. It is conceivable that ill-matured ECV/MVBs/late endosomes would be incompetent to properly fuse with lysosomes. Thus, if PIKfyve and PtdIns(3,5)P_2 affect late endosome-lysosome fusion as claimed by others (Tsuruta et al. 2009), this is most likely secondary to the ECV/MVBs formation/maturation, a notion supported by data that lysosomes are devoid of PIKfyve even when the latter is greatly overexpressed (Cabezas et al. 2006; Ikonomov et al. 2006; Rutherford et al. 2006).

Intra-Endosomal Dynamics

Intralumenal vesicles (ILVs) are detached membranous structures that form from the limiting membrane of endosomes by inward budding and fission, a process typically (but not necessarily) associated with cargo sorting in the lysosomal pathway for degradation (Ikonomov et al. 2003a; van der Goot and Gruenberg 2006; Woodman and Futter 2008; Falguieres et al. 2009; Stuffers et al. 2009). In fact, ILVs within MVBs represent a quite heterogeneous population of vesicles, with some, involved in transport of signaling receptors to lysosomes, whereas others, undergoing selective fusion with the limiting membrane, a process known as back-fusion, that returns ILV proteins and lipids to the limiting membrane for further export to other cellular destinations (van der Goot and Gruenberg 2006; Falguieres et al. 2009). ILVs appear to undergo a constant cycle of fission and back-fusion at the limiting membrane, forming a highly dynamic membrane network within the late endosome lumen. Ultrastructural data in an HEK293 cell line stably expressing dominant-negative kinase-deficient PIKfyve^{K1831E}, demonstrating profoundly dilated MVB structures containing a significantly lower number of ILVs, are consistent with PIKfyve regulating ILV dynamics (Ikonomov et al. 2003a). Concordantly, fewer ILVs within profoundly dilated endosomes and lysosomes are also observed upon Fab1/PIKfyve inactivation in *S. cerevisiae* and *Drosophila* (Gary et al. 1998; Rusten et al. 2006). This apparent inability of MVBs to store ILVs under PIKfyve perturbation could be viewed as an additional mechanism contributing to membrane expansion of dilated MVBs. However, whether PIKfyve dysfunction reduces ILV fission or accelerates ILV back-fusion is yet to be determined.

Ion Flux in the Control of Inter- and Intra-Endosomal PIKfyve Functions

Ion flux into and out of endolysosomal compartments has long been considered to be a direct determinant of both inter- and intra-endosomal operations, but our mechanistic knowledge of its functionality is only fragmented (Pryor and Luzio 2009; Scott and Gruenberg 2011). A recent study implicates TRPML1, a member of the mucolipin transient receptor potential Ca^{2+} channels localized on late endosomes/lysosomes, as a candidate downstream effector of PtdIns(3,5)P_2-regulated plasticity of endo-lysosomal compartments (Dong et al. 2010). Thus, enlarged cytoplasmic inclusion bodies are formed in cells lacking TRPML1 although they are less severe than those under PIKfyve dysfunction. Ectopic expression of TRPML1 in ArPIKfyve$^{-/-}$ mouse fibroblasts with reduced PtdIns(3,5)P_2/PtdIns5P production suppresses the vacuolation phenotype. TRPML1 binds PtdIns(3,5)P_2 in vitro and the PtdIns(3,5)P_2 delivery into cells accelerates the TRPML1-dependent Ca^{2+} release (Dong et al. 2010). Although how membrane TRPML1 and PtdIns(3,5)P_2 meet one another remains unclear, these data link the PAS complex and PtdIns(3,5)P_2 -regulated inter-endososomal dynamics with intracellular Ca^{2+} homeostasis.

Besides TRPML1 and $Ca_V1.2$, discussed above, PIKfyve and/or PtdIns(3,5)P_2 are implicated in trafficking and/or transport activity of other Ca^{2+} channels distributed to various organelles, promoting Ca^{2+} entry and/or intracellular release. These include the cell surface TRPV6 and the ryanodine receptors, high-conductance Ca^{2+} channels localized on endoplasmic reticulum (Shen et al. 2009; Sopjani et al. 2010; Touchberry et al. 2010; Silswal et al. 2011). In addition, PIKfyve appears to also affect channels that control Cl^- and K^+ influx (Seebohm et al. 2007; Gehring et al. 2009a; Klaus et al. 2009b). Whereas mechanistic insights are yet to be provided, it is clear that, as in the case with other PIs (Balla 2006), PtdIns(3,5)P_2 and PtdIns5P regulate several ion fluxes that in turn, could have a major impact on endosome plasticity. PIKfyve-dependent regulation of inter- and intra-endosomal dynamics could also be linked with the endosome acidification, determined by the vacuolar type ATPase and the flux of ions other than protons. Indeed, both *Drosophila* and *C. elegans* PIKfyve null models exhibit defective acidification (Nicot et al. 2006; Rusten et al. 2006) but the situation in mammalian cells is still elusive. Thus, PIKfyve pharmacological inhibition appears to insignificantly affect the endosome acidification in several mammalian cell types (Jefferies et al. 2008; de Lartigue et al. 2009); the knock-down or knockout conditions are yet to be explored.

PIKfyve functioning in endosome processing, outlined above, suggests that the numerous defects in cellular trafficking pathways observed under PIKfyve dysfunction may be a consequence of perturbed inter- and intra-endosomal dynamics. In addition, PtdIns(3,5)P_2 deficiency and abnormal endomembrane vacuolation are often difficult to uncouple and, thus, at least some PIKfyve effects could be secondary to the defective morphology. With these inherent limitations of the data interpretation, in the subsequent sections I discuss in detail the multitude of cell biological processes shown to be dependent on PIKfyve and/or its lipid products (Fig. 7.4).

Endocytic Traffic Progression to Lysosomes

Endocytosis of Membrane and Fluid-Phase Cargoes

Endocytosis is typically divided into early events that include budding and internalization of vesicles from the cell surface, often resulting in recycling of endocytosed material to the plasma membrane, and later degradative events that involve transport of internalized materials through the endolysosomal system (Pryor and Luzio 2009). Receptor internalization, recycling, and degradation in the lysosomes are uninterrupted by PIKfyve suppression through dominant-negative mutation or PIKfyve protein knockdown (Ikonomov et al. 2003a; Rutherford et al. 2006; de Lartigue et al. 2009). More sensitive appears to be the lysosomal degradation of the $Ca_V1.2$ channels that physically associate with PIKfyve (see $Ca_V1.2$). For example, PIKfyve protein depletion delays glutamate-induced $Ca_V1.2$ traffic through and degradation in endo-lysosomal compartments, resulting in increased Ca^{2+} influx and neuronal susceptibility to excitotoxicity (Tsuruta et al. 2009). The fluid-phase endocytosis appears to proceed normally at the initial stage of internalization but its progression through the degradative endocytic arm is markedly impaired as revealed by biochemical evaluation of HRP uptake in a PIKfyve^{K1831E} expressing HEK293 cell line (Ikonomov et al. 2003a). Concordantly, microscopy analyses in other mammalian cell types with PIKfyve suppression, indicating accumulation of a number of fluid-phase markers around the limiting membrane of the dilated vacuoles (Ikonomov et al. 2003a; Jefferies et al. 2008; de Lartigue et al. 2009), are consistent with a post-endosomal defect in the endocytic traffic progression of solutes. The observations that PIKfyve is critical in fluid-phase post-endosomal traffic might reflect greater dependence on requirements for PtdIns(3,5)P_2 and/or PtdIns5P.

Autophagy and Macropinocytosis

In autophagy, a process used by higher eukaryotes to turn over organelles, protein aggregates, and cytoplasmic constituents, the double-membraned autophagosome fuses with endosomes or lysosomes to form structures known as amphisomes and autolysosomes (Chen and Klionsky 2011). Whereas autophagy in yeast seems to proceed independently of Fab1, in *Drosophila* intact PIKfyve/Fab1 activity is required for progression of amphisomes to degradative autolysosomes (Rusten et al. 2007; Dove et al. 2009). That PIKfyve may regulate mammalian autophagy is suggested by data demonstrating a mild accumulation of markers for autophagosomal structures upon PIKfyve pharmacological inhibition in several cell types (Jefferies et al. 2008; de Lartigue et al. 2009). More specific means for PIKfyve suppression is yet to be explored. Two groups have also examined the autophagic activity in tissues and cells from the pale tremor mice that are Sac3-null but reached contradictory conclusions (Ferguson et al. 2009; Katona et al. 2011) with yet-to-be determined causes of discrepancy. Thus, whereas the PIKfyve pathway may generate molecular signals for autophagosome maturation and progression of autophagy, its contribution is still unclear.

Numerous infectious pathogens exploit macropinocytosis to invade host cells. PIKfyve appears to be required in this process as exemplified by arrested replication and intracellular survival of *Salmonella* upon PIKfyve cell suppression by various means. Under these conditions, *Salmonella*-containing vacuoles fail to mature and fuse with late endosomes and/or lysosomes (Kerr et al. 2010). It is conceivable that impaired ECV/MVB formation/maturation due to PIKfyve/PtdIns(3,5)P_2 dysfunction may result in endocytic compartments that are incapacitated to maintain the *Salmonella* cycle.

Trafficking out of the Endolysosomal System

To the Trans-Golgi network

A set of internalized transmembrane proteins, including intracellular sorting receptors, enzymes, and toxins, exits the endosome compartments and are transported to the TGN through the retrograde route (Bonifacino and Rojas 2006; Johannes and Popoff 2008). The endosome-to-TGN trafficking could occur from early, recycling, or late endosomes, following roads characterized by different sets of machinery. Several cargoes, including the acid-hydrolase-sorting receptor CI-MPR, the transmembrane protein TGN38 or Shiga toxin, traffic from early endosomes in a manner that requires the evolutionarily conserved proteins of the retromer complex serving as an endosomal coat. Conversely, the late endosome-to-TGN trafficking, the best-characterized cargo being the endopeptidase furin, is retromer-independent but requires intact microtubules and the Rab9GTPase (Chia et al. 2011). Several studies exploring all three means of PIKfyve suppression, i.e., dominant interfering mutation, RNA interference or acute pharmacological inhibition, implicate PIKfyve in supporting late endosome-to-TGN trafficking of furin (Rutherford et al. 2006; de Lartigue et al. 2009; Ikonomov et al. 2009a). Our lab has further demonstrated that microtubule-dependent furin trafficking also requires physical interaction between PIKfyve and the kinesin adaptor JLP (Ikonomov et al. 2009a). For instance, inhibitory peptides disrupting the JLP association with either PIKfyve or kinesin profoundly delay the furin arrival at the TGN. Importantly, furin trafficking to the TGN is also arrested by dominant-negative kinase-deficient $PIKfyve^{K1831E}$ (Ikonomov et al. 2009a). These data are consistent with the idea that PIKfyve enzymatic products are required but whether they support the association of PIKfyve with JLP or function independently within the same segment of the endosome processing step is unknown. Consistent with the reported requirement for Rab9 in the retromer-independent retrograde transport (Chia et al. 2011), PIKfyve is found to bind the Rab9 effector p40 (see Rab9 Effector p40 and JLP Adapter). How the interaction of PIKfyve with Rab9p40 is coordinated with that with JLP and coupled to microtubule tracks remains to be elucidated. Unraveling the mechanistic link between PIKfyve, Rab9p40, and JLP becomes particularly significant in the light of observations that the retroviral replication and release of mature viral particles is dependent on intact endosome-to-TGN trafficking machinery, and proper Rab9, Rab9p40, PIKfyve and TIP47 functioning (Murray et al. 2005; Jefferies et al. 2008).

Several studies have examined the role of PIKfyve in retromer-dependent trafficking but reached contradictory conclusions. Thus, under PIKfyve protein depletion, the arrival of CI-MPR to the TGN is delayed but that of TGN38 appears unaffected (Rutherford et al. 2006; Ikonomov et al. 2009a). Under acute PIKfyve inhibition, the retrograde trafficking of Shiga toxin but not that of CI-MPR is affected (Rutherford et al. 2006; de Lartigue et al. 2009). In addition to different cell types and means of PIKfyve suppression (chronic vs. acute) causing the discrepancies, it is also worth considering the different transport requirements for Rab9 effector p40. Reportedly, the latter supports late endosome-to-TGN trafficking of CI-MPR, which may be preferentially sensitive to PIKfyve protein depletion rather than lipid kinase inhibition (Diaz et al. 1997; Ikonomov et al. 2003b). Consistent with this, knockdown of the retromer subunits in several mammalian cell types does not induce cytoplasmic vacuolation, suggesting that the retromer-dependent endosome-to-TGN-retrieval is either a minor aspect of PIKfyve functionality or proceeds independently of the PIKfyve-catalyzed $PtdIns(3,5)P_2$ production (Shisheva 2008b). Concordantly, the yeast sorting receptor Vps10 that functions in this pathway is unaffected in *Δfab1* yeast strain (Dove et al. 2009).

To the Plasma Membrane

Regulated exocytosis of specialized vesicles, containing transporters, ion channels, or neurotransmitters, relies on a well-orchestrated series of sorting events in the endosomal system, leading to vesicle outward trafficking and fusion with the plasma membrane; cargo is thereby released or inserted into the plasma membrane. The role of PIKfyve in exocytosis has been first recognized in the context of the cell-surface translocation of GLUT4 (gene symbol, SLC2A4), the insulin-regulated fat/muscle-specific glucose transporter that is responsible for postprandial glucose uptake. GLUT4 is in a constant dynamic cycle between the plasma membrane and several intracellular compartments (Huang and Czech 2007; Foley et al. 2011). Under basal conditions, the TGN and endosomal GLUT4 pools are sequestered into a specialized post-endosomal GLUT4 storage compartment (GSV) that is mobilized to the plasma membrane in response to insulin. Following insulin challenge, cell-surface GLUT4 is quickly internalized via the early endosomal system. Intriguingly, PIKfyve suppression by negative-interfering mutation, protein depletion, or pharmacological inhibition reduces the insulin response of GLUT4 cell-surface translocation in 3T3L1 adipocytes (Ikonomov et al. 2002b; 2007; 2009d). This is paralleled by decreased glucose entry. Several lines of evidence are consistent with the role of the PAS complex in these effects. For example, combined loss of PIKfyve and ArPIKfyve proteins leads to greater inhibition of insulin-stimulated glucose uptake, correlating with a greater reduction in the intracellular PtdIns(3,5)P_2 pool (Ikonomov et al. 2007). Similarly, perturbation of the PAS complex integrity by ArPIKfyve inhibitory peptides arrests GLUT4 plasma membrane translocation (Sbrissa et al. 2008). Conversely, partial depletion of the Sac3 protein (to $\sim$40 % of control) or expression of a phosphatase-inactive point mutant enhances insulin-dependent gain in surface GLUT4 and glucose uptake (Ikonomov et al. 2009c). Concordantly, acute insulin action is associated with robust increases in intracellular levels of PtdIns(3,5)P_2 and PtdIns5P through a mechanism that involves both membrane recruitment of cytosolic subpopulations of PAS complexes and concomitant inhibition of the Sac3 phosphatase activity, presumably, through conformational changes resulting from oxidation of the catalytic Cys residue (Shisheva et al. 2001; Sbrissa et al. 2004b; Ikonomov et al. 2009c). These data indicate that the PAS complex, PtdIns(3,5)P_2 and/or PtdIns5P are positive regulators of GLUT4-mediated glucose entry. Although the precise cellular mechanism is not yet known, we maintain the view that robust PIKfyve-catalyzed PtdIns(3,5)P_2/PtdIns5P production is required in GLUT4 vesicle budding/fission from early endosomes *en route* to GSVs for replenishing the readily releasable and depleting GSV pool during and after insulin challenge (Shisheva 2008a). Strikingly, the key importance of PIKfyve as a physiological regulator in whole-body glucose homeostasis has recently emerged by observations for peripheral insulin resistance in mice with *pikfyve* haploinsufficiency or muscle-specific knockout (Ikonomov et al. 2011; Ikonomov, Sbrissa and Shisheva, in preparation).

In a series of reports exploring the *Xenopus* oocyte heterologous coexpression system, Lang and co-workers have shown that PIKfyve facilitates the exocytosis and cell-surface activity of a plethora of ion channels and excitatory amino acid or metabolite transporters. These include the voltage-gated K + channel KCNQ1/KCNE1, Cl^- channels ClC-2, and cystic fibrosis transmembrane conductance regulator, transient receptor potential Ca^{2+} channel TRPV6, Na +/glucose co-transporter SLGT-1, and the EAAT2-4 glutamate, and SLC6A8 creatine transporters (Seebohm et al. 2007; Shojaiefard et al. 2007; Strutz-Seebohm et al. 2007; Gehring et al. 2009a, b; Klaus et al. 2009b; Alesutan et al. 2010; Sopjani et al. 2010). All these transport systems appear to be activated by the serum- and glucocorticoid-induced serine/threonine kinase 1 (SGK1), whose expression and activity is controlled by a number of extracellular stimuli, including cell shrinkage, insulin, and other growth factors, just to name a few. Ser318 in PIKfyve is considered to be a selective target for the SGK1-dependent phosphorylation, which increases the cell surface abundance and/or activity of all channels/transporters listed above. How the Ser318 phosphorylation affects PIKfyve activity and functionality in vivo is unclear. In one of the studies, the authors claim that phosphorylation-deficient $PIKfyve^{S318A}$ compromises the intracellular supply of $PtdIns(3,5)P_2$ (Alesutan et al. 2010). Whereas this conclusion might be correct, it stems solely from the usage of anti-$PtdIns(3,5)P_2$ antibody and thus requires validation by quantitative and specific HPLC analyses. Therefore, whether exocytosis of all these cargoes is potentiated by the PIKfyve-catalyzed supply of $PtdIns(3,5)P_2$, PtdIns5P, both or none, is yet to be demonstrated. Interestingly, based on a comparative genomic approach of PI enzymes and regulators, PtdIns5P is postulated to play a role in exocytosis (Lecompte et al. 2008). Consistent with such a function, PtdIns5P is implicated in filamentous actin remodeling (Niebuhr et al. 2002; Sbrissa et al. 2004b), an event required for proper exocytosis of many cargoes. A more direct link between the PIKfyve-synthesized PtdIns5P pool and complex actin remodeling in the course of exocytic cargo delivery to the cell surface remains to be shown.

Contrary to the above observations implicating PIKfyve as a positive regulator in exocytosis, one study finds PIKfyve catalytic activity to negatively affect Ca^{2+}-dependent secretion (Osborne et al. 2008). Thus, PIKfyve suppression potentiates depolarization-induced secretion of catecholamine and hGH in bovine chromaffin or rat neuroendocrine PC12 cells, respectively. Recent findings for the role of PIK3C2 in promoting secretion of insulin granules in pancreatic β-cells may provide some mechanistic insights into the data in neurosecretory cells (Dominguez et al. 2011). Thus, clearance of this PIK3C2-controlled PtdIns3P pool by PIKfyve may impinge on exocytosis, a putative mechanism that might be significant in cells where cargo secretion is executed by large dense-core granules. Importantly, ectopically expressed GLUT4 or endogenous IRAP, a GLUT4-vesicle reporter protein, are both excluded from the secretory granules in either

PC12 or pancreatic β-cells (Herman et al. 1994; Thorens and Roth 1996; Thoidis and Kandror 2001), an observation that sheds mechanistic insights into the positive role of PIKfyve in GLUT4 exocytosis.

Nuclear Trafficking, Transcriptional Regulation and Cell Cycle Progression

Two recent studies in mammals and yeast implicate the function of PIKfyve and Fab1 in transmitting signaling into the nucleus to regulate gene transcription. Thus, in mammalian cells, receptor tyrosine kinases may transit to the nucleus where they regulate gene expression (Carpenter and Liao 2009). Intriguingly, PIKfyve appears to assist the nuclear trafficking of EGFR, as shown in human bladder cancer cells abundant in HB-EGF (Kim et al. 2007). Cell stimulation with HB-EGF enhances the EGFR association with PIKfyve (see EGFR and Class IA PI3K), which activates receptor nuclear trafficking, binding to the cyclin D1 promoter, and cell cycle progression. Conversely, PIKfyve depletion inhibits EGFR nuclear localization and chromatin association, resulting in cell cycle arrest at the Go-G1 phase. The study in yeast provides evidence for highly specific interaction of the endosomal PtdIns(3,5)P_2 pool with two novel factors, Tup1 and Cti6, which control multiple transcriptional regulators, determining an activated or repressed chromatin state of numerous genes (Han and Emr 2011). It is unclear whether PIKfyve and/or Fab1 enter the nucleus. As outlined above, the nuclear PtdIns5P pool regulates gene expression and cell cycle progression in both mammalian and plant cells, but the contribution of PIKfyve to this pool is questionable.

PIKfyve's role in mitogenesis has been suggested by earlier observations for inhibition of insulin/serum-induced DNA synthesis in a HEK293 cell line stably expressing PIKfyve^{K1831E}, as determined by BrdU or [^{3}H]thymidine incorporation (Ikonomov et al. 2002b). Further evidence supporting this function is provided by similar analyses in *pikfyve*-null MEFs derived from recently generated PIKfyve KO mice (Ikonomov et al. 2011) (see below). These data from mammalian cells corroborate the pioneering observation for impaired nuclear division in the *Δfab1* yeast strain, characterized by an aploid and binucleate phenotype, hence the name (Yamamoto et al. 1995). However, in neither case could a secondary defect on cell division due to massive vacuolation be ruled out. On the other hand, a recent report failed to relate PIKfyve protein depletion to suppression of cell proliferation in an NPM-ALK transformed cell line that exhibits a high content of PIKfyve-related PtdIns5P (Dupuis-Coronas et al. 2011). Thus, PIKfyve's role in cell cycle progression will require further clarification.

Physiological Consequences of Disrupted PIKfyve and PtdIns(3,5)P_2/PtdIns5P Production

Metazoan and Mouse Knockout Models

Unlike in yeast, where *fab1* deletion is dispensable for life, *PIKFYVE* knockout in multicellular organisms causes developmental defects and early death. For example, a complete loss of PIKfyve in *C. elegans* induces embryonic lethality and partial loss of function causes growth retardation (Nicot et al. 2006). *Drosophila* null or close to null mutants die at the pupal stage displaying an overgrowth phenotype, but without tumor-like overproliferation of tissues (Rusten et al. 2006). In both model organisms the endo-lysosomal structures are grossly enlarged, exhibiting impaired acidification, but besides these defects, there are no other morphological aberrations. Thus, the cause of early lethality in *C. elegans* or *Drosophila* PIKfyve null models remains unclear.

The first PIKfyve KO mouse model, recently generated in our laboratory, reveals that as in metazoans, PIKfyve carries essential and nonredundant functions in mammals, supporting cell viability and early embryonic development (Ikonomov et al. 2011). Indeed, almost all PIKfyve$^{KO/KO}$ embryos die during preimplantation development—at the 32–64-cell stage—consistent with the requirement of PIKfyve function as early as the first rounds of divisions in the fertilized egg. Although cell proliferation defects are highly likely as outlined above, the exact cause of PIKfyve$^{-/-}$ embryonic death requires more investigation. Unlike with metazoans, the PIKfyve heterozygous mice develop into old adults without ostensible defects, exhibiting only ~60 % of the normal PtdIns(3,5)P_2 and PtdIns5P levels. This indicates that in mice, a single PIKfyve allele is sufficient to support growth and development to late adulthood despite the apparent lack of functional compensation (Ikonomov et al. 2011).

It is worth comparing the phenotypes in PIKfyve$^{WT/KO}$ mice with those manifested in pale tremor mice that are Sac3-null or in mice with ArPIKfyve genetic deletion. Both ArPIKfyve- and Sac3-null mice exhibit early lethality and widespread peripheral and central neuropathy (Chow et al. 2007; Zhang et al. 2007; Katona et al. 2011). The pale tremor mice also display diluted pigmentation and hydrocephalus (Lenk et al. 2011) but these defects could be Sac3-unrelated as no specific rescue is established (see further). In any case, the lifespan of these mouse models is quite different: ArPIKfyve$^{-/-}$ mice die at postnatal day 1–2, whereas the Sac3-nulls, around postnatal day 42 (Chow et al. 2007; Zhang et al. 2007). Strikingly, these are the Sac3-null mice that exhibit more severe reduction in steady-state levels of PtdIns(3,5)P_2 rather than the ArPIKfyve$^{-/-}$ mice; 28 and 50 % of the control levels, respectively. This inconsistent correlation between lifespan and PtdIns(3,5)P_2 content in ArPIKfyve$^{-/-}$ *versus* Sac3-null mice is unclear and could be related to functions independent of regulating PIKfyve-catalyzed lipid production. Different genetic backgrounds of the mouse models might also contribute to some of the variations. Furthermore, the normal development of the

PIKfyve$^{WT/KO}$ at 60–65 % of control PtdIns(3,5)P_2 and PtdIns5P levels, without signs of neurodegeneration, suggests that this 10–15 % gain over ArPIKfyve$^{-/-}$ is quite significant, resulting in healthy life versus perinatal death (Ikonomov et al. 2011). Clearly, whereas the threshold PtdIns(3,5)P_2/PtdIns5P levels that are life-nonpermissible require further research, the presumed absence of these lipids under *pikfyve* homozygous deletion in mice causes unconditional early embryonic lethality (Ikonomov et al. 2011).

Human Diseases

The observations for early embryonic lethality in mice with *pikfyve* homozygous deletion yet lack of known severe human diseases associated with *PIKFYVE* mutations are consistent with the idea that many variants of the *pikfyve* gene would be incompatible with human life. It is conceivable that key mutations lead to the demise of embryos and fetuses before birth, an event estimated to occur to ~60 % of human embryos (Copp 1995). An asymptomatic disorder linked to heterozygous mutations in *PIKFYVE* is Francois–Neetens fleck corneal dystrophy (CFD), an autosomal dominant syndrome characterized by numerous small white flecks scattered in all layers of the stroma (Li et al. 2005). CFD subjects carry primarily protein-terminating nonsense or frame-shift heterozygous mutations within the *PIKFYVE* gene, all localized within the region spanning residues 705–1217 (Li et al. 2005). The enlarged/swollen vesicles within keratocytes likely arise from abnormal PIKfyve-dependent ECV/MVB fission/maturation presumably due to (see Inter-Endosomal Dynamics) reduced PtdIns(3,5)P_2 levels . The latter might result from both haploinsufficiency of the PIKfyve catalytic activity and enhanced presence of the Sac3 phosphatase, delivered to key locations by the PIKfyve truncated fragments, that retain some binding affinity for Sac3-ArPIKfyve (Ikonomov et al. 2011).

PIKfyve could also play a role in certain human cancers. Experimental support for such a connection are the data for PIKfyve-assisted EGFR nuclear trafficking in human bladder carcinoma cells, promoting gene regulation and cell cycle progression, as outlined above. Furthermore, pre-publication data finds PIKfyve somatic mutations in all of the examined seven ovarian adenocarcinoma samples (http://www.sanger.ac.uk). These findings are corroborated by a recent study indicating that the invasive capacity of the NPM-ALK oncogene, a constitutively active chimeric tyrosine kinase found in anaplastic large cell lymphomas, is enhanced upon interaction with PIKfyve (Dupuis-Coronas et al. 2011). This effect might be related to PIKfyve-catalyzed PtdIns5P production, as inferred by findings for an elevated PIKfyve-dependent PtdIns5P pool in human NPM-ALK-positive lymphoma cells (Coronas et al. 2008). However, this is yet to be characterized as is the contribution of PtdIns(3,5)P_2 and the ArPIKfyve/Sac3 subcomplex in cell invasiveness.

Mutations in the Sac3 phosphatase are linked with hereditary neurodegenerative disorders, such as Charcot-Marie-Tooth 4 J (CMT4 J) and amyotrophic lateral sclerosis (ALS). In the case of CMT4 J, a disease characterized by motor and sensory neuropathy, muscle weakness, and amyotrophy, patients are compound heterozygotes carrying one (Fig. 7.4)/*SAC3* null allele in combination with an I^{41}-to-T variant (Chow et al. 2007; Zhang et al. 2008). Our lab has recently provided mechanistic insights into the underlying cause of CMT4 J disease, associated with disrupted ability of ArPIKfyve to stabilize $Sac3^{I41T}$, leading to robust proteosome-dependent degradation of the phosphatase (Ikonomov et al. 2010). Because without Sac3, ArPIKfyve cannot efficiently associate with PIKfyve, the CMT4 J pathogenic mechanism is likely related to a failure of PIKfyve to be activated and produce proper amounts of PtdIns(3,5)P_2 (Ikonomov et al. 2010). Our observations and conclusions have been recently supported by findings for undetectable levels of the Sac3 protein in fibroblasts from CMT4 J patients (Lenk et al. 2011). Strikingly, the disrupted ability of ArPIKfyve to protect $Sac3^{I41T}$ from degradation occurs without substantial disturbance in the association between the two proteins as identified by protein co-immunoprecipitation (Ikonomov et al. 2010). These findings are corroborated by recent evidence in a mouse model that recapitulates the CMT4 J genetic defect carrying the $Sac3^{I41T}$ transgene on a Sac3-null background (Lenk et al. 2011). The ameliorated neurodegeneration and life expectancy by only 2-fold expression of $Sac3^{I41T}$ are consistent with functional ArPIKfyve-$Sac3^{I41T}$ complexes being readily restored. ALS is another severe neurological disorder, characterized by neurodegeneration in the spinal cord, brainstem, and cortex. Various Sac3 heterozygous mutations are reported in 1–2 % of the tested patients, but as ALS is a polygenic disease, the precise definition of the Sac3 contribution is currently unclear (Chow et al. 2009).

Conclusions and Perspectives

There is no doubt that in the dozen years after PIKfyve's discovery, the field has advanced tremendously. It is now clear that PIKfyve and its lipid products regulate key cellular processes such as membrane trafficking, stress/hormone-induced signaling, ion channel activity, cytoskelatal dynamics, nuclear transport, gene transcription, and cell cycle progression. Their dysfunction has wide-ranging consequences on disease-related conditions, including pathogen colonization, neuronal excitotoxicity, insulin resistance, and tumor cell invasiveness. The identification of the PAS complex that holds two active yet antagonistic enzymes together with the aid of a scaffolding component, has unraveled the critical requirement for a tight homeostatic control of PtdIns(3,5)P_2 levels in the course of dynamic endosome membrane remodeling and performance of several trafficking pathways. Although the mammalian PIKfyve pathway shares similarities with the orthologous system in yeast, there are some striking differences, one of the foremost being the vital function of PIKfyve in survival. Genetic mouse models that are null for any constituent of the PAS core complex exhibit early embryonic or perinatal/juvenile lethality, underscoring the critical need for each of the three proteins. Several human genetic disorders associated with the PAS complex dysfunction have already emerged.

Despite the considerable advances, many challenges remain. Among those is our lack of knowledge on the spatial/temporal regulation and coordination of the individual enzymatic activities within the PAS complex. Furthermore, it is unclear whether all PIKfyve functions are mediated through the PAS complex or some could proceed independently. Does PIKfyve participate in other multiprotein complexes to support these other functions? Does the PAS complex control synthesis and turnover of only PtdIns(3,5)P_2 and can such a regulation also govern PtdIns5P homeostasis? Are the two lipids equally essential in supporting mammalian life? A major roadblock to the successful development of a more coherent picture is, in part, the lack of high-quality reagents such as specific bioreporters or antibodies for intracellular detection of localized PtdIns(3,5)P_2 and PtdIns5P production. Furthermore, despite the tremendous efforts, downstream effectors of PtdIns(3,5)P_2/PtdIns5P remain largely uncertain. Even the one that was thought to fulfill such a mission, i.e., the membrane TRPML1 whose dysregulation causes mucolipidosis type IV, challenges our concept of the downstream effector paradigm involving recruitment of cytosolic proteins. Finally, structural insights into the PAS complex would certainly shed further light. Thus, there is an exciting multifaceted avenue for future research that will broaden our knowledge, providing a better comprehension of disease mechanisms associated with dysfunction of the PIKfyve pathway.

Acknowledgments I am grateful to the past and current members of my laboratory, and particularly to Drs. Ogi Ikonomov and Diego Sbrissa for their excellent work and stimulating discussions. I thank Dr. Steven Cala for his insightful comments and Violeta Shisheva for her many years of support. The work described from my laboratory was funded by ADA, JDFI, and NIH.

References

Alesutan IS, Ureche ON, Laufer J, Klaus F, Zurn A, Lindner R, Strutz-Seebohm N, Tavare JM, Boehmer C, Palmada M, Lang UE, Seebohm G, Lang F (2010) Regulation of the glutamate transporter EAAT4 by PIKfyve. Cell Physiol Biochem 25:187–194

Al-Qusairi L, Weiss N, Toussaint A, Berbey C, Messaddeq N, Kretz C, Sanoudou D, Beggs AH, Allard B, Mandel JL, Laporte J, Jacquemond V, Buj-Bello A (2009) T-tubule disorganization and defective excitation-contraction coupling in muscle fibers lacking myotubularin lipid phosphatase. Proc Natl Acad Sci U S A 106:18763–18768

Andrade MA, Petosa C, O'Donoghue SI, Muller CW, Bork P (2001) Comparison of ARM and HEAT protein repeats. J Mol Biol 309:1–18

Backer JM (2008) The regulation and function of class III PI3Ks: novel roles for Vps34. Biochem J 410:1–17

Backer JM (2010) New methods for capturing the mystery lipid, PtdIns5P. Biochem J 428:1–2

Balla T (2006) Phosphoinositide-derived messengers in endocrine signaling. J Endocrinol 188:135–153

Banfic H, Downes CP, Rittenhouse SE (1998) Biphasic activation of PKBalpha/Akt in platelets. Evidence for stimulation both by phosphatidylinositol 3,4-bisphosphate, produced via a novel pathway, and by phosphatidylinositol 3,4,5-trisphosphate. J Biol Chem 273:11630–11637

Berger P, Bonneick S, Willi S, Wymann M, Suter U (2002) Loss of phosphatase activity in myotubularin-related protein 2 is associated with Charcot-Marie-Tooth disease type 4B1. Hum Mol Genet 11:1569–1579

Berwick DC, Dell GC, Welsh GI, Heesom KJ, Hers I, Fletcher LM, Cooke FT, Tavare JM (2004) Protein kinase B phosphorylation of PIKfyve regulates the trafficking of GLUT4 vesicles. J Cell Sci 117:5985–5993

Blagoveshchenskaya A, Mayinger P (2009) SAC1 lipid phosphatase and growth control of the secretory pathway. Mol BioSyst 5:36–42

Bolis A, Coviello S, Visigalli I, Taveggia C, Bachi A, Chishti AH, Hanada T, Quattrini A, Previtali SC, Biffi A, Bolino A (2009) Dlg1, Sec8, and Mtmr2 regulate membrane homeostasis in Schwann cell myelination. J Neurosci 29:8858–8870

Bonifacino JS, Rojas R (2006) Retrograde transport from endosomes to the trans-Golgi network. Nat Rev Mol Cell Biol 7:568–579

Cabezas A, Pattni K, Stenmark H (2006) Cloning and subcellular localization of a human phosphatidylinositol 3-phosphate 5-kinase, PIKfyve/Fab1. Gene 371:34–41

Carlton JG, Cullen PJ (2005) Coincidence detection in phosphoinositide signaling. Trends Cell Biol 15:540–547

Carpenter G, Liao HJ (2009) Trafficking of receptor tyrosine kinases to the nucleus. Exp Cell Res 315:1556–1566

Casamayor A, Snyder M (2003) Molecular dissection of a yeast septin: distinct domains are required for septin interaction, localization, and function. Mol Cell Biol 23:2762–2777

Chen Y, Klionsky DJ (2011) The regulation of autophagy—unanswered questions. J Cell Sci 124:161–170

Chia PZ, Gasnereau I, Lieu ZZ, Gleeson PA (2011) Rab9-dependent retrograde transport and endosomal sorting of the endopeptidase furin. J Cell Sci 124:2401–2413

Chow CY, Zhang Y, Dowling JJ, Jin N, Adamska M, Shiga K, Szigeti K, Shy ME, Li J, Zhang X, Lupski JR, Weisman LS, Meisler MH (2007) Mutation of Fig. 4 causes neurodegeneration in the pale tremor mouse and patients with CMT4 J. Nature 448:68–72

Chow CY, Landers JE, Bergren SK, Sapp PC, Grant AE, Jones JM, Everett L, Lenk GM, McKenna-Yasek DM, Weisman LS, Figlewicz D, Brown RH, Meisler MH (2009) Deleterious variants of Fig. 4, a phosphoinositide phosphatase, in patients with ALS. Am J Hum Genet 84:85–88

Clarke JH, Letcher AJ, D'Santos CS, Halstead JR, Irvine RF, Divecha N (2001) Inositol lipids are regulated during cell cycle progression in the nuclei of murine erythroleukaemia cells. Biochem J 357:905–910

Copp AJ (1995) Death before birth: clues from gene knockouts and mutations. Trends Genet 11:87–93

Coronas S, Lagarrigue F, Ramel D, Chicanne G, Delsol G, Payrastre B, Tronchere H (2008) Elevated levels of PtdIns5P in NPM-ALK transformed cells: implication of PIKfyve. Biochem Biophys Res Commun 372:351–355

D'Andrea LD, Regan L (2003) TPR proteins: the versatile helix. Trends Biochem Sci 28:655–662

de Lartigue J, Polson H, Feldman M, Shokat K, Tooze SA, Urbe S, Clague MJ (2009) PIKfyve regulation of endosome-linked pathways. Traffic 10:883–893

Dhanasekaran DN, Kashef K, Lee CM, Xu H, Reddy EP (2007) Scaffold proteins of MAP-kinase modules. Oncogene 26:3185–3202

Di Paolo G, De Camilli P (2006) Phosphoinositides in cell regulation and membrane dynamics. Nature 443:651–657

Diaz E, Schimmoller F, Pfeffer SR (1997) A novel Rab9 effector required for endosome-to-TGN transport. J Cell Biol 138:283–290

Dominguez V, Raimondi C, Somanath S, Bugliani M, Loder MK, Edling CE, Divecha N, da Silva-Xavier G, Marselli L, Persaud SJ, Turner MD, Rutter GA, Marchetti P, Falasca M, Maffucci T (2011) Class II phosphoinositide 3-kinase regulates exocytosis of insulin granules in pancreatic beta cells. J Biol Chem 286:4216–4225
Dong XP, Shen D, Wang X, Dawson T, Li X, Zhang Q, Cheng X, Zhang Y, Weisman LS, Delling M, Xu H (2010) PI(3,5)P(2) controls membrane trafficking by direct activation of mucolipin Ca(2+) release channels in the endolysosome. Nat Commun 1:38
Dove SK, Cooke FT, Douglas MR, Sayers LG, Parker PJ, Michell RH (1997) Osmotic stress activates phosphatidylinositol-3,5-bisphosphate synthesis. Nature 390:187–192
Dove SK, Dong K, Kobayashi T, Williams FK, Michell RH (2009) Phosphatidylinositol 3,5-bisphosphate and Fab1p/PIKfyve underPPIn endo-lysosome function. Biochem J 419:1–13
Dupuis-Coronas S, Lagarrigue F, Ramel D, Chicanne G, Saland E, Gaits-Iacovoni F, Payrastre B, Tronchere H (2011) The nucleophosmin-anaplastic lymphoma kinase oncogene interacts, activates, and uses the kinase PIKfyve to increase invasiveness. J Biol Chem 286:32105–32114
Falasca M, Maffucci T (2009) Rethinking phosphatidylinositol 3-monophosphate. Biochim Biophys Acta 1793:1795–1803
Falasca M, Hughes WE, Dominguez V, Sala G, Fostira F, Fang MQ, Cazzolli R, Shepherd PR, James DE, Maffucci T (2007) The role of phosphoinositide 3-kinase C2alpha in insulin signaling. J Biol Chem 282:28226–28236
Falguieres T, Luyet PP, Gruenberg J (2009) Molecular assemblies and membrane domains in multivesicular endosome dynamics. Exp Cell Res 315:1567–1573
Ferguson CJ, Lenk GM, Meisler MH (2009) Defective autophagy in neurons and astrocytes from mice deficient in PI(3,5)P2. Hum Mol Genet 18:4868–4878
Foley K, Boguslavsky S, Klip A (2011) Endocytosis, recycling, and regulated exocytosis of glucose transporter 4. Biochemistry 50:3048–3061
Gary JD, Wurmser AE, Bonangelino CJ, Weisman LS, Emr SD (1998) Fab1p is essential for PtdIns(3)P 5-kinase activity and the maintenance of vacuolar size and membrane homeostasis. J Cell Biol 143:65–79
Gehring EM, Lam RS, Siraskar G, Koutsouki E, Seebohm G, Ureche ON, Ureche L, Baltaev R, Tavare JM, Lang F (2009a) PIKfyve upregulates CFTR activity. Biochem Biophys Res Commun 390:952–957
Gehring EM, Zurn A, Klaus F, Laufer J, Sopjani M, Lindner R, Strutz-Seebohm N, Tavare JM, Boehmer C, Palmada M, Lang UE, Seebohm G, Lang F (2009b) Regulation of the glutamate transporter EAAT2 by PIKfyve. Cell Physiol Biochem 24:361–368
Gillooly DJ, Morrow IC, Lindsay M, Gould R, Bryant NJ, Gaullier JM, Parton RG, Stenmark H (2000) Localization of phosphatidylinositol 3-phosphate in yeast and mammalian cells. EMBO J 19:4577–4588
Gozani O, Karuman P, Jones DR, Ivanov D, Cha J, Lugovskoy AA, Baird CL, Zhu H, Field SJ, Lessnick SL, Villasenor J, Mehrotra B, Chen J, Rao VR, Brugge JS, Ferguson CG, Payrastre B, Myszka DG, Cantley LC, Wagner G, Divecha N, Prestwich GD, Yuan J (2003) The PHD finger of the chromatin-associated protein ING2 functions as a nuclear phosphoinositide receptor. Cell 114:99–111
Gruenberg J, Stenmark H (2004) The biogenesis of multivesicular endosomes. Nat Rev Mol Cell Biol 5:317–323
Han BK, Emr SD (2011) Phosphoinositide [PI(3,5)P2] lipid-dependent regulation of the general transcriptional regulator Tup1. Genes Dev 25:984–995
Herman GA, Bonzelius F, Cieutat AM, Kelly RB (1994) A distinct class of intracellular storage vesicles, identified by expression of the glucose transporter GLUT4. Proc Natl Acad Sci U S A 91:12750–12754
Hill EV, Hudson CA, Vertommen D, Rider MH, Tavare JM (2010) Regulation of PIKfyve phosphorylation by insulin and osmotic stress. Biochem Biophys Res Commun 397:650–655
Hnia K, Tronchere H, Tomczak KK, Amoasii L, Schultz P, Beggs AH, Payrastre B, Mandel JL, Laporte J (2011) Myotubularin controls desmin intermediate filament architecture and mitochondrial dynamics in human and mouse skeletal muscle. J Clin Invest 121:70–85

Huang S, Czech MP (2007) The GLUT4 glucose transporter. Cell Metab 5:237–252

Hudson AM, Cooley L (2008) Phylogenetic, structural and functional relationships between WD- and Kelch-repeat proteins. Subcell Biochem 48:6–19

Hughes WE, Cooke FT, Parker PJ (2000) Sac phosphatase domain proteins. Biochem J 350 (Pt 2):337–352

Ikonomov OC, Sbrissa D, Shisheva A (2001) Mammalian cell morphology and endocytic membrane homeostasis require enzymatically active phosphoinositide 5-kinase PIKfyve. J Biol Chem 276:26141–26147

Ikonomov OC, Sbrissa D, Mlak K, Kanzaki M, Pessin J, Shisheva A (2002a) Functional dissection of lipid and protein kinase signals of PIKfyve reveals the role of PtdIns 3,5–P2 production for endomembrane integrity. J Biol Chem 277:9206–9211

Ikonomov OC, Sbrissa D, Mlak K, Shisheva A (2002b) Requirement for PIKfyve enzymatic activity in acute and long-term insulin cellular effects. Endocrinology 143:4742–4754

Ikonomov OC, Sbrissa D, Foti M, Carpentier JL, Shisheva A (2003a) PIKfyve controls fluid phase endocytosis but not recycling/degradation of endocytosed receptors or sorting of procathepsin D by regulating multivesicular body morphogenesis. Mol Biol Cell 14:4581–4591

Ikonomov OC, Sbrissa D, Mlak K, Deeb R, Fligger J, Soans A, Finley RL Jr, Shisheva A (2003b) Active PIKfyve associates with and promotes the membrane attachment of the late endosome-to-trans-Golgi network transport factor Rab9 effector p40. J Biol Chem 278:50863–50871

Ikonomov OC, Sbrissa D, Shisheva A (2006) Localized PtdIns 3,5–P2 synthesis to regulate early endosome dynamics and fusion. Am J Physiol Cell Physiol 291:393–404

Ikonomov OC, Sbrissa D, Dondapati R, Shisheva A (2007) ArPIKfyve-PIKfyve interaction and role in insulin-regulated GLUT4 translocation and glucose transport in 3T3-L1 adipocytes. Exp Cell Res 313:2404–2416

Ikonomov OC, Fligger J, Sbrissa D, Dondapati R, Mlak K, Deeb R, Shisheva A (2009a) Kinesin adapter JLP links PIKfyve to microtubule-based endosome-to-trans-Golgi network traffic of furin. J Biol Chem 284:3750–3761

Ikonomov OC, Sbrissa D, Fenner H, Shisheva A (2009b) PIKfyve-ArPIKfyve-Sac3 core complex: contact sites and their consequence for Sac3 phosphatase activity and endocytic membrane homeostasis. J Biol Chem 284:35794–35806

Ikonomov OC, Sbrissa D, Ijuin T, Takenawa T, Shisheva A (2009c) Sac3 is an insulin-regulated phosphatidylinositol 3,5-bisphosphate phosphatase: gain in insulin responsiveness through Sac3 down-regulation in adipocytes. J Biol Chem 284:23961–23971

Ikonomov OC, Sbrissa D, Shisheva A (2009d) YM201636, an inhibitor of retroviral budding and PIKfyve-catalyzed PtdIns(3,5)P2 synthesis, halts glucose entry by insulin in adipocytes. Biochem Biophys Res Commun 382:566–570

Ikonomov OC, Sbrissa D, Fligger J, Delvecchio K, Shisheva A (2010) ArPIKfyve regulates Sac3 protein abundance and turnover: disruption of the mechanism by Sac3I41T mutation causing Charcot-Marie-Tooth 4 J disorder. J Biol Chem 285:26760–26764

Ikonomov OC, Sbrissa D, Delvecchio K, Xie Y, Jin JP, Rappolee D, Shisheva A (2011) The Phosphoinositide Kinase PIKfyve Is Vital in Early Embryonic Development: PREIMPLANTATION LETHALITY OF PIKfyve-/- EMBRYOS BUT NORMALITY OF PIKfyve $\pm$ MICE. J Biol Chem 286:13404–13413

Ivetac I, Munday AD, Kisseleva MV, Zhang XM, Luff S, Tiganis T, Whisstock JC, Rowe T, Majerus PW, Mitchell CA (2005) The type Ialpha inositol polyphosphate 4-phosphatase generates and terminates phosphoinositide 3-kinase signals on endosomes and the plasma membrane. Mol Biol Cell 16:2218–2233

Jefferies HB, Cooke FT, Jat P, Boucheron C, Koizumi T, Hayakawa M, Kaizawa H, Ohishi T, Workman P, Waterfield MD, Parker PJ (2008) A selective PIKfyve inhibitor blocks PtdIns(3,5)P(2) production and disrupts endomembrane transport and retroviral budding. EMBO Rep 9:164–170

Johannes L, Popoff V (2008) Tracing the retrograde route in protein trafficking. Cell 135:1175–1187

Johnson EE, Overmeyer JH, Gunning WT, Maltese WA (2006) Gene silencing reveals a specific function of hVps34 phosphatidylinositol 3-kinase in late versus early endosomes. J Cell Sci 119:1219–1232

Jones DR, Gonzalez-Garcia A, Diez E, Martinez AC, Carrera AC, Merida I (1999) The identification of phosphatidylinositol 3,5-bisphosphate in T-lymphocytes and its regulation by interleukin-2. J Biol Chem 274:18407–18413

Jones DR, Bultsma Y, Keune WJ, Halstead JR, Elouarrat D, Mohammed S, Heck AJ, D'Santos CS, Divecha N (2006) Nuclear PtdIns5P as a transducer of stress signaling: an in vivo role for PIP4Kbeta. Mol Cell 23:685–695

Katona I, Zhang X, Bai Y, Shy ME, Guo J, Yan Q, Hatfield J, Kupsky WJ, Li J (2011) Distinct pathogenic processes between Fig. 4-deficient motor and sensory neurons. Eur J Neurosci 33:1401–1410

Kerr MC, Wang JT, Castro NA, Hamilton NA, Town L, Brown DL, Meunier FA, Brown NF, Stow JL, Teasdale RD (2010) Inhibition of the PtdIns(5) kinase PIKfyve disrupts intracellular replication of Salmonella. EMBO J 29:1331–1347

Kim J, Jahng WJ, Di Vizio D, Lee JS, Jhaveri R, Rubin MA, Shisheva A, Freeman MR (2007) The phosphoinositide kinase PIKfyve mediates epidermal growth factor receptor trafficking to the nucleus. Cancer Res 67:9229–9237

Klaus F, Gehring EM, Zurn A, Laufer J, Lindner R, Strutz-Seebohm N, Tavare JM, Rothstein JD, Boehmer C, Palmada M, Gruner I, Lang UE, Seebohm G, Lang F (2009a) Regulation of the Na(+)-coupled glutamate transporter EAAT3 by PIKfyve. Neurochem Int 54:372–377

Klaus F, Laufer J, Czarkowski K, Strutz-Seebohm N, Seebohm G, Lang F (2009b) PIKfyve-dependent regulation of the Cl- channel ClC-2. Biochem Biophys Res Commun 381:407–411

Kong AM, Horan KA, Sriratana A, Bailey CG, Collyer LJ, Nandurkar HH, Shisheva A, Layton MJ, Rasko JE, Rowe T, Mitchell CA (2006) Phosphatidylinositol 3-phosphate [PtdIns3P] is generated at the plasma membrane by an inositol polyphosphate 5-phosphatase: endogenous PtdIns3P can promote GLUT4 translocation to the plasma membrane. Mol Cell Biol 26:6065–6081

Lecompte O, Poch O, Laporte J (2008) PtdIns5P regulation through evolution: roles in membrane trafficking? Trends Biochem Sci 33:453–460

Lemmon MA (2008) Membrane recognition by phospholipid-binding domains. Nat Rev Mol Cell Biol 9:99–111

Lenk GM, Ferguson CJ, Chow CY, Jin N, Jones JM, Grant AE, Zolov SN, Winters JJ, Giger RJ, Dowling JJ, Weisman LS, Meisler MH (2011) Pathogenic mechanism of the Fig. 4 mutation responsible for Charcot-Marie-Tooth disease CMT4 J. PLoS Genet 7:e1002104

Li S, Tiab L, Jiao X, Munier FL, Zografos L, Frueh BE, Sergeev Y, Smith J, Rubin B, Meallet MA, Forster RK, Hejtmancik JF, Schorderet DF (2005) Mutations in PIP5K3 are associated with Francois-Neetens mouchetee fleck corneal dystrophy. Am J Hum Genet 77:54–63

Lindmo K, Stenmark H (2006) Regulation of membrane traffic by phosphoinositide 3-kinases. J Cell Sci 119:605–614

Liu Y, Bruzik KS, Ananthanarayanan B, Cho W (2003) New aspects of formation of 1,2-cyclic phosphates by phospholipase C-delta1. Bioorg Med Chem 11:2471–2475

Majerus PW, York JD (2009) Phosphoinositide phosphatases and disease. J Lipid Res 50(Suppl):249–254

McEwen RK, Dove SK, Cooke FT, Painter GF, Holmes AB, Shisheva A, Ohya Y, Parker PJ, Michell RH (1999) Complementation analysis in PtdInsP kinase-deficient yeast mutants demonstrates that Schizosaccharomyces pombe and murine Fab1p homologues are phosphatidylinositol 3-phosphate 5-kinases. J Biol Chem 274:33905–33912

Meijer HJ, Berrie CP, Iurisci C, Divecha N, Musgrave A, Munnik T (2001) Identification of a new polyphosphoinositide in plants, phosphatidylinositol 5-monophosphate (PtdIns5P), and its accumulation upon osmotic stress. Biochem J 360:491–498

Michell RH, Heath VL, Lemmon MA, Dove SK (2006) Phosphatidylinositol 3,5-bisphosphate: metabolism and cellular functions. Trends Biochem Sci 31:52–63

Morris JB, Hinchliffe KA, Ciruela A, Letcher AJ, Irvine RF (2000) Thrombin stimulation of platelets causes an increase in phosphatidylinositol 5-phosphate revealed by mass assay. FEBS Lett 475:57–60

Munday AD, Norris FA, Caldwell KK, Brown S, Majerus PW, Mitchell CA (1999) The inositol polyphosphate 4-phosphatase forms a complex with phosphatidylinositol 3-kinase in human platelet cytosol. Proc Natl Acad Sci U S A 96:3640–3645

Murray JW, Wolkoff AW (2003) Roles of the cytoskeleton and motor proteins in endocytic sorting. Adv Drug Deliv Rev 55:1385–1403

Murray JL, Mavrakis M, McDonald NJ, Yilla M, Sheng J, Bellini WJ, Zhao L, Le Doux JM, Shaw MW, Luo CC, Lippincott-Schwartz J, Sanchez A, Rubin DH, Hodge TW (2005) Rab9 GTPase is required for replication of human immunodeficiency virus type 1, filoviruses, and measles virus. J Virol 79:11742–11751

Nicot AS, Fares H, Payrastre B, Chisholm AD, Labouesse M, Laporte J (2006) The phosphoinositide kinase PIKfyve/Fab1p regulates terminal lysosome maturation in caenorhabditis elegans. Mol Biol Cell 17:3062–3074

Niebuhr K, Giuriato S, Pedron T, Philpott DJ, Gaits F, Sable J, Sheetz MP, Parsot C, Sansonetti PJ, Payrastre B (2002) Conversion of PtdIns(4,5)P(2) into PtdIns(5)P by the S.flexneri effector IpgD reorganizes host cell morphology. EMBO J 21:5069–5078

Ooms LM, Fedele CG, Astle MV, Ivetac I, Cheung V, Pearson RB, Layton MJ, Forrai A, Nandurkar HH, Mitchell CA (2006) The inositol polyphosphate 5-phosphatase, PIPP, Is a novel regulator of phosphoinositide 3-kinase-dependent neurite elongation. Mol Biol Cell 17:607–622

Osborne SL, Wen PJ, Boucheron C, Nguyen HN, Hayakawa M, Kaizawa H, Parker PJ, Vitale N, Meunier FA (2008) PIKfyve negatively regulates exocytosis in neurosecretory cells. J Biol Chem 283:2804–2813

Pryor PR, Luzio JP (2009) Delivery of endocytosed membrane proteins to the lysosome. Biochim Biophys Acta 1793:615–624

Rameh LE, Tolias KF, Duckworth BC, Cantley LC (1997) A new pathway for synthesis of phosphatidylinositol-4,5-bisphosphate. Nature 390:192–196

Rink J, Ghigo E, Kalaidzidis Y, Zerial M (2005) Rab conversion as a mechanism of progression from early to late endosomes. Cell 122:735–749

Roberts HF, Clarke JH, Letcher AJ, Irvine RF, Hinchliffe KA (2005) Effects of lipid kinase expression and cellular stimuli on phosphatidylinositol 5-phosphate levels in mammalian cell lines. FEBS Lett 579:2868–2872

Rudge SA, Anderson DM, Emr SD (2004) Vacuole size control: regulation of PtdIns(3,5)P2 levels by the vacuole-associated Vac14-Fig. 4 complex, a PtdIns(3,5)P2-specific phosphatase. Mol Biol Cell 15:24–36

Rusten TE, Rodahl LM, Pattni K, Englund C, Samakovlis C, Dove S, Brech A, Stenmark H (2006) Fab1 phosphatidylinositol 3-phosphate 5-kinase controls trafficking but not silencing of endocytosed receptors. Mol Biol Cell 17:3989–4001

Rusten TE, Vaccari T, Lindmo K, Rodahl LM, Nezis IP, Sem-Jacobsen C, Wendler F, Vincent JP, Brech A, Bilder D, Stenmark H (2007) ESCRTs and Fab1 regulate distinct steps of autophagy. Curr Biol 17:1817–1825

Rutherford AC, Traer C, Wassmer T, Pattni K, Bujny MV, Carlton JG, Stenmark H, Cullen PJ (2006) The mammalian phosphatidylinositol 3-phosphate 5-kinase (PIKfyve) regulates endosome-to-TGN retrograde transport. J Cell Sci 119:3944–3957

Sarkes D, Rameh LE (2010) A novel HPLC-based approach makes possible the spatial characterization of cellular PtdIns5P and other phosphoinositides. Biochem J 428:375–384

Sasaki T, Takasuga S, Sasaki J, Kofuji S, Eguchi S, Yamazaki M, Suzuki A (2009) Mammalian phosphoinositide kinases and phosphatases. Prog Lipid Res 48:307–343

Sbrissa D, Ikonomov OC, Shisheva A (1999) PIKfyve, a mammalian ortholog of yeast Fab1p lipid kinase, synthesizes 5-phosphoinositides. Effect of insulin. J Biol Chem 274:21589–21597

Sbrissa D, Ikonomov OC, Shisheva A (2000) PIKfyve lipid kinase is a protein kinase: downregulation of 5'-phosphoinositide product formation by autophosphorylation. Biochemistry 39:15980–15989
Sbrissa D, Ikonomov O, Shisheva A (2001) Selective insulin-induced activation of class I(A) phosphoinositide 3-kinase in PIKfyve immune complexes from 3T3-L1 adipocytes. Mol Cell Endocrinol 181:35–46
Sbrissa D, Ikonomov OC, Deeb R, Shisheva A (2002a) Phosphatidylinositol 5-phosphate biosynthesis is linked to PIKfyve and is involved in osmotic response pathway in mammalian cells. J Biol Chem 277:47276–47284
Sbrissa D, Ikonomov OC, Shisheva A (2002b) Phosphatidylinositol 3-phosphate-interacting domains in PIKfyve. Binding specificity and role in PIKfyve. Endomenbrane localization. J Biol Chem 277:6073–6079
Sbrissa D, Ikonomov OC, Strakova J, Dondapati R, Mlak K, Deeb R, Silver R, Shisheva A (2004a) A mammalian ortholog of Saccharomyces cerevisiae Vac14 that associates with and up-regulates PIKfyve phosphoinositide 5-kinase activity. Mol Cell Biol 24:10437–10447
Sbrissa D, Ikonomov OC, Strakova J, Shisheva A (2004b) Role for a novel signaling intermediate, phosphatidylinositol 5-phosphate, in insulin-regulated F-actin stress fiber breakdown and GLUT4 translocation. Endocrinology 145:4853–4865
Sbrissa D, Ikonomov OC, Shisheva A (2005) Analysis of potential binding of the recombinant Rab9 effector p40 to phosphoinositide-enriched synthetic liposomes. Methods Enzymol 403:696–705
Sbrissa D, Shisheva A (2005) Acquisition of unprecedented phosphatidylinositol 3,5-bisphosphate rise in hyperosmotically stressed 3T3-L1 adipocytes, mediated by ArPIKfyve-PIKfyve pathway. J Biol Chem 280:7883–7889
Sbrissa D, Ikonomov OC, Fu Z, Ijuin T, Gruenberg J, Takenawa T, Shisheva A (2007) Core protein machinery for mammalian phosphatidylinositol 3,5-bisphosphate synthesis and turnover that regulates the progression of endosomal transport. Novel Sac phosphatase joins the ArPIKfyve-PIKfyve complex. J Biol Chem 282:23878–23891
Sbrissa D, Ikonomov OC, Fenner H, Shisheva A (2008) ArPIKfyve homomeric and heteromeric interactions scaffold PIKfyve and Sac3 in a complex to promote PIKfyve activity and functionality. J Mol Biol 384:766–779
Sbrissa D. Ikonomov OC Filios C, Delvecchio K, Shisheva A (2012) Functional dissociation between PIKfyve-synthesized PtdIns5P and PtdIns(3,5)P2 by means of the PIKfyve inhibitor YM201636. Am J Physiol Cell Physiol 303:C436–46
Schnapp BJ (2003) Trafficking of signaling modules by kinesin motors. J Cell Sci 116:2125–2135
Scott CC, Gruenberg J (2011) Ion flux and the function of endosomes and lysosomes: pH is just the start: the flux of ions across endosomal membranes influences endosome function not only through regulation of thc luminal pH. BioEssays 33:103–110
Seebohm G, Strutz-Seebohm N, Birkin R, Dell G, Bucci C, Spinosa MR, Baltaev R, Mack AF, Korniychuk G, Choudhury A, Marks D, Pagano RE, Attali B, Pfeufer A, Kass RS, Sanguinetti MC, Tavare JM, Lang F (2007) Regulation of endocytic recycling of KCNQ1/KCNE1 potassium channels. Circ Res 100:686–692
Shen J, Yu WM, Brotto M, Scherman JA, Guo C, Stoddard C, Nosek TM, Valdivia HH, Qu CK (2009) Deficiency of MIP/MTMR14 phosphatase induces a muscle disorder by disrupting Ca(2+) homeostasis. Nat Cell Biol 11:769–776
Shin HW, Hayashi M, Christoforidis S, Lacas-Gervais S, Hoepfner S, Wenk MR, Modregger J, Uttenweiler-Joseph S, Wilm M, Nystuen A, Frankel WN, Solimena M, De Camilli P, Zerial M (2005) An enzymatic cascade of Rab5 effectors regulates phosphoinositide turnover in the endocytic pathway. J Cell Biol 170:607–618
Shisheva A, Sbrissa D, Ikonomov O (1999) Cloning, characterization, and expression of a novel Zn2+-binding FYVE finger-containing phosphoinositide kinase in insulin-sensitive cells. Mol Cell Biol 19:623–634
Shisheva A (2001) PIKfyve: the road to PtdIns 5-P and PtdIns 3,5-P(2). Cell Biol Int 25:1201–1206

Shisheva A, Rusin B, Ikonomov OC, DeMarco C, Sbrissa D (2001) Localization and insulin-regulated relocation of phosphoinositide 5-kinase PIKfyve in 3T3-L1 adipocytes. J Biol Chem 276:11859–11869

Shisheva A, DeMarco C, Ikonomov O, Sbrissa D (2002) PIKfyve and acute insulin actions. In: Sima A, Shafrir E e (eds) Insulin signaling: from cultured cells to animal models. 189–206

Shisheva A (2008a) Phosphoinositides in insulin action on GLUT4 dynamics: not just PtdIns(3,4,5)P3. Am J Physiol Endocrinol Metab 295:536–544

Shisheva A (2008b) PIKfyve: Partners, significance, debates and paradoxes. Cell Biol Int 32:591–604

Shojaiefard M, Strutz-Seebohm N, Tavare JM, Seebohm G, Lang F (2007) Regulation of the Na(+), glucose cotransporter by PIKfyve and the serum and glucocorticoid inducible kinase SGK1. Biochem Biophys Res Commun 359:843–847

Silswal N, Parelkar NK, Wacker MJ, Brotto M, Andresen J (2011) Phosphatidylinositol 3,5-bisphosphate increases intracellular free Ca2+ in arterial smooth muscle cells and elicits vasocontraction. Am J Physiol Heart Circ Physiol 300:2016–2026

Sopjani M, Kunert A, Czarkowski K, Klaus F, Laufer J, Foller M, Lang F (2010) Regulation of the Ca(2+) channel TRPV6 by the kinases SGK1, PKB/Akt, and PIKfyve. J Membr Biol 233:35–41

Strauss HM, Keller S (2008) Pharmacological interference with protein-protein interactions mediated by coiled-coil motifs. Handb Exp Pharmacol 186:461–482

Strutz-Seebohm N, Shojaiefard M, Christie D, Tavare J, Seebohm G, Lang F (2007) PIKfyve in the SGK1 mediated regulation of the creatine transporter SLC6A8. Cell Physiol Biochem 20:729–734

Stuffers S, Sem Wegner C, Stenmark H, Brech A (2009) Multivesicular endosome biogenesis in the absence of ESCRTs. Traffic 10:925–937

Thoidis G, Kandror KV (2001) A Glut4-vesicle marker protein, insulin-responsive aminopeptidase, is localized in a novel vesicular compartment in PC12 cells. Traffic 2:577–587

Thorens B, Roth J (1996) Intracellular targeting of GLUT4 in transfected insulinoma cells: evidence for association with constitutively recycling vesicles distinct from synaptophysin and insulin vesicles. J Cell Sci 109(Pt 6):1311–1323

Touchberry CD, Bales IK, Stone JK, Rohrberg TJ, Parelkar NK, Nguyen T, Fuentes O, Liu X, Qu CK, Andresen JJ, Valdivia HH, Brotto M, Wacker MJ (2010) Phosphatidylinositol 3,5-bisphosphate (PI(3,5)P2) potentiates cardiac contractility via activation of the ryanodine receptor. J Biol Chem 285:40312–40321

Tronchere H, Laporte J, Pendaries C, Chaussade C, Liaubet L, Pirola L, Mandel JL, Payrastre B (2004) Production of phosphatidylinositol 5-phosphate by the phosphoinositide 3-phosphatase myotubularin in mammalian cells. J Biol Chem 279:7304–7312

Tsien RW, Ellinor PT, Horne WA (1991) Molecular diversity of voltage-dependent Ca2 + channels. Trends Pharmacol Sci 12:349–354

Tsujita K, Itoh T, Ijuin T, Yamamoto A, Shisheva A, Laporte J, Takenawa T (2004) Myotubularin regulates the function of the late endosome through the gram domain-phosphatidylinositol 3,5-bisphosphate interaction. J Biol Chem 279:13817–13824

Tsuruta F, Green EM, Rousset M, Dolmetsch RE (2009) PIKfyve regulates $Ca_V1.2$ degradation and prevents excitotoxic cell death. J Cell Biol 187:279–294

van der Goot FG, Gruenberg J (2006) Intra-endosomal membrane traffic. Trends Cell Biol 16:514–521

Vanhaesebroeck B, Guillermet-Guibert J, Graupera M, Bilanges B (2010) The emerging mechanisms of isoform-specific PI3 K signalling. Nat Rev Mol Cell Biol 11:329–341

Vicinanza M, D'Angelo G, Di Campli A, De Matteis MA (2008) Function and dysfunction of the PI system in membrane trafficking. EMBO J 27:2457–2470

Walker DM, Urbe S, Dove SK, Tenza D, Raposo G, Clague MJ (2001) Characterization of MTMR3. An inositol lipid 3-phosphatase with novel substrate specificity. Curr Biol 11:1600–1605

Whiteford CC, Brearley CA, Ulug ET (1997) Phosphatidylinositol 3,5-bisphosphate defines a novel PI 3-kinase pathway in resting mouse fibroblasts. Biochem J 323(Pt 3):597–601

Woodman PG, Futter CE (2008) Multivesicular bodies: co-ordinated progression to maturity. Curr Opin Cell Biol 20:408–414

Yamamoto A, DeWald DB, Boronenkov IV, Anderson RA, Emr SD, Koshland D (1995) Novel PI(4)P 5-kinase homologue, Fab1p, essential for normal vacuole function and morphology in yeast. Mol Biol Cell 6:525–539

Yuan Y, Gao X, Guo N, Zhang H, Xie Z, Jin M, Li B, Yu L, Jing N (2007) rSac3, a novel Sac domain phosphoinositide phosphatase, promotes neurite outgrowth in PC12 cells. Cell Res 17:919–932

Zhang Y, Zolov SN, Chow CY, Slutsky SG, Richardson SC, Piper RC, Yang B, Nau JJ, Westrick RJ, Morrison SJ, Meisler MH, Weisman LS (2007) Loss of Vac14, a regulator of the signaling lipid phosphatidylinositol 3,5-bisphosphate, results in neurodegeneration in mice. Proc Natl Acad Sci U S A 104:17518–17523

Zhang X, Chow CY, Sahenk Z, Shy ME, Meisler MH, Li J (2008) Mutation of Fig. 4 causes a rapidly progressive, asymmetric neuronal degeneration. Brain 131:1990–2001

Zhang J, Guan Z, Murphy AN, Wiley SE, Perkins GA, Worby CA, Engel JL, Heacock P, Nguyen OK, Wang JH, Raetz CR, Dowhan W, Dixon JE (2011) Mitochondrial phosphatase PTPMT1 is essential for cardiolipin biosynthesis. Cell Metab 13:690–700

Zou J, Marjanovic J, Kisseleva MV, Wilson M, Majerus PW (2007) Type I phosphatidylinositol-4,5-bisphosphate 4-phosphatase regulates stress-induced apoptosis. Proc Natl Acad Sci U S A 104:16834–16839

Chapter 8
Class I Phosphoinositide 3-Kinases in Normal and Pathologic Hematopoietic Cells

François Vergez, Christian Recher and Bernard Payrastre

Abstract Class I phosphoinositide 3-kinases which produce the D3-phosphoinositide second messenger phosphatidylinositol 3,4,5-trisphosphate in response to membrane receptors activation play a critical role in cell proliferation, survival, metabolism, and motility. These lipid kinases and the phosphatases regulating the level of D3-phosphoinositides have been an intense area of research these last two decades. The class I phosphoinositide 3-kinases signaling is found aberrantly activated in numerous human cancers, including in malignant hemopathies, and are important therapeutic targets for cancer therapy. Haematopoiesis is an ongoing process which generates the distinct blood cell types from a common hematopoietic stem cell through the action of a variety of cytokines. In the human adult hematopoiesis occurs primarily in the bone marrow, and defects in hematopoiesis result in diseases, such as anemia, thrombocytopenia, myeloproliferative syndromes, or leukemia. Here we give a brief overview of the role of class I phosphoinositide 3-kinases in hematopoietic stem cells, in hematopoietic lineage development and in leukemia, particularly in acute myeloid leukemia and summarize the potential therapeutic implications.

B. Payrastre (✉)
Inserm, U1048, I2MC, Université de Toulouse 3 and Laboratoire d'Hématologie du CHU de Toulouse, Hôpital de Toulouse, 31024 Toulouse, France
e-mail: bernard.payrastre@inserm.fr

F. Vergez
Inserm, U1037, Cancer Research Center of Toulouse and Laboratoire d'Hématologie, Hôpital de Toulouse, 31024 Toulouse, France

C. Recher
Inserm, U1037, Cancer Research Center of Toulouse and Service d'Hématologie, Hôpital de Toulouse, 31024 Toulouse, France

B. Payrastre
Inserm, U1048, University of Toulouse 3, I2MC and Laboratoire d'Hématologie, Hôpital de Toulouse, 1 Avenue Jean Poulhès BP84225, 31432 Toulouse Cedex 04, France

M. Falasca (ed.), *Phosphoinositides and Disease*, Current Topics in Microbiology and Immunology 362, DOI: 10.1007/978-94-007-5025-8_8,

Introduction

Phosphoinositides are a family of phospholipids implicated in the regulation of a variety of highly dynamic cellular processes including spatiotemporal organization of signaling pathways, actin cytoskeleton rearrangements, establishment of cell polarity, and intracellular vesicle trafficking (Di Paolo and De Camilli 2006; Payrastre et al. 2001). Polyphosphoinositides can be rapidly synthesized and degraded through different metabolic pathways involving specific 3, 4, or 5 -kinases and -phosphatases (Sasaki et al. 2010). Mutations in several phosphoinositide metabolizing enzymes take part in the development of human pathologies including cancer and genetic diseases and gene knockout (KO)/knock-in (KI) studies in mice have shown disease-related phenotypes in many cases (Pendaries et al. 2003; McCrea and De Camilli 2009). Moreover, human pathogens have evolved different strategies to subvert the phosphoinositide metabolism for pathogenicity and virulence (Payrastre et al. 2012). Through sequential phosphorylations of phosphatidylinositol, the 4 and 5 -kinases produce phosphatidylinositol 4,5 bisphosphate ($PtdIns(4,5)P_2$), which can be hydrolyzed by phospholipases C (PLC) to generate the second messengers diacylglycerol (DAG), an activator of protein kinase C (PKC), and inositol 1,4,5 *tris*phosphate ($InsP_3$), initiating calcium release from the endoplasmic reticulum. $PtdIns(4,5)P_2$ is also the substrate of class I phosphoinositide 3-kinases (PI 3-kinases) which generate the lipid second messenger $PtdIns(3,4,5)P_3$. Polyphosphoinositides can interact with protein domains including PH, FYVE, PX, PHD, or ENTH (Lemmon 2008; Carlton and Cullen 2005) leading to specific relocalization of proteins but also in some cases their activation or changes in conformation. $PtdIns(3,4,5)P_3$ and $PtdIns(3,4)P_2$ are well-known to recruit the serine/threonine kinase Akt (protein kinase B) at the plasma membrane via its PH domain allowing activation of the kinase by phosphorylation on Thr308 by PDK1 and Ser473 by mTORC2 and/or possibly other kinases such as ILK or DNA-PK. Active Akt then goes to the cytosol, to some organelles and to the nucleus to phosphorylate and modulate the activity of a number of downstream effectors playing direct or indirect roles in cell growth, proliferation, differentiation, or survival. $PtdIns(3,4,5)P_3$ phosphatases are important negative regulators of this pathway. Among them, the 3-phosphatase PTEN and the polyphosphate 5-phosphatases such as the type IV 5-phosphatase, SKIP, and the SH2-containing inositol polyphosphate 5-phosphatases 1 and 2 (SHIP1 and SHIP2) are best characterized. Class I PI 3-kinases are heterodimers consisting of a catalytic subunit (class IA: p110 α, β, or δ and class IB: p110 γ) and a regulatory subunit (p85α, p85β, p55α, p55γ, and p50α for the class IA isoforms and p101, p87, or p84 for the class IB). Class IA PI 3-kinases are activated by receptor tyrosine kinases through recruitment of the regulatory subunits by specific tyrosine phosphorylated motifs via their SH2 domains. Class IB is activated by heterotrimeric G protein β/γ-subunits, often from Gi. Recently, the class IA PI 3-kinase β has been shown to be activated by heterotrimeric-G protein coupled receptors rather than tyrosine kinase receptors or by a contribution of both (Vanhaesebroeck et al. 2010; Gratacap et al. 2011).

Besides being lipid kinases, PI 3-kinases have a protein kinase activity which remains hill defined and have also scaffolding functions (Costa and Hirsch 2010). Deregulation of the class I PI 3-kinase signaling is sufficient to induce cellular transformation and aberrant activation of this pathway has been described in numerous human cancers, including hematopoietic malignancies.

Various experimental approaches have demonstrated the critical role of several phosphoinositide-kinases, -phosphatases, and -phospholipases in the normal hematopoietic system but so far the PI 3-kinase signaling and the nuclear phospholipase Cβ1 (PLCβ1) are the major pathways found deregulated in human hematopoietic disorders.

Hematopoiesis is a tightly regulated process leading to the production of all blood lineages from hematopoietic stem cells (HSCs). HSCs have self-renewal capacity and the ability to sustain all blood cell lineages. They can reconstitute the entire blood system of a recipient. It is admitted that HSCs give rise to common lymphoid and myeloid multipotent progenitors which will proliferate and differentiate. In adults, billions of mature blood cells are generated every day and this system must adapt to pathophysiological demands. Within the bone marrow, the developing hematopoietic cells are retained until they have maturated and are then released into the vascular system. Broadly acting or lineage-specific cytokines play a pivotal role in the regulation of hematopoiesis and many of them act through receptors able to stimulate the class I PI 3-kinase signaling. Several reports have demonstrated the critical role of the class I PI 3-kinase/Akt and of the mTOR pathways in hematopoietic stem cells maintenance and function. In general, hematopoietic cells express all four class I PI 3-kinase isoforms (α, β, δ, and γ). While PI 3-kinase α and β are ubiquitous, PI 3-kinase δ and γ are expressed predominantly, but not exclusively, in hematopoietic cells. In hematopoietic progenitor cells p110δ appears to contribute for the largest fraction of class I PI 3-kinase activity as assessed by in vitro lipid kinase assays (Foukas et al. 2010). Consistent with this, p110δ is the predominant PI 3-kinase isoform in leukocytes (Geering et al. 2007). Mice deprived of PI 3-kinase δ or γ expression or function are viable, fertile, and have a normal life span. However, invalidation of PI 3-kinase δ activity leads to defects in B cell development and functions and signaling deficiencies are found in multiple cell types including B, T, and mast cells. In the absence of PI 3-kinase γ mice have a significant defect in leukocyte chemotaxis and to a lesser extent in B and T cell development and activation. In all cases PI 3-kinase δ or γ inactivation leads to a reduction but not a complete abrogation of leukocyte functions, possibly due to the overlapping functions with other PI 3-kinase members present in these cells. The p110α isoform seems to account for approximately 20 % of PI 3-kinase activity measured in vitro in hematopoietic progenitors cells (Foukas et al. 2010) but in vivo it may significantly contribute to PtdIns(3,4,5)P_3 production at least under certain conditions. P110α KO mice are embryonically lethal very early due to proliferation defects. P110α is considered as the main isoform involved in insulin action but its role in hematopoietic tissues is still hill defined. A study reports its implication in phagocytosis and fluid-phase pinocytosis in macrophages (Tamura et al. 2009). Despite a high degree of homology with PI 3-kinase α, PI 3-kinase β, which is also ubiquitously expressed, appears to have specific regulatory mechanisms and functions. While p110β KO mice are embryonically lethal at the blastocyst stage, and the embryos fail to implant, germline KI of a kinase dead allele of p110β is only partially lethal (Ciraolo et al. 2008) possibly due to putative scaffold functions of p110β. In the hematopoietic system, pharmacological inhibitors and mice models have shown a role for PI 3-kinase β essentially in platelet activation processes, particularly in adhesion/aggregation under high shear stress (Jackson et al. 2005; Martin et al. 2010), and in neutrophiles (Kulkarni et al. 2011).

PtdIns(3,4,5)P_3 phosphatases are important negative regulators of the PI 3-kinase pathway. Among them, the 3-phosphatase PTEN and the polyphosphate 5-phosphatases, such as the type IV 5-phosphatase, SKIP, and the SH2-containing inositol polyphosphate 5-phosphatases 1 and 2 (SHIP1 and SHIP2) are best characterized. PTEN and the 5-phosphatases SHIP1 and SHIP2 are largely expressed in hematopoietic cells (PTEN and SHIP2 are ubiquitous while SHIP1 is predominantly expressed in hematopoietic tissues). As discussed below, mice models and in vitro knockdown approaches have demonstrated the important functions of PTEN and SHIP1 as regulators of normal hematopoiesis.

In this chapter, we will cover several aspects of our current knowledge on the role of class I PI 3-kinases in normal HSCs and their malignant counterpart. Malignant hematological diseases comprise a diverse array of disorders. The clinical and biological diversity of hematological malignancies is linked to the cellular origin and stage of their ontogenic differentiation supported by genetic and molecular perturbation. Here we will particularly focus on acute myeloid leukemia (AML), the predominant acute leukemia in adults. This clonal disorder is characterized by an accumulation of immature leukemic cells in the bone marrow leading to marrow failure. It is thought that the leukemic clone arises from acquired mutations occurring either in HSCs or in a myeloid progenitor which can re-acquire self-renewal properties. These leukemic stem cells (LSCs) or leukemic initiating cells (LICs) can enter a proliferation and differentiation process leading to the generation of a more mature pool of highly cycling clonogenic progenitors which will form a terminally differentiated pool of leukemic cells arrested at a given stage of granulo-monocytic differentiation. It is thought that two mutational events are required for genesis of AML; class I mutations affecting key actors of cell proliferation and survival and class II mutations targeting transcription factors leading to impaired cell differentiation. Despite recent progress new treatments of AML are strongly expected as most patients relapse and die of the disease. In this context, we will discuss potential therapeutic implications of PI 3-kinase pathway inhibition.

PI 3-Kinase Signaling in Hematopoietic and Leukemic Stem Cells

A large part of HSCs are maintained in an undifferentiated quiescent sate (Go phase of the cell cycle) in the bone marrow/endosteal niche (Arai et al. 2004). Only a small fraction of HSCs is dividing (around 5 %), mainly in the vascular niche located at the bone marrow sinusoids which is permissive for proliferation and differentiation. HSCs and their progeny are surrounded by stromal cells in the bone marrow niche including chondrocytes, osteoblasts, fibroblasts, adipocytes, and endothelial cells. These cells modulate the maintenance and expansion of HSCs and have a critical influence on hematopoiesis (Zhang et al. 2003; Calvi et al. 2003; Scadden 2006; Kiel and Morrison 2008; Kobayashi et al. 2010). Many studies have correlated a loss of control of HSCs proliferation to their impaired long-term functions. In quiescent HSCs, the PI 3-kinase/Akt pathway is not found activated but following stimulation by various factors such as pleiotrophin (Himburg et al. 2010) or SDF-1 (Chabanon et al. 2008), this pathway becomes activated and contributes to the expansion of HSCs. Manipulating this pathway in vivo in mice has provided important information concerning its function in HSC. For instance, the PtdIns(3,4,5)P_3 3-phosphatase PTEN has been found to regulate HSCs maintenance through restriction of HSCs proliferation (Yilmaz et al. 2006; Zhang et al. 2006) while Evi1 (ecotropic viral integration site 1), which is mandatory for HSCs proliferation, directly represses PTEN expression (Goyama et al. 2008; Yoshimi et al. 2011). Conditional depletion of PTEN in HSCs in mice promotes short-term expansion followed by long-term depletion indicating that normal HSCs are unable to maintain themselves without PTEN. As a consequence, these mice develop a myeloproliferative disorder within days and leukemia within weeks. Interestingly, rapamycin treatment can revert the phenotype of PTEN-deficient HSCs (Yilmaz et al. 2006) indicating that mTORC1 is a key effector of the overactivated PI 3-kinase signaling in HSCs (Fig. 8.1).

SHIP1 deficient mice have also demonstrated the implication of this lipid phosphatase in normal hematopoiesis as well as in bone maintenance and in the immune system (Helgason et al. 1998; Helgason et al. 2003). SHIP1 is indeed involved in the negative regulation of B lymphocytes, macrophages, mast cells, osteoclasts, and also of HSCs. As a consequence, SHIP1 KO mice develop a myeloproliferative disease. However, the effect of SHIP1 depletion on the homeostasis and regeneration of HSCs is possibly due, at least in part, to its role in cells comprising the HSC niche (Helgason et al. 2003; Hazen et al. 2009). Moreover, SHIP1 KO/PTEN heterozygous mice have a myelodysplasia with a marked normocytic anemia (hemoglobin: 83.4 ± 5.0 vs 137.7 ± 3.5 g/L in wild-type), a thrombocytopenia (platelets: 106.2 ± 27.0 vs 662.2 ± 110.0 G/L in wild-type), and a significant leukocytosis (white blood cell count: 18.8 ± 3.0 vs 3.6 ± 0.3 G/L in wild-type) (Moody et al. 2004). Those mice have reduced number of marrow-derived granulocyte-erythrocyte-macrophage-megakaryocyte (GEMM) progenitors but show extramedullar hematopoiesis (Moody et al. 2004).

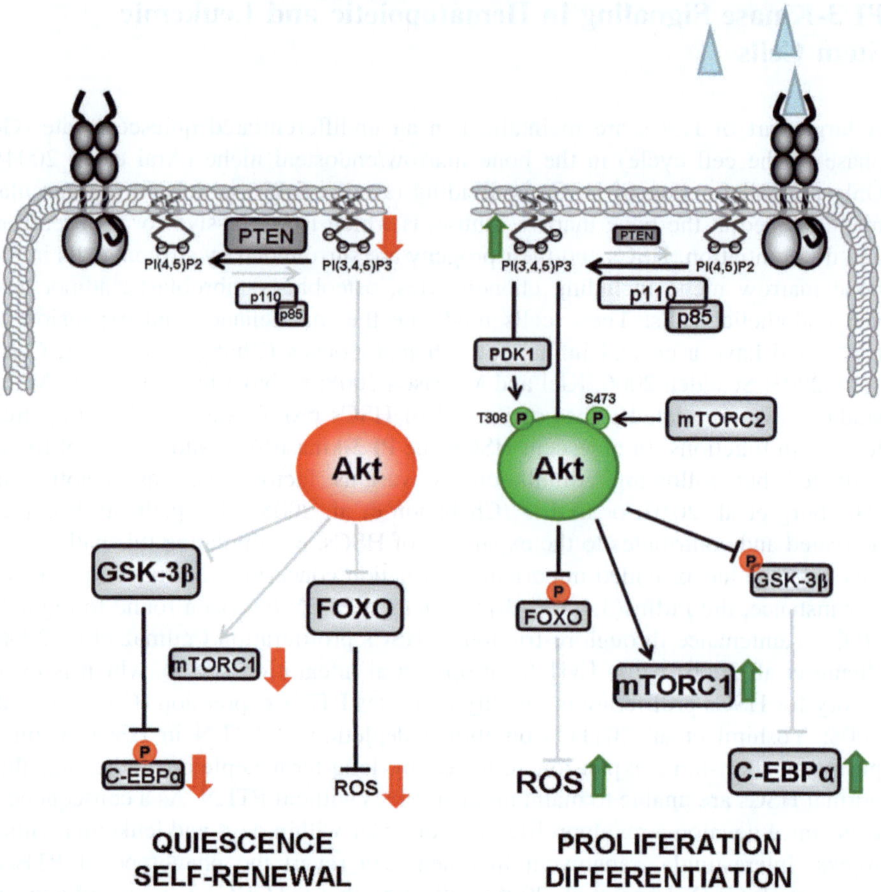

Fig. 8.1 *The Hematopoietic Stem Cell balance between self-renewal and proliferation is dependant of the PI 3-kinase/Akt pathway*. PTEN is the major regulator of HSCs maintenance. It inhibits Akt localization to the membrane and activation, by dephosphorylating PtdIns(3,4,5)P_3. When PI 3-kinases are activated (or PTEN inactivated), Akt can be phosphorylated on Ser473 by PDK1 and Thr308 by mTORC2. Its downstream targets, mTORC1, FOXO family, and GSK3-β, participate to the induction of HSCs entry in proliferation and maturation, in particular, via elevation of intracellular ROS levels and activation of CEBPα

Using a mouse model in which HSCs were stably transduced with a constitutively active form of Akt Kharas et al. (Kharas et al. 2010) have shown that Akt-activated HSCs are capable of homing normally to the bone marrow. However, self-renewal properties are no longer observed in these cells. In particular, the HSCs short-term repopulating activity is impaired because of an increased cell cycle entry and an expansion of more differentiated progenitors (Kharas et al. 2010). Again, the mTORC1 inhibitor rapamycin partially rescued Akt-activated HSCs functions (Kharas et al. 2010). Conversely, Akt deficient HSCs persist in the G0 phase of the cell cycle (Juntilla et al. 2010). Overall, these studies demonstrate that an activation of the PI 3-kinase/Akt and mTORC1 axis in HSCs reduces cell quiescence and self-renewal capacity and increases the rate at which HSCs exit the stem cell pool (Fig. 8.1).

Besides mTORC1, other important effectors downstream of PI 3-kinase/Akt are members of the FOXO family of transcription factors (FOXO1, 3, and 4 are substrates of Akt). Several studies report the role of FOXO3a in HSCs and propose a role of Akt in the export of FOXO3a from the nucleus to the cytoplasm (Chabanon et al. 2008; Miyamoto et al. 2007). Pharmacological inhibition of the PI 3-kinase/Akt pathway leads to the cytoplasm-to-nucleus translocation of FOXO3a as assessed by cellular-imaging assays (Zanella et al. 2008). Phosphorylated cytoplasmic FOXO3a is inactivated by facilitating its association with 14-3-3 proteins and its cytoplasmic retention and degradation. Concomitant deletion of FOXO1, 3, and 4 in the adult hematopoietic tissue leads to a reduction of HSCs number and function in response to a physiological oxidative stress (Miyamoto et al. 2007; Tothova et al. 2007). FOXO-deficient HSCs show an increased ROS level due to a reduced expression of ROS metabolism genes (including superoxide dismutase genes Sod1 and Sod3) (Tothova et al. 2007). Accordingly, antioxidant treatment of FOXO-deficient HSCs can rescue HSC functions. This is an important pathway since low intracellular ROS levels control the long-term self-renewal ability of HSCs (Jang and Sharkis 2007). Moreover, it has been shown in multiple cellular models that high intracellular ROS induces inactivating phosphorylations of PTEN and consequently further activation of the PI3-kinase/Akt pathway (Lee et al. 2011; Chetram et al. 2011).

These studies indicate that the balance between quiescence and proliferation/differentiation of HSCs is strongly affected by manipulation of the PI 3-kinase/Akt/mTORC1 and the PI 3-kinase/Akt/FOXO3a pathways (Fig. 8.1).

It is important to note that the PI 3-kinase/Akt and mTORC1 pathways are also important in the regulation and functions of the stroma cells surrounding HSCs and their progeny. For instance, activation of the PI3-kinase/Akt/mTORC1 pathway in mice endothelial cells of the niche contributes to self-renewal of HSCs by inducing the expression of membrane-bound and soluble angiocrine factors (Kobayashi et al. 2010). In particular, the expression of Notch ligands by endothelial cells promotes maintenance of HSCs (Butler et al. 2010) and expression of IGFBP2 or FGF2 induces their expansion (Kobayashi et al. 2010). Conversely, activation of the MAP-kinase pathway in endothelial cells rather stimulates the expression of factors promoting lineage-specific differentiation of HSCs (Kobayashi et al. 2010). Thus, endothelial cells of the bone marrow niche participate to the fine regulation of the balance between self-renewal and differentiation of HSCs thanks to the production of angiocrine factors and the PI 3-kinase/Akt and MAP-kinase pathways contribute to this regulation.

A critical cell population in AML, responsible for the relapse after conventional chemotherapy, is LSCs or LICs. These rare cells have the ability to transfer disease upon transplantation into irradiated mice. Few years ago it has been demonstrated that CD34 + CD38- leukemic cells, but not the more mature CD34 + CD38 + progenitors, could generate leukemia in immunocompromised

NOD/SCID mice (Bonnet and Dick 1997). It supports the hierarchical model of leukemia's where leukemia is initiated and maintained in vivo by a small fraction of rare LSCs or LICs, derived from HSCs. More recently, it has been shown that other less primitive fractions of leukemic cells, including CD34 + CD38 + cells, can generate leukemia in the more immunocompromised NSG mice model, supporting the notion that LICs can also arise from hematopoietic progenitors (Sarry et al. 2011; Dick 2009).

In a mouse model for AML, engineered to overexpress genes important for self-renewal like HoxA9 and Meis1 homeobox genes (Kroon et al. 1998), three immunophenotypic heterogenous LICs were identified and able to generate leukemia in irradiated wild-type mice (Gibbs et al. 2012). Furthermore, each immunophenotype of LIC from primary leukemia could recapitulate all the LIC immunophenotypes in secondary recipients. In this model, LICs represent malignant hematopoietic progenitors that are immunophenotypically heterogeneous but interestingly, all three distinct LICs have largely conserved signaling networks. They all exhibit activation of MAP-kinase, DNA methyltransferase, tyrosine kinase receptors, and PI 3-kinase/Akt pathways (Gibbs et al. 2012).

In human leukemia's, it is also difficult to identify immunophenotypic markers that strictly characterize all LICs. For example, two recent studies propose that LICs are hematopoietic progenitors with several different immunophenotypes (Sarry et al. 2011; Goardon et al. 2011) nevertheless the different immunophenotypic LICs can generate each other in vivo.

Self-renewal of LICs is dependent on quite similar mechanisms as HSCs. In the murine model of MLL-AF9-induced myeloid leukemia, which closely phenocopies human AML (Krivtsov et al. 2006), the LIC-enriched subpopulation show a repressed Akt activity and a nuclear localization of FOXOs which are active and required to inhibit myeloid maturation (Sykes et al. 2011). Importantly, c-JUN activity can counteract Akt activation or FOXOs inhibition to maintain long-term LIC properties (Sykes et al. 2011) indicating that, contrary to HSCs, new pathways can be developed in LICs to preserve their self-renewal capacities.

PI 3-Kinase Signaling in Hematopoietic Lineage Development

Several groups have investigated the role of PI 3-kinase in the hematopoietic lineage development. To give rise to all hematopoietic lineages, HSCs first differentiate to multipotent progenitors. Using the mouse embryonic stem cells model, Bone and Welham (Bone and Welham 2007) observed that inhibition of PI 3-kinase during earliest stage of development has no significant impact on the generation of progeny with hematopoietic potential but affect their expansion. However, PI 3-kinase signaling is required at a stage following hematopoietic progenitor formation for the development of mature hematopoietic lineages

(Bone and Welham 2007). Deletion of class I PI 3-kinase subunits in mouse models has highlighted the important role of p85α, p110γ, and p110δ for lymphocytes development and proliferation (reviewed in (Vanhaesebroeck et al. 2005)). CD19 is a major regulator of PI 3-kinase activity in B cells (Tuveson et al. 1993) and CD19 deficient mice lack some of the B cell subpopulations (CD19 + CD5 + and Marginal Zone B cells) suggesting that PI 3-kinase activation through CD19 signaling is important in B cell differentiation.

In erythroid progenitors PI 3-kinase is involved together with several tyrosine-phosphorylated proteins (IRS2, SHIP, Gab1, and Epo-R) in response to Epo (Bouscary et al. 2003). Activation of PI 3-kinase by Epo leads to the phosphorylation of Akt which is important to sustain erythropoiesis. Epo stimulation enhances the level of cyclins D1 and D3 in a PI 3-kinase-dependant manner. In addition, PI 3-kinase activation downregulates p27kip1 level in erythroid progenitors via proteasome degradation (Bouscary et al. 2003). Among PI 3-kinases subunits, the p85α regulatory subunit is of special interest for erythropoiesis as p85α-deficient mouse embryos lack early and late erythroid progenitors (Huddleston et al. 2003).

Development of megakaryocytes from hematopoietic progenitors involves the PI 3-kinase/Akt pathway downstream of the NOTCH pathway, especially activated by its ligand Delta-like 1 (Watts et al. 2010; Cornejo et al. 2011). However, the PI 3-kinase/Akt pathway does not seem to be indispensable to NOTCH induced megakaryocyte development at the mouse HSC level (Cornejo et al. 2011). In fact, the exact role of the different class I PI 3-kinases in megakaryocyte differentiation remain poorly documented. Nevertheless, expression of a kinase-dead form of Akt induces the depletion of the megakaryocyte pool whereas its constitutive activation or the expression of a dominant-negative form of FOXO3a enhance megakaryocyte development (Nakao et al. 2008). Existent mouse models of class I PI 3-kinases deficiency indicate that independent invalidation of p110δ, p110γ, and p110β has no significant impact on platelet production at study state. These observations suggest that either compensatory mechanisms are taking place between these PI 3-kinases or p110α is the most important isoform in megakaryocyte maturation. Mouse models showing invalidation of p110α specifically in the megakaryocytes and platelets are currently under development and should clarify this point. Following megakaryocyte fragmentation platelets are delivered in the bloodstream in bone marrow sinusoids. All class I PI 3-kinases are expressed in platelets. The p110β isoform has been shown to play an important role in platelet functions. It is activated downstream of immune tyrosine-based activation motifs (ITAM) signaling, G protein-coupled receptors, and αIIbβ3 integrin in human and mouse platelets. A high level of PI3-kinase β activation appears to involve a synergistic action of tyrosine kinases and Gβ/γ subunits of heterotrimeric G-proteins, a condition often observed in vivo because platelets receive parallel stimulations upon vascular injury. This PI 3-kinase is also important to mediate integrin αIIbβ3 signaling under high shear stress conditions. Thus, PI 3-kinase β may be considered as an integrator amplifying concurrent signals. Accordingly, mouse exhibiting p110β inactivation selectively in megakaryocytes/platelets are resistant to thromboembolism induced by carotid injury (Martin et al. 2010) and PI 3-kinase β has been proposed as a target for the development of antithrombotic drugs (Jackson et al. 2005). Besides the critical role of PI 3-kinases β, the α and γ isoforms are likely to be involved to a lesser extent in platelet activation while the δ isoform is dispensable (Gratacap et al. 2011).

Granulopoiesis is also strongly affected by PI 3-kinase or Akt inhibition at the hematopoietic progenitor stages (Buitenhuis et al. 2008). In fact, eosinophil differentiation is enhanced by Akt inactivation whereas the development of neutrophils and monocytes is decreased (Buitenhuis et al. 2008). Akt activation during granulopoiesis leads to the inhibition of GSK3β phosphorylation which in turn can no longer phosphorylate CEBPα resulting in neutrophil development. Conversely, GSK3β activation following Akt silencing during granulopoiesis leads to CEBPα activation and eosinophil development (Buitenhuis et al. 2008). Thus, the Akt/GSK3β/CEBPα axis contributes to granulopoietic lineage choice decisions (Fig. 8.1).

Several studies have addressed the question of the relative implication of the different isoforms of class I PI 3-kinases in normal or pathologic hematopoietic cells. They have revealed isoform specific roles but also some plasticity of the system. For instance, in macrophages, p110α controls DNA synthesis while p110β and p110δ are rather implicated in the regulation of actin cytoskeleton changes and chemotaxis (Vanhaesebroeck et al. 1999). In primary macrophages, p110δ is the major isoform regulating the PI 3-kinase/Akt pathway (Papakonstanti et al. 2008). Nevertheless, when macrophages are immortalized, Akt phosphorylation becomes dependent on p110α while p110δ has a minimal impact on its activation (Papakonstanti et al. 2008). Similarly, knockdown of p110α in the chronic myeloid leukemia (CML) cell line K562 is sufficient to downregulate Akt phosphorylation (Hui et al. 2008). In B cell lymphoma cell lines, the role of p110δ becomes less important compared to primary untransformed B cells (Bilancio et al. 2006). Similarly, p110α inhibitors are more efficient than p110δ ones to diminish Akt Ser473 phosphorylation in chronic lymphocytic leukemia (CLL), a disorder of B lymphocytes (Niedermeier et al. 2009). These data suggest that the relative contribution of p110 isoforms to signaling can differ significantly from normal to transformed cells and this seems to be independent of the modifications of their expression levels (Papakonstanti et al. 2008).

Deregulation of Class I PI 3-Kinases in Hematological Malignancies

The relatively ready access to primary blast cells from patients with hematological malignancies has allowed a number of studies related to the characterization of signaling pathways in these tumor cells. Genomic studies coupled to biochemical analysis and flow cytometry approaches have clearly shown deregulations of the PI 3-kinase pathway in many hematologic tumors. In contrast to solid tumors, the frequency of mutation in PI 3-kinase α in hematological malignancies is very low. Only 1 % (1/88) of activating mutations in p110α have been detected in leukemia (Karakas et al. 2006), no mutation (0/51) in myelodyspasia (Machado-Neto et al. 2011) and 4 % (2/44) in pediatric T-acute lymphocytic leukemia (T-ALL) (Gutierrez et al. 2009).

Mutations/deletions of PTEN or SHIP1 are also less frequent than in solid tumors. In AML only 3 % (1/30) of SHIP mutations have been observed (Luo et al. 2003) and no (0/62) or rare (1/59) mutations of PTEN have been reported (Liu et al. 2000; Aggerholm et al. 2000). PTEN alterations are more frequent in T-ALL with deletions in 8 % (4/46) and mutations in 27 % (12/44) of the cases (Gutierrez et al. 2009). In pediatric T-ALL, patients harboring PTEN gene alterations have a tendency toward inferior overall survival and event-free survival than those devoid of these genetic lesions (Jotta et al. 2010). Furthermore, primary T-ALL harbors frequent (33/61) deletions, insertions, and splicing variants of SHIP1 leading to translational inactivation and generation of shorter forms (Lo et al. 2009). It is however important to note that inactivation of PTEN and SHIP1 can occur via postraductional modifications. For instance, inactivation of PTEN can result from inhibitory phosphorylations through mechanisms involving CK2 or ROS-dependant oxidation in T-ALL cells (Silva et al. 2008).

It is noteworthy that Akt mutations have not been reported yet in leukemia (Tibes et al. 2008).

Besides these relatively rare cases of mutations, PI 3-kinase activation in hematological malignancies is mostly due to upstream signaling abnormalities at the level of tyrosine kinases or Ras. In AML, the PI 3-kinase/Akt signaling is activated in a high proportion of cases and the level of PtdIns(3,4,5)P_3 has been found elevated in a small series of primary AML cells (Xu et al. 2003; Sujobert et al. 2005; Recher et al. 2005). Except in one report (Tamburini et al. 2007), most studies in AML show that the constitutive Akt phosphorylation in blasts cells is significantly associated with shorter overall survival (Min et al. 2003; Kornblau et al. 2006). When analyzed by flow cytometry, the intensity of Akt Thr308 phosphorylation appeared significantly correlated with the cytogenetic risk, particularly with complex karyotype, and a decreased activity of PP2A (Gallay et al. 2009; Ruvolo et al. 2011). Conversely, this phosphorylation is quite low in the core binding factor AML group which has a better prognosis (Gallay et al. 2009). Moreover, constitutive Akt phosphorylation is found in mononuclear cells of high risk myelodysplasic syndromes (i.e patients with a high risk to develop AML) but not in low risk MDS or normal bone marrow (Nyakern et al. 2006).

The constitutive activation of FMS-like tyrosine kinase 3 by internal tandem duplication (Flt3-ITD) or point mutation observed in 30 % of AML cases (Brandts et al. 2005) and the much less frequent activating mutations in c-Kit (Hashimoto et al. 2003) contribute to PI 3-kinase/Akt signaling activation. It is noteworthy that this is not the case of all deregulated tyrosine kinase receptors in AML. The c-Met-related receptor tyrosine kinase Ron is especially expressed in normal HSCs in which it is physiologically activated by its ligand, the macrophage-stimulating protein. The hypermethylation of the RON gene can lead to the synthesis of a

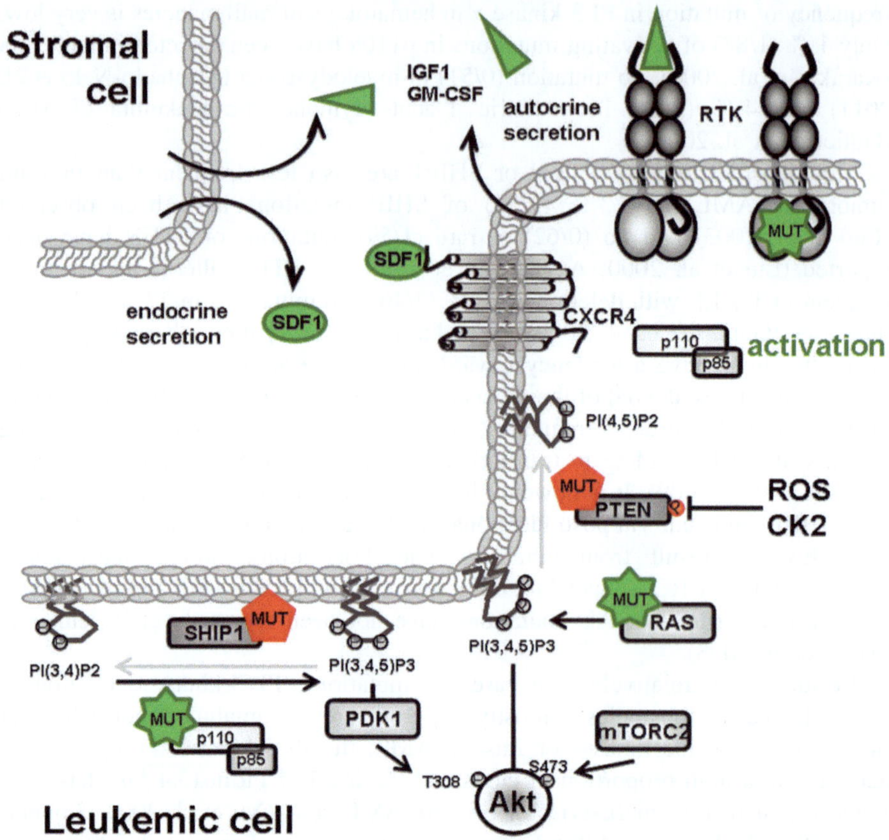

Fig. 8.2 *Deregulation of PI 3-kinase/Akt pathway leads to leukemogenesis*. PI 3-kinase activation in leukemia is mostly due to upstream signaling abnormalities at the level of receptors (RTK mutations, autocrine/endocrine secretion of growth factors) or Ras (mutated in around 15 % of AML). In only rare cases it can be found activating mutations of PI 3-kinases and inhibiting mutations of SHIP or PTEN. Inhibiting phosphorylation of PTEN can be induced by high ROS levels or CK2

constitutively active short-form of Ron, sfRon (Angeloni et al. 2007). Recently, we observed the expression of SfRon in 43 % (37/86) of AMLs while it is undetectable in HSCs (Fialin et al. 2012). Although Ron is a receptor known to strongly activate PI 3-kinase/Akt, we found that sfRon does not account for the activation of this pathway in AML cells. However, sfRon modulates the mTORC1 pathway which is important for clonogenicity and survival of AML cells (Recher et al. 2005). In AML, activation of mTORC1 is largely independent of Akt, but uses a pathway involving the Src-kinase Lyn, which is constitutively activated in a large proportion of AML cells (Dos Santos et al. 2008). sfRon appears to interact with Lyn in a multiprotein complex involved in the control of mTORC1 signaling independently of PI 3-kinase/Akt.

Besides aberrant activation of tyrosine kinase receptors, mutations in KRAS, NRAS, and HRAS found in 5 %, 10 %, and 0.3 % of AML cases, respectively (see Catalog Of Somatic Mutations In Cancer, http://www.sanger.ac.uk/genetics/CGP/cosmic/), likely contribute to class I PI 3-kinase activation (Rodriguez-Viciana et al. 1994; Hu et al. 2003). Finally, the autocrine production of growth factors such as IGF1 and GM-CSF also impacts on the activation of PI 3-kinase in AML (Doepfner et al. 2007; Tazzari et al. 2007; Tamburini et al. 2008; Chapuis et al. 2010a), (Fig. 8.2).

Targeting PI 3-Kinase As a Therapeutic Option

Different studies have characterized the pattern of expression of class I PI3-kinases in primary AML cells. The p110δ catalytic subunits and the p85 regulatory subunits are expressed in all AML cases. The expression of other class IA PI 3-kinase catalytic subunits, p110α and β, are variable (Sujobert et al. 2005; Billottet et al. 2006). Using selective inhibitors it has been proposed that a large part of the constitutive activation of PI 3-kinase in AML cells is attributable to p110δ (Sujobert et al. 2005; Billottet et al. 2006) and targeting this kinase affects the proliferative and clonogenic properties of AML cells ex vivo.

In these conditions, could a selective PI 3-kinase δ inhibition be sufficient to treat AML? This strategy would have the advantage to target hematopoietic cells thus reducing potential toxicity to non-hematopoietic tissues. In this context, clinical trials are ongoing to evaluate its effect on AML and other hematological malignancies. However, ex vivo, in primary AML cells, a selective inhibitor of PI 3-kinase δ reduces proliferation but weakly increases apoptosis. Thus, it is unlikely that this single inhibitor will prove sufficient to exert long-term disease control by eradicating LICs. To note, in primary acute promyelocytic leukemia cells, the PI 3-kinase δ inhibitor IC87114 can trigger apoptosis in the presence or in the absence of the differentiating agent ATRA suggesting that it could be envisaged to treat this subclass of AML which, in contrary to other subtypes of AML, is now efficiently treated by a combination of ATRA and conventional chemotherapy (Billottet et al. 2009).

It is important to note that p110δ is a major regulator of inflammatory responses and its inhibition is considered an excellent potential therapeutic strategy for reducing inflammation (Ji et al. 2007; Rommel et al. 2007). Accordingly, p110δ-selective drugs are currently under investigation as a treatment for rheumatoid arthritis or asthma. Therefore, inhibition of PI 3-kinase δ may decrease the impact of immune cells in eliminating tumor cells, including the cytolytic capacity of NK cells (Zebedin et al. 2008).

Moreover, in some conditions, transformed cells may compensate a loss of PI 3-kinase δ activity by using other class I PI3-kinases. For instance, as mentioned above, in primary macrophages stimulated with CSF1, p110δ selectively controls Akt activation; but under immortalized conditions, p110α take up a more prominent role in CSF1-receptor signaling (Papakonstanti et al. 2008). Similarly, in immortalized HSCs, p110β alone can sustain proliferation, despite the fact that p110α and p110δ are the major HSCs class I isoforms (Foukas et al. 2010). Furthermore, our recent results suggest that, in some cases, PI 3-kinase α is also contributing to AML (FV personal observation). Indeed, we observed that treatment of the AML cell lines KG1, KG1a, and HL60 with the p110α inhibitor PIK75 at relatively low doses decreases cell proliferation and induces a massive apoptosis. Conversely, the p110β (TGX221), p110γ (AS252424), and p110δ (IC87114) inhibitors have only a slight effect on cell proliferation or apoptosis in KG1 and HL60 cells. The effects of PIK75 were also observed in a significant number of primary AML cases but not in normal HSCs. These results suggest that such a compound could be of interest in AML treatment, however, it remains to be established that PI 3-kinase α is the only target of PIK75 in these cells.

Sensitization of AML cells, particularly LICs, to conventional chemotherapy by PI 3-kinase inhibition is an attractive option to develop innovative therapeutic treatments. Drug resistance is considered to be a multifactorial phenomenon involving several mechanisms such as increased DNA damage repair, reduced apoptosis, and increased energy-dependant drug efflux (McLornan et al. 2007). PTEN expression has been shown to restore the chemosensitivity of AML cell lines to anti-cancer agents (Tabellini et al. 2005; Zhou et al. 2003) whereas its inhibition by miR-21 induces anthracycline resistance in K562 leukemic cells (Bai et al. 2011).

The drug efflux pump MRP1, a member of the ATP-binding cassette (ABC) membrane transporters, is expressed in about 30 % of AML (Tazzari et al. 2007; Walter et al. 2003). Its expression, as well as the capacity of leukemic blasts to extrude viable dye, is downregulated by addition of pan PI 3-kinase inhibitors (Tazzari et al. 2007) suggesting that in leukemic cells, the PI 3-kinase signaling promotes, at least in part, chemoresistance via drug efflux. Consistent with these observations inhibition of PI 3-kinase δ by a selective inhibitor in primary AML cells, efficiently sensitizes tumor cells to the topoisomerase 2 inhibitor etoposide/VP16 (Billottet et al. 2006).

As mentioned above, molecules secreted by neighboring cells are critical for the maintenance and expansion of pathologic cells but also for drug resistance. CXCL12 (or Stromal cell-Derived Factor 1) is secreted by marrow stroma cells and activates the CXCR4 receptor pathway and downstream pathways including PI 3-kinase/Akt. CXCL12 and its receptor CXCR4 have been shown to contribute to the mechanism of cell adhesion-mediated drug resistance possibly via PI 3-kinases (Burger et al. 2005; Liesveld et al. 2007; Zeng et al. 2006).

Combining PI 3-kinase inhibition with other therapeutic agents targeting signal transduction actors may improve efficacy. For example, the tyrosine kinase Lyn is constitutively activated in most AML cases (Dos Santos et al. 2008) and contributes to mTORC1 pathway activation in a PI 3-kinase independent way. Inhibition of mTORC1 may amplify the activation of the PI 3-kinase pathway via IGF1 autocrine regulation (Tamburini et al. 2008). These observations provide rational reasons to inhibit both pathways simultaneously. Accordingly, association of PI 3-kinase and mTORC1 inhibitors is more effective in inducing apoptosis of AML cells than the drugs used independently. Concomitant inhibition of PI 3-kinase and mTORC1 pathways by RAD001 plus IC87114 (Tamburini et al. 2008) or by rapamycin plus LY294002 (Mise et al. 2011) induces additive anti-proliferative effects in AML and T-ALL primary samples. The dual inhibitor PI-103 targeting mTORC1 and class I PI 3-kinases inhibits leukemic proliferation, clonogenicity of leukemic progenitors, and induces apoptosis, especially in the compartment containing LICs (Park et al. 2008). Furthermore, PI-103 has additive proapoptotic effects with etoposide/VP16 (Park et al. 2008). Similarly, the dual mTORC1 and class I PI 3-kinases inhibitor NVP-BEZ235 induces a marked inhibition of protein translation and potent antileukemic activities in AML without affecting normal hematopoiesis ex vivo (Chapuis et al. 2010b).

Inhibiting simultaneously class I PI 3-kinases/Akt and mTORC1 is expected to induce undesirable metabolic effects, particularly increased blood insulin and/or glucose levels. These effects seem to be relatively modest, as suggested by early clinical trials results (Martelli et al. 2009). Overall these data suggest that combining PI 3-kinase inhibitors with mTORC1 pathways inhibitors or with conventional chemotherapy could be an attractive therapeutic option.

Conclusions

A number of studies using in vitro and in vivo approaches have shown the important role of the class I PI 3-kinase pathway in the control of normal hematopoiesis. While the role of PI 3-kinase α in normal hematopoietic cells remains hill identified, it is well established that PI 3-kinases δ and γ are particularly involved in the development and function of B cells, T cells, and NK cells. PI 3-kinase β is rather implicated in the regulation of platelet functions. At the level of HSCs, the implication of the PI 3-kinase signaling in the control of the balance between quiescence and proliferation has been demonstrated in vivo in mice. Therefore, it is not surprising that the class I PI 3-kinase signaling is found aberrantly activated in a high proportion of primary cells from patients with hematological malignancies. In many cases, including in AML, activation of this pathway is mainly due to the deregulation of upstream regulators as mutations on PI 3-kinases, PTEN, and Akt are relatively rare in these tumors. Since the PI 3-kinase pathway influences survival, proliferation, and drug resistance of leukemic cells, it appears as an attractive target for therapy. In a large proportion of AML cases, PI 3-kinase δ is constitutively activated. Ex vivo, its inhibition decreases their proliferation suggesting that this isoform could be a good candidate for therapeutic targeting. However, the use of a PI 3-kinase δ inhibitor as a single agent is unlikely to exert long-term effects in AML as it will be insufficient to eradicate LICs. Therapeutic options combining PI 3-kinase inhibitors and cytotoxic drugs or molecules targeting other signaling molecules deregulated in AML may be of interest. Molecular diagnosis of AML cells and mice models with xenografted human AML cells should be useful to evaluate the sensitivity to PI

3-kinase inhibitors in combination with other targeted drugs and to identify subsets of patients that are likely to benefit from these combinations. This may be an important preclinical step before conducting clinical trials with the novel targeted agents that become now available, including PI 3-kinase inhibitors or dual inhibitors targeting PI 3-kinase/Akt and mTORC1.

References

Aggerholm A, Gronbaek K, Guldberg P, Hokland P (2000) Mutational analysis of the tumour suppressor gene MMAC1/PTEN in malignant myeloid disorders. Eur J Haematol 65:109–113

Angeloni D, Danilkovitch-Miagkova A, Ivanova T, Braga E, Zabarovsky E, Lerman MI (2007) Hypermethylation of Ron proximal promoter associates with lack of full-length Ron and transcription of oncogenic short-Ron from an internal promoter. Oncogene 26:4499–4512

Arai F, Hirao A, Ohmura M, Sato H, Matsuoka S, Takubo K, Ito K, Koh GY, Suda T (2004) Tie2/angiopoietin-1 signaling regulates hematopoietic stem cell quiescence in the bone marrow niche. Cell 118:149–161

Bai H, Xu R, Cao Z, Wei D, Wang C (2011) Involvement of miR-21 in resistance to daunorubicin by regulating PTEN expression in the leukaemia K562 cell line. FEBS Lett 585:402–408

Bilancio A, Okkenhaug K, Camps M, Emery JL, Ruckle T, Rommel C, Vanhaesebroeck B (2006) Key role of the p110delta isoform of PI3 K in B-cell antigen and IL-4 receptor signaling: comparative analysis of genetic and pharmacologic interference with p110delta function in B cells. Blood 107:642–650

Billottet C, Grandage VL, Gale RE, Quattropani A, Rommel C, Vanhaesebroeck B, Khwaja A (2006) A selective inhibitor of the p110delta isoform of PI 3-kinase inhibits AML cell proliferation and survival and increases the cytotoxic effects of VP16. Oncogene 25:6648–6659

Billottet C, Banerjee L, Vanhaesebroeck B, Khwaja A (2009) Inhibition of class I phosphoinositide 3-kinase activity impairs proliferation and triggers apoptosis in acute promyelocytic leukemia without affecting atra-induced differentiation. Cancer Res 69:1027–1036

Bone HK, Welham MJ (2007) Phosphoinositide 3-kinase signalling regulates early development and developmental haemopoiesis. J Cell Sci 120:1752–1762

Bonnet D, Dick JE (1997) Human acute myeloid leukemia is organized as a hierarchy that originates from a primitive hematopoietic cell. Nat Med 3:730–737

Bouscary D, Pene F, Claessens YE, Muller O, Chretien S, Fontenay-Roupie M, Gisselbrecht S, Mayeux P, Lacombe C (2003) Critical role for PI 3-kinase in the control of erythropoietin-induced erythroid progenitor proliferation. Blood 101:3436–3443

Brandts CH, Sargin B, Rode M, Biermann C, Lindtner B, Schwable J, Buerger H, Muller-Tidow C, Choudhary C, McMahon M et al (2005) Constitutive activation of Akt by Flt3 internal tandem duplications is necessary for increased survival, proliferation, and myeloid transformation. Cancer Res 65:9643–9650

Buitenhuis M, Verhagen LP, van Deutekom HW, Castor A, Verploegen S, Koenderman L, Jacobsen SE, Coffer PJ (2008) Protein kinase B (c-akt) regulates hematopoietic lineage choice decisions during myelopoiesis. Blood 111:112–121

Burger M, Hartmann T, Krome M, Rawluk J, Tamamura H, Fujii N, Kipps TJ, Burger JA (2005) Small peptide inhibitors of the CXCR4 chemokine receptor (CD184) antagonize the activation, migration, and antiapoptotic responses of CXCL12 in chronic lymphocytic leukemia B cells. Blood 106:1824–1830

Butler JM, Nolan DJ, Vertes EL, Varnum-Finney B, Kobayashi H, Hooper AT, Seandel M, Shido K, White IA, Kobayashi M et al (2010) Endothelial cells are essential for the self-renewal and repopulation of Notch-dependent hematopoietic stem cells. Cell Stem Cell 6:251–264

Calvi LM, Adams GB, Weibrecht KW, Weber JM, Olson DP, Knight MC, Martin RP, Schipani E, Divieti P, Bringhurst FR et al (2003) Osteoblastic cells regulate the haematopoietic stem cell niche. Nature 425:841–846

Carlton JG, Cullen PJ (2005) Coincidence detection in phosphoinositide signaling. Trends Cell Biol 15:540–547

Chabanon A, Desterke C, Rodenburger E, Clay D, Guerton B, Boutin L, Bennaceur-Griscelli A, Pierre-Louis O, Uzan G, Abecassis L et al (2008) A cross-talk between stromal cell-derived factor-1 and transforming growth factor-beta controls the quiescence/cycling switch of CD34(+) progenitors through FoxO3 and mammalian target of rapamycin. Stem Cells 26:3150–3161

Chapuis N, Tamburini J, Cornillet-Lefebvre P, Gillot L, Bardet V, Willems L, Park S, Green AS, Ifrah N, Dreyfus F et al (2010a) Autocrine IGF-1/IGF-1R signaling is responsible for constitutive PI3 K/Akt activation in acute myeloid leukemia: therapeutic value of neutralizing anti-IGF-1R antibody. Haematologica 95:415–423

Chapuis N, Tamburini J, Green AS, Vignon C, Bardet V, Neyret A, Pannetier M, Willems L, Park S, Macone A et al (2010b) Dual inhibition of PI3K and mTORC1/2 signaling by NVP-BEZ235 as a new therapeutic strategy for acute myeloid leukemia. Clin Cancer Res 16:5424–5435

Chetram MA, Don-Salu-Hewage AS, Hinton CV (2011) ROS enhances CXCR4-mediated functions through inactivation of PTEN in prostate cancer cells. Biochem Biophys Res Commun 410:195–200

Ciraolo E, Iezzi M, Marone R, Marengo S, Curcio C, Costa C, Azzolino O, Gonella C, Rubinetto C, Wu H, et al (2008) Phosphoinositide 3-kinase p110beta activity: key role in metabolism and mammary gland cancer but not development. Sci Signal 1:ra3

Cornejo MG, Mabialah V, Sykes SM, Khandan T, Lo Celso C, Lopez CK, Rivera-Munoz P, Rameau P, Tothova Z, Aster JC et al (2011) Crosstalk between NOTCH and AKT signaling during murine megakaryocyte lineage specification. Blood 118:1264–1273

Costa C, Hirsch E (2010) More than just kinases: the scaffolding function of PI3 K. Curr Top Microbiol Immunol 346:171–181

Di Paolo G, De Camilli P (2006) Phosphoinositides in cell regulation and membrane dynamics. Nature 443:651–657

Dick JE (2009) Looking ahead in cancer stem cell research. Nat Biotechnol 27:44–46

Doepfner KT, Spertini O, Arcaro A (2007) Autocrine insulin-like growth factor-I signaling promotes growth and survival of human acute myeloid leukemia cells via the phosphoinositide 3-kinase/Akt pathway. Leukemia 21:1921–1930

Dos Santos C, Demur C, Bardet V, Prade-Houdellier N, Payrastre B, Recher C (2008) A critical role for Lyn in acute myeloid leukemia. Blood 111:2269–2279

Fialin C, Larrue C, Vergez F, Sarry JE, Bertoli S, Mansat- De Mas V, Demur C, Delabesse E, Payrastre B, Manenti S, Roche S, Recher C (2012) The short form of RON is expressed in acute myeloid leukemia and sensitizes leukemic cells to cMET inhibitors. Leukemia (in press)

Foukas LC, Berenjeno IM, Gray A, Khwaja A, Vanhaesebroeck B (2010) Activity of any class IA PI3 K isoform can sustain cell proliferation and survival. Proc Natl Acad Sci U S A 107:11381–11386

Gallay N, Dos Santos C, Cuzin L, Bousquet M, Simmonet Gouy V, Chaussade C, Attal M, Payrastre B, Demur C, Recher C (2009) The level of AKT phosphorylation on threonine 308 but not on serine 473 is associated with high-risk cytogenetics and predicts poor overall survival in acute myeloid leukaemia. Leukemia 23:1029–1038

Geering B, Cutillas PR, Nock G, Gharbi SI, Vanhaesebroeck B (2007) Class IA phosphoinositide 3-kinases are obligate p85–p110 heterodimers. Proc Natl Acad Sci U S A 104:7809–7814

Gibbs KD Jr, Jager A, Crespo O, Goltsev Y, Trejo A, Richard CE, Nolan GP (2012) Decoupling of tumor-initiating activity from stable immunophenotype in HoxA9-Meis1-Driven AML. Cell Stem Cell 10:210–217

Goardon N, Marchi E, Atzberger A, Quek L, Schuh A, Soneji S, Woll P, Mead A, Alford KA, Rout R et al (2011) Coexistence of LMPP-like and GMP-like leukemia stem cells in acute myeloid leukemia. Cancer Cell 19:138–152

Goyama S, Yamamoto G, Shimabe M, Sato T, Ichikawa M, Ogawa S, Chiba S, Kurokawa M (2008) Evi-1 is a critical regulator for hematopoietic stem cells and transformed leukemic cells. Cell Stem Cell 3:207–220

Gratacap MP, Guillermet-Guibert J, Martin V, Chicanne G, Tronchere H, Gaits-Iacovoni F, Payrastre B (2011) Regulation and roles of PI3 Kbeta, a major actor in platelet signaling and functions. Adv Enzyme Regul 51:106–116

Gutierrez A, Sanda T, Grebliunaite R, Carracedo A, Salmena L, Ahn Y, Dahlberg S, Neuberg D, Moreau LA, Winter SS et al (2009) High frequency of PTEN, PI3 K, and AKT abnormalities in T-cell acute lymphoblastic leukemia. Blood 114:647–650

Hashimoto K, Matsumura I, Tsujimura T, Kim DK, Ogihara H, Ikeda H, Ueda S, Mizuki M, Sugahara H, Shibayama H et al (2003) Necessity of tyrosine 719 and phosphatidylinositol 3'-kinase-mediated signal pathway in constitutive activation and oncogenic potential of c-kit receptor tyrosine kinase with the Asp814Val mutation. Blood 101:1094–1102

Hazen AL, Smith MJ, Desponts C, Winter O, Moser K, Kerr WG (2009) SHIP is required for a functional hematopoietic stem cell niche. Blood 113:2924–2933

Helgason CD, Damen JE, Rosten P, Grewal R, Sorensen P, Chappel SM, Borowski A, Jirik F, Krystal G, Humphries RK (1998) Targeted disruption of SHIP leads to hemopoietic perturbations, lung pathology, and a shortened life span. Genes Dev 12:1610–1620

Helgason CD, Antonchuk J, Bodner C, Humphries RK (2003) Homeostasis and regeneration of the hematopoietic stem cell pool are altered in SHIP-deficient mice. Blood 102:3541–3547

Himburg HA, Muramoto GG, Daher P, Meadows SK, Russell JL, Doan P, Chi JT, Salter AB, Lento WE, Reya T et al (2010) Pleiotrophin regulates the expansion and regeneration of hematopoietic stem cells. Nat Med 16:475–482

Hu L, Shi Y, Hsu JH, Gera J, Van Ness B, Lichtenstein A (2003) Downstream effectors of oncogenic ras in multiple myeloma cells. Blood 101:3126–3135

Huddleston H, Tan B, Yang FC, White H, Wenning MJ, Orazi A, Yoder MC, Kapur R, Ingram DA (2003) Functional p85alpha gene is required for normal murine fetal erythropoiesis. Blood 102:142–145

Jackson SP, Schoenwaelder SM, Goncalves I, Nesbitt WS, Yap CL, Wright CE, Kenche V, Anderson KE, Dopheide SM, Yuan Y et al (2005) PI 3-kinase p110beta: a new target for antithrombotic therapy. Nat Med 11:507–514

Jang YY, Sharkis SJ (2007) A low level of reactive oxygen species selects for primitive hematopoietic stem cells that may reside in the low-oxygenic niche. Blood 110:3056–3063

Ji H, Rintelen F, Waltzinger C, Bertschy Meier D, Bilancio A, Pearce W, Hirsch E, Wymann MP, Ruckle T, Camps M et al (2007) Inactivation of PI3 Kgamma and PI3 Kdelta distorts T-cell development and causes multiple organ inflammation. Blood 110:2940–2947

Jotta PY, Ganazza MA, Silva A, Viana MB, da Silva MJ, Zambaldi LJ, Barata JT, Brandalise SR, Yunes JA (2010) Negative prognostic impact of PTEN mutation in pediatric T-cell acute lymphoblastic leukemia. Leukemia 24:239–242

Juntilla MM, Patil VD, Calamito M, Joshi RP, Birnbaum MJ, Koretzky GA (2010) AKT1 and AKT2 maintain hematopoietic stem cell function by regulating reactive oxygen species. Blood 115:4030–4038

Karakas B, Bachman KE, Park BH (2006) Mutation of the PIK3CA oncogene in human cancers. Br J Cancer 94:455–459

Kharas MG, Okabe R, Ganis JJ, Gozo M, Khandan T, Paktinat M, Gilliland DG, Gritsman K (2010) Constitutively active AKT depletes hematopoietic stem cells and induces leukemia in mice. Blood 115:1406–1415

Kiel MJ, Morrison SJ (2008) Uncertainty in the niches that maintain haematopoietic stem cells. Nat Rev Immunol 8:290–301

Kobayashi H, Butler JM, O'Donnell R, Kobayashi M, Ding BS, Bonner B, Chiu VK, Nolan DJ, Shido K, Benjamin L et al (2010) Angiocrine factors from Akt-activated endothelial cells balance self-renewal and differentiation of haematopoietic stem cells. Nat Cell Biol 12:1046–1056

Kornblau SM, Womble M, Qiu YH, Jackson CE, Chen W, Konopleva M, Estey EH, Andreeff M (2006) Simultaneous activation of multiple signal transduction pathways confers poor prognosis in acute myelogenous leukemia. Blood 108:2358–2365

Krivtsov AV, Twomey D, Feng Z, Stubbs MC, Wang Y, Faber J, Levine JE, Wang J, Hahn WC, Gilliland DG et al (2006) Transformation from committed progenitor to leukaemia stem cell initiated by MLL-AF9. Nature 442:818–822

Kroon E, Krosl J, Thorsteinsdottir U, Baban S, Buchberg AM, Sauvageau G (1998) Hoxa9 transforms primary bone marrow cells through specific collaboration with Meis1a but not Pbx1b. EMBO J 17:3714–3725

Kulkarni S, Sitaru C, Jakus Z, Anderson KE, Damoulakis G, Davidson K, Hirose M, Juss J, Oxley D, Chessa TA, et al (2011) PI3 Kbeta plays a critical role in neutrophil activation by immune complexes. Sci Signal 4:ra23

Lee EJ, Kim N, Kang KH, Kim JW (2011) Phosphorylation/inactivation of PTEN by Akt-independent PI3 K signaling in retinal pigment epithelium. Biochem Biophys Res Commun 414:384–389

Lemmon MA (2008) Membrane recognition by phospholipid-binding domains. Nat Rev Mol Cell Biol 9:99–111

Liesveld JL, Bechelli J, Rosell K, Lu C, Bridger G, Phillips G 2nd, Abboud CN (2007) Effects of AMD3100 on transmigration and survival of acute myelogenous leukemia cells. Leuk Res 31:1553–1563

Liu TC, Lin PM, Chang JG, Lee JP, Chen TP, Lin SF (2000) Mutation analysis of PTEN/MMAC1 in acute myeloid leukemia. Am J Hematol 63:170–175

Lo TC, Barnhill LM, Kim Y, Nakae EA, Yu AL, Diccianni MB (2009) Inactivation of SHIP1 in T-cell acute lymphoblastic leukemia due to mutation and extensive alternative splicing. Leuk Res 33:1562–1566

Luo JM, Yoshida H, Komura S, Ohishi N, Pan L, Shigeno K, Hanamura I, Miura K, Iida S, Ueda R et al (2003) Possible dominant-negative mutation of the SHIP gene in acute myeloid leukemia. Leukemia 17:1–8

Machado-Neto JA, Traina F, Lazarini M, Campos Pde M, Pagnano KB, Lorand-Metze I, Costa FF, Saad ST (2011) Screening for hotspot mutations in PI3 K, JAK2, FLT3 and NPM1 in patients with myelodysplastic syndromes. Clinics (Sao Paulo) 66:793–799

Martelli AM, Evangelisti C, Chiarini F, Grimaldi C, Manzoli L, McCubrey JA (2009) Targeting the PI3K/AKT/mTOR signaling network in acute myelogenous leukemia. Expert Opin Investig Drugs 18:1333–1349

Martin V, Guillermet-Guibert J, Chicanne G, Cabou C, Jandrot-Perrus M, Plantavid M, Vanhaesebroeck B, Payrastre B, Gratacap MP (2010) Deletion of the p110beta isoform of phosphoinositide 3-kinase in platelets reveals its central role in Akt activation and thrombus formation in vitro and in vivo. Blood 115:2008–2013

McCrea HJ, De Camilli P (2009) Mutations in phosphoinositide metabolizing enzymes and human disease. Physiology (Bethesda) 24:8–16

McLornan DP, McMullin MF, Johnston P, Longley DB (2007) Molecular mechanisms of drug resistance in acute myeloid leukaemia. Expert Opin Drug Metab Toxicol 3:363–377

Min YH, Eom JI, Cheong JW, Maeng HO, Kim JY, Jeung HK, Lee ST, Lee MH, Hahn JS, Ko YW (2003) Constitutive phosphorylation of Akt/PKB protein in acute myeloid leukemia: its significance as a prognostic variable. Leukemia 17:995–997

Mise J, Dembitz V, Banfic H, Visnjic D (2011) Combined inhibition of PI3K and mTOR exerts synergistic antiproliferative effect, but diminishes differentiative properties of rapamycin in acute myeloid leukemia cells. Pathol Oncol Res 17:645–656

Miyamoto K, Araki KY, Naka K, Arai F, Takubo K, Yamazaki S, Matsuoka S, Miyamoto T, Ito K, Ohmura M et al (2007) Foxo3a is essential for maintenance of the hematopoietic stem cell pool. Cell Stem Cell 1:101–112

Moody JL, Xu L, Helgason CD, Jirik FR (2004) Anemia, thrombocytopenia, leukocytosis, extramedullary hematopoiesis, and impaired progenitor function in Pten ± SHIP-/- mice: a novel model of myelodysplasia. Blood 103:4503–4510

Nakao T, Geddis AE, Fox NE, Kaushansky K (2008) PI3 K/Akt/FOXO3a pathway contributes to thrombopoietin-induced proliferation of primary megakaryocytes in vitro and in vivo via modulation of p27(Kip1). Cell Cycle 7:257–266

Niedermeier M, Hennessy BT, Knight ZA, Henneberg M, Hu J, Kurtova AV, Wierda WG, Keating MJ, Shokat KM, Burger JA (2009) Isoform-selective phosphoinositide 3'-kinase inhibitors inhibit CXCR4 signaling and overcome stromal cell-mediated drug resistance in chronic lymphocytic leukemia: a novel therapeutic approach. Blood 113:5549–5557

Nyakern M, Tazzari PL, Finelli C, Bosi C, Follo MY, Grafone T, Piccaluga PP, Martinelli G, Cocco L, Martelli AM (2006) Frequent elevation of Akt kinase phosphorylation in blood marrow and peripheral blood mononuclear cells from high-risk myelodysplastic syndrome patients. Leukemia 20:230–238

Papakonstanti EA, Zwaenepoel O, Bilancio A, Burns E, Nock GE, Houseman B, Shokat K, Ridley AJ, Vanhaesebroeck B (2008) Distinct roles of class IA PI3 K isoforms in primary and immortalised macrophages. J Cell Sci 121:4124–4133

Park S, Chapuis N, Bardet V, Tamburini J, Gallay N, Willems L, Knight ZA, Shokat KM, Azar N, Viguie F et al (2008) PI-103, a dual inhibitor of Class IA phosphatidylinositide 3-kinase and mTOR, has antileukemic activity in AML. Leukemia 22:1698–1706

Payrastre B, Missy K, Giuriato S, Bodin S, Plantavid M, Gratacap M (2001) Phosphoinositides: key players in cell signalling, in time and space. Cell Signal 13:377–387

Payrastre B, Gaits-Iacovoni F, Sansonetti P, Tronchere H (2012) Phosphoinositides and cellular pathogens. Subcell Biochem 59:363–388

Pendaries C, Tronchere H, Plantavid M, Payrastre B (2003) Phosphoinositide signaling disorders in human diseases. FEBS Lett 546:25–31

Recher C, Beyne-Rauzy O, Demur C, Chicanne G, Dos Santos C, Mas VM, Benzaquen D, Laurent G, Huguet F, Payrastre B (2005) Antileukemic activity of rapamycin in acute myeloid leukemia. Blood 105:2527–2534

Rodriguez-Viciana P, Warne PH, Dhand R, Vanhaesebroeck B, Gout I, Fry MJ, Waterfield MD, Downward J (1994) Phosphatidylinositol-3-OH kinase as a direct target of Ras. Nature 370:527–532

Rommel C, Camps M, Ji H (2007) PI3 K delta and PI3 K gamma: partners in crime in inflammation in rheumatoid arthritis and beyond? Nat Rev Immunol 7:191–201

Ruvolo PP, Qui YH, Coombes KR, Zhang N, Ruvolo VR, Borthakur G, Konopleva M, Andreeff M, Kornblau SM (2011) Low expression of PP2A regulatory subunit B55alpha is associated with T308 phosphorylation of AKT and shorter complete remission duration in acute myeloid leukemia patients. Leukemia 25:1711–1717

Sarry JE, Murphy K, Perry R, Sanchez PV,. Secreto A, Keefer C, Swider CR, Strzelecki AC, Cavelier C, Recher C et al (2011) Human acute myelogenous leukemia stem cells are rare and heterogeneous when assayed in NOD/SCID/IL2Rgammac-deficient mice. J Clin Invest 121:384–395

Sasaki J, Kofuji S, Itoh R, Momiyama T, Takayama K, Murakami H, Chida S, Tsuya Y, Takasuga S, Eguchi S (2010) The PtdIns(3,4)P(2) phosphatase INPP4A is a suppressor of excitotoxic neuronal death. Nature 465:497–501

Scadden DT (2006) The stem-cell niche as an entity of action. Nature 441:1075–1079

Silva A, Yunes JA, Cardoso BA, Martins LR, Jotta PY, Abecasis M, Nowill AE, Leslie NR, Cardoso AA, Barata JT (2008) PTEN posttranslational inactivation and hyperactivation of the PI3 K/Akt pathway sustain primary T cell leukemia viability. J Clin Invest 118:3762–3774

Sujobert P, Bardet V, Cornillet-Lefebvre P, Hayflick JS, Prie N, Verdier F, Vanhaesebroeck B, Muller O, Pesce F, Ifrah N et al (2005) Essential role for the p110delta isoform in phosphoinositide 3-kinase activation and cell proliferation in acute myeloid leukemia. Blood 106:1063–1066

Sykes SM, Lane SW, Bullinger L, Kalaitzidis D, Yusuf R, Saez B, Ferraro F, Mercier F, Singh H, Brumme KM et al (2011) AKT/FOXO signaling enforces reversible differentiation blockade in myeloid leukemias. Cell 146:697–708

Tabellini G, Cappellini A, Tazzari PL, Fala F, Billi AM, Manzoli L, Cocco L, Martelli AM (2005) Phosphoinositide 3-kinase/Akt involvement in arsenic trioxide resistance of human leukemia cells. J Cell Physiol 202:623–634

Tamburini J, Elie C, Bardet V, Chapuis N, Park S, Broet P, Cornillet-Lefebvre P, Lioure B, Ugo V, Blanchet O et al (2007) Constitutive phosphoinositide 3-kinase/Akt activation represents a favorable prognostic factor in de novo acute myelogenous leukemia patients. Blood 110:1025–1028

Tamburini J, Chapuis N, Bardet V, Park S, Sujobert P, Willems L, Ifrah N, Dreyfus F, Mayeux P, Lacombe C et al (2008) Mammalian target of rapamycin (mTOR) inhibition activates phosphatidylinositol 3-kinase/Akt by up-regulating insulin-like growth factor-1 receptor signaling in acute myeloid leukemia: rationale for therapeutic inhibition of both pathways. Blood 111:379–382

Tamura N, Hazeki K, Okazaki N, Kametani Y, Murakami H, Takaba Y, Ishikawa Y, Nigorikawa K, Hazeki O (2009) Specific role of phosphoinositide 3-kinase p110alpha in the regulation of phagocytosis and pinocytosis in macrophages. Biochem J 423:99–108

Tazzari PL, Tabellini G, Bortul R, Papa V, Evangelisti C, Grafone T, Martinelli G, McCubrey JA, Martelli AM (2007a) The insulin-like growth factor-I receptor kinase inhibitor NVP-AEW541 induces apoptosis in acute myeloid leukemia cells exhibiting autocrine insulin-like growth factor-I secretion. Leukemia 21:886–896

Tazzari PL, Cappellini A, Ricci F, Evangelisti C, Papa V, Grafone T, Martinelli G, Conte R, Cocco L, McCubrey JA et al (2007b) Multidrug resistance-associated protein 1 expression is under the control of the phosphoinositide 3 kinase/Akt signal transduction network in human acute myelogenous leukemia blasts. Leukemia 21:427–438

Tibes R, Kornblau SM, Qiu Y, Mousses SM, Robbins C, Moses T, Carpten JD (2008) PI3 K/AKT pathway activation in acute myeloid leukaemias is not associated with AKT1 pleckstrin homology domain mutation. Br J Haematol 140:344–347

Tothova Z, Kollipara R, Huntly BJ, Lee BH, Castrillon DH, Cullen DE, McDowell EP, Lazo-Kallanian S, Williams IR, Sears C et al (2007) FoxOs are critical mediators of hematopoietic stem cell resistance to physiologic oxidative stress. Cell 128:325–339

Tuveson DA, Carter RH, Soltoff SP, Fearon DT (1993) CD19 of B cells as a surrogate kinase insert region to bind phosphatidylinositol 3-kinase. Science 260:986–989

Vanhaesebroeck B, Jones GE, Allen WE, Zicha D, Hooshmand-Rad R, Sawyer C, Wells C, Waterfield MD, Ridley AJ (1999) Distinct PI(3)Ks mediate mitogenic signalling and cell migration in macrophages. Nat Cell Biol 1:69–71

Vanhaesebroeck B, Ali K, Bilancio A, Geering B, Foukas LC (2005) Signalling by PI3 K isoforms: insights from gene-targeted mice. Trends Biochem Sci 30:194–204

Vanhaesebroeck B, Guillermet-Guibert J, Graupera M, Bilanges B (2010) The emerging mechanisms of isoform-specific PI3 K signalling. Nat Rev Mol Cell Biol 11:329–341

Walter RB, Raden BW, Hong TC, Flowers DA, Bernstein ID, Linenberger ML (2003) Multidrug resistance protein attenuates gemtuzumab ozogamicin-induced cytotoxicity in acute myeloid leukemia cells. Blood 102:1466–1473

Watts KL, Delaney C, Humphries RK, Bernstein ID, Kiem HP (2010) Combination of HOXB4 and Delta-1 ligand improves expansion of cord blood cells. Blood 116:5859–5866

Xu Q, Simpson SE, Scialla TJ, Bagg A, Carroll M (2003) Survival of acute myeloid leukemia cells requires PI3 kinase activation. Blood 102:972–980

Yilmaz OH, Valdez R, Theisen BK, Guo W, Ferguson DO, Wu H, Morrison SJ (2006) Pten dependence distinguishes haematopoietic stem cells from leukaemia-initiating cells. Nature 441:475–482

Yoshimi A, Goyama S, Watanabe-Okochi N, Yoshiki Y, Nannya Y, Nitta E, Arai S, Sato T, Shimabe M, Nakagawa M et al (2011) Evi1 represses PTEN expression and activates PI3 K/AKT/mTOR via interactions with polycomb proteins. Blood 117:3617–3628

Zanella F, Rosado A, Garcia B, Carnero A, Link W (2008) Chemical genetic analysis of FOXO nuclear-cytoplasmic shuttling by using image-based cell screening. Chem Bio Chem 9:2229–2237

Zebedin E, Simma O, Schuster C, Putz EM, Fajmann S, Warsch W, Eckelhart E, Stoiber D, Weisz E, Schmid JA et al (2008) Leukemic challenge unmasks a requirement for PI3Kdelta in NK cell-mediated tumor surveillance. Blood 112:4655–4664

Zeng Z, Samudio IJ, Munsell M, An J, Huang Z, Estey E, Andreeff M, Konopleva M (2006) Inhibition of CXCR4 with the novel RCP168 peptide overcomes stroma-mediated chemoresistance in chronic and acute leukemias. Mol Cancer Ther 5:3113–3121

Zhang J, Niu C, Ye L, Huang H, He X, Tong WG, Ross J, Haug J, Johnson T, Feng JQ et al (2003) Identification of the haematopoietic stem cell niche and control of the niche size. Nature 425:836–841

Zhang J, Grindley JC, Yin T, Jayasinghe S, He XC, Ross JT, Haug JS, Rupp D, Porter-Westpfahl KS, Wiedemann LM et al (2006) PTEN maintains haematopoietic stem cells and acts in lineage choice and leukaemia prevention. Nature 441:518–522

Zhou M, Gu L, Findley HW, Jiang R, Woods WG (2003) PTEN reverses MDM2-mediated chemotherapy resistance by interacting with p53 in acute lymphoblastic leukemia cells. Cancer Res 63:6357–6362

Chapter 9
The Diverse Functions of Phosphatidylinositol Transfer Proteins

Shamshad Cockcroft

Abstract Phosphatidylinositol transfer proteins (PITPs), comprising five members in the human genome are implicated in the non-vesicular traffic of phosphatidylinositol (PI) between intracellular membranes and the plasma membrane. Three members of the PITP family (PITPα, PITPβ, and RdgBβ (retinal degeneration type B) alt. name PITPNC1) are present as single domain proteins and two (RdgBαI and RdgBαII alt. name PITPNM1 and PITPNM2) are present as multi-domain proteins with the PITP domain located at the N-terminus. The hallmark of PITP proteins is to extract PI molecules from a membrane, sequester in its binding pocket and deposit the lipid to membranes. PITPs regulate the synthesis of phosphoinositides (PPIs) either by delivery of the substrate, PI to specific membrane compartments or by potentiating the activities of the lipid kinases, or both. In the light of recent studies, we propose that PITPs are regulators of phosphoinositide pathways by recruitment to membranes through specific protein interactions to promote molecular exchange between closely opposed membranes i.e., at membrane contact sites. Individual PITP proteins play highly specific roles in many biological processes including neurite outgrowth, membrane traffic, cytokinesis, and sensory transduction in mammals as well as in the model organisms, *Drosophila, Caenorhabditis elegans,* and zebrafish. The common requirement for the diverse functions for all PITPs is their ability to bind PI and coupling its function to phosphoinositide-dependent pathways.

S. Cockcroft (✉)
Department of Neuroscience, Physiology and Pharmacology, Division of Biosciences, University College London, London, WC1E 6JJ, UK
e-mail: s.cockcroft@ucl.ac.uk

M. Falasca (ed.), *Phosphoinositides and Disease*, Current Topics in Microbiology and Immunology 362, DOI: 10.1007/978-94-007-5025-8_9,

Abbreviations

PPIs	Phosphoinositides
PI	Phosphatidylinositol
PITP	Phosphatidylinositol transfer protein
RdgB	Retinal degeneration type B
PC	Phosphatidylcholine
$PI(4,5)P_2$	Phosphatidylinositol (4,5) bisphosphate
PA	Phosphatidic acid
HAD	Haloacid dehalogenase
PCTP	PC transfer protein
HL60	Human promyelocytic leukemia cells
DCC	Deleted in colon cancer
COP-1	Coat protein complex-1
SRC	Subrhabdomeric cisternae
PIS	PI synthase
CDS	CDP-DAG synthase
DAG	Diacylglycerol
ERG	Electroretinogram
ATRAP (alt. name AGTRAP)	Angiotensin II receptor-associated protein
AT1R	Angiotensin II Type 1 receptor

Introduction

Phosphoinositides (PPIs) are the phosphorylated derivatives of phosphatidylinositol (PI) that regulate fundamental biological processes including cell signaling, cell division, membrane trafficking, and cytoskeletal dynamics. PPIs comprise less than 1 % of cell lipids and are derived by phosphorylation of the inositol headgroup of PI by a variety of lipid kinases present in specific target organelles. The inositol headgroup of PI can be reversibly phosphorylated at three positions, 3, 4, and 5, and all seven PI derivatives, (PI(3)P, PI(4)P, PI(5)P), $PI(3,4)P_2$, $PI(3,5)P_2$, $PI(4,5)P_2$, and $PI(3,4,5)P_3$) are present in mammalian cells. The levels and turnover of PPIs are tightly controlled by a large set of lipid kinases and phosphatases as well as by phospholipase Cs. Phospholipase C cleaves mainly $PI(4,5)P_2$ and replenishment of $PI(4,5)P_2$ requires the synthesis of PI, which occurs at the endoplasmic reticulum. Trafficking of PI to different organelles can be accomplished by vesicular and non-vesicular mechanisms. As PI is largely water insoluble, their non-vesicular transport between membranes is predicted to require phosphatidylinositol transfer proteins (PITPs), which extract a PI molecule from a donor membrane, and after enclosing the bound PI molecule in a shielded pocket, transfer the PI to an acceptor membrane. Efficient exchange can be accomplished when transfer occurs across two membranes which are in close proximity.

PI transfer proteins are characterized by the presence of the PITP domain; a compact structure with a hydrophobic cavity which can be occupied by either one molecule of PI or phosphatidylcholine (PC) (Fig. 9.1). The first PITP protein, PITPα, was identified as a 35 kDa cytosolic protein that could facilitate the energy-independent transfer of PI or PC between membrane compartments *in vitro*. PITPα is a single domain protein and such proteins have been identified in both multicellular and unicellular organisms (Cockcroft and Garner 2011). However, in addition to being present as a solo domain protein, the PITP domain can also be found in combination with other domains (Fig. 9.2).

The human genome encodes five PITPs, all of which contain their PITP domain at the N-terminus (Fig 9.2). Three of the members of the PITP family are small, soluble proteins (PITPα, PITPβ, and RdgBβ (Retinal degeneration type B) alt name PITPNC1), whereas the RdgBα (alt. name PITPNM or Nir) proteins are large, multi-domain, membrane-associated proteins characterized by the presence of the DDHD and LNS2 domains. The DDHD domain is also found in PA-PLA$_1$ and Sec23 interacting protein (Higgs et al. 1998; Nakajima et al. 2002; Mizoguchi et al. 2000; Tani et al. 1999) whilst the LNS2 domain is found in the lipins which code for PA phosphatases. The LNS2 domain in lipins contains the halo acid dehalogenase (HAD) catalytic site (DIDGT motif) that is required for activity. This motif is not conserved in the RdgBα proteins suggesting that they are unlikely to possess PA phosphatase activity. RdgBαIII, which shares similarity with the RdgBα proteins, is also included here although it lacks the PITP domain.

The PITPs are categorized according to their sequence identity into class I (PITPα and PITPβ), and class II proteins (RdgB proteins). Class II is further subdivided into A and B; Class IIA comprise RdgBα proteins whilst Class IIB consists of RdgBβ (Allen-Baume et al. 2002). RdgBα proteins are named after the retinal degeneration phenotype observed in *Drosophila*. RdgBα mutant flies show light–induced retinal degeneration as well as an abnormal light response. There are two mammalian homologs, RdgBαI and II, of which only RdgBαI can fully rescue the retinal degeneration phenotype and the abnormal light response in flies (Chang et al. 1997). In contrast, RdgBαII only rescues the retinal degeneration phenotype but fails to fully restore the electrophysiological light response (Lu et al. 1999). The PITP domain of RdgBα proteins shares $\sim$40 % identity with Class I PITPs. The last PITP to be cloned was RdgBβ; the PITP domain of RdgBβ shares $\sim$40 % identity with both Class I PITPs (α and β) and with Class 2A RdgBα proteins.

Lipid-Binding Properties of the Hydrophobic Cavity of the PITP Domain

The PITP domain belongs to a large superfamily of proteins with a common fold, the Bet v 1 fold (Radauer et al. 2008). This family is named after Bet v 1, the major birch pollen antigen which is one of many plant pathogenesis-related proteins

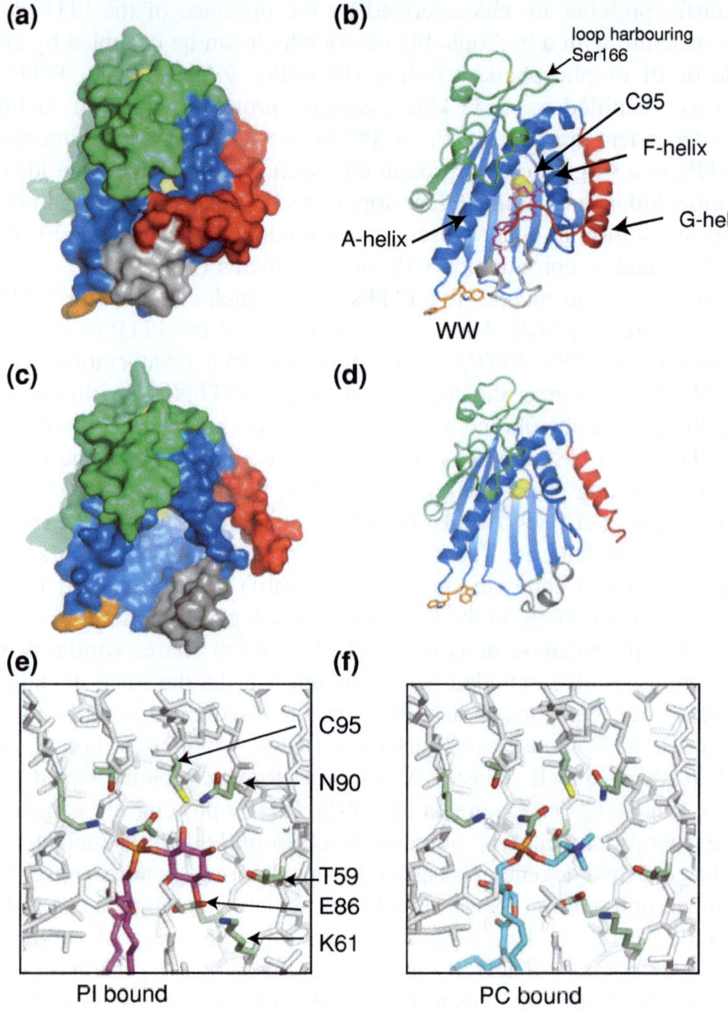

Fig. 9.1 **a** Surface representation and **b** cartoon representation of PITPα with PC-bound (PDB: IT27); **c** Surface representation and **d** cartoon representation of PITPα in the apo-form (PDB: 1kcm). The lipid binding core residues are *colored blue*, the G-helix and the extended 11 amino acids at the C-terminus that forms the lid are *colored red*, the regulatory loop is *colored green* and the lipid exchange loop is *colored gray*. C95 is depicted as balls and is *colored yellow*. The side chains of W203 and W204 are *colored orange*. In the apo-structure **c** and **d**, the lipid exchange loop and the G-helix have moved to the open configuration and the C-terminal region is disordered. **e** A stick model showing a PI molecule (carbon atoms *colored magenta*) and the functionally important inositol-binding residues K61, N90, T59 and E86 in the lipid binding cavity of PITPα are indicated (PDB code: 1UW5). The location of C95 in the lipid binding pocket is also indicated. **f** A stick model showing a PC molecule (carbon atoms *labelled blue-green*) in the lipid binding cavity of PITPα (PDB code: IT27). The diagrams were generated using the pymol software

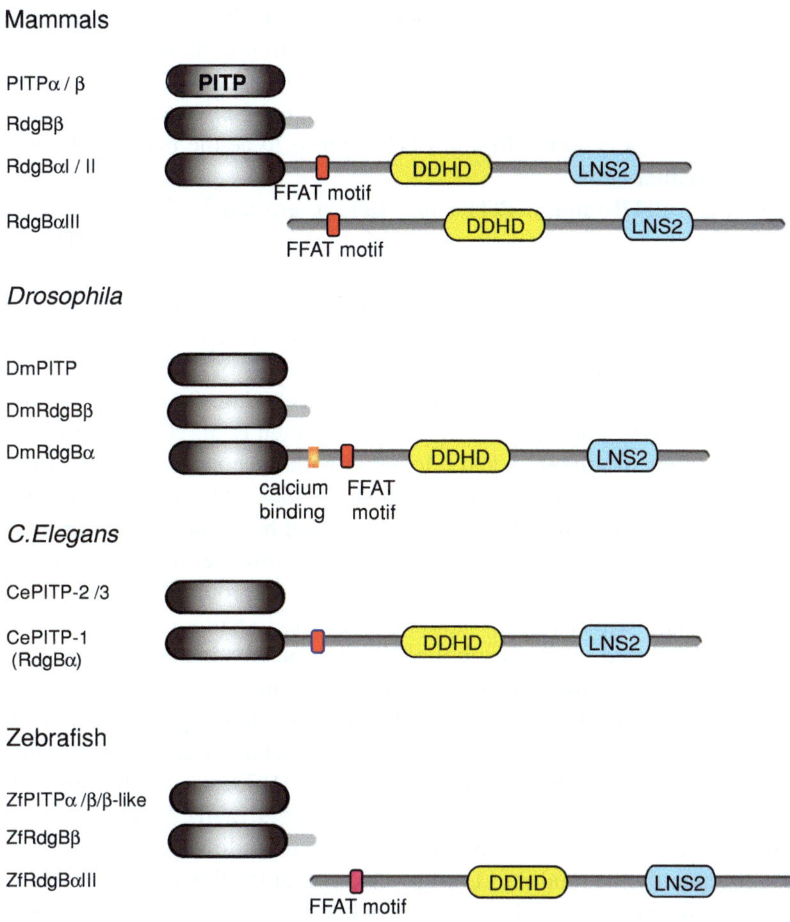

Fig. 9.2 Representative structural organization of the PITP families in mammals, flies (*Drosophila*), worms (*Caenorhabditis elegans*) and fish (zebrafish). The PITP domain is located at the N-terminus. FFAT motif, LNS2 and DDHD domain are indicated. See Table 1 for alternative names and additional comments

(PR-10). PR-10 is cytoplasmic proteins of 15–17 kDa. This fold has a distinctive topology determined by the position of the secondary structure elements, β-sheets and α-helices, in the sequence. This gives rise to a three-dimensional structure consisting of a large hydrophobic cavity, which is open to the exterior and functions as a ligand binding site. The ubiquitous distribution of proteins with a Bet v1 fold among all superkingdoms suggests that a Bet v 1-like protein was already present in the last common ancestor. During evolution, it is thought that this protein diversified into numerous families with low sequence similarity but with a common fold that succeeded as a versatile scaffold for binding of bulky ligands

(Radauer et al. 2008). This superfamily includes the START and PITP families which have been shown to bind lipid molecules including cholesterol, ceramide, PC, and PI. START family proteins include StAR (binds cholesterol), CERT (binds ceramide), and PC transfer protein (PCTP) (binds PC). No START proteins have been identified that can bind PI and PITPs appear to be the major PI binding proteins in mammalian cells.

Members of the PITP family, PITPα and PITPβ, are unique in that they can bind either PC or PI. X-ray crystal structures of three distinct forms of PITPα are available; PITPα bound to PI or to PC, and one that is devoid of lipid (the 'apo'structure) (Fig. 9.1). The structures of the PI- and PC-loaded PITPα are near identical, excluding the possibility of conformational changes in the structure regulated by the identity of the lipid bound. PITPα consists of eight β-strands that forms the concave sheet, together with seven α-helices (Fig. 9.1b) (Tilley et al. 2004; Yoder et al. 2001; Vordtriede et al. 2005; Schouten et al. 2002; Wirtz et al. 2006). Two of the long α-helices, A and F together with the concave sheet form the lipid-binding cavity (Fig. 9.1b). A single phospholipid molecule occupies this hydrophobic cavity. The entrance to the cavity is closed by a 'lid' formed from the C-terminal α-helix (the G-helix, colored red), which is followed by the 11 a.a. C-terminal random coil (colored red). The phospholipid headgroup is accommodated deep inside the cavity whilst the *sn*1 and *sn*2 fatty acyl chains point toward the 'lid'. With the lid closed, the phospholipid is completely enclosed within the protein (Fig. 9.1a). In the structure of the lipid-free form, the G-helix has swung outwards from the main structure (Fig. 9.1c and d), allowing passage of the phospholipid into or out of the cavity. From these structures coupled with mutational analysis of two tryptophan residues (WW203/204) located at the tip of the loop between beta strand 8 and alpha helix F (see Fig. 9.1a–d), it is conjectured that insertion of these residues into the membrane could indirectly perturb the G-helix, and consequently the random 11 amino acid coil at the C-terminus providing the driving force for opening of the cavity (Cockcroft 2007; Shadan et al. 2008).

Lipid binding and transfer properties have been extensively studied for Class I PITPs but not in great depth for Class II PITPs (Tilley et al. 2004). From the studies on PITPα and PITPβ, both proteins have been shown to bind either PI or PC and facilitate intermembrane transfer (Segui et al. 2002). PITPα has a 16-fold higher affinity for binding to PI than it does to PC (de Brouwer et al. 2002) compensating for the low concentrations of PI (5–8 % of total cellular phospholipids). In cells, approximately 50 % of the PITPα molecules are bound to PI and the rest to PC (Helmkamp et al. 1974; Thomas et al. 1993), and the same is true for PITPβ (Morgan et al. 2006). Comparison of the structures of the PI-bound and PC-bound PITPα reveals that both lipids occupy the same space in the lipid-binding cavity, and the main difference is that the inositol headgroup makes more hydrogen bonds with specific amino acid residues than does the choline moiety. Four residues (T59, E86, K61, and N90) in the lipid-binding cavity make hydrogen bond contact with the inositol headgroup (Fig. 9.1e) and mutation of any one of these residues results in loss of PI binding (de Brouwer et al. 2002; Carvou et al. 2010). The common feature shared by all members of the PITP family is the conservation of the inositol headgroup binding site. These four amino acid residues are not contiguous but reside on two separate β-strands and are conserved in all PITP proteins and can be regarded as the hallmark that identifies a PITP protein.

For PC binding and transfer, cysteine 95 in PITPα and in PITPβ is important. Cysteine 95 lies in the lipid-binding cavity (see Fig. 9.1) and forms a hydrogen bond with the headgroup of PC indirectly via a water molecule and mutation of this residue to either threonine or alanine reduces PC transfer, without affecting PI transfer (Shadan et al. 2008; Carvou et al. 2010). Cysteine 95 is only conserved in Class I PITPs but not Class II PITPs where it is replaced with a threonine raising the possibility that PC binding and transfer may not be a prominent feature of Class II PITPs.

Another residue that affects PC transfer is phenylalanine 225. PC transfer activity is robust when residue 225 is a phenylalanine but when it is replaced with a leucine residue, as in rodent PITPβ, PC transfer is lower (Vordtriede et al. 2005). In human PITPβ this residue is a phenylalanine and shows robust PC transfer activity. Residue 225 is variable in different PITPs. Human RdgBα has alanine at this location whilst RdgBβ has a glycine; both are small aliphatic residues similar to leucine and thus would contribute to low PC transfer activity in RdgB proteins.

PITPs Potentiate PI Kinase Activities

PITPα was first purified as a reconstituting factor that potentiated the hydrolysis of PI(4,5)P_2 by receptor-regulated phospholipase C at the plasma membrane (Thomas et al. 1993; Cunningham et al. 1995; Kauffmann-Zeh et al. 1995). PITPα was also purified, together with a PIP 5-kinase, for reconstituting Ca^{2+}-activated exocytosis of secretory granules from neuroendocrine cells (Hay and Martin 1993; Hay et al. 1995). Reconstitution of exocytosis was also dependent on PITPα in cytosol-depleted human promyelocytic leukemia cells (HL60 cells) and mast cells and was required for maintaining a pool of PI(4,5)P_2 (Fensome et al. 1996; Way et al. 2000). Subsequent studies also showed that PITPα and β were also able to restore G-protein-stimulated PI(3,4,5)P_3 production in permeabilized HL60 cells (Kular et al. 1997, 2002). In these reconstituted systems, thus the common theme was the ability of PITP to stimulate the synthesis of PPIs, both PI(4,5)P_2 and PI(3,4,5)P_3.

Studies using in vitro lipid kinase assays showed that human Class III PI 3-kinase which phosphorylates PI to PI(3)P was greatly enhanced in the presence of either PITPα, β or Sec14p, the unrelated yeast PITP (Panaretou et al. 1997). The yeast PITP (Sec14p) bears no sequence or structural homology to mammalian PITPs (Phillips et al. 2006). Despite this, Sec14p is able to replace mammalian PITPs in these reconstitution assays including activation of phospholipase C activity and exocytosis. Furthermore, PITPα and PITPβ are both able to replace Sec14p in yeast in stimulating PI(4)P synthesis. Yeast ablated for the lipid phosphatase, Sac1, accumulates PI(4)P and this is reduced when Sec14p is inactivated. Class I PITP proteins from diverse sources including PITPα and PITPβ are able to reconstitute PI(4)P synthesis (Ile et al. 2010).

From these studies, a clear concept emerges: PITP proteins promote the synthesis of phosphorylated derivatives of PI. The underlying requirement is a functional PI transfer protein that is able to bind and transfer PI. Mutants devoid of PI binding and transfer are inactive. The molecular mechanism, however, remains unclear. Two models have been entertained: the PI delivery model and the PI presentation model. In the PI delivery model, PI synthesized at the ER is redistributed by a rapid non-vesicular transport mechanism. The membranes involved must be close to equilibrium with respect to PI levels. The conversion of PI to the phosphorylated forms in target membranes would effectively lower the PI concentration allowing for vectorial transfer. The accumulation of phosphorylated derivatives of PI in specific membrane compartments would be determined by the activity of the lipid kinases. For example, the combined activity of PI 4-kinase and PIP 5-kinase accounts for the fact that $PI(4,5)P_2$ is more concentrated at the plasma membrane than elsewhere in the cell. In the PI presentation model, the PI bound to PITP gets preferentially phosphorylated by the lipid kinases. In this case, PITPs would be recruited for lipid delivery on demand. Recruitment of PITPα by the netrin-1 receptor, deleted in colon cancer (DCC) has been reported (Xie et al. 2005). In neurons, PITPα is the major PITP specifically localized in the axons and stimulation of phospholipase Cγ by BDNF or netrin-1 is dependent on PITPα (Xie et al. 2006). Netrin-1 and BDNF act as guidance cues for axonal outgrowth during development and the activity of both PLCγ and PI 3-kinase is required.

To obtain a complete understanding of the molecular mechanisms of PITP function, identification of the specific phosphoinositide-requiring systems that each PITP participates in, and the level of redundancy between the different PITPs, needs to be explored. Thus, to identify the specific functions of PITPs, analysis of the expression patterns of individual PITP proteins in different tissues and cell-types, their subcellular localisation and their binding partners will provide important clues. The idea that PITPs have highly specific functions is reinforced by the observations that when individual PITP proteins are ablated in model organisms, specific phenotypes are observed (Ile et al. 2010; Xie et al. 2005; Hamilton et al. 1997; Iwata et al. 2011; Milligan et al. 1997). Studies in mice, *C. elegans* (worms), zebrafish and *Drosophila* (flies) each reveal the rich repertoire of functions for the PITP proteins. At a molecular level, many of these functions are linked to phosphoinositide metabolism. In some cases, PITPs appear to be regulators of diacylglycerol (DAG) levels (Iwata et al. 2011; Litvak et al. 2005). In several studies, where the importance of PI binding has been examined making use of mutants that are deficient in PI binding, it is clear that PI binding is essential for function (Carvou et al. 2010; Alb et al. 2007).

As discussed above, PITPs enhance the phosphorylation of PI by lipid kinases in vitro. In the structure of the soluble form of PITPα, the inositol headgroup of PI is buried deep inside its binding pocket and is not immediately accessible for phosphorylation. The structure of PITP when associated with the membrane is not known but one possibility is that proteins that interact with the regulatory loop may promote access to the headgroup of PI (Cockcroft and Garner 2011; Yoder et al. 2001) (Fig. 9.1). Since PITPs are unable to bind the phosphorylated forms of PI in their soluble conformation, exchange of the phosphorylated lipid with PI or PC from the membrane would allow PITP to regain its soluble compact closed form. Evidence that supports this model comes from phosphorylation studies of PITPα. The regulatory domain (colored green in Fig. 9.1a–d) harbors a short loop containing a serine residue (Ser166) which is conserved in almost all PITP sequences. It lies in a consensus site for phosphorylation by protein kinase C and this is confirmed by experimental evidence (van Tiel et al. 2000; Morgan et al. 2004). Ser166, however, is not exposed to solvent in any of the PITP structures and therefore a dramatic conformational change is required to expose this residue for phosphorylation. This concept of a major structural change when PITPs are at the membrane is supported by other members of the Bet v 1 superfamily. Both Steroidogenic acute regulatory protein (StAR) and Bet v 1 protein undergoes a loss in tertiary structure to form a highly-helical 'molten globule' structure upon interaction with membranes (Miller 2007; Mogensen et al. 2007). Therefore, an analogous conformational change would provide a mechanism that enables the lipid kinases to access the PI head group while still bound to PITP.

In this model, PITP would face the membrane from the opposite end to where the 'lid' is located. This dilemma could be easily resolved if PITPs engage two separate membranes at opposite ends thus bridging two membranes. For PITP proteins that contain additional domains that anchor the protein at a membrane, the PITP domain could still bridge two membranes. Thus, PITP could present PI to the lipid kinases as suggested in the original studies (Cunningham et al. 1995). Phosphorylation of PI would therefore occur locally in a specific organelle membrane where specific PI kinases are localized. The endoplasmic reticulum is the site for PI resynthesis and is known to make close contacts with many compartments including the plasma membrane, Golgi, endosomes, and mitochondria allowing for vectorial transport from the ER to other compartments. The presence of the FFAT motif in the RdgBα proteins provides additional support. The FFAT motif binds to the ER –localized integral membrane proteins, VAP-A, and VAP-B (Amarilio et al. 2005; Loewen et al. 2003)

Function of Class I PITPs

Mammalian PITPα and PITPβ comprise Class I PITPs and have different tissue distributions and subcellular localizations. Mice present highly specific phenotypes when their proteins levels are reduced or completely absent. Ablation of the PITPβ gene is embryonic lethal while PITPα shows dosage effects where partial reduction of PITPα (~80 % reduction) in the naturally occurring *vibrator* mice cause death within 30 days of birth (Hamilton et al. 1997) while mice which have no PITPα dies within 24 h (Alb et al. 2003). Genetic ablation of PITPα in mice results in spinocerebellar neurodegeneration, accumulation of fat (steatosis) in enterocytes and liver, and hypoglycemia (Alb et al. 2007). Only spinocerebellar neurodegeneration is observed in the *vibrator* mice. Deficiency greater than 90 % in PITPα is required to observe accumulation of fat in liver and intestine (enterocytes) and hypoglycemia. Moreover, the intestinal steatosis is independent of liver steatosis as tissue-specific expression of PITPα in the intestine does not rescue liver steatosis and also partially restores hypoglycemia. Hypoglycemia appears to be a consequence of intestinal and liver steatosis.

Mammalian PITPα in Neuronal Function

The highest expression of PITPα is found in the brain and specifically in the axons of the neurons in the cerebellum and the hippocampus (Xie et al. 2005; Cosker et al. 2008). The naturally-occurring *vibrator* mice have severe action tremor and suffers progressive neurodegeneration in the brain and spinal cord. Degenerative changes in neurons are observed in the brain stem, cerebellum, spinal cord, and dorsal root ganglia. Affected neurons display progressive intracellular vacuolation suggestive of defective ER. Isolated hippocampal and cortical neurons have been used to identify the molecular function of PITPα. PITPα levels steadily increase during the maturation of hippocampal neurons in vitro, and during the first few days of development, PITPα is found specifically in the axonal processes, including their growth cones (Cosker et al. 2008). This discrete localisation in neurons contrasts significantly with the localisation of PITPα in cultured cells such as COS-7, which is throughout the cytosol and nucleus (Larijani et al. 2003). This suggests that PITPα has specific binding partners in axons which determine its discrete localisation. Two binding partners that have been identified for PITPα is neogenin and DCC. DCC is a receptor for netrin 1, which is released by floor plate cells in the developing spinal cord, attracting spinal commissural axons to cross the midline. The netrin-1 receptor, DCC binds PITPα via its C-terminus in response to netrin-1 binding (Xie et al. 2005, 2006), and commissural axons, expressing a mutant PITPα lacking 5 amino acids at the C-terminus, are unable to cross to the other side in a chick embryo. This mutant can bind DCC but is impaired for PI transfer and is unable to support phospholipase C signaling (Hara et al. 1997). Furthermore, when morpholinos to ablate PITPα are injected into zebrafish embryos, a variable loss of spinal cord neurons is observed, and when present, defects in motor-axon outgrowth, indicating the importance of PITPα in axon guidance (Xie et al. 2005).

Growth cone guidance by NGF, BDNF, netrin-1, or myelin-associated glycoprotein (MAG) requires co-activation of the PLCγ and PI 3-kinase (Ming et al. 1999), and stimulation of PLCγ by netrin-1 is impaired in neurons from the vibrator mouse (Cosker et al. 2008). In addition to netrin-1, laminin also enhances axonal outgrowth. Over-expression of PITPα makes even longer axons while a decrease in PITPα using RNAi results in shorter axons (Cosker et al. 2008).

The effects of PITPα over-expression are sensitive to the PI 3-kinase inhibitor, wortmannin, suggesting that PITPα effects are mediated via PI 3-kinase.

Mammalian PITPβ is Required for Retrograde Traffic From the Golgi to ER

PITPβ shares 77 % identity with PITPα and deletion of the PITPβ gene, *PITPNB*, is embryonic lethal in mice, indicating an essential function (Alb et al. 2002). Depletion of PITPβ by RNAi in HeLa cells shows a number of phenotypes; these include a compacted Golgi apparatus, a misshapen nucleus and a defect in retrograde traffic from the Golgi to the ER mediated by coat protein complex I (COPI)-coated vesicles (Carvou et al. 2010). The KDEL receptor, important for retrieval of ER-resident proteins from the Golgi in COPI-coated vesicles, is arrested at the Golgi in PITPβ-knockdown cells, and ERGIC-53, which cycles between the ER-Golgi intermediate compartment (ERGIC) and the ER in COPI-coated vesicles, is similarly arrested at the ERGIC in PITPβ-knockdown cells. Accompanying these phenotypes is a reduction in PI(4)P by ~40 %. The retrograde trafficking defect can be rescued with wild type PITPβ but not mutant PITPβ deficient in PI or PC transfer, or lacking the membrane-interacting WW203/204 residues. These observations support the idea that both PI and PC binding contribute to function, very likely for maintaining a specific pool of PI(4)P. Here, PITPβ can be seen as presenting PI to PI4KIIIα, to produce PI(4)P, required for the assembly of COPI-coated vesicles at the Golgi apparatus. The function of PI(4)P at the Golgi remains is unclear but could be required to form the COP1 complex or to regulate the actin cytoskeleton (Cockcroft and Garner 2011; Carvou et al. 2010; Dumaresq-Doiron et al. 2010).

Drosophila Class I PITP is Required for Cytokinesis

Mammalian cells possess two Class I PITPs, PITPα and PITPβ. In flies, the situation is simplified as they possess a single Class I PITP which cannot be classified as either PITPα or PITPβ (see Table 1). Dm-PITP is encoded by *Gio* and is required for cytokinesis in spermatocytes (Gatt and Glover 2006; Giansanti et al. 2006). In addition to *Gio, fwd*, the gene coding for Dm-PI4KIIIβ, is required for cytokinesis. *Gio* and *fwd* are specifically required for the regulation of the actomyosin ring constriction and cleavage furrow ingression (Gatt and Glover 2006; Giansanti et al. 2006). Membrane delivery in the form of Golgi-derived vesicles is required and mutations in *Gio* or in *Fwd* results in the abnormal localisation of Golgi-derived vesicles at the cell equator. Failure of fusion of these Golgi-derived vesicles with the invaginating furrow appears to be responsible for the defects in

Table 1 PITP proteins found in humans (*H.sapiens*), flies *(D.melanogaster)*, worms (*C.Elegans*) and fish (*D.rerio)*

Protein name (Species)	Gene name	Other names	Comments	Reference
H.sapiens				
PITPα	PITPNA		C-terminus binds to neogenin and DCC	(Xie et al. 2005)
PITP*β*	PITPNB		Two splice variants which differ at the C-terminus	(Morgan et al. 2006)
RdgB*β*	PITPNC1		Two splice variants which differ at the C-terminus	(Takano et al. 2003)
RdgBαI	PITPNM1	Nir2		
RdgBαII	PITPNM2	Nir3		
RdgBαIII	PITPNM3	Nir1	Lacks the PITP domain	
D.melanogaster				
PITP(vib)	Gio	Giotto	Required for cytokinesis in spermatocytes	(Gatt and Glover 2006; Giansanti et al. 2006)
RdgB*β*	CG17818			
RdgBα	RdgBα		Phototransduction	(Milligan et al. 1997)
C.Elegans				
CE46531	Y54F10AR.1	CE46531	Interacts with Vps34	(Lee et al. 2008)
CE26250	Y71G12B.17	CE26250		
PITP-1 (RdgBα)	MO1F1.7	CE33673	Required for sensory transduction	(Iwata et al. 2011)
D.rerio				
PITPα	PITPNA			(Ile et al. 2010; Xie et al. 2005)
PITP*β*	PITPNB			(Ile et al. 2010)
PITP*β*-like	PITPNG			(Ile et al. 2010)
RdgB*β*	PITPNC1			
RdgBαIII	pI-RdgB		Lacks the PITP domain	(Elagin et al. 2000)

cytokinesis. In *Drosophila* spermatocytes, Rab11 regulates membrane addition to the advancing cleavage furrow (Giansanti et al. 2007; Polevoy et al. 2009). The phenotypes observed with *Gio* and *Fwd* resembles *rab11* mutants and these include incomplete constriction of the actin ring and accumulation of Golgi-derived vesicles at the cell equator. Rab11 associates with Golgi-derived vesicles and this vesicle association depends on *Gio* and *Fwd*. PITP and PI4KIII*β* are thus upstream to Rab11 and probably functions in PI(4)P generation (Polevoy et al. 2009). In mammalian cells, an interaction between Rab11 and PI4KIII*β* is also observed and in this case, PI4KIII*β* is also required for Rab11 localisation at the Golgi (de Graaf et al. 2004). Whether a PITP is required for Rab11 localisation is not known in mammalian cells but both PITP*β* and RdgBαI (Nir2) are potential candidates as they localize to the cleavage furrow during cytokinesis (Cockcroft and Carvou 2007; Litvak et al. 2004). In mammalian cells, a requirement for $PI(4,5)P_2$ at the midzone has been also identified where it recruits anillin, required for furrow ingression (Liu et al. 2012).

Functions of Class IIA PITPs

Class II PITPs comprise the RdgB family of PITPs. The RdgB proteins are subdivided into Class IIA (RdgBα) and IIB (RdgBβ). (Alternative names for RdgBα proteins are PITPNM and Nir proteins.) Class IIA proteins are multi-domain proteins and their function in model organisms including *Drosophila* and *C. elegans* indicate specific roles in sensory transduction. Mammals contain two genes and their function is less well-understood. Studies in *Drosophila* and *C.elegans* indicate that they function in phosphoinositide signaling.

Drosophila **RdgBα is required for Phototransduction**

The *rdgB* mutant was one of the first *Drosophila* retinal degeneration mutants identified and mutation of the gene causes an abnormal termination of the light response in the photoreceptor cells as well as retinal degeneration (Trivedi and Padinjat 2007). RdgB encodes for a multi-domain protein that contains a PITP domain at its N-terminus (Fig. 9.2). Dm-RdgBα is the founding member of the Class IIA PITP family. The apical domain of the *Drosophila* photoreceptor cell is a specialized light-sensing organelle, the rhabdomere that consists of the closely-packed photoreceptive microvilli. The light-sensing protein, rhodopsin, together with its G-protein, Gq, and phospholipase C-β (NorpA) are highly concentrated in these microvilli while Dm-RdgBα localizes to the region immediately beneath the light-sensing rhabdomeric membrane, the subrhabdomeric cisternae (SRC), as well as in the adjacent plasma membrane (Vihtelic et al. 1993). In the *rdgB* mutant, the rhabdomere membrane undergoes vesiculation, is internalized and lost resulting in reduction in the size of the apical membrane. Expression of the N-terminal PITP domain of Dm-RdgBα in *rdgB* mutant flies is sufficient to completely restore the wild-type electrophysiological light response and prevent degeneration. Expression of mammalian PITPα is unable to restore function implying differences in the PITP domains. Moreover, the function of the rest of the RdgBα protein appears to be non-essential. This may be misleading as expression levels of the PITP domain are likely to be higher than that of the full length protein.

$PI(4,5)P_2$ is the critical lipid required at the rhabdomere. Light induces robust activation of phospholipase C resulting in the consumption of $PI(4,5)P_2$. To reset the system, replenishment of PI by Dm-RdgBα to the rhabdomere from the adjacent membrane, the subrhabdomere offers an attractive mechanism. The subrhabdomere is an extension of the ER, where the enzymes for PI synthesis are located. The retinal degeneration phenotype of Dm-RdgBα is light-dependent in that it can be both protected by rearing flies in the dark and accelerated by rearing flies in a light/dark cycle. (In the absence of stimulation by light, PI(4,5)P_2 hydrolysis would be prevented.) Such protection is also observed in mutants that have reduced levels of the rhodopsin receptor or have been reared in vitamin A-deficient medium that reduces the levels of functional rhodopsin receptors (Stark and Sapp 1987), and in *norpA* mutants that lack a functional phospholipase Cβ (Paetkau et al. 1999). In contrast, retinal degeneration is accelerated in flies that express the constitutively active Gq. Retinas of double mutants of Gq with mutant *rdgB* degenerate even in the dark and is phospholipase C-dependent (Lee et al. 1994). Phosphatidylinositol synthase (PIS) is required for a key step during PI(4,5)P_2 regeneration, the production of PI. Overexpression of PIS suppresses the retinal degeneration resulting from two other mutations affecting PI(4,5)P_2 cycling, *rdgB* (retinal degeneration B) and CDS (CDP-DAG synthase) (CDP-diacylglycerol synthase) (Wang and Montell 2006).

Together these data provide compelling genetic evidence to support the idea that the *rdgB* mutant phenotype is due to the inability to replenish PI(4,5)P_2 levels following stimulation by light. This conclusion is supported by another study which used an electrophysiological biosensor, the PI(4,5)P_2-sensitive Kir channel. The channel was targeted to the microvillar membranes. Depletion of PI(4,5)P_2 and its recovery during phototransduction was measured indirectly by recording the activity of the Kir channel. Comparison of wild-type and *rdgB* mutant flies showed that microvillar PI(4,5)P_2 levels in wild-type flies had recovered by $\sim$30 s after stimulation yet in *rdgB* mutant flies a recovery of no more than 50 % wild-type levels was observed after several minutes (Hardie et al. 2001). This study provides the most direct evidence supporting a role for RdgBα in PI(4,5)P_2 resynthesis during cell signaling.

An additional role for PPIs is in the recycling of arrestin. Arrestin trafficking is impaired in *rdgB* and CDS mutants (Lee et al. 2003). The C-terminal domain of arrestin binds to PI(3,4,5)P_3 in vitro, and mutation of this site delays arrestin shuttling and results in defects in the termination of the light response, which is normally accelerated by prior exposure to light. Disruption of the arrestin/PI(3,4,5)P_3 interaction also suppresses retinal degeneration caused by excessive endocytosis of rhodopsin/arrestin complexes. These findings indicate that light-dependent trafficking of arrestin is regulated by direct interaction with PPIs and is required for light adaptation. Since phospholipase C activity is required for activation of *Drosophila* phototransduction, these data point to a dual role of PPIs in phototransduction (Lee et al. 2003). Both functions depend on RdgBα.

While the function of the single *Drosophila* RdgBα in phototransduction has been extensively characterized, mammals do not share the same signal transduction cascade for phototransduction. However, in recent studies, the intrinsically-photosensitive retinal ganglion cells (pRGCs) have been shown to respond to light. The photopigment is melanopsin that transduces its signals via a G-protein coupled to phospholipase Cβ4 (Graham et al. 2008; Moldrup et al. 2010; Bailes and Lucas 2010). These cells modulate behavioral responses to light including synchronization of the circadian clock to light/dark cycles, regulation of pupil size, sleep propensity and pineal melatonin production. Thus, this signaling cascade is similar to phototransduction in flies and therefore a mammalian RdgBα might work in this pathway. RdgBαII expression is restricted to the retina and the dentate gyrus and could potentially function in the melanopsin signal transduction pathway in a manner similar to flies. The availability of the knockout mice for RdgBαII, which show no gross phenotype should now be examined for defects in light-sensitive behavioral responses such as regulation of pupil size to test this prediction (Lu et al. 2001).

Caenorhabditis Elegans RdgBα is Required for Sensory Transduction in Specific Neurons

PPIs have been implicated in neural and behavioral plasticity in the worm, *C.elegans* PI(3,4,5)P_3 and phospholipase C/diacylglycerol signaling play pivotal role in the plasticity of salt chemotaxis. *C.elegans* is attracted to salt (NaCl) but learn to avoid salt after exposure to salt in the absence of food. The nervous system of *C.elegans* consists of 302 neurons and ASER is the gustatory neuron that senses NaCl. When ASER senses salt, it mediates attraction behavior, but elicits avoidance behavior after worms are exposed to salt under starvation conditions. *C. elegans* express three PITP proteins one of which is the multi-domain Class IIA PITP in addition to two Class I PITPs (Fig. 9.2 and Table 1). The Class IIA PITP (analogous to RdgBα), referred to as PITP-1 localizes to sensory neurons and plays an essential role in the regulation of salt chemotaxis. Both the attraction and plasticity after exposure to salt are lost when PITP-1 is mutated. Previous studies showed that PI(3,4,5)P_3 signaling in the ASER salt-sensing neuron promotes the learning, leading to avoidance of NaCl. In contrast, DAG signaling in ASER counteracts PI(3,4,5)P_3 signaling and promotes attraction behavior towards NaCl (Tomioka et al. 2006).

PITP-1 is expressed broadly in the nervous system and the defects in chemotaxis and learning are completely rescued when the *pitp-1* cDNA is expressed in all the neurons. The attraction defect is partially rescued by expression of *pitp-1* cDNA in the ASER and ASEL neurons, but not by expression in either one alone. In contrast, the plasticity defect is partially rescued by expression in ASER alone. No rescue is seen when expressed in ASEL, but expression in both is better at rescuing plasticity. Interestingly, no morphological changes are observed in the ASER neuron in *pitp-1* mutants, in contrast to *Drosophila*, where mutations in RdgBα lead to the light-induced retinal degeneration.

The requirement for PITP-1 is also noted for other sensory modalities including olfaction and osmo-sensation. The AWC olfactory neuron senses benzaldehyde and this is lost in *Pitp-1* mutants. Re-expression of PITP-1 cDNA in AWC but not in ASER rescues the defect. Similarly, osmotic behavior is sensed by the ASH neuron and defects are observed in *pitp-1* mutants which are rescued by re-expression of *pitp-1* cDNA in ASH. Although PITP-1 is essential for a number of sensory behaviors, locomotion and egg-laying are unaffected indicating a highly specific requirement for PITP-1. Locomotion is dependent on phospholipase C and therefore the possibility that a different PITP is required needs investigation.

In ASER, a PITP-1-GFP fusion protein shows a punctate localization mainly in the cell body and the axon, suggesting that PITP-1 functions at the presynaptic regions. In the presynapses of ASER and AWC, DAG signaling promotes attraction to salt and odor, possibly by enhancing synaptic release from the sensory neurons and DAG can be produced from PI(4,5)P_2 by hydrolysis by phospholipase C. In support of this idea, the reduced NaCl attraction to mutant PITP-1, could be counteracted by mutations in DAG kinase, indicating that DAG signaling is decreased. However, the defective learning phenotype of mutant PITP-1 was not suppressed by DAG kinase, implying a role for other PPIs in learning. These results suggest that a PITP regulates the production of DAG via phospholipase C as well as PPIs to dynamically regulate sensory behaviors. The PITP domain is sufficient to rescue the behavioral phenotypes supporting the notion that PI delivery by the PITP domain is the major functional activity. Rescue with PITP mutants unable to bind PI would provide strong support for this notion.

Mammalian RdgBα Proteins and Their Functions in Signalling and Membrane Traffic

In *Drosophila* and *C.elegans*, the single RdgBα has specific roles in light and sensory transduction respectively in neurons. In mammals there are two genes encoding for RdgBα –I and –II. RdgBαI is expressed ubiquitously while RdgBαII expression is restricted to the hippocampus and the retina (Lu et al. 1999). It is tempting to speculate whether this orthologue has a role in learning and memory. Mice ablated of RdgBαII show no gross phenotype but have not been examined for learning and memory. As discussed above, this PITP could be required for light – sensitive behavioral responses.

Mice ablated of RdgBαI are embryonic lethal and therefore studies have been mainly conducted in tissue cultured cells. Depletion of RdgBαI in cultured HeLa cells using RNAi reveals disorganized Golgi with protein export to the plasma membrane arrested (Litvak et al. 2005). A decrease in DAG levels is thought to be responsible for the defects in membrane traffic. DAG levels are decreased in RdgBαI-knockdown cells due to its enhanced utilization for PC synthesis. They speculate that RdgBα inhibits the CDP-choline pathway for PC synthesis. Since re-expression of the PITP domain is sufficient to rescue the membrane trafficking defect, further analysis using mutants of the PITP domain which are defective in PI and PC binding may provide clues as to how this PITP regulates PC metabolism.

Function of RdgBαIII, the Protein Lacking a PITP Domain

In many of the studies described above for RdgBα, the PITP domain is sufficient to rescue the phenotypes observed in *Drosophila*, *C.elegans* and in Hela cells. In mammalian cells and in fish, a RdgBα protein, RdgBαIII (alt names: PITPNM3/Nir1), is found that lacks a PITP domain. Recent studies have identified that RdgBαIII is mutated in two large Swedish families with autosomal dominant cone dystrophy. These patients show defective color vision, low visual acuity, and abnormal cone response when tested with an electroretinogram (ERG). The mutation (Q626H) is located in a conserved region in the C-terminal part of RdgBαIII which is the region that interacts with the tyrosine kinase, PYK2. A homologue of RdgBαIII (pI-RdgB) has also been identified in zebrafish where it is found to be exclusively localized in the retina to the cone cell inner segments (Elagin et al. 2000). When used to restore the defective light response and retinal degeneration in *rdgB* mutant *Drosophila*, retinal degeneration was slowed without restoring the electrophysiological light response. These results from human patients and from zebrafish suggest that the other domains (LNS2 and DDHD) found in RdgB proteins possess an activity that promotes photoreceptor viability which is independent of the PITP domain. This conclusion is reinforced by the observation that RdgBαII can also rescue photoreceptor degeneration in RdgB mutant flies, without fully restoring the ERG light response.

A recent study has also identified RdgBαIII (PITPNM3) as a functional receptor for a cytokine CCL18 (Chen et al. 2011). CCL18 is produced in abundance by macrophages associated with breast tumors which promote the invasiveness of tumor cells by enhancing their adhesion to the extracellular matrix and migration. The receptor for CCL18 on the breast cancer cells was identified as PITPNM3. An antibody to PITPNM3 identified that CCL18 co-localized with PITPNM3 at the plasma membrane of breast cancer cells, MDA-MB-231. CCL18 also induced tyrosine phosphorylation of PITPNM3, and a stable HEK cell-line expressing PITPNM3 were used to demonstrate that CCL18 bound to PITPNM3 and stimulate an increase in cytosol Ca^{2+}, phosphorylation of PLCγ1 and PKC-zeta and also induce directional migration. These results suggest that PITPNM3 is an integral membrane protein (Chen et al. 2011). This conclusion is at variance with the biochemical analysis of RdgB proteins.

When the Dm-RdgBα was initially cloned, it was thought to be an integral protein based on the presence of 6 hydrophobic segments. The human RdgBαII was subsequently found to associate with the membrane pellet after subcellular fractionation, but the protein remained in the pellet even after solubilisation of the membranes with detergents suggesting that the protein is associated with a large protein complex likely to be the cytoskeleton (Lu et al. 1999). Similar analysis with the zebrafish PITPNM3 (pI-RdgB/Nir1) (Elagin et al. 2000) and with human RdgBαI (Nir2) gave identical results (Litvak et al. 2002). In all these studies, RdgBα was present in the particulate fractions and could be solubilised after treatment with a denaturing agent or alkaline membrane extraction, but not with detergents. Given the high conservation of the amino acid sequences of the RdgB family members and their conserved hydrophobic stretches, it is likely that RdgB family members are peripheral proteins and exhibit similar protein topology. Thus, for PITPNM3 to work as a receptor for CCL18, it would need to be an integral membrane protein with a binding site for CCL18 facing outwards. Further work is required to reconcile these contradictory observations.

RdgBβ Binds to 14-3-3 Proteins and ATRAP and Enhances Metastasis of Breast Cancer Cells

RdgBβ is the third soluble PITP found in mammalian cells, *Drosophila* and zebrafish but not in *C.elegans* (Fig. 9.2). The gene encoding RdgBβ is *PITPNC1*. The PITP domain shares ~50 % identity with Class I and Class IIA PITPs making it distinct from either family. Unlike PITPα and PITPβ, it is also the least studied. Our recent work has identified two binding partners for human RdgBβ (Garner et al. 2011). Unlike PITPα and β which have compact structures, RdgBβ has a region of disorder of 80 amino acids at the C-terminus following its N-terminal PITP domain. Two serine residues, Ser^{274} and Ser^{299} are phosphorylated and form a docking site for 14-3-3 proteins. A splice variant of RdgBβ (sp2) lacks the region for 14-3-3 binding and is found both in the cytosol and the nucleus unlike RdgBβ which is excluded from the nucleus (Takano et al. 2003). Binding of 14-3-3 proteins to the C-terminus shields two PEST sequences, regions rich in proline, glutamic acid, serine and threonine. PEST sequences are found in proteins that are rapidly degraded by proteolysis (Rechsteiner and Rogers 1996) and RdgBβ is indeed rapidly degraded with a half-life of 4 h and mutants that cannot bind 14-3-3 proteins are degraded even faster with a half-life of ~2 h.

Whilst PITPα and PITPβ are ubiquitous and have been detected in most tissues and in cultured cell lines such as Hela and COS-7 cells (Morgan et al. 2006), RdgBβ expression is either too low or undetectable using available antibodies and the only tissue that has substantial amounts of endogenous RdgBβ is the heart (Garner et al. 2011). These results suggest that RdgBβ expression is tightly regulated. Human dermal fibroblasts when stimulated to proliferate with serum, EGF, FGF, and PDGF show increased expression of *PITPNC1* transcripts together with increased expression of 14-3-3γ (Gu and Iyer 2006). Recent studies indicate that RdgBβ expression can also be regulated by the microRNA-126 (Png et al. 2012). MicroRNAs (miRNAs) are a family of small non-coding RNAs (approximately 21–25 nucleotides long) that are capable of targeting genes for either degradation of mRNA or inhibition of translation. MiR-126 down-regulates the expression of three genes, *IGFBP2, MERTK* and *PITPNC1* and the loss of miR-126 expression from cancer cells has been reported in colon cancer, gastric cancer and breast cancer. Thus, silencing of miR-126 in MDA-231 breast cancer cells significantly increases metastatic colonization and this could be recapitulated by increased expression of the individual genes, *IGFBP2, MERTK* and *PITPNC1*. IGFBP2 is a secreted protein which requires RdgBβ for its secretion. Metastatic cells over-expressing these genes are able to efficiently establish endothelial interactions by the modulation of the IGF1 signaling pathway. IGFBP2 secreted by the cancer cells aided by RdgBβ assists in IGF1 receptor activation on endothelial cells. miR-126 is also expressed in endothelial cells during development and is a promoter of normal developmental angiogenesis. However, in this case, miR-126 targets other genes and not *PITPNC1, MERTK* and *IGFBP2*. miRNAs appear to have cell-type specific effects using distinct molecular mechanisms to regulate physiologic or pathological processes (Png et al. 2012).

While the function of RdgBβ in enhancing the metastatic potential of breast cancer cells in pathology is evident from the study described above, the physiologic functions of RdgBβ remain to be identified. We have identified that RdgBβ is highly expressed in cardiomyocytes. Furthermore, we have identified that the PITP domain of RdgBβ binds to Angiotensin II receptor-associated protein (ATRAP). ATRAP is an integral membrane protein of 18 kDa, which is composed of three transmembrane domains and an extended hydrophilic cytoplasmic C-terminal tail (108–159 a.a.). ATRAP interacts with the angiotensin II type I receptor (AT1R) and ATRAP residues in the cytoplasmic C-terminus 110–120 a.a are required for this interaction (Daviet et al. 1999). Although identified as binding partner for AT1R, ATRAP is more widely expressed than AT1R. In addition, a splice variant of 152 amino acids is also expressed that lacks the residues, 116–121 a.a which is part of the region that interacts with the receptor (Daviet et al. 1999). Both splice variants of ATRAP are widely expressed in cultured cell lines including HL60 cells and COS-7 cells (our unpublished observations). The wide-spread expression of ATRAP suggests that its function is unlikely to be restricted to AT1R signaling. ATRAP also binds to several other proteins including calcium modulating cyclophilin ligand (CAML) (Guo et al. 2005) and receptor for activated protein C kinase (RACK1) (Wang et al. 2002). CAML regulates nuclear factor of activated T cell (NFAT) and studies suggest that ATRAP can inactivate the NFAT pathway (Guo et al. 2005; Min et al. 2009). The presence of RdgBβ and ATRAP in cardiomyocytes implies a function for RdgBβ in the heart, potentially in angiotensin II signaling.

Concluding Remarks

Proteins with a PITP domain display a rich repertoire of function ranging from cytokinesis, phototransduction, neurotransmission, behavioral plasticity at sensory neurons, membrane traffic, and each individual PITP appears to play specific roles. The theme that unites the PITPs is their ability to bind PI and all PITP proteins contain a signature set of residues that are non-linear which constitute the inositol headgroup binding site. Binding of PI coupled with the potentiation of PI phosphorylation by numerous lipid kinases is the major function of this class of proteins. Since PPIs play such widespread roles, it is not surprising that PITPs are required for so many different functions. Understanding the specific roles of the individual PITPs in mammalian systems are revealing some surprises. The recent discovery of PITPNC1 expression regulated by miR-126 is a case in point (Png et al. 2012). The tumor-suppressive effects of mir-126 offer these miRNAs as novel cancer therapeutics and in breast cancer cells, an association between loss of miR-126 and poor prognosis due to metastases has been made. How the increased expression of PITPNC1 in breast cancer cells contributes to metastases remains to be examined.

References

Alb JG Jr, Phillips SE, Rostand K, Cui X, Pinxteren J, Cotlin L, Manning TGS, York JD, Sontheimer JF, Collawn JF, Bankaitis VA (2002) Genetic ablation of phosphatidylinositol transfer protein function in murine embryonic stem cells. Mol Biol Cell 13:739–754

Alb JG Jr, Cortese JD, Phillips SE, Albin RL, Nagy TR, Hamilton BA, Bankaitis VA (2003) Mice lacking phosphatidylinositol transfer protein alpha exhibit spinocerebellar degeneration, intestinal and hepatic steatosis, and hypoglycemia. J Biol Chem 278:33501–33518

Alb JG Jr, Phillips SE, Wilfley LR, Philpot BD, Bankaitis VA (2007) The pathologies associated with functional titration of phosphatidylinositol transfer protein alpha activity in mice. J Lipid Res 48:1857–1872

Allen-Baume V, Segui B, Cockcroft S (2002) Current thoughts on the phosphatidylinositol transfer protein family. FEBS Lett 531:74–80

Amarilio R, Ramachandran S, Sabanay H, Lev S (2005) Differential regulation of endoplasmic reticulum structure through VAP-Nir protein interaction. J Biol Chem 280:5934–5944

Bailes HJ, Lucas RJ (2010) Melanopsin and inner retinal photoreception. Cell Mol Life Sci 67:99–111

Carvou N, Holic R, Li M, Futter C, Skippen A, Cockcroft S (2010) Phosphatidylinositol- and phosphatidylcholine- transfer activity of PITPβ is essential for COP1-mediated retrograde transport from the Golgi to the endoplasmic reticulum. J Cell Sci 123:1262–1273

Chang JT, Milligan S, Li Y, Chew CE, Wiggs J, Copeland NG, Jenkins NA, Campochiaro PA, Hyde DR, Zack DJ (1997) Mammalian homolog of *Drososphila retinal degeneration B* rescues the mutant fly phenotype. J Neurosci 17:5881–5890

Chen J, Yao Y, Gong C, Yu F, Su S, Chen J, Liu B, Deng H, Wang F, Lin L, Yao H, Su F, Anderson KS, Liu Q, Ewen ME, Yao X, Song E (2011) CCL18 from tumor-associated macrophages promotes breast cancer metastasis via PITPNM3. Cancer Cell 19:541–555

Cockcroft S (2007) Trafficking of phosphatidylinositol by phosphatidylinositol transfer proteins. Biochem Soc Symp 74:259–271

Cockcroft S, Carvou N (2007) Biochemical and biological functions of class I phosphatidylinositol transfer proteins. Biochim Biophys Acta 1771:677–691

Cockcroft S, Garner K (2011) Function of the phosphatidylinositol transfer protein gene family: is phosphatidylinositol transfer the mechanism of action? Crit Rev Biochem Mol Biol 46:89–117

Cosker KE, Shadan S, van Diepen M, Morgan C, Li M, Allen-Baume V, Hobbs C, Doherty P, Cockcroft S, Eickholt BJ (2008) Regulation of PI3 K signalling by the phosphatidylinositol transfer protein PITP{alpha} during axonal extension in hippocampal neurons. J Cell Sci 121:796–803

Cunningham E, Thomas GMH, Ball A, Hiles I, Cockcroft S (1995) Phosphatidylinositol transfer protein dictates the rate of inositol trisphosphate production by promoting the synthesis of PIP_2. Current Biol 5:775–783

Daviet L, Lehtonen JY, Tamura K, Griese DP, Horiuchi M, Dzau VJ (1999) Cloning and characterization of ATRAP, a novel protein that interacts with the angiotensin II type 1 receptor. J Biol Chem 274:17058–17062

de Brouwer AP, Versluis C, Westerman J, Roelofsen B, Heck AJ, Wirtz KW (2002) Determination of the stability of the noncovalent phospholipid transfer protein-lipid complex by electrospray time-of-flight mass spectrometry. Biochemistry 41:8013–8018

de Graaf P, Zwart WT, van Dijken RA, Deneka M, Schulz TK, Geijsen N, Coffer PJ, Gadella BM, Verkleij AJ, van der SP, van Bergen en Henegouwen PM (2004) Phosphatidylinositol 4-kinasebeta is critical for functional association of rab11 with the Golgi complex. Mol Biol Cell 15:2038–2047

Dumaresq-Doiron K, Savard MF, Akam S, Costantino S, Lefrancois S (2010) The phosphatidylinositol 4-kinase PI4KIIIalpha is required for the recruitment of GBF1 to Golgi membranes. J Cell Sci 123:2273–2280

Elagin VA, Elagina RB, Doro CJ, Vihtelic TS, Hyde DR (2000) Cloning and tissue localization of a novel zebrafish RdgB homolog that lacks a phospholipid transfer domain. Vis Neurosci 17:303–311

Fensome A, Cunningham E, Prosser S, Tan SK, Swigart P, Thomas G, Hsuan J, Cockcroft S (1996) ARF and PITP restore GTPγS-stimulated protein secretion from cytosol-depleted HL60 cells by promoting PIP_2 synthesis. Current Biol 6:730–738

Garner K, Li M, Ugwuanya N, Cockcroft S (2011) The phosphatidylinositol transfer protein, RdgBβ binds 14–3-3 via its unstructured C-terminus, whereas its lipid binding domain interacts with the integral membrane protein, ATRAP (Angiotensin II Type I receptor-associated protein). Biochem J 439:97–111

Gatt MK, Glover DM (2006) The *Drosophila* phosphatidylinositol transfer protein encoded by vibrator is essential to maintain cleavage-furrow ingression in cytokinesis. J Cell Sci 119:2225–2235

Giansanti MG, Bonaccorsi S, Kurek R, Farkas RM, Dimitri P, Fuller MT, Gatti P (2006) The class I PITP Giotto is required for *Drosophila* cytokinesis. Curr Biol 16:195–201

Giansanti MG, Belloni G, Gatti M (2007) Rab11 Is required for membrane trafficking and actomyosin ring constriction in meiotic cytokinesis of *Drosophila* males. Mol Biol Cell 18:5034–5047

Graham DM, Wong KY, Shapiro P, Frederick C, Pattabiraman K, Berson DM (2008) Melanopsin ganglion cells use a membrane-associated rhabdomeric phototransduction cascade. J Neurophysiol 99:2522–2532

Gu J, Iyer VR (2006) PI3 K signaling and miRNA expression during the response of quiescent human fibroblasts to distinct proliferative stimuli. Genome Biol 7:R42

Guo S, Lopez-Ilasaca M, Dzau VJ (2005) Identification of calcium-modulating cyclophilin ligand (CAML) as transducer of angiotensin II-mediated nuclear factor of activated T cells (NFAT) activation. J Biol Chem 280:12536–12541

Hamilton BA, Smith DJ, Mueller KL, Kerrebrock AW, Bronson RT, Berkel V, Daly V, Kroglyak MJ, Reeve L, Nernhauser MP, Hawkins JL, Rubin TL, M E, Lander ES (1997) The *vibrator* mutation causes neurogeneration via reduced expression of PITPα: positional complementation cloning and extragenic suppression. Neuron 18:711–722

Hara S, Swigart P, Jones D, Cockcroft S (1997) The first 5 amino acids of the carboxy terminus of phosphatidylinositol transfer protein α (PITPα) play a critical role in inositol lipid signalling: transfer activity of PITP is essential but not sufficient for restoration of phospholipase C signalling. J Biol Chem 272:14909–14913

Hardie RC, Raghu P, Moore S, Juusola M, Baines RA, Sweeney ST (2001) Calcium influx via TRP channels is required to maintain PIP2 levels in *Drosophila* photoreceptors. Neuron 30:149–159

Hay JC, Martin TFJ (1993) Phosphatidylinositol transfer protein required for ATP-dependent priming of Ca^{2+}-activated secretion. Nature 366:572–575

Hay JC, Fisette PL, Jenkins GH, Fukami K, Takenawa T, Anderson RE, Martin TFJ (1995) ATP-dependent inositide phosphorylation required for Ca^{2+}-activated secretion. Nature 374:173–177

Helmkamp GM Jr, Harvey MS, Wirtz KWA, van Deenen LLM (1974) Phospholipid exchange between membranes. Purification of bovine brain proteins that preferentially catalyze the transfer of phosphatidylinositol. J Biol Chem 249:6382–6389

Higgs HN, Han MH, Johnson GE, Glomset JA (1998) Cloning of a phosphatidic acid-preferring phospholipase A1 from bovine testis. J Biol Chem 273:5468–5477

Ile KE, Kassen S, Cao C, Vihtehlic T, Shah SD, Mousley CJ, Alb JG Jr, Huijbregts RP, Stearns GW, Brockerhoff SE, Hyde DR, Bankaitis VA (2010) zebrafish class 1 phosphatidylinositol transfer proteins: PITPbeta and double cone cell outer segment integrity in retina. Traffic 11:1151–1167

Iwata R, Oda S, Kunitomo H, Iino Y (2011) Roles for class IIA phosphatidylinositol transfer protein in neurotransmission and behavioural plasticity at the sensory neuron synapses of *Caenorhabditis elegans*. Proc Natl Acad Sci U S A 108:7589–7594

Kauffmann-Zeh A, Thomas GMH, Ball A, Prosser S, Cunningham E, Cockcroft S, Hsuan JJ (1995) Requirement for phosphatidylinositol transfer protein in epidermal growth factor signalling. Science 268:1188–1190

Kular G, Loubtchenkov M, Swigart P, Whatmore J, Ball A, Cockcroft S, Wetzker R (1997) Co-operation of phosphatidylinositol transfer protein with phosphoinositide 3-kinase(gamma) in the formylmethionyl-leucylphenylalanine-dependent production of phosphatidylinositol 3,4,5 trisphosphate in human neutrophils. Biochem J 325:299–301

Kular GS, Chaudhary A, Prestwich G, Swigart P, Wetzker R, Cockcroft S (2002) Co-operation of phosphatidylinositol transfer protein with phosphoinositide 3-kinase γ in vitro. Adv Enz Reg 42:53–61

Larijani B, Allen-Baume V, Morgan CP, Li M, Cockcroft S (2003) EGF regulation of PITP dynamics is blocked by inhibitors of phospholipase C and of the Ras-MAP kinase pathway. Curr Biol 13:78–84

Lee YJ, Shah S, Suzuki E, Zars T, O'Day PM, Hyde DR (1994) The *Drosophila* dgq gene encodes a G alpha protein that mediates phototransduction. Neuron 13:1143–1157

Lee SJ, Xu H, Kang LW, Amzel LM, Montell C (2003) Light adaptation through phosphoinositide-regulated translocation of *Drosophila* visual arrestin. Neuron 39:121–132

Lee I, Lehner B, Crombie C, Wong W, Fraser AG, Marcotte EM (2008) A single gene network accurately predicts phenotypic effects of gene perturbation in *Caenorhabditis elegans*. Nat Genet 40:181–188

Litvak V, Shaul YD, Shulewitz M, Amarilio R, Carmon S, Lev S (2002) Targeting of Nir2 to lipid droplets is regulated by a specific threonine residue within its PI-transfer domain. Curr Biol 12:1513–1518

Litvak V, Argov R, Dahan N, Ramachandran S, Amarilio R, Shainskaya A, Lev S (2004) Mitotic phosphorylation of the peripheral Golgi protein Nir2 by Cdk1 provides a docking mechanism for Plk1 and affects cytokinesis completion. Mol Cell 14:319–330

Litvak V, Dahan N, Ramachandran S, Sabanay H, Lev S (2005) Maintenance of the diacylglycerol level in the Golgi apparatus by the Nir2 protein is critical for Golgi secretory function. Nat Cell Biol 7:225–234

Liu J, Fairn GD, Ceccarelli DF, Sicheri F, Wilde A (2012) Cleavage furrow organization requires PIP(2)-mediated recruitment of anillin. Curr Biol 22:64–69

Loewen CJ, Roy A, Levine TP (2003) A conserved ER targeting motif in three families of lipid binding proteins and in Opi1p binds VAP. EMBO J 22:2025–2035

Lu C, Vihtelic TS, Hyde DR, Li T (1999) A neuronal-specific mammalian homolog of the *Drosophila* retinal degeneration B gene with expression to the retina and dentate gyrus. J Neurosci 19:7317–7325

Lu C, Peng YW, Shang J, Pawlyk BS, Yu F, Li T (2001) The mammalian retinal degeneration B2 gene is not required for photoreceptor function and survival. Neuroscience 107:35–41

Miller WL (2007) Steroidogenic acute regulatory protein (StAR), a novel mitochondrial cholesterol transporter. Biochim Biophys Acta 1771:663–676

Milligan SC, Alb JG, Elagina RB, Bankaitis VA, Hyde DR (1997) The phosphatidylinositol transfer protein domain of *Drosophila* retinal degeneration B protein is essential for photoreceptor cell survival and recovery from light stimulation. J Cell Biol 139:351–363

Min LJ, Mogi M, Tamura K, Iwanami J, Sakata A, Fujita T, Tsukuda K, Jing F, Iwai M, Horiuchi M (2009) Angiotensin II type 1 receptor-associated protein prevents vascular smooth muscle cell senescence via inactivation of calcineurin/nuclear factor of activated T cells pathway. J Mol Cell Cardiol 47:798–809

Ming G, Song H, Berninger B, Inagaki N, Tessier-Lavigne M, Poo M (1999) Phospholipase C-gamma and phosphoinositide 3-kinase mediate cytoplasmic signalling in nerve growth cone guidance. Neuron 23:139–148

Mizoguchi T, Nakajima K, Hatsuzawa K, Nagahama M, Hauri HP, Tagaya M, Tani K (2000) Determination of functional regions of p125, a novel mammalian Sec23p-interacting protein. Biochem Biophys Res Commun 279:144–149

Mogensen JE, Ferreras M, Wimmer R, Petersen SV, Enghild JJ, Otzen DE (2007) The major allergen from birch tree pollen, Bet v 1, binds and permeabilizes membranes. Biochemistry 46:3356–3365

Moldrup ML, Georg B, Falktoft B, Mortensen R, Hansen JL, Fahrenkrug J (2010) Light induces Fos expression via extracellular signal-regulated kinases 1/2 in melanopsin-expressing PC12 cells. J Neurochem 112:797–806

Morgan CP, Skippen A, Segui B, Ball A, Allen-Baume V, Larijani B, Murray-Rust J, McDonald N, Sapkota G, Morrice NA, Cockcroft S (2004) Phosphorylation of a distinct structural form of phosphatidylinositol transfer protein α at Ser^{166} by protein kinase C disrupts receptor-mediated phospholipase C signalling by inhibiting delivery of phosphatidylinositol to membranes. J Biol Chem 279:47159–47171

Morgan CP, Allen-Baume V, Radulovic M, Li M, Skippen AJ, Cockcroft S (2006) Differential expression of a C-terminal splice variant of PITPβ lacking the constitutive-phosphorylated Ser262 that localises to the Golgi Compartment. Biochem J 398:411–421

Nakajima K, Sonoda H, Mizoguchi T, Aoki J, Arai H, Nagahama M, Tagaya M, Tani K (2002) A novel phospholipase A1 with sequence homology to a mammalian Sec23p-interacting protein, p125. J Biol Chem 277:11329–11335

Paetkau DW, Elagin VA, Sendi LM, Hyde DR (1999) Isolation and characterization of *Drosophila* retinal degeneration B suppressors. Genetics 151:713–724

Panaretou C, Domin J, Cockcroft S, Waterfield MD (1997) Characterization of p150, an adaptor protein for the human phosphatidylinositol (PtdIns) 3-kinase. Substrate presentation by phosphatidylinositol transfer protein to the p150-PtdIns 3-kinase complex. J Biol Chem 272:2477–2485

Phillips SE, Vincent P, Rizzieri KE, Schaaf G, Bankaitis VA, Gaucher EA (2006) The diverse biological functions of phosphatidylinositol transfer proteins in eukaryotes. Crit Rev Biochem Mol Biol 41:21–49

Png KJ, Halberg N, Yoshida M, Tavazoie SF (2012) A microRNA regulon that mediates endothelial recruitment and metastasis by cancer cells. Nature 481:190–194

Polevoy G, Wei HC, Wong R, Szentpetery Z, Kim YJ, Goldbach P, Steinbach SK, Balla T, Brill JA (2009) Dual roles for the *Drosophila* PI 4-kinase four wheel drive in localizing Rab11 during cytokinesis. J Cell Biol 187:847–858

Radauer C, Lackner P, Breiteneder H (2008) The Bet v 1 fold: an ancient, versatile scaffold for binding of large, hydrophobic ligands. BMC Evol Biol 8:286

Rechsteiner M, Rogers SW (1996) PEST sequences and regulation by proteolysis. Trends Biochem Sci 21:267–271

Schouten A, Agianian B, Westerman J, Kroon J, Wirtz KW, Gros P (2002) Structure of apo-phosphatidylinositol transfer protein alpha provides insight into membrane association. EMBO J 21:2117–2121

Segui B, Allen-Baume V, Cockcroft S (2002) Phosphatidylinositol transfer protein-beta displays minimal sphingomyelin transfer activity and is not required for biosynthesis and trafficking of sphingomyelin. Biochem J 366:23–34

Shadan S, Holic R, Carvou N, Ee P, Li M, Murray-Rust J, Cockcroft S (2008) Dynamics of lipid transfer by phosphatidylinositol transfer proteins in cells. Traffic 9:1743–1756

Stark WS, Sapp R (1987) Ultrastructure of the retina of *Drosophila* melanogaster: the mutant ora (outer rhabdomeres absent) and its inhibition of degeneration in rdgB (retinal degeneration-B). J Neurogenet 4:227–240

Takano N, Owada Y, Suzuki R, Sakagami H, Shimosegawa T, Kondo H (2003) Cloning and characterization of a novel variant (mM-rdgBbeta1) of mouse M-rdgBs, mammalian homologs of *Drosophila* retinal degeneration B gene proteins, and its mRNA localization in mouse brain in comparison with other M-rdgBs. J Neurochem 84:829–839

Tani K, Mizoguchi T, Iwamatsu A, Hatsuzawa K, Tagaya M (1999) p125 is a novel mammalian Sec23p-interacting protein with structural similarity to phospholipid-modifying proteins. J Biol Chem 274:20505–20512

Thomas GMH, Cunningham E, Fensome A, Ball A, Totty NF, Troung O, Hsuan JJ, Cockcroft S (1993) An essential role for phosphatidylinositol transfer protein in phospholipase C-mediated inositol lipid signalling. Cell 74:919–928

Tilley SJ, Skippen A, Murray-Rust J, Swigart P, Stewart A, Morgan CP, Cockcroft S, McDonald NQ (2004) Structure-function analysis of human phosphatidylinositol transfer protein alpha bound to phosphatidylinositol. Structure 12:317–326

Tomioka M, Adachi T, Suzuki H, Kunitomo H, Schafer WR, Iino Y (2006) The insulin/PI 3-kinase pathway regulates salt chemotaxis learning in *Caenorhabditis elegans*. Neuron 51:613–625

Trivedi D, Padinjat R (2007) RdgB proteins: functions in lipid homeostasis and signal transduction. Biochim Biophys Acta 1771:692–699

van Tiel CM, Westerman J, Paasman M, Wirtz KWA, Snoek GT (2000) The protein kinase C-dependent phosphorylation of serine-166 is controlled by the phospholipid species bound to the phosphatidylinositol transfer protein α. J Biol Chem 275:21532–21538

Vihtelic TS, Goebl M, Milligan S, O'Tousa SE, Hyde DR (1993) Localization of *Drosophila* retinal degeneration B, a membrane- associated phosphatidylinositol transfer protein. J Cell Biol 122:1013–1022

Vordtriede PB, Doan CN, Tremblay JM, Helmkamp GM Jr, Yoder MD (2005) Structure of PITPβ in complex with phosphatidylcholine: comparison of structure and lipid transfer to other PITP isoforms. Biochemistry 44:14760–14771

Wang T, Montell C (2006) A phosphoinositide synthase required for a sustained light response. J Neurosci 26:12816–12825

Wang W, Huang Y, Zhou Z, Tang R, Zhao W, Zeng L, Xu M, Cheng C, Gu S, Ying K, Xie Y, Mao Y (2002) Identification and characterization of AGTRAP, a human homolog of murine angiotensin II receptor-associated protein (Agtrap). Int J Biochem Cell Biol 34:93–102

Way G, O'Luanaigh N, Cockcroft S (2000) Activation of exocytosis by cross-linking of the IgE receptor is dependent on ARF-regulated phospholipase D in RBL-2H3 mast cells: evidence that the mechanism of activation is via regulation of PIP_2 synthesis. Biochem J 346:63–70

Wirtz KW, Schouten A, Gros P (2006) Phosphatidylinositol transfer proteins: from closed for transport to open for exchange. Adv Enzyme Regul 46:301–311

Xie Y, Ding YQ, Hong Y, Feng Z, navarre S, Xi CX, Wang CL, Zhu XJ, Ackerman SL, Kozlowski D, Mei L, Xiong WC (2005) Role of phosphatidylinositol transfer protein α in netrin-1-induced PLC signalling and neurite outgrowth. Nature Cell Biol 7:1124–1132

Xie Y, Hong Y, Ma XY, Ren XR, Ackerman S, Mei L, Xiong WC (2006) DCC-dependent phospholipase C signaling in netrin-1-induced neurite elongation. J Biol Chem 281:2605–2611

Yoder MD, Thomas LM, Tremblay JM, Oliver RL, Yarbrough LR, Helmkamp GM Jr (2001) Structure of a multifunctional protein. Mammalian phosphatidylinositol transfer protein complexed with phosphatidylcholine. J Biol Chem 276:9246–9252

Chapter 10
Myotubularin Phosphoinositide Phosphatases in Human Diseases

Leonela Amoasii, Karim Hnia and Jocelyn Laporte

Abstract The level and turnover of phosphoinositides (PIs) are tightly controlled by a large set of PI-specific enzymes (PI kinases and phosphatases). Mammalian PI phosphatases are conserved through evolution and among this large family the dual-specificity phosphatase (PTP/DSP) are metal-independent enzymes displaying the amino acid signature Cys-X5-Arg-Thr/Ser (CX5RT/S) in their active site. Such catalytic site characterizes the myotubularin 3-phosphatases that dephosphorylate PtdIns3*P* and PtdIns(3,5)P_2 and produce PtdIns5*P*. Substrates of myotubularins have been implicated in endocytosis and membrane trafficking while PtdIns5*P* may have a role in signal transduction. As a paradox, 6 of the 14 members of the myotubularin family lack enzymatic activity and are considered as dead phosphatases. Several myotubularins have been genetically linked to human diseases: MTM1 is mutated in the congenital myopathy X-linked centronuclear or myotubular myopathy (XLCNM) and MTMR14 (JUMPY) has been linked to an autosomal form of such disease, while MTMR2 and MTMR13 are mutated in Charcot-Marie-Tooth

Authors: Leonela Amoasii, Karim Hnia are equally contributed.

L. Amoasii · K. Hnia · J. Laporte (✉)
Department of Translational Medicine and Neurogenetics, IGBMC (Institut de Génétique et de Biologie Moléculaire et Cellulaire), Illkirch, France
e-mail: jocelyn@igbmc.fr

L. Amoasii · K. Hnia · J. Laporte
U964, Institut National de la Santé et de la Recherche Médicale, Illkirch, France

L. Amoasii · K. Hnia · J. Laporte
UMR7104, Centre National de la Recherche Scientifique, Illkirch, France

L. Amoasii · K. Hnia · J. Laporte
Université de Strasbourg, Strasbourg, France

L. Amoasii · K. Hnia · J. Laporte
Chaire de Génétique Humaine, Collège de France, Illkirch, France

M. Falasca (ed.), *Phosphoinositides and Disease*, Current Topics in Microbiology and Immunology 362, DOI: 10.1007/978-94-007-5025-8_10,

(CMT) neuropathies. Furthermore, recent evidences from genetic association studies revealed that several other myotubularins could be associated to chronic disorders such as cancer and obesity, highlighting their importance for human health. Here, we discuss cellular and physiological roles of myotubularins and their implication in human diseases, and we present potential pathological mechanisms affecting specific tissues in myotubularin-associated diseases.

Abbreviations

AAV	Adeno-associated-virus
BIN1	Bridging integrator 1 gene coding fo amphiphysin 2
CC	Coiled-coil
CMT	Charcot-Marie-Tooth neuropathies
CMTX	X-linked inherited CMT
CNM	Centronuclear myopathy
Dlg1	Disks large 1
DI-CMT	Dominant inherited CMT
DRM	Desmin related myopathies
DRCM	Desmin related cardiomyopathies
PTP/DSP	Protein Tyrosine Phosphatase/Dual-specificity phosphatase
DNM2	Dynamin 2
EC	Excitation–contraction
EGF	Epidermal growth factor
GEFS$^+$	Generalized epilepsy with febrile seizures plus
ING2	Inhibitor of growth 2
KO	Knockout
MTM1	Myotubularin
MTMR	Myotubularin-related
NCV	Nerve conduction velocity
NF-L	Neurofilament light chain protein
Pi3K68D	PI 3-kinase class II
PIs	Phosphoinositides
PtdIns	Phosphatidylinositol
PHD	Plant homeodomain
PH-GRAM	Pleckstrin homology, glucosyltransferases, rab-like GTPase activators and myotubularins
RID	Rac-induced recruitment domain
RYR1	Ryanodine receptor
SAP97	Synapse associated 97
SID	Set-interacting domain
XLCNM	X-linked recessive form of centronuclear myopathy

Introduction

Phosphoinositide (PI) lipids are essential components of eukaryotic cell membranes. PIs, the phosphorylated derivatives of phosphatidylinositol (PtdIns), regulate fundamental biological processes including cell growth and survival, membrane trafficking, and cytoskeletal dynamics (Di Paolo and De Camilli 2006). PIs comprise less than 1 % of cell lipids, yet they play very important roles in major signal transduction pathways, serving as docking sites for signaling effectors and as precursors of secondary messengers. The inositol head group of PtdIns can be reversibly phosphorylated at three positions, 3, 4, and 5, and all seven PI isoforms, including three monophosphorylated (PtdIns3*P*, PtdIns4*P*, and PtdIns5*P*), three bisphosphorylated [PtdIns(3,4)P_2, PtdIns(3,5)P_2, and PtdIns(4,5)P_2], and one tris-phosphorylated [PtdIns(3,4,5)P_3] species have been identified in eukaryotic cells (Fig. 10.1). The level and turnover of PIs are tightly controlled by a large set of PI-specific enzymes (PI-kinases and phosphatases). Found on the cytosolic side of membrane bilayers, PIs are readily accessible to PI-kinases and phosphatases capable of attaching and removing phosphate groups, respectively, and to phospholipases that cleave the lipids. PIs species are enriched in distinct intracellular membranes and thus each form of PIs may serve as a marker of membrane subdomains or organelles. However, the dynamics of PIs more than their absolute concentration at a given subdomain is probably underlying their specific functions. For example, the PtdIns3*P* is the most represented PI signature on early endosomes and a ligand for a large number of endosomal proteins (Gillooly et al. 2000; Simonsen et al. 1998). The bulk of PtdIns3*P* pool is generated directly on endosomes by the phosphorylation of PtdIns at the 3-position by PI 3-kinases (Murray et al. 2002). Subsequently, phosphorylation of PtdIns3*P* into PtdIns(3,5)P_2 by PIKfyve 5-kinase on endosomes is thought to generate docking sites for the recruitment of cytosolic factors that control outgoing traffic from early and late endosomes and multivesicular bodies formation (Odorizzi et al. 1998, 2000). The localization of PtdIns5*P*, a low abundance PI species that can be generated by either phosophorylation of PtdIns by PIKfyve or dephosphorylation of PtdIns(3,5)P_2 or PtdIns(4,5)P_2 by myotubularins or 4-phosphatases, respectively, remains under investigation (Coronas et al. 2007; Tronchere et al. 2004; Ungewickell et al. 2005). However, specific mass assays and imaging using high affinity probes like the plant homeodomain (PHD) of the inhibitor of growth 2 (ING2) lead several groups to suggest that PtdIns5*P* may act as a second messenger both in the cytoplasm and the nucleus (Gozani et al. 2003, 2005; Morris et al. 2000). The amount of PtdIns5*P* can be increased by different stimuli such as thrombin in platelets, insulin in 3T3-L1 cells, T cell receptor engagement, increased tyrosine kinase activity, and stress signals (Grainger et al. 2011; Morris et al. 2000; Pizarro-Cerda and Cossart 2004; Sbrissa et al. 2001, 2002). In addition, an increase in PtdIns5*P* was reported during the G1 phase of the cell cycle (Clarke et al. 2001). Infection by the intracellular pathogens *Shigella flexneri* or *Salmonella Typhimurium* also results in increased amounts of PtdIns5*P* in host cells supporting that pathogens divert the PIs metabolism to

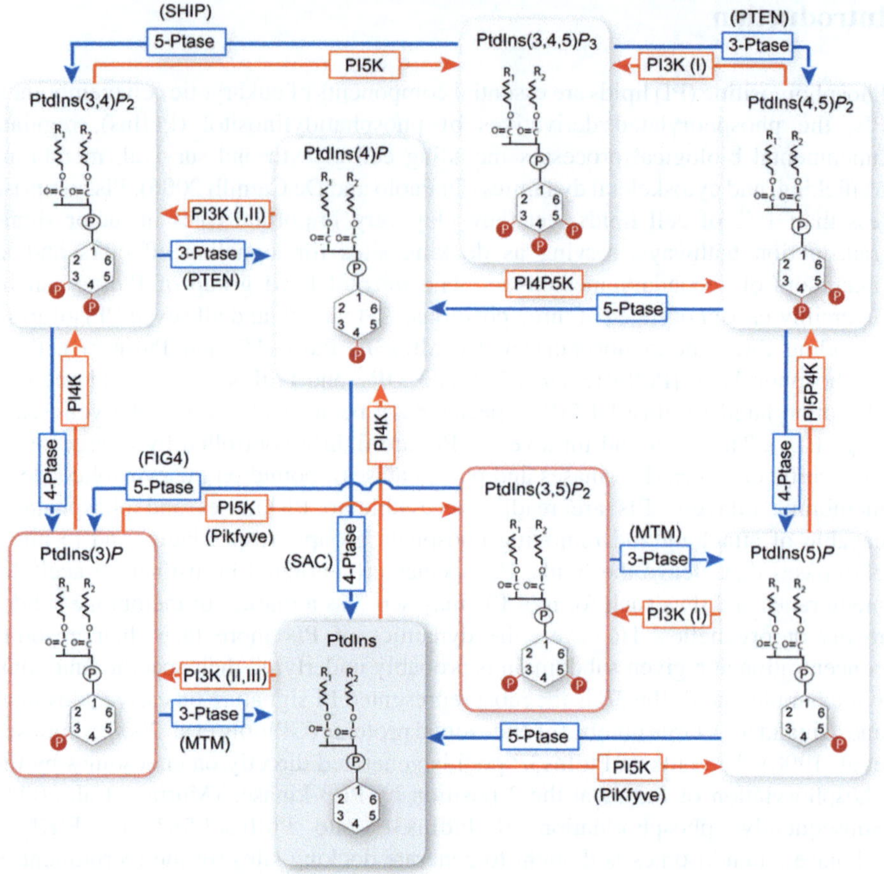

Fig. 10.1 Integration of myotubularins in the PI metabolism and PI-metabolizing reactions. The phosphatidylinositol (*PtdIns*) is an acidic (anionic) phospholipid that in essence consists of a phosphatidic acid backbone, linked via a phosphate group to inositol ring. PtdIns is subjected to reversible phosphorylation on three of the five free hydroxyl groups on the inositol ring (D3, D4 and D5), and this generates seven possible PI species. The PI metabolism is spatially and temporally regulated in the cell through controlled recruitment and activation of the different PI kinases (*red arrow*) and phosphatases (*blue arrow*). The majority of these metabolizing-enzymes are implicated in a variety of human diseases such as the 3-phosphatase (3-Ptase) PTEN, the 5-phosptahases SHIP and FIG4, and the 5-kinase (PI5 K) PIKfyve. Myotubularins are 3-phosphatase that dephosphorylate both PtdIns3,5-bisphosphate (PtdIns3,5P_2) and the PtdIns3P to PtdIns5P and PtdIns, respectively. Three catalytically active myotubularin-like proteins have been implicated in neuromuscular diseases; MTM1 and MTMR14 in centronuclear myopathies and MTMR2 is mutated in the Charcot-Marie Tooth neuropathy

promote host cell survival and their subsequent dissemination (Pizarro-Cerda and Cossart 2004; Ramel et al. 2011). Other PIs species have been discussed in detail in other reviews (Di Paolo and De Camilli 2006; Itoh and Takenawa 2002; Lecompte et al. 2008; Martin 1998; Michell et al. 2006; Payrastre 2004).

Mammalian PI phosphatases are conserved through evolution and classified into two superfamilies: the protein tyrosine phosphatase and dual-specificity phosphatase (PTP/DSP) superfamily, and the inositide polyphosphate phosphatase superfamily. PI phosphatases of the PTP/DSP superfamily are metal-independent enzymes that exhibit the signature Cys-X5-Arg-Thr/Ser (CX5RT/S) in their active site implicated in the various chemical steps of the enzymatic reaction. First, the PO_3 moiety from the phosphoinositol headgroup is accepted by the nucleophilic cysteine residue to generate a phosphocysteine intermediate. The invariant arginine of the CX5RT/S motif stabilizes the transition-state, and the PO_3 moiety is transferred to a water molecule. Finally, a conserved aspartic acid, usually from an adjacent structural loop, completes the catalytic cycle by protonating the leaving oxygen and reconstituting an uncharged hydroxyl group on the PI that is released (Guan and Dixon 1991; Fauman and Saper 1996). The PI phosphatases of the PTP/DSP superfamily fall into four primary classes: the PI 4-phosphatases that either transform PtdIns(3,4)P_2 into PtdIns3P or PtdIns(4,5)P_2 to PtdIns5P; the PtdIns(3,4,5)P_3 3-phosphatase PTEN tumor suppressor; the PI 5-phosphatases like SHIP, synaptojanins, OCRL, or FIG4; and the myotubularin PtdIns3P and PtdIns(3,5)P_2 3-phosphatases (Chow et al. 2007; Gary et al. 2002; Maehama and Dixon 1998; Norris et al. 1997; Nystuen et al. 2001; Ungewickell et al. 2005). Here, we highlight the role of myotubularins in several cellular pathways, their implication in human diseases and discuss the potential pathological mechanisms.

The Myotubularin Phosphoinositides Phosphatases and Inactive Homologs

Myotubularins are only found in eukaryotes and define a large protein family highly conserved from yeast to human (Laporte et al. 1998; Lecompte et al. 2008). Humans express 14 myotubularin members distributed into 6 phylogenetic subgroups (Fig. 10.2a). Three subgroups have PI 3-phosphatase activity: MTM1 and the myotubularin-related (MTMR) proteins MTMR1 and MTMR2, MTMR3 and MTMR4, MTMR6 together with MTMR7 and MTMR8. The catalytically active myotubularins preferentially dephosphorylate PtdIns3P and PtdIns(3,5)P_2 and is one of the main pathways leading to PtdIns5P production (Blondeau et al. 2000; Chaussade et al. 2003; Taylor et al. 2000b; Tronchere et al. 2004). Noteworthy, the other three sub-groups contain naturally-occurring missense substitutions that involve the Cys and Arg residues of the canonical CX5R(S/T) motif and lack enzymatic activity: MTMR5 and MTMR13, MTMR9, MTMR10 with MTMR11 and MTMR12. In addition, the MTMR14 protein, also called JUMPY or egg-derive tyrosine phosphatase, has a similar CX5R(S/T) motif to myotubularins and share their substrate specificity; however, phylogenetic studies and protein domain composition suggested it defines a separated family (Lecompte et al. 2008; Tosch et al. 2006; Yamaguchi et al. 1999). The MTMR15 protein, once defined as part of

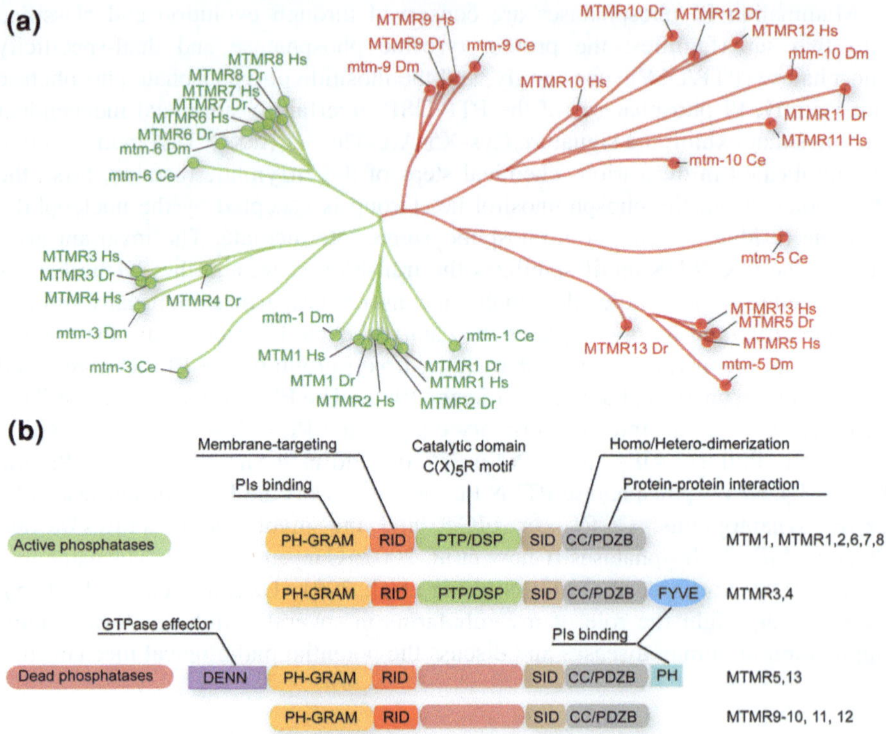

Fig. 10.2 Phylogenetic relationship and protein domains of active and inactive myotubularin phosphatases. (**a**) A phylogeny of myotubularins from human (*Homo sapiens, Hs*), zebrafish (*Danio rerio, Dr*), fruit fly (*Drosophila melanogaster, Dm*), and nematode (*Caenorhabditis elegans, Ce*). Catalytically active myotubularins are in green, catalytically inactive in red. Myotubularins were identified by using human sequences as queries for BLAST searches of the indicated genomes. The phylogenetic tree was adapted from Robinson and Dixon (2006). (**b**) Drawing of myotubularin domains and motifs in the human myotubularins. All myotubularins have a pleckstrin homology-glucosyltransferases rab-like GTPase activators and myotubularins (*PH-GRAM*), a rac-induced recruitment domain (*RID*), the protein tyrosine phosphatase/dual-specificity phosphatase (*PTP/DSP domain*), the set-interacting domain (*SID*) and a coiled-coil (*CC*) domain followed by a PDZ-binding motif (Postsynaptic Density-95, disks large and Zona occludens-1). Besides, MTMR3 and MTMR4 have a FYVE domain, while MTMR5 and MTMR13 contain DENN and PH domains

the myotubularin inactive members, does not share extensive homology with myotubularins and was renamed FAN1, a nuclease implicated in DNA repair (Alonso et al. 2004; Liu et al. 2010; Patzko and Shy 2011).

The so-called dead phosphatase myotubularins are probably not acting as substrate-traps but several studies converged to show that they unexpectedly heterodimerize with their active homologs (Kim et al. 2003; Mochizuki and Majerus 2003; Nandurkar et al. 2003). Doing so, they may either be critical for the localization of their active homolog, for appropriate presentation of the PI substrate to the catalytically active subunit in a cellular context, or for a direct activation of its catalytic activity (Berger et al. 2006b; Taylor et al. 2000b; Zou et al. 2009). The dead phosphatases may thus indirectly regulate the phosphorylation status of PIs. This original regulation does not hold true for all species, as for example *Saccharomyces cerevisiae* expresses a single active MTM1 homolog in yeast (Ymr1), and the yeast genome does not encode for a dead-phosphatase myotubularin (Lecompte et al. 2008; Robinson and Dixon 2006). Homodimerization of myotubularins has also been reported, for example in the case of MTM1 where it could be triggered by PtdIns5*P* and induces an increase in enzymatic activity based on an allosteric mechanism (Schaletzky et al. 2003).

The domain structure common to myotubularins includes a Pleckstrin homology, glucosyltransferases, rab-like GTPase activators and myotubularins(PH-GRAM), a rac-induced recruitment (RID) domain also called MTMR domain (IPR010569 in Interpro database; http://wwwdev.ebi.ac.uk/interpro/), the PTP/DSP phosphatase-like domain, the set-interacting domain (SID) and a coiled-coil (CC) domain followed by a PDZ-binding motif (Fig. 10.2b). The GRAM domain, which overlaps with a PH motif, appears to bind to PIs although a more recent study did not confirm this affinity (Begley et al. 2003; Choudhury et al. 2006; Doerks et al. 2000; Tsujita et al. 2004). The RID is a membrane-targeting motif necessary for myotubularin recruitment to Rac1-induced membrane ruffles, whereas the SID and the PDZ-binding motif domain mediate protein–protein interactions (Bolino et al. 2004; Laporte et al. 2002; Previtali et al. 2003). The CC domain is critical for homo and/or hetero-dimerization of myotubularins, and the PH-GRAM has also been implicated in this process (Berger et al. 2003; Lorenzo et al. 2005; Nandurkar et al. 2003; Robinson and Dixon 2005). The MTMR3/4 and MTMR5/13 subclasses contain additional conserved modules, including FYVE, PH, and DENN domains (Robinson and Dixon 2006; Laporte et al. 2001). While the FYVE and PH domain are domains found to bind PtdIns3*P* and PIs in other proteins and sustain the implication of myotubularins in PIs regulation, the DENN domain is found in several Rab GTPase exchange factors, suggesting a role for MTMR5 and MTMR13 in Rab regulation (Yoshimura et al. 2010).

Studies in cultured cells showed that several myotubularins are associated to endosomal compartments and other intramembranes i.e. profiles consistent with involvements of these proteins in the control of PtdIns3*P* and PtdIns(3,5)P_2 signaling to regulate protein trafficking at the endosomal system. Indeed, PtdIns3*P* and PtdIns(3,5)P_2 are present on endosomal compartments where PtdIns3*P* predominates on early endosomes and PtdIns(3,5)P_2 on late endosomes (Cao et al. 2007, 2008; Laporte et al. 2002; Mochizuki and Majerus 2003; Zhao et al. 2001). Additional studies using yeasts supported a role of myotubularin in vesicle homeostasis (Blondeau et al. 2000; Taylor et al. 2000a).

Myotubularin-Related Diseases

Myotubularin (MTM1), the pioneer member of myotubularins, was discovered by positional cloning as the gene mutated in X-linked centronuclear myopathy (XLCNM) (Laporte et al. 1996). Later, MTMR2 and MTMR13 were found mutated in demyelinating Charcot-Marie-Tooth neuropathy types 4B1 and 4B2, respectively (Azzedine et al. 2003; Bolino et al. 2000; Senderek et al. 2003). In addition, several myotubularins have been recently linked to multifactorial diseases. Hence, the interest in myotubularins is stoked by their implication in human diseases.

X-Linked Centronuclear/Myotubular Myopathy

Mutations in the myotubularin (MTM1) gene on chromosome Xq28 have been identified in patients with the X-linked recessive form of centronuclear myopathy (XLCNM) , also called myotubular myopathy (XLCNM, OMIM 310400) (Laporte et al. 1996). XLCNM is resulting in a very severe and generalized muscle weakness and hypotonia, external ophthalmoplegia, and respiratory distress (Jungbluth et al. 2008). The disease incidence is about 1:50,000 in newborn males. Signs of antenatal onset comprise reduced fetal movements and polyhydramnios. Death occurs usually during the first year of life from respiratory failure, although a few milder cases were described to survive to adulthood (Herman et al. 1999). Mutations in the MTM1 gene have been now identified in more than 90 % of affected male patients and about 200 MTM1 mutations widespread through the gene have been reported so far (Laporte et al. 2000). On muscle biopsy, centronuclear myopathy (CNM) is characterized by centrally placed nuclei surrounded by a perinuclear halo devoid of myofilaments and occupied by mitochondrial and glycogen aggregates (Romero 2010). The characteristic central nuclei are seen in all muscles, including extraocular muscles, and may affect up to 90 % of fibers (Ringel et al. 1979; Romero 2010). Predominance of type I fibers was also frequently observed in muscle patients. Immunohistochemical studies have demonstrated consistent but nonspecific abnormalities: persistent fetal expression pattern of various proteins including the cell surface protein N-CAM, myosin, vimentin, and desmin have been reported in male infants with the X-linked form, but more recent immunohistochemical studies on sequential biopsies in long-term survivors suggest that the expression of developmentally regulated proteins eventually decreases as in healthy individuals (Dorchies et al. 2001; Fidzianska et al. 1994; Grainger et al. 2011; Misra et al. 1992; Van der Ven et al. 1995a; Yu et al. 2003). Other proteins abnormally expressed in XLCNM include laminin and collagen components (Van der Ven et al. 1995b). Observations in an *Mtm1*-null mouse model suggest a muscle-specific role of myotubularin in muscle fiber maintenance but not in myogenesis (Buj-Bello et al. 2002). A recent gene expression profiling study in muscles with different MTM1 mutations revealed upregulation of transcripts for cytoskeleton and extracellular matrix proteins, indicating that remodeling of cytoskeleton and extracellular architecture plays a role in the atrophy and intracellular disorganization observed in XLCNM (Noguchi et al. 2005). Prolonged expression but eventual decrease of developmentally regulated proteins in muscle from affected infants suggests maturational delay rather than complete developmental arrest in this condition. Recent data in *Mtm1* knockout mice and zebrafish morphants suggested structural anomalies of the triad, a connection of one T-tubule and two sarcoplasmic reticulum saccules underlying the excitation–contraction (EC) coupling (Al-Qusairi et al. 2009; Dowling et al. 2009) (Fig. 10.3a). These morphological defects lead to disruption of EC-coupling machinery, which affects normal Ca2+ turnover and muscle contraction (Al-Qusairi et al. 2009). However, the precise localization and role of MTM1 in skeletal muscle remain to be determined. Interestingly, defects in triad structure have also been found in autosomal forms of centronuclear myopathies due to mutations in BIN1 (amphiphysin 2) and DNM2 (dynamin 2), both regulators of membrane remodeling and trafficking (Bitoun et al. 2005, 2007, 2009; Claeys et al. 2010; Nicot et al. 2007; Toussaint et al. 2011). Thus, an intriguing hypothesis is emerging of a common pathway involving MTM1, DMN2, and BIN1, which should function in the maintenance of the triad structure by regulating specialized membrane compartment and/or cytoskeleton rearrangement.

MTMR14/hJUMPY Implication in Centronuclear Myopathy

The MTMR14 gene is located on chromosome 3p25.3 and is conserved down to flesh flies and drosophila. It is expressed in different tissues and particularly at a high level in skeletal muscle and, as for *Mtm1*, *Mtmr14* expression level increases during myoblast differentiation to myotubes in vitro. MTMR14 is also able to efficiently dephosphorylate PtdIns3*P* and PtdIns(3,5)P_2 at position 3 of the inositol ring (Tosch et al. 2006) .

Screening of CNM patients with neonatal, childhood and adult onsets, without mutation in the MTM1 genes, revealed amino acid changes in MTMR14/JUMPY in two cases with early onset (Tosch et al. 2006). One patient had a heterozygous missense p.Arg336Glu. This 12-year-old boy had a neonatal hypotonia and hypoxia and a stable or slowly progressive disease with general hypotonia and diffuse weakness, affecting predominantly proximal portions of the limbs, together with ophthalmoparesis. The arginine at position 336 is conserved in all the MTMR14 protein orthologs and in all PTP/DSP phosphatases, as it is one of the amino acids required for the enzymatic activity in the CX5RT/S signature (Tosch et al. 2006). The second patient had a non-conservative heterozygous p.Tyr462Cys missense variation. This 36-year-old female had a neonatal hypotonia followed by a stable muscle involvement and also presented ophthalmoparesis. The tyrosine 462 lies outside the PTP/DSP domain and is conserved at least in mammals and birds. Importantly, the p.Arg336Glu variant and, to a lesser extent, the p.Tyr462Cys variant impaired the enzymatic activity of MTMR14. One of the patients also had a mutation in the *DNM2* gene, suggesting that MTMR14 defect may modulate the phenotype due to the *DNM2* mutation but not directly induces CNM.

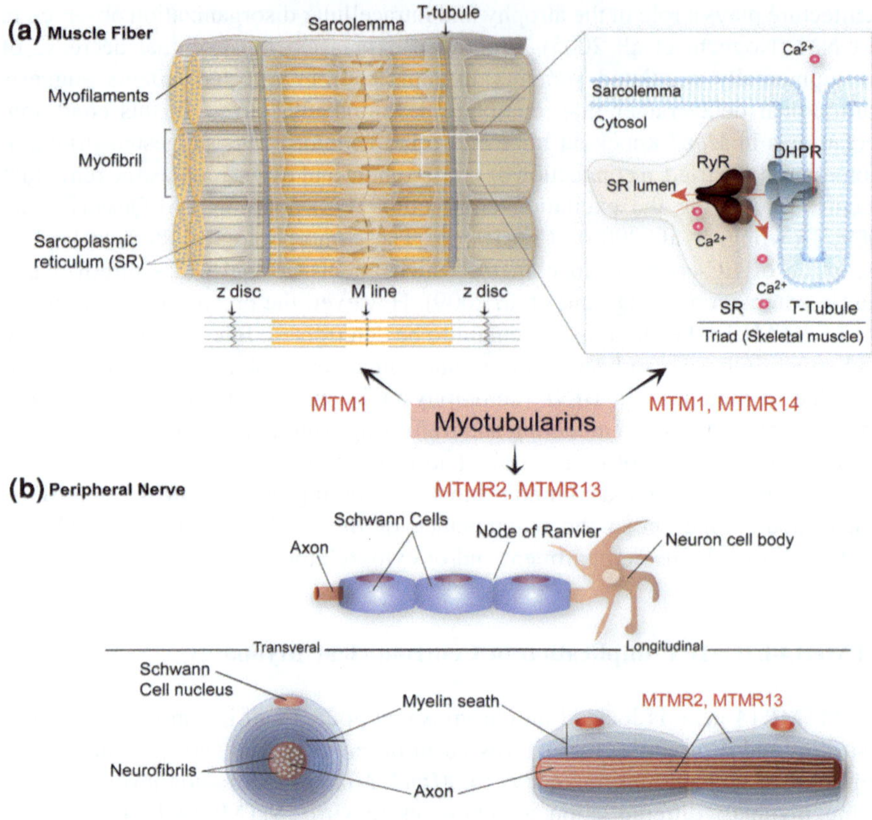

Fig. 10.3 Tissue-specific role of myotubularins and pathological relevance. (**a**) MTM1 and MTMR14 are implicated in centronuclear myopathies (X-linked form and sporadic cases, respectively). Both proteins are implicated in skeletal muscle maintenance and Ca2+ homeostasis. The latter mechanism involves the Ca2+ entry from the extracellular space via the transmembrane channel complex DHPR while a second channel, the Ryanodine receptor (*RyR1*) on the internal compartment storage called the sarcoplasmic reticulum (*SR*), is in charge of the fine regulation of the cytoplasmic level of Ca2+ (*squared*). In centronuclear myopathies involving MTM1 and MTMR14, defects in the triad (junction between two sarcoplasmic reticulum and one T-tubule) was observed, paralleling defects in RyR1 function. (**b**) Both MTMR2 and MTMR13 have been associated to Charcot-Marie-Tooth neuropathies (CMT4B1 and CMT4B2, respectively) characterized by myelin outfoldings. Myelin in the peripheral nervous system is generated and maintained by myelinating Schwann cells. This highly specialized cell type enwraps segments of axons with multiple layers of its plasma membrane, and segments are connected through nodes of Ranvier. MTMR2 and MTMR13 can form a heterodimer and MTMR13 could regulate MTMR2 phosphatase activity in vitro suggesting an important role of the MTMR2/MTMR13 complex in CMT physiopathology

Vergne et al. reported a role for MTMR14/JUMPY in the regulation of autophagy. They discovered that MTMR14 associated with early autophagosomes and negatively regulated progression through the steps of autophagy (Vergne et al. 2009).

Knockdown of *MTMR14* in cultured cells resulted in increased autophagy, whereas overexpression inhibited this process. MTMR14 phosphatase activity was required for its role as a regulator of autophagy, a finding consistent with the known importance of PtdIns3*P* in the promotion of the early steps of autophagy (Vergne et al. 2009). In agreement, an increase in lipidated LC3 was noted in *Mtmr14* KO mouse muscle and in muscles from *Mtmr14* zebrafish morphants (Dowling et al. 2010; Hnia et al. 2011a). Shen et al. (2009) reported on mice with a targeted deletion in *Mtmr14* that resulted in a mild phenotype of early onset exercise intolerance and late-onset muscle wasting. They also observed a reduction in force production in muscle from the knockout animals, and demonstrated that this change was associated with calcium leakage from the sarcoplasmic reticulum. These changes in calcium homeostasis were most probably due to a direct activation of the skeletal muscle ryanodine receptor (RYR1) by excess of PtdIns(3,5)P_2 (Shen et al. 2009). Similarly, morpholino-based experiments in zebrafish suggested that MTMR14 is important for excitation—contraction coupling as morphants failed to respond to high-frequency electrical stimulation (Dowling et al. 2010). In both zebrafish and mice, downregulation or deletion of MTMR14 did not reproduce the centralization of nuclei observed in patients with CNM; however, structural defects of the triad were observed by electron microscopy (Dowling et al. 2010; Hnia et al. 2011a). Moreover, decreased *Mtmr14* in an *Mtm1* knockdown background potentiated the CNM-like phenotype in zebrafish. As defect in calcium homeostasis and structural anomalies of the triads were also reported as common signs of different forms of centronuclear myopathies, it suggests that MTMR14 impairment is associated to muscle weakness in such diseases (Toussaint et al. 2011).

Charcot-Marie-Tooth Peripheral Neuropathies Type 4B

Charcot-Marie-Tooth disease (CMT) denotes a heterogeneous group of genetic diseases that affect peripheral nerves, leading to pronounced muscular atrophy and weakness of distal limbs. Electrophysiological and neuropathological data differentiate sensory and motor neuropathies. In some classes, only motor or sensory neurons might be affected, and in others, both cell types may be involved. Thus, axonal CMT2 diseases can be dominant or recessive, and similarly demyelinating CMT are classified depending on their inheritance: CMT1 for dominant and CMT4 for autosomal recessive. The intermediate forms are mainly following a dominant inheritance (DI-CMT) or X-linked inheritance (CMTX) (Patzko and Shy 2011). Nerve conduction velocity (NCV) measurements are used to differentiate between demyelinating and axonal forms: uniformly slow NCV less than 38 m/s in the arms is characteristic of demyelinating CMT. Intermediate CMT have NCV values in the range 25–45 m/s.

Autosomal recessive forms of demyelinating CMT are collectively designated CMT4 and focally folded myelin sheaths around peripheral nerves are characteristic of CMT4B sub-type (Dubourg et al. 2006; Gambardella et al. 1997; Zappia et al. 1997). Myelin in the peripheral nervous system is generated and maintained by myelinating Schwann cells. This highly specialized cell type enwraps segments of axons with multiple layers of its plasma membrane, and segments are connected through nodes of Ranvier. MTMR2 was found mutated in CMT4B (CMT4B1, OMIM 601382) (Bolino et al. 2000). Mutations in the MTMR13/SBF2 gene, which encodes a catalytically inactive member of the myotubularin family, cause CMT4B2 (OMIM 604563), which has almost similar pathological features to CMT4B1 (Azzedine et al. 2003; Senderek et al. 2003). MTMR2 and MTMR13 can form heterotetramers and MTMR13 could regulate MTMR2 phosphatase activity in vitro (Berger et al. 2006b; Robinson and Dixon 2005). The implication of both proteins in similar disease was the first proof that myotubularin heterodimerization has important roles at the physiological level. However, it remains an open question how their interaction could impact on the physiopathology of CMT4B1 and CMT4B2. MTMR2 and MTMR13 deficiency exclusively in Schwann cells in mouse models is sufficient to generate myelin outfoldings as seen in CMT4B1 and CMT4B2 (Bonneick et al. 2005; Robinson et al. 2008; Tersar et al. 2007). Such aberrant structures contain redundant myelin membranes that originate during late development of myelinating Schwann cells mainly at the paranodes and Schmidt-Lanterman incisures, regions that contain Schwann cell cytoplasm and intracellular organelles such as endosomes and lysosomes (Fig. 10.3b). Paranodes and Schmidt-Lanterman incisures are also thought to be the main regions of the myelin sheath where addition and turnover of myelin membranes and its protein components occur. These findings, together with the known functions of the PI substrates of MTMR2 on intramembranes suggest that MTMR2/MTMR13 couple could be involved somehow in the regulation of membrane addition and/or remodeling. Altered membrane trafficking, recycling, or impaired degradation may lead to the observed phenotype. Noteworthy, other genes mutated in different forms of demyelinating CMT encode proteins implicated in endosome/lysosome function like FIG4, a PtdIns(3,5)P_2 phosphatase, and LITAF, an integral membrane protein of lysosome and late endosome (Patzko and Shy 2011; Zhang et al. 1997). Indeed, recent elegant crosses in mice by Vaccari and colleagues have shown a genetic interaction between MTMR2 and FIG4 both in Schwann cells and neurons (Vaccari et al. 2011). In addition, reduction of FIG4 rescued *Mtmr2*-null myelin outfoldings in vitro and in vivo, suggesting some possible therapeutic approaches and supporting that PIs misregulation is a cause of CMT4 neuropathies.

Potential Link between Myotubularins and Multifactorial Diseases

Cancer: Over the past 2 decades, PI effectors and several metabolizing enzymes have been investigated in cancer. In particular, PTEN has blossomed into one of the featured molecules in cancer biology. Hundreds of published reports focused on PTEN structure, function, and mutations in the context of various human diseases. After P53, PTEN is the most commonly mutated gene in human cancer, and the most deleted/mutated phosphatase in human sporadic and hereditary cancer syndromes (Liaw et al. 1997; Nelen et al. 1997; Stiles et al. 2004; Suzuki et al. 2009). More recently, alteration of several other PTPs/DSPs has been implicated in oncogenic transformation (Wang et al. 2004). By analyzing a public genomic database, a recent study reported several MTMR3 mutations in colorectal and gastric cancers (Song et al. 2010). MTMR3 genomic region is also linked to lung cancer susceptibility (Hu et al. 2011). The suspected role of MTMR3 in autophagy might be linked to the pathology, although a direct link remains to be found (Taguchi-Atarashi et al. 2010). In addition, expression of the dead phosphatase MTMR11 is also altered following modulation of the oncogene HER2 in cells, suggesting a potential link between myotubularins and oncogenic program in cancer (Lucci et al. 2010).

Epilepsy: Generalized epilepsy with febrile seizures plus ($GEFS^+$) is a familial condition in which patients have febrile seizures variably associated with epilepsy. Febrile seizures are the most common convulsive event in humans, affecting 2–5 % of children. The $GEFS^+$ context is inherited as an autosomal dominant trait with incomplete penetrance. It is genetically heterogeneous and several ion channel genes have been implicated in $GEFS^+$. Recently, a sixth locus for $GEFS^+$ was identified on chromosome 8p23-p21 (Baulac et al. 2008). No ion channel genes are located in this interval. Two of the candidate genes located in the genetic interval are MTMR7 and MTMR9, although no mutations have been identified in these genes yet. The identification of the responsible gene will probably uncover a new mechanism of pathogenesis for $GEFS^+$.

Obesity: Obesity is a genetically heterogeneous disorder and its genetic susceptibility is likely to differ among various ethnic groups. Common obesity arises when an individual's genetic background is susceptible to an environment that promotes energy intake over energy expenditure. Many cases of monogenic obesity (obesity associated with a single-gene mutation) have been reported. Most of the genes causing monogenic obesity are expressed in the hypothalamus and have been indicated to have important roles in the regulation of food intake; therefore, genes expressed in the hypothalamus are likely to be good candidates for susceptibility to obesity. Recent reports, performed through an association studies using a large number of gene-based SNPs, have reported that the MTMR protein-9 (MTMR9) gene is associated with obesity (Hotta et al. 2011; Yanagiya et al. 2007). MTMR9 is expressed in the lateral hypothalamic area, paraventricular nucleus and accurate nucleus of the hypothalamus, the center for food intake regulation, and MTMR9 expression was regulated by diet. These data suggest that MTMR9 is likely to contribute to genetic susceptibility to obesity.

Connecting Molecular Pathways and Physiopathology in Myotubularinopathies

Phosphoinositide Metabolism and Membrane Trafficking

The myotubularin PI 3-phosphatase family has been partially explored in vitro and ex vivo. As for MTMR2, MTM1 was proposed to bind to PtdIns(3,5)P_2 through the N-terminal PH-GRAM domain (Berger et al. 2003; Tsujita et al. 2004). MTM1 and MTMR2 localizes to early and late endosomal compartments, respectively, where PtdIns3*P* and PtdIns(3,5)P_2 are particularly enriched (Cao et al. 2007, 2008). Overexpression of MTM1 in cultured cells altered the trafficking of EGF receptor from the late endosome to lysosome and subsequently its degradation. However, the PH-GRAM domain was not mediating membrane localization, as mutations in this domain did not affect MTM1 endosomal localization. The role of the PH-GRAM domain of several myotubularins is likely to mediate their PtdIns3*P*/PtdIns5*P*-dependent oligomerization and a PtdIns5*P*-specific activation rather than targeting the concerned myotubularin in specific compartments. Indeed, addition of PtdIns5*P* to a catalytically inactive mutant of MTM1 induced its oligomerization as a heptameric ring (Schaletzky et al. 2003). Cao et al. (2007) proposed that activated Rab5 and Rab7 bind to P150 (VPS15), the regulatory subunit of PI 3-kinase class III (VPS34) leading to activation of VPS34, elevation of PtdIns3*P* and recruitment of EEA1 and Hrs effectors on early and late endosomes, respectively. The subsequent interaction of MTM1 with p150 should then lead to a decrease of PtdIns3*P* levels and a decreased level of EEA1 and Hrs on endosomes. Such tripartite complex, consisting of a PI kinase, a PI phosphatase and a regulatory protein, may also exist for other PI transformation reactions controlling the level of PtdIns5*P* and PtdIns(3,5)P_2 (Lecompte et al. 2008).

Similarly, using HEK293 cells, Franklin et al. have shown a localization of MTMR2 at endosomes (Franklin et al. 2011). They reported that MTMR2, not phosphorylated at residue Ser58, co-localized with Rab5 on early endosomes, where it depleted pools of PtdIns3*P* in response to extracellular stimuli likely activating serine phosphatases. Conversely, when Ser58 is phosphorylated in steady-state conditions, MTMR2 preferentially localizes in the cytosol where it would probably not access any PI pools. In addition, downregulation of MTMR2 mRNA in A431 cells has been shown to alter EGF receptor trafficking at the level of late endosome/lysosomes, where MTMR2 has been proposed to function in complex with P150 and Rab7 (Cao et al. 2008). Moreover, the increase of PtdIns(3,5)P_2 levels upon hypo-osmotic shock in COS7 cells led to the recruitment of MTMR2 at membranes of vacuoles formed under these conditions (Berger et al. 2006b). Berger et al. showed that membrane association of MTMR2 in hypo-osmotic condition requires both PH-GRAM and coiled coil domains and does not depend on phosphatase activity. However, unlike MTM1, MTMR2 vacuolar localization is dependent on its PH-GRAM domain underlying differential PI-dependent regulatory mechanisms within the myotubularin family.

Drosophila *mtm*, the ortholog of both human MTM1 and MTMR2, acts on a sub-pool of PtdIns3*P* generated by the class II PI 3-kinase (Pi3K68D), likely at the plasma membrane or on early endosomes (Velichkova et al. 2010). Velichkova et al. elegantly showed that, by controlling this PtdIns3*P* sub-pool, *mtm* regulates cortical remodeling, downregulates endosomal influx and promotes membrane efflux, thus maintaining endosomal homeostasis in the endolysosomal axis. By performing in vivo studies on *mtm* mutants, the same authors also reported that *mtm* is required for integrin-mediated myofiber attachment (Ribeiro et al. 2011). *Mtm* depletion increased integrin turnover at the sarcolemma, an accumulation of integrin on PtdIns3*P* enriched endosomes and disruption of T-tubules. Depletion of Pi3K68D rescued integrin accumulation, thus revealing the existence of a pathway that controls integrin recycling in muscle. However, Pi3K68D depletion did not rescue the defect of T-tubules caused by *mtm* deplection, suggesting a distinct Pi3K68D-dependent and PI-dependent and independent *mtm* functions in muscle. Similar defects in integrin localization and T-tubule disruption were also observed in XLCNM biopsies, thus supporting a role for integrin adhesion and triad maintenance in the pathogenesis of the XLCNM (Ribeiro et al. 2011; Toussaint et al. 2011). Furthermore, ectopic overexpression of MTM1 by adeno-associated-virus (AAV) in wild type skeletal muscle lead to membrane stacks accumulation at subsarcolemmal regions, suggesting that depletion or/and over-production of the MTM1 PIs substrates and products is likely to affects membrane formation and remodeling in muscle cells (Buj-Bello et al. 2008).

Similarly, MTMR2 acts on membrane homeostasis. *Mtmr2*-null nerves showed myelin outfoldings, which can be considered as a model of impaired membrane formation preferentially arising at nodal/paranodal regions, known sites of membrane remodeling (Dang et al. 2004). MTMR2 has been shown to interact with the Dlg1 (disks large 1)/synapse associated 97 (SAP97) scaffolding protein, which is involved in polarized membrane trafficking and membrane addition (Bolino et al. 2004). Interestingly, Dlg1 localization is altered at paranodal regions in *Mtmr2*-null peripheral nerves, suggesting that the MTMR2/Dlg1 complex might be relevant for the CMT4B1 pathogenesis (Bolino et al. 2004; Bolis et al. 2005). Bolis et al. (2009) have reported Dlg1 interaction with kinesin 13B (kif13B) and sec8, which are involved in vesicle transport and membrane tethering in polarized cells, respectively. They proposed that the interaction of Dlg1 with Sec8 promotes membrane formation whereas *Mtmr2* negatively regulates membrane addition through its interaction with Dlg1. Defects in MTMR2/Dlg1-mediated control of membrane formation during Schwann cell myelination might lead to excessive myelin with redundant folds (Bolis et al. 2009). The role of MTMR2 in the regulation of endocytic trafficking might be physiologically more relevant in specialized cell types as for example in neurons. Recently, downregulation of *Mtmr2* mRNA in cultured hippocampus neurons suggested that MTMR2, through PSD95 interaction, might downregulate endocytic events important for spine maintenance (Gambardella et al. 1997; Lee et al. 2010).

Interactors and Tissue-specific Regulation

While myotubularins appear rather ubiquitously expressed, mutations in several of them lead to diseases affecting different and specific tissues. Thus, there should be some tissue-specific regulatory mechanisms that could be mediated by the type of extracellular stimuli, specificity of PIs turnover in different cell types, or specific interactors. This hypothesis is supported by the recent discovery of interactors for specific myotubularins.

Initially, in vitro studies suggesting that a myotubularin could form a heterodimer with another member within the same family have supported that such regulation could have a major role in the cellular function of myotubularins. Such molecular configuration implicated an active/dead-phosphatase couple of myotubularins. Specific heterodimers exist, although all possibilities were not investigated yet, and such specificity may sustain tissue-targeted functions of myotubularins (Lorenzo et al. 2005). The best characterized heterodimers are MTM1 with MTMR12, MTMR2 with MTMR5 or MTMR13, and MTMR9 with MTMR6 or MTMR7 (Azzedine et al. 2003; Kim et al. 2003; Mochizuki and Majerus 2003; Robinson and Dixon 2005; Taylor et al. 2000b). A first example is the complex between MTM1 and MTMR12 (3-PAP: 3-phosphatase associated protein). The MTM1-MTMR12 heterodimer was biochemically discovered before the identification of the first myotubularin gene sequence as a protein complex with PtdIns3*P* phosphatase activity purified from rat brain extracts (Caldwell et al. 1991). Subsequent sequencing of this protein complex matched the discovered myotubularins and also suggested a role in human platelets (Nandurkar et al. 2001, 2003). Overexpressed MTM1 localized to the plasma membrane, causing extensive filopodia formation (Laporte et al. 2002). Co-expression of MTMR12 with MTM1 led to attenuation of filopodia formation and relocalization of MTM1 to the cytosol, toward the localization of MTMR12 (Nandurkar et al. 2003). This indicates that MTMR12 functions as an "adapter" for MTM1, regulating its intracellular location and thereby altering the phenotype resulting from MTM1 overexpression. MTMR12, and more generally dead-phosphatase myotubularins, may thus direct the enzymatically active myotubularins to specific PI pools.

MTMR2 interacts with MTMR5, another catalytically inactive myotubularin via its CC domain and mutations in the CC domain of either MTMR2 or MTMR5 abrogate this interaction. Through this interaction, MTMR5 increased the enzymatic activity of MTMR2 and modulated its subcellular localization (Kim et al. 2003). In addition, MTMR2 binds to MTMR13 (Sbf2), and both proteins are mutated in CMT4B (Previtali et al. 2003). MTMR2 homodimers interacted with MTMR13 homodimers to form tetrameric complexes (Berger et al. 2006a; Robinson and Dixon 2005). This association dramatically increased the enzymatic activity of MTMR2 toward PtdIns3*P* and PtdIns(3,5)P_2. MTMR2 and MTMR13 are mostly co-localized in the cytoplasm when exogenously expressed in cells. On membranes of large vesicles formed under hypo-osmotic conditions, MTMR13 favorably competed with MTMR2 for binding sites. These observations suggested that MTMR2 activity is tightly regulated, being high in the complex, moderate if

MTMR2 is not associated with MTMR13 or functionally blocked through competition with MTMR13 for membrane-binding sites (Berger et al. 2006b; Robinson and Dixon 2005). Given the fact that MTMR2/MTMR13 are interaction partners in Schwann cells in vitro and in vivo (Berger et al. 2006a), defect in this interaction is most likely the primary cause for the similar phenotypes observed in CMT4B1 and CMT4B2, indicating that these proteins act in concert in a common pathway essential for myelin maintenance.

The dead phosphatase MTMR9 interacts with MTMR6 and with MTMR7 (Mochizuki and Majerus 2003; Zou et al. 2009). MTMR9 binding increased the binding of MTMR6 to PIs and its catalytic activity. Moreover, co-expression of MTMR6 with MTMR9 decreased etoposide-induced apoptosis, suggesting that formation of this heterodimer has important cellular impact.

An unexpected link between MTM1/MTMR2 and intermediate filaments was reported. Intermediate filaments are important structural components of living cells and are essential for normal tissue structure and function; they provide physical resilience for cells to withstand the mechanical stresses of the host tissue (Herrmann et al. 2009). The Neurofilament light chain protein (NF-L), mutated in CMT disease including axonal, intermediate, and demyelinating forms, was found to interact with MTMR2 (Liem and Messing 2009; Previtali et al. 2003; Szaro and Strong 2010). Myelin outfoldings are due to loss of MTMR2 in Schwann cells, since conditional ablation in mice of *Mtmr2* in Schwann cells is both sufficient and necessary to provoke the myelin defect hallmark of CMT4B1 (Bolis et al. 2005). In the peripheral nervous system, NF-L is expressed mainly in neurons, whereas Schwann cells express NF-L mRNA only upon damage. Thus, it is likely that the MTMR2 and NF-L interaction is physiologically relevant in neurons, but the functional insight of this partnership and its impact on the physiopathology of CMT4B1 is not yet identified.

The second example is the interaction of MTM1 with the type III intermediate filaments, desmin, a muscle-specific protein (Hnia et al. 2011b). MTM1 bound directly to desmin and regulated filament assembly and architecture independently of its enzymatic activity, suggesting a crucial role for MTM1 in the regulation of the desmin network in skeletal muscle. Knockout or knockdown of MTM1 expression and disruption of the MTM1-desmin complex promoted desmin aggregation. Both MTM1 and desmin are implicated in muscle disorders and desmin mutations are associated with myofibrillar myopathies and cardiomyopathies (DRM, Desmin Related Myopathies, and DRCM, Desmin Related Cardiomyopathies). XLCNM mutations as well as DRM mutations abolished MTM1/desmin interaction and could not re-establish normal intermediate filaments network in *Mtm1* KO muscle cells. Accordingly, these data suggested a common pathophysiological mechanism between centronuclear and myofibrillar myopathies and underlined the importance of myotubularins in the regulation of intermediate filaments in different tissues (Hnia et al. 2011b). In addition, MTM1/desmin complex was found in the mitochondrial fraction and MTM1 depletion from muscle cells affected mitochondrial dynamics and function. These effects appeared both dependent and independent of desmin interaction as several MTM1 mutations, particularly those affecting enzymatic activity, did not interfere with desmin interaction but affected mitochondrial dynamics (Hnia et al. 2011b). In zebrafish *mtm1* morphant muscles, the presence of dysmorphic and swollen mitochondria in the perinuclear region also supported the link between MTM1 and mitochondrial homeostasis (Dowling et al. 2009). Indeed, mitochondrial collapse around central nuclei is a sign of XLCNM found on muscle biopsies from affected patients (Romero 2010).

Concluding Remarks and Future Directions

Myotubularins have raised a high interest since the discovery of their implication in human diseases. In addition to that consideration, myotubularins appear as key players in the PIs metabolism. Although the lipid phosphatase activity of myotubularins was largely investigated in vitro and ex vivo, little is known about its importance in vivo. How and in which biological context the PI-related function of myotubularins is required will be an important query in the future. One may wonder why there are so many myotubularins in higher eukaryotes, most of them being rather ubiquitously expressed. The growing data obtained during the last few years support the idea that myotubularins functions are differentially regulated from one tissue to another. The emergence of newly identified protein partners supports the hypothesis that myotubularins regulate different PIs subpools in tissues and that they also have phosphatase independent functions. Recent finding using cell and animal models suggested that these PI phosphatases can also regulate many processes including cell proliferation and differentiation, survival, autophagy, cytokinesis, cytoskeletal, and cell junction dynamics. How myotubularins regulate these processes is barely understood, and whether defects in these pathways are primary causes of the "myotubularinopathies" remains to be investigated.

References

Alonso A, Sasin J, Bottini N, Friedberg I, Friedberg I, Osterman A, Godzik A, Hunter T, Dixon JE, Mustelin T (2004) Protein tyrosine phosphatases in the human genome. Cell 117:699–711

Al-Qusairi L, Weiss N, Toussaint A, Berbey C, Messaddeq N, Kretz C, Sanoudou D, Beggs AH, Allard B, Mandel JL, Laporte J, Jacquemond V, Buj-Bello A (2009) T-tubule disorganization and defective excitation-contraction coupling in muscle fibers lacking myotubularin lipid phosphatase. Proc Natl Acad Sci U S A 106:18763–18768

Azzedine H, Bolino A, Taieb T, Birouk N, Di Duca M, Bouhouche A, Benamou S, Mrabet A, Hammadouche T, Chkili T, Gouider R, Ravazzolo R, Brice A, Laporte J, LeGuern E (2003) Mutations in MTMR13, a new pseudophosphatase homologue of MTMR2 and Sbf1, in two families with an autosomal recessive demyelinating form of Charcot-Marie-Tooth disease associated with early-onset glaucoma. Am J Hum Genet 72:1141–1153

Baulac S, Gourfinkel-An I, Couarch P, Depienne C, Kaminska A, Dulac O, Baulac M, LeGuern E, Nabbout R (2008) A novel locus for generalized epilepsy with febrile seizures plus in French families. Arch Neurol 65:943–951

Begley MJ, Taylor GS, Kim SA, Veine DM, Dixon JE, Stuckey JA (2003) Crystal structure of a phosphoinositide phosphatase, MTMR2: insights into myotubular myopathy and Charcot-Marie-Tooth syndrome. Mol Cell 12:1391–1402

Berger P, Schaffitzel C, Berger I, Ban N, Suter U (2003) Membrane association of myotubularin-related protein 2 is mediated by a pleckstrin homology-GRAM domain and a coiled-coil dimerization module. Proc Natl Acad Sci U S A 100:12177–12182

Berger P, Niemann A, Suter U (2006a) Schwann cells and the pathogenesis of inherited motor and sensory neuropathies (Charcot-Marie-Tooth disease). Glia 54:243–257

Berger P, Berger I, Schaffitzel C, Tersar K, Volkmer B, Suter U (2006b) Multi-level regulation of myotubularin-related protein-2 phosphatase activity by myotubularin-related protein-13/set-binding factor-2. Hum Mol Genet 15:569–579

Bitoun M, Maugenre S, Jeannet PY, Lacene E, Ferrer X, Laforet P, Martin JJ, Laporte J, Lochmuller H, Beggs AH, Fardeau M, Eymard B, Romero NB, Guicheney P (2005) Mutations in dynamin 2 cause dominant centronuclear myopathy. Nat Genet 37:1207–1209

Bitoun M, Bevilacqua JA, Prudhon B, Maugenre S, Taratuto AL, Monges S, Lubieniecki F, Cances C, Uro-Coste E, Mayer M, Fardeau M, Romero NB, Guicheney P (2007) Dynamin 2 mutations cause sporadic centronuclear myopathy with neonatal onset. Ann Neurold 62: 666–670

Bitoun M, Durieux AC, Prudhon B, Bevilacqua JA, Herledan A, Sakanyan V, Urtizberea A, Cartier L, Romero NB, Guicheney P (2009) Dynamin 2 mutations associated with human diseases impair clathrin-mediated receptor endocytosis. Hum Mutat 30:1419–1427

Blondeau F, Laporte J, Bodin S, Superti-Furga G, Payrastre B, Mandel JL (2000) Myotubularin, a phosphatase deficient in myotubular myopathy, acts on phosphatidylinositol 3-kinase and phosphatidylinositol 3-phosphate pathway. Hum Mol Genet 9:2223–2229

Bolino A, Muglia M, Conforti FL, LeGuern E, Salih MA, Georgiou DM, Christodoulou K, Hausmanowa-Petrusewicz I, Mandich P, Schenone A, Gambardella A, Bono F, Quattrone A, Devoto M, Monaco AP (2000) Charcot-Marie-Tooth type 4B is caused by mutations in the gene encoding myotubularin-related protein-2. Nat Genet 25:17–19

Bolino A, Bolis A, Previtali SC, Dina G, Bussini S, Dati G, Amadio S, Del Carro U, Mruk DD, Feltri ML, Cheng CY, Quattrini A, Wrabetz L (2004) Disruption of Mtmr2 produces CMT4B1-like neuropathy with myelin outfolding and impaired spermatogenesis. J Cell Biol 167:711–721

Bolis A, Coviello S, Bussini S, Dina G, Pardini C, Previtali SC, Malaguti M, Morana P, Del Carro U, Feltri ML, Quattrini A, Wrabetz L, Bolino A (2005) Loss of Mtmr2 phosphatase in Schwann cells but not in motor neurons causes Charcot-Marie-Tooth type 4B1 neuropathy with myelin outfoldings. J Neurosci 25:8567–8577

Bolis A, Coviello S, Visigalli I, Taveggia C, Bachi A, Chishti AH, Hanada T, Quattrini A, Previtali SC, Biffi A, Bolino A (2009) Dlg1, Sec8, and Mtmr2 regulate membrane homeostasis in Schwann cell myelination. J Neurosci 29:8858–8870

Bonneick S, Boentert M, Berger P, Atanasoski S, Mantei N, Wessig C, Toyka KV, Young P, Suter U (2005) An animal model for Charcot-Marie-Tooth disease type 4B1. Hum Mol Genet 14:3685–3695

Buj-Bello A, Laugel V, Messaddeq N, Zahreddine H, Laporte J, Pellissier JF, Mandel JL (2002) The lipid phosphatase myotubularin is essential for skeletal muscle maintenance but not for myogenesis in mice. Proc Natl Acad Sci U S A 99:15060–15065

Buj-Bello A, Fougerousse F, Schwab Y, Messaddeq N, Spehner D, Pierson CR, Durand M, Kretz C, Danos O, Douar AM, Beggs AH, Schultz P, Montus M, Denefle P, Mandel JL (2008) AAV-mediated intramuscular delivery of myotubularin corrects the myotubular myopathy phenotype in targeted murine muscle and suggests a function in plasma membrane homeostasis. Hum Mol Genet 17:2132–2143

Caldwell KK, Lips DL, Bansal VS, Majerus PW (1991) Isolation and characterization of two 3-phosphatases that hydrolyze both phosphatidylinositol 3-phosphate and inositol 1,3-bisphosphate. J Biol Chem 266:18378–18386

Cao C, Laporte J, Backer JM, Wandinger-Ness A, Stein MP (2007) Myotubularin lipid phosphatase binds the hVPS15/hVPS34 lipid kinase complex on endosomes. Traffic 8: 1052–1067
Cao C, Backer JM, Laporte J, Bedrick EJ, Wandinger-Ness A (2008) Sequential actions of myotubularin lipid phosphatases regulate endosomal PI(3)P and growth factor receptor trafficking. Mol Biol Cell 19:3334–3346
Chaussade C, Pirola L, Bonnafous S, Blondeau F, Brenz-Verca S, Tronchère H, Portis F, Rusconi S, Payrastre B, Laporte J, Van Obberghen E (2003) Expression of myotubularin by a novel adenoviral vector demonstrates its function as a PtdIns(3)P phosphatase in muscle cell lines. Involvement of PtdIns(3)P in insulin-stimulated glucose transport. Mol Endocrinol 17: 2448–2460
Choudhury P, Srivastava S, Li Z, Ko K, Albaqumi M, Narayan K, Coetzee WA, Lemmon MA, Skolnik EY (2006) Specificity of the myotubularin family of phosphatidylinositol-3-phosphatase is determined by the PH/GRAM domain. J Biol Chem 281:31762–31769
Chow CY, Zhang Y, Dowling JJ, Jin N, Adamska M, Shiga K, Szigeti K, Shy ME, Li J, Zhang X, Lupski JR, Weisman LS, Meisler MH (2007) Mutation of FIG4 causes neurodegeneration in the pale tremor mouse and patients with CMT4J. Nature 448:68–72
Claeys KG, Maisonobe T, Bohm J, Laporte J, Hezode M, Romero NB, Brochier G, Bitoun M, Carlier RY, Stojkovic T (2010) Phenotype of a patient with recessive centronuclear myopathy and a novel BIN1 mutation. Neurology 74:519–521
Clarke JH, Letcher AJ, D'Santos CS, Halstead JR, Irvine RF, Divecha N (2001) Inositol lipids are regulated during cell cycle progression in the nuclei of murine erythroleukaemia cells. Biochem J 357:905–910
Coronas S, Ramel D, Pendaries C, Gaits-Iacovoni F, Tronchere H, Payrastre B (2007) PtdIns5P: a little phosphoinositide with big functions? Biochem Soc Symp 74:117–128
Dang H, Li Z, Skolnik EY, Fares H (2004) Disease-related myotubularins function in endocytic traffic in caenorhabditis elegans. Mol Biol Cell 15:189–196
Di Paolo G, De Camilli P (2006) Phosphoinositides in cell regulation and membrane dynamics. Nature 443:651–657
Doerks T, Strauss M, Brendel M, Bork P (2000) GRAM, a novel domain in glucosyltransferases, myotubularins and other putative membrane-associated proteins. Trends Biochem Sci 25:483–485
Dorchies OM, Laporte J, Wagner S, Hindelang C, Warter JM, Mandel JL, Poindron P (2001) Normal innervation and differentiation of X-linked myotubular myopathy muscle cells in a nerve-muscle coculture system. Neuromuscul Disord 11:736–746
Dowling JJ, Vreede AP, Low SE, Gibbs EM, Kuwada JY, Bonnemann CG, Feldman EL (2009) Loss of myotubularin function results in T-tubule disorganization in zebrafish and human myotubular myopathy. PLoS Genet 5:e1000372
Dowling JJ, Low SE, Busta AS, Feldman EL (2010) Zebrafish MTMR14 is required for excitation-contraction coupling, developmental motor function and the regulation of autophagy. Hum Mol Genet 19:2668–2681
Dubourg O, Azzedine H, Verny C, Durosier G, Birouk N, Gouider R, Salih M, Bouhouche A, Thiam A, Grid D, Mayer M, Ruberg M, Tazir M, Brice A, LeGuern E (2006) Autosomal-recessive forms of demyelinating Charcot-Marie-Tooth disease. NeuroMol Med 8:75–86
Fauman EB, Saper MA (1996) Structure and function of the protein tyrosine phosphatases. Trends Biochem Sci 21:413–417
Fidzianska A, Warlo I, Goebel HH (1994) Neonatal centronuclear myopathy with N-CAM decorated myotubes. Neuropediatrics 25:158–161
Franklin NE, Taylor GS, Vacratsis PO (2011) Endosomal targeting of the phosphoinositide 3-phosphatase MTMR2 is regulated by an N-terminal phosphorylation site. J Biol Chem 286:15841–15853
Gambardella A, Muglia M, Quattrone A (1997) Hereditary demyelinating neuropathy of infancy: a genetically complex syndrome. Brain: J Neurol 120(Pt 11):2113–2115

Gary JD, Sato TK, Stefan CJ, Bonangelino CJ, Weisman LS, Emr SD (2002) Regulation of Fab1 phosphatidylinositol 3-phosphate 5-kinase pathway by Vac7 protein and Fig4, a polyphosphoinositide phosphatase family member. Mol Biol Cell 13:1238–1251

Gillooly DJ, Morrow IC, Lindsay M, Gould R, Bryant NJ, Gaullier JM, Parton RG, Stenmark H (2000) Localization of phosphatidylinositol 3-phosphate in yeast and mammalian cells. EMBO J 19:4577–4588

Gozani O, Karuman P, Jones DR, Ivanov D, Cha J, Lugovskoy AA, Baird CL, Zhu H, Field SJ, Lessnick SL, Villasenor J, Mehrotra B, Chen J, Rao VR, Brugge JS, Ferguson CG, Payrastre B, Myszka DG, Cantley LC, Wagner G, Divecha N, Prestwich GD, Yuan J (2003) The PHD finger of the chromatin-associated protein ING2 functions as a nuclear phosphoinositide receptor. Cell 114:99–111

Gozani O, Field SJ, Ferguson CG, Mahlke C, Cantley LC, Prestwich GD, Yuan J (2005) Modification of protein sub-nuclear localization by synthetic phosphoinositides: evidence for nuclear phosphoinositide signaling mechanisms. Adv Enzyme Reg. 45:171–185

Grainger DL, Tavelis C, Ryan AJ, Hinchliffe KA (2011) Involvement of phosphatidylinositol 5-phosphate in insulin-stimulated glucose uptake in the L6 myotube model of skeletal muscle. Pflugers Arch 462:723–732

Guan KL, Broyles SS, Dixon JE (1991) A Tyr/Ser protein phosphatase encoded by vaccinia virus. Nature 350:359–362

Herman GE, Finegold M, Zhao W, de Gouyon B, Metzenberg A (1999) Medical complications in long-term survivors with X-linked myotubular myopathy. J Pediatr 134:206–214

Herrmann H, Strelkov SV, Burkhard P, Aebi U (2009) Intermediate filaments: primary determinants of cell architecture and plasticity. J Clin Invest 119:1772–1783

Hnia K, Kretz C, Amoasii L, Boehm J, Liu X, Messaddeq N, Qu C, Laporte J, (2011a) Primary T-tubule and autophagy defects in the phosphoinositide phosphatase jumpy/MTMR14 knockout mice muscle. Adv Enzyme Regul (in press)

Hnia K, Tronchere H, Tomczak KK, Amoasii L, Schultz P, Beggs AH, Payrastre B, Mandel JL, Laporte J (2011b) Myotubularin controls desmin intermediate filament architecture and mitochondrial dynamics in human and mouse skeletal muscle. J Clin Invest 121:70–85

Hotta K, Kitamoto T, Kitamoto A, Mizusawa S, Matsuo T, Nakata Y, Kamohara S, Miyatake N, Kotani K, Komatsu R, Itoh N, Mineo I, Wada J, Yoneda M, Nakajima A, Funahashi T, Miyazaki S, Tokunaga K, Masuzaki H, Ueno T, Hamaguchi K, Tanaka K, Yamada K, Hanafusa T, Oikawa S, Yoshimatsu H, Sakata T, Matsuzawa Y, Nakao K, Sekine A (2011) Association of variations in the FTO, SCG3 and MTMR9 genes with metabolic syndrome in a Japanese population. J Hum Genet 56:647–651

Hu Z, Wu C, Shi Y, Guo H, Zhao X, Yin Z, Yang L, Dai J, Hu L, Tan W, Li Z, Deng Q, Wang J, Wu W, Jin G, Jiang Y, Yu D, Zhou G, Chen H, Guan P, Chen Y, Shu Y, Xu L, Liu X, Liu L, Xu P, Han B, Bai C, Zhao Y, Zhang H, Yan Y, Ma H, Chen J, Chu M, Lu F, Zhang Z, Chen F, Wang X, Jin L, Lu J, Zhou B, Lu D, Wu T, Lin D, Shen H (2011) A genome-wide association study identifies two new lung cancer susceptibility loci at 13q12.12 and 22q12.2 in Han Chinese. Nat Genet 43:792–796

Itoh T, Takenawa T (2002) Phosphoinositide-binding domains: Functional units for temporal and spatial regulation of intracellular signalling. Cell Signal 14:733–743

Jungbluth H, Wallgren-Pettersson C, Laporte J (2008) Centronuclear (myotubular) myopathy. Orphanet J Rare Dis 3:26

Kim SA, Vacratsis PO, Firestein R, Cleary ML, Dixon JE (2003) Regulation of myotubularin-related (MTMR)2 phosphatidylinositol phosphatase by MTMR5, a catalytically inactive phosphatase. Proc Natl Acad Sci U S A 100:4492–4497

Laporte J, Hu LJ, Kretz C, Mandel JL, Kioschis P, Coy JF, Klauck SM, Poustka A, Dahl N (1996) A gene mutated in X-linked myotubular myopathy defines a new putative tyrosine phosphatase family conserved in yeast. Nat Genet 13:175–182

Laporte J, Blondeau F, Buj-Bello A, Tentler D, Kretz C, Dahl N, Mandel JL (1998) Characterization of the myotubularin dual specificity phosphatase gene family from yeast to human. Hum Mol Genet 7:1703–1712

Laporte J, Biancalana V, Tanner SM, Kress W, Schneider V, Wallgren-Pettersson C, Herger F, Buj-Bello A, Blondeau F, Liechti-Gallati S, Mandel JL (2000) MTM1 mutations in X-linked myotubular myopathy. Hum Mutat 15:393–409

Laporte J, Blondeau F, Buj-Bello A, Mandel JL (2001) The myotubularin family: from genetic disease to phosphoinositide metabolism. Trends Genet 17:221–228

Laporte J, Blondeau F, Gansmuller A, Lutz Y, Vonesch JL, Mandel JL (2002) The PtdIns3P phosphatase myotubularin is a cytoplasmic protein that also localizes to Rac1-inducible plasma membrane ruffles. J Cell Sci 115:3105–3117

Lecompte O, Poch O, Laporte J (2008) PtdIns5P regulation through evolution: roles in membrane trafficking? Trends Biochem Sci 33:453–460

Lee HW, Kim Y, Han K, Kim H, Kim E (2010) The phosphoinositide 3-phosphatase MTMR2 interacts with PSD-95 and maintains excitatory synapses by modulating endosomal traffic. J Neurosci 30:5508–5518

Liaw D, Marsh DJ, Li J, Dahia PL, Wang SI, Zheng Z, Bose S, Call KM, Tsou HC, Peacocke M, Eng C, Parsons R (1997) Germline mutations of the PTEN gene in Cowden disease, an inherited breast and thyroid cancer syndrome. Nat Genet 16:64–67

Liem RK, Messing A (2009) Dysfunctions of neuronal and glial intermediate filaments in disease. J Clin Investig 119:1814–1824

Liu T, Ghosal G, Yuan J, Chen J, Huang J (2010) FAN1 acts with FANCI-FANCD2 to promote DNA interstrand cross-link repair. Science 329:693–696

Lorenzo O, Urbe S, Clague MJ (2005) Analysis of phosphoinositide binding domain properties within the myotubularin-related protein MTMR3. J Cell Sci 118:2005–2012

Lucci MA, Orlandi R, Triulzi T, Tagliabue E, Balsari A, Villa-Moruzzi E (2010) Expression profile of tyrosine phosphatases in HER2 breast cancer cells and tumors. Cellular Oncology: Official J Int Soc Cell Oncol 32:361–372

Maehama T, Dixon JE (1998) The tumor suppressor, PTEN/MMAC1, dephosphorylates the lipid second messenger, phosphatidylinositol 3,4,5-trisphosphate. J Biol Chem 273:13375–13378

Martin TF (1998) Phosphoinositide lipids as signaling molecules: common themes for signal transduction, cytoskeletal regulation, and membrane trafficking. Annu Rev Cell Dev Biol 14:231–264

Michell RH, Heath VL, Lemmon MA, Dove SK (2006) Phosphatidylinositol 3,5-bisphosphate: metabolism and cellular functions. Trends Biochem Sci 31:52–63

Misra AK, Menon NK, Mishra SK (1992) Abnormal distribution of desmin and vimentin in myofibers in adult onset myotubular myopathy. Muscle Nerve 15:1246–1252

Mochizuki Y, Majerus PW (2003) Characterization of myotubularin-related protein 7 and its binding partner, myotubularin-related protein 9. Proc Natl Acad Sci U S A 100:9768–9773

Morris JB, Hinchliffe KA, Ciruela A, Letcher AJ, Irvine RF (2000) Thrombin stimulation of platelets causes an increase in phosphatidylinositol 5-phosphate revealed by mass assay. FEBS Lett 475:57–60

Murray JT, Panaretou C, Stenmark H, Miaczynska M, Backer JM (2002) Role of Rab5 in the recruitment of hVps34/p150 to the early endosome. Traffic 3:416–427

Nandurkar HH, Caldwell KK, Whisstock JC, Layton MJ, Gaudet EA, Norris FA, Majerus PW, Mitchell CA (2001) Characterization of an adapter subunit to a phosphatidylinositol (3)P 3-phosphatase: identification of a myotubularin-related protein lacking catalytic activity. Proc Natl Acad Sci U S A 98:9499–9504

Nandurkar HH, Layton M, Laporte J, Selan C, Corcoran L, Caldwell KK, Mochizuka Y, Majerus PW, Mitchell CA (2003) Identification of myotubularin as the lipid phosphatase catalytic subunit associated with the 3-phosphatase adapter protein, 3-PAP. Proc Natl Acad Sci U S A 100:8660–8665

Nelen MR, van Staveren WC, Peeters EA, Hassel MB, Gorlin RJ, Hamm H, Lindboe CF, Fryns JP, Sijmons RH, Woods DG, Mariman EC, Padberg GW, Kremer H (1997) Germline mutations in the PTEN/MMAC1 gene in patients with Cowden disease. Hum Mol Genet 6:1383–1387

Nicot AS, Toussaint A, Tosch V, Kretz C, Wallgren-Pettersson C, Iwarsson E, Kingston H, Garnier JM, Biancalana V, Oldfors A, Mandel JL, Laporte J (2007) Mutations in amphiphysin

2 (BIN1) disrupt interaction with dynamin 2 and cause autosomal recessive centronuclear myopathy. Nat Genet 39:1134–1139

Noguchi S, Fujita M, Murayama K, Kurokawa R, Nishino I (2005) Gene expression analyses in X-linked myotubular myopathy. Neurology 65:732–737

Norris FA, Atkins RC, Majerus PW (1997) The cDNA cloning and characterization of inositol polyphosphate 4-phosphatase type II. Evidence for conserved alternative splicing in the 4-phosphatase family. J Biol Chem 272:23859–23864

Nystuen A, Legare ME, Shultz LD, Frankel WN (2001) A null mutation in inositol polyphosphate 4-phosphatase type I causes selective neuronal loss in weeble mutant mice. Neuron 32: 203–212

Odorizzi G, Babst M, Emr SD (1998) Fab1p PtdIns(3)P 5-kinase function essential for protein sorting in the multivesicular body. Cell 95:847–858

Odorizzi G, Babst M, Emr SD (2000) Phosphoinositide signaling and the regulation of membrane trafficking in yeast. Trends Biochem Sci 25:229–235

Patzko A, Shy ME (2011) Update on Charcot-Marie-Tooth disease. Curr Neurol Neurosci Rep 11:78–88

Payrastre B (2004) Phosphoinositides: lipid kinases and phosphatases. Methods Mol Biol 273:201–212

Pizarro-Cerda J, Cossart P (2004) Subversion of phosphoinositide metabolism by intracellular bacterial pathogens. Nat Cell Biol 6:1026–1033

Previtali SC, Zerega B, Sherman DL, Brophy PJ, Dina G, King RHM, Salih MM, Feltri L, Quattrini A, Ravazzolo R, Wrabetz L, Monaco AP, Bolino A (2003) Myotubularin-related 2 protein phosphatase and neurofilament light chain protein, both mutated in CMT neuropathies, interact in peripheral nerve. Hum Mol Genet 12:1713–1723

Ramel D, Lagarrigue F, Pons V, Mounier J, Dupuis-Coronas S, Chicanne G, Sansonetti PJ, Gaits-Iacovoni F, Tronchere H, Payrastre B (2011) Shigella flexneri infection generates the lipid PI5P to alter endocytosis and prevent termination of EGFR signaling. Sci Signal 4:ra61

Ribeiro I, Yuan L, Tanentzapf G, Dowling JJ, Kiger A (2011) Phosphoinositide regulation of integrin trafficking required for muscle attachment and maintenance. PLoS Genet 7:e1001295

Ringel SP, Wilson WB, Barden MT (1979) Extraocular muscle biopsy in chronic progressive external ophthalmoplegia. Ann Neurol 6:326–339

Robinson FL, Dixon JE (2005) The phosphoinositide-3-phosphatase MTMR2 associates with MTMR13, a membrane-associated pseudophosphatase also mutated in type 4B Charcot-Marie-Tooth disease. J Biol Chem 280:31699–31707

Robinson FL, Dixon JE (2006) Myotubularin phosphatases: policing 3-phosphoinositides. Trends Cell Biol 16:403–412

Robinson FL, Niesman IR, Beiswenger KK, Dixon JE (2008) Loss of the inactive myotubularin-related phosphatase Mtmr13 leads to a Charcot-Marie-Tooth 4B2-like peripheral neuropathy in mice. Proc Natl Acad Sci U S A 105:4916–4921

Romero NB (2010) Centronuclear myopathies: a widening concept. Neuromuscul Disord 20: 223–228

Sbrissa D, Ikonomov O, Shisheva A (2001) Selective insulin-induced activation of class I(A) phosphoinositide 3-kinase in PIKfyve immune complexes from 3T3-L1 adipocytes. Mol Cell Endocrinol 181:35–46

Sbrissa D, Ikonomov OC, Deeb R, Shisheva A (2002) Phosphatidylinositol 5-phosphate biosynthesis is linked to PIKfyve and is involved in osmotic response pathway in mammalian cells. J Biol Chem 277:47276–47284

Schaletzky J, Dove SK, Short B, Lorenzo O, Clague MJ, Barr FA (2003) Phosphatidylinositol-5-phosphate activation and conserved substrate specificity of the myotubularin phosphatidylinositol 3-phosphatases. Curr Biol 13:504–509

Senderek J, Bergmann C, Weber S, Ketelsen UP, Schorle H, Rudnik-Schoneborn S, Buttner R, Buchheim E, Zerres K (2003) Mutation of the SBF2 gene, encoding a novel member of the myotubularin family, in Charcot-Marie-Tooth neuropathy type 4B2/11p15. Hum Mol Genet 12:349–356

Shen J, Yu WM, Brotto M, Scherman JA, Guo C, Stoddard C, Nosek TM, Valdivia HH, Qu CK (2009) Deficiency of MIP/MTMR14 phosphatase induces a muscle disorder by disrupting Ca(2 +) homeostasis. Nat Cell Biol 11:769–776

Simonsen A, Lippe R, Christoforidis S, Gaullier JM, Brech A, Callaghan J, Toh BH, Murphy C, Zerial M, Stenmark H (1998) EEA1 links PI(3)K function to Rab5 regulation of endosome fusion. Nature 394:494–498

Song SY, Kang MR, Yoo NJ, Lee SH (2010) Mutational analysis of mononucleotide repeats in dual specificity tyrosine phosphatase genes in gastric and colon carcinomas with microsatellite instability. APMIS: Acta Pathologica, Microbiologica, et Immunologica Scandinavica 118:389–393

Stiles B, Wang Y, Stahl A, Bassilian S, Lee WP, Kim YJ, Sherwin R, Devaskar S, Lesche R, Magnuson MA, Wu H (2004) Liver-specific deletion of negative regulator PTEN results in fatty liver and insulin hypersensitivity. Proc Natl Acad Sci USA 101:2082–2087

Suzuki N, Tsumoto K, Hajicek N, Daigo K, Tokita R, Minami S, Kodama T, Hamakubo T, Kozasa T (2009) Activation of leukemia-associated RhoGEF by Galpha13 with significant conformational rearrangements in the interface. J Biol Chem 284:5000–5009

Szaro BG, Strong MJ (2010) Post-transcriptional control of neurofilaments: New roles in development, regeneration and neurodegenerative disease. Trends Neurosci 33:27–37

Taguchi-Atarashi N, Hamasaki M, Matsunaga K, Omori H, Ktistakis NT, Yoshimori T, Noda T (2010) Modulation of local PtdIns3P levels by the PI phosphatase MTMR3 regulates constitutive autophagy. Traffic 11:468–478

Taylor GS, Maehama T, Dixon JE (2000a) Inaugural article: myotubularin, a protein tyrosine phosphatase mutated in myotubular myopathy, dephosphorylates the lipid second messenger, phosphatidylinositol 3-phosphate. Proc Natl Acad Sci U S A 97:8910–8915

Taylor GS, Maehama T, Dixon JE (2000b) Myotubularin, a protein tyrosine phosphatase mutated in myotubular myopathy, dephosphorylates the lipid second messenger, phosphatidylinositol 3-phosphate. Proc Nat Acad Sci U S A 97:8910–8915

Tersar K, Boentert M, Berger P, Bonneick S, Wessig C, Toyka KV, Young P, Suter U (2007) Mtmr13/Sbf2-deficient mice: an animal model for CMT4B2. Hum Mol Genet 16:2991–3001

Tosch V, Rohde HM, Tronchere H, Zanoteli E, Monroy N, Kretz C, Dondaine N, Payrastre B, Mandel JL, Laporte J (2006) A novel PtdIns3P and PtdIns(3,5)P2 phosphatase with an inactivating variant in centronuclear myopathy. Hum Mol Genet 15:3098–3106

Toussaint A, Cowling BS, Hnia K, Mohr M, Oldfors A, Schwab Y, Yis U, Maisonobe T, Stojkovic T, Wallgren-Pettersson C, Laugel V, Echaniz-Laguna A, Mandel JL, Nishino I, Laporte J (2011) Defects in amphiphysin 2 (BIN1) and triads in several forms of centronuclear myopathies. Acta Neuropathol 121:253–266

Tronchere H, Laporte J, Pendaries C, Chaussade C, Liaubet L, Pirola L, Mandel J-L, Payrastre B (2004) Production of phosphatidylinositol 5-phosphate by the phosphoinositide 3-phosphatase myotubularin in mammalian cells. J Biol Chem 279:7304–7312

Tsujita K, Itoh T, Ijuin T, Yamamoto A, Shisheva A, Laporte J, Takenawa T (2004) Myotubularin regulates the function of late endosome through the GRAM domain-PtdIns(3,5)P2 interaction. J Biol Chem 279:13817–13824

Ungewickell A, Hugge C, Kisseleva M, Chang SC, Zou J, Feng Y, Galyov EE, Wilson M, Majerus PW (2005) The identification and characterization of two phosphatidylinositol-4,5-bisphosphate 4-phosphatases. Proc Natl Acad Sci U S A 102:18854–18859

Vaccari I, Dina G, Tronchere H, Kaufman E, Chicanne G, Cerri F, Wrabetz L, Payrastre B, Quattrini A, Weisman LS, Meisler MH, Bolino A (2011) Genetic interaction between MTMR2 and FIG4 phospholipid phosphatases involved in Charcot-Marie-Tooth neuropathies. PLoS Genet 7:e1002319

Van der Ven PF, Jap PH, Barth PG, Sengers RC, Ramaekers FC, Stadhouders AM (1995a) Abnormal expression of intermediate filament proteins in X-linked myotubular myopathy is not reproduced in vitro. Neuromuscul Disord 5:267–275

Van der Ven PF, Jap PH, ter Laak HJ, Nonaka I, Barth PG, Sengers RC, Stadhouders AM, Ramaekers FC (1995b) Immunophenotyping of congenital myopathies: disorganization of sarcomeric, cytoskeletal and extracellular matrix proteins. J Neurol Sci 129:199–213

Velichkova M, Juan J, Kadandale P, Jean S, Ribeiro I, Raman V, Stefan C, Kiger AA (2010) Drosophila Mtm and class II PI3 K coregulate a PI(3)P pool with cortical and endolysosomal functions. J Cell Biol 190:407–425

Vergne I, Roberts E, Elmaoued RA, Tosch V, Delgado MA, Proikas-Cezanne T, Laporte J, Deretic V (2009) Control of autophagy initiation by phosphoinositide 3-phosphatase jumpy. EMBO J 28:2244–2258

Wang Y, Li Q, Xu J, Liu Q, Wang W, Lin Y, Ma F, Chen T, Li S, Shen Y (2004) Mutation analysis of five candidate genes in Chinese patients with hypospadias. Eur J Hum Genet 12:706–712

Yamaguchi S, Homma K, Natori S (1999) A novel egg-derived tyrosine phosphatase, EDTP, that participates in the embryogenesis of Sarcophaga peregrina (flesh fly). Eur J Biochem 259:946–953

Yanagiya T, Tanabe A, Iida A, Saito S, Sekine A, Takahashi A, Tsunoda T, Kamohara S, Nakata Y, Kotani K, Komatsu R, Itoh N, Mineo I, Wada J, Masuzaki H, Yoneda M, Nakajima A, Miyazaki S, Tokunaga K, Kawamoto M, Funahashi T, Hamaguchi K, Tanaka K, Yamada K, Hanafusa T, Oikawa S, Yoshimatsu H, Nakao K, Sakata T, Matsuzawa Y, Kamatani N, Nakamura Y, Hotta K (2007) Association of single-nucleotide polymorphisms in MTMR9 gene with obesity. Hum Mol Genet 16:3017–3026

Yoshimura S, Gerondopoulos A, Linford A, Rigden DJ, Barr FA (2010) Family-wide characterization of the DENN domain Rab GDP-GTP exchange factors. J Cell Biol 191:367–381

Yu S, Manson J, White S, Bourne A, Waddy H, Davis M, Haan E (2003) X-linked myotubular myopathy in a family with three adult survivors. Clin Genet 64:148–152

Zappia M, Colao R, Montesanti R, Rizzo M, Aguglia U, Gambardella A, Oliveri RL, Quattrone A (1997) Long-duration response to levodopa influences the pharmacodynamics of short-duration response in Parkinson's disease. Ann Neurol 42:245–248

Zhang X, Loijens JC, Boronenkov IV, Parker GJ, Norris FA, Chen J, Thum O, Prestwich GD, Majerus PW, Anderson RA (1997) Phosphatidylinositol-4-phosphate 5-kinase isozymes catalyze the synthesis of 3-phosphate-containing phosphatidylinositol signaling molecules. J Biol Chem 272:17756–17761

Zhao R, Qi Y, Chen J, Zhao ZJ (2001) FYVE-DSP2, a FYVE domain-containing dual specificity protein phosphatase that dephosphorylates phosphotidylinositol 3-phosphate. Exp Cell Res 265:329–338

Zou J, Chang SC, Marjanovic J, Majerus PW (2009) MTMR9 increases MTMR6 enzyme activity, stability, and role in apoptosis. J Biol Chem 284:2064–2071

Chapter 11
Nuclear PI-PLC $\beta 1$ and Myelodysplastic Syndromes: From Bench to Clinics

Sara Mongiorgi, Matilde Y. Follo, Cristina Clissa, Roberto Giardino, Milena Fini, Lucia Manzoli, Giulia Ramazzotti, Roberta Fiume, Carlo Finelli and Lucio Cocco

Abstract Myelodysplastic syndromes (MDS), clonal hematopoietic stem-cell disorders mainly affecting older adult patients, show ineffective hematopoiesis in one or more of the lineages of the bone marrow. A number of MDS progresses to acute myeloid leukemia (AML) with the involvement of genetic and epigenetic mechanisms affecting PI-PLC $\beta 1$. The molecular mechanisms underlying the MDS evolution to AML are still unclear, even though it is now clear that the nuclear signaling elicited by PI-PLC $\beta 1$, Cyclin D3, and Akt plays an important role in the control of the balance between cell cycle progression and apoptosis in both normal and pathologic conditions. Moreover, a correlation between other PI-PLCs, such as PI-PLC $\beta 3$, kinases and phosphatases has been postulated in MDS pathogenesis. Here, we review the findings hinting at the role of nuclear lipid signaling pathways in MDS, which could become promising therapeutic targets.

The authors 'Sara Mongiorgi and Matilde Y. Follo' are equally contributed to this work.

S. Mongiorgi · M. Y. Follo · L. Manzoli · G. Ramazzotti · R. Fiume · L. Cocco (✉)
Cellular Signalling Laboratory, Department of Human Anatomical Sciences,
University of Bologna, via Irnerio 48, 40126 Bologna, Italy
e-mail: lucio.cocco@unibo.it

C. Clissa · C. Finelli
Department of Hematology and Medical Oncology L. e A. Seràgnoli, University of Bologna,
via Massarenti 9, 40138 Bologna, Italy

R. Giardino · M. Fini
Laboratory of Preclinical and Surgical Studies, Rizzoli Orthopedic Institute, Bologna, Italy

M. Falasca (ed.), *Phosphoinositides and Disease*, Current Topics in Microbiology and Immunology 362, DOI: 10.1007/978-94-007-5025-8_11,

Introduction

The regulations of a great variety of cellular processes, both in the cytoplasm and in the plasma membrane, are mediated by phosphoinositides (PIs). Over the last years it has been established that these molecules are present not only in the plasma membrane, but also in the inner part of the nucleus, within the nuclear speckles, thus hinting at new roles in cellular signaling (Martelli et al. 1992; Faenza et al. 2008).

Indeed, nuclear inositides are now considered essential co-factors for several nuclear processes, including DNA repair, transcription regulation, and RNA dynamics. Their metabolism changes during cell growth and differentiation (Martelli et al. 1992; Faenza et al. 2008) and it is regulated independently from their plasma membrane counterpart, suggesting that the nucleus constitutes a functionally distinct compartment of inositol lipids metabolism (Martelli et al. 2005).

Among the enzymes of the nuclear PI cycle, phosphoinositide-specific phospholipase C (PI-PLC) $\beta 1$, whose hydrolysis generates diacylglycerol and inositol 1,4,5-trisphosphate as second messengers, appears to play a fundamental role as a checkpoint in the G1 phase of the cell cycle (Faenza et al. 2000), mainly targeting cyclin D3 (Faenza et al. 2007), as well as in the G2/M transition (Fiume et al. 2009).

Moreover, PI-PLC $\beta 1$ is implicated in the hematopoietic system, playing a role in proliferation and cell survival (Suh et al. 2008) and, more importantly, in the early stages of the hematopoietic differentiation, which show a high expression of PI-PLC $\beta 1$ (Furukawa 2002; Cooper et al. 2006).

The myelodysplastic syndromes (MDS) are a heterogeneous group of bone marrow diseases characterized by an ineffective differentiation of the hematopoietic stem cell that causes anemia, neutropenia, bleeding problems, and infections (Tefferi and Vardiman 2009). The disease can result in a slow decrease in blood cell counts, but it may also evolve toward a severe cytopenia or, in about 30 % of all the patients, transformation into acute myeloid leukemia (AML).

The MDS diagnosis is currently based on morphological evaluations, according to either the French–American–British (FAB) (Bennett et al. 1982) or World Health Organization (WHO) (Vardiman et al. 2009) classification, as well as following two more complex systems, based on the percentage of marrow blasts, number of cytopenias, and bone marrow cytogenetic findings, useful for the assessment of the risk of evolution into AML, i.e., the International Prognostic Scoring System (IPSS) (Greenberg et al. 1997) and/or the WHO classification-based Scoring System (WPSS) (Malcovati et al. 2007). The identification of the MDS risk is essential because it may change the therapeutic approach. In low-risk MDS cases the therapy aims at the improvement of peripheral cytopenia and quality of life (Jabbour et al. 2008), while high-risk MDS patients need to increase survival and delay the AML evolution (Morgan and Reuter 2006).

Nuclear PI-PLC β1 and MDS: Genetics

High-risk MDS patients have demonstrated, in comparison with healthy donors, an altered level of expression of both PI-PLC β1 splicing variants: PI-PLC β1a, the nuclear and cytoplasmatic isoform and PI-PLC β1b, the nuclear isoform (Follo et al. 2006). PI-PLC β1a mRNA shows a strong decrease in all cases analyzed, while the levels of PI-PLC β1b mRNA are low in the majority of the patients. Interestingly, MDS cells always expressed higher levels of PI-PLC β1b mRNA, as if the nuclear isoform should be preserved, with an imbalance of the nuclear versus the cytoplasmatic signaling which, in turn, could affect the cell cycle progression mechanisms of MDS cells. This is the key that could explain why PI-PLC β1 could regulate fundamental roles in MDS cells and seems to be involved in the progression toward AML, both at a genetic and epigenetic level (Follo et al. 2009a, b).

By fluorescence in situ hybridization (FISH) analysis it has been demonstrated that PI-PLC β1 locus gene is mapped on chromosome 20p12 (Peruzzi et al. 2000), and that a cryptic and interstitial mono-allelic deletion of the gene is present in some MDS cases, both at high and low-risk. Interestingly, the deletion has been associated with a higher risk of AML evolution (Fig. 11.1). Indeed, 80 cases belonging to all of the IPSS risk groups have been tested for the evaluation of the allelic status of PI-PLC β1, and it has been studied the correlation with the MDS progression toward AML: 30 % of the patients had the mono-allelic deletion and these cases evolved more rapidly into AML as compared with cases without deletion. This is very important because it suggests not only that PI-PLC β1 mono-allelic deletion is associated with disease progression, but also that it could have a prognostic role, through the identification of a sub-group of patients at a poorer prognosis also among low-risk cases, usually considered as patients with a better outcome.

Nuclear PI-PLC β1 and MDS: Epigenetics

Epigenetic mechanisms, that affect chromatin structure, possibly contribute to regulate gene expression and assure the inheritance of DNA information, which are essential for the expression of key regulatory genes in healthy cells, tissues, and organs. Among epigenetic processes, DNA hypermethylation and histone modifications are the most studied. In particular, promoter DNA hypermethylation is a common hallmark of cancer and, opposite to deletions or mutations in tumor suppressor genes, that are irreversible, this mechanism can be reversed by the epigenetic therapy with demethylating agents.

In the past few years there has been an advancement in supportive care and several agents have been tested for the treatment of MDS. For instance, demethylating agents

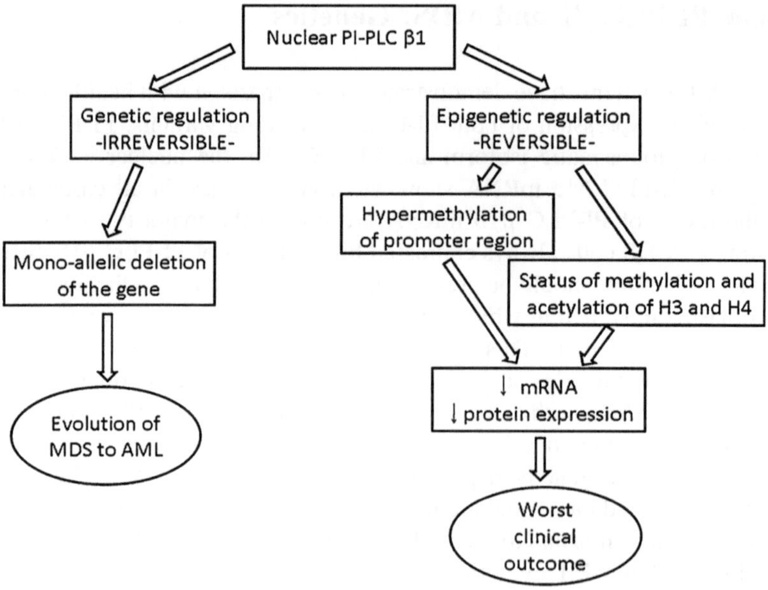

Fig. 11.1 *Role of nuclear PI-PLC β1 in MDS: genetic and epigenetic regulation*. The presence of a PI-PLC β1 mono-allelic deletion in MDS patients is an irreversible event that can be useful for the identification of a subgroup of patients at higher risk of evolution into AML. On the other hand, both the presence of a PI-PLC β1 promoter hypermethylation and histone methylation/acetylation are reversible phenomena which can be linked to PI-PLC β1 epigenetic regulation

(azacytidine, decitabine), and Histone Deacetylase (HDAC) Inhibitors [Valproic Acid (VPA) and Vorinostat] have been successfully tested in MDS therapy (Kaminskas et al. 2005; Fenaux et al. 2007, 2009; Griffiths and Gore 2008; Park et al. 2008; Sekeres et al. 2008; Perl et al. 2009). However, these agents benefit a minority of patients and the overall outcomes are still unsatisfactory (Srinivasan and Schiffer 2008). That is why clinical trials are now testing combination therapies, also because until now the only curative option for MDS seems to be allogeneic stem cell transplantation that is not suitable for all MDS, in that the majority of patients are not eligible because of older age and other medical problems.

Azacitidine is a DNA methyltransferase (DNMT) inhibitor currently approved for the treatment of MDS (Kaminskas et al. 2005; Silverman and Mufti 2005) and under experimental evaluation for other hematologic malignancies (Quintas-Cardama et al. 2008). Besides showing response rates of 50–80 % in high-risk MDS, azacitidine has been reported to have a significant impact on the overall survival and progression toward AML (Fenaux et al. 2009). Nevertheless, the molecular mechanisms underlying this drug are not completely understood, even though it is clear how the DNMT inhibitors can induce the re-expression of methylated silenced gene products (Griffiths and Gore 2008): after incorporation of demethylating agents into DNA, the methyltransferases are inhibited, but complete demethylation occurs only after several cycles of replication, thus accounting for time to response to these drugs (Stresemann and Lyko 2008). Low-dose regimens with azacitidine have been assumed to act by reversing the epigenetic silencing of target genes involved in the control of cell growth and differentiation. For instance, demethylation of a hypermethylated p15/INK4B gene, as well as of other genes, such as p21WAF/Cip1 and p73, has been demonstrated in MDS patients treated with demethylating therapy (Daskalakis et al. 2002; Raj et al. 2007).

Recently, it has been demonstrated that azacitidine specifically targets PI-PLC $\beta 1$ (Follo et al. 2009b). PI-PLC $\beta 1$ promoter methylation and gene expression were quantified in high-risk MDS patients during azacitidine administration and compared to the expression in patients treated with only best supportive care as well as healthy subjects. Interestingly, promoter methylation and gene expression had an opposite trend, with PI-PLC $\beta 1$ mRNA levels following and anticipating the clinical outcome (Fig. 11.1). In fact, the variations in PI-PLC $\beta 1$ expression, increase or decrease, were detectable prior to the clinical improvement or worsening, respectively. This is particulary appealing, since usually some cycles of azacitidine are needed in order to asses the clinical response.

As previously stated, in the past few years it has emerged the importance of combining different drugs, with the purpose of increasing the efficacy of the single agents, therefore improving response rates, prolonging response duration and decreasing the toxicities associated with the treatment (Minucci and Pelicci 2006). That is why several combination therapies are beginning to be tested, either as chemosensitizing agents in association with standard chemotherapy or combination of epigenetic drugs, aiming at reverting the hypermethylation of clonal cells (Braiteh et al. 2008; Mercurio et al. 2010).

HDAC inhibitors can modulate the chromatin structure that is under control of both histone acetyl-transferase and deacetylases. This class of enzyme can induce in biochemical changes that modify the chromatin-associated histone proteins, such as acetylation of histone H3 or H4, thus resulting in differential gene transcription and inducing other biological effects, such as apoptosis in leukemogenesis (Neff and Armstrong 2009).

It has been recognized that the combination of DNMT inhibitors, such as azacitidine, with HDAC inhibitors, like VPA, in high-risk MDS patients, may have a synergic effect in enhancing their demethylating effect on PI-PLC $\beta 1$ (Follo et al. 2011). Indeed, the combination of azacitidine and VPA induce a major demethylation of PI-PLC $\beta 1$ promoter and an increased reactivation of both gene and protein expression in responder patients, as compared with azacitidine alone. The synergistic effect of the combined therapy might offer better efficacy by modulating the methylation and acetylation states of silenced genes (Fenaux et al. 2009).

Furthermore, this therapeutic strategy increases Cyclin D3 mRNA and protein expression. Cyclin D3 is a downstream target of PI-PLC $\beta 1$ signaling (Faenza et al. 2000; O'Carroll et al. 2009) and this suggests a role for the PI-PLC $\beta 1$ and Cyclin D3 in hematopoietic differentiation and could be useful for a better comprehension of the molecular mechanisms underling the epigenetic therapy.

Possible Correlation Between PI-PLCs, Kinases, and Phosphatases in MDS

As previously described, the modulation of PI-PLC $\beta 1$ at a nuclear level is implicated in the progression of MDS to AML. Interestingly, other studies have elucidated that there is an inverse correlation between PI-PLC $\beta 1$ and activation of Akt. This is particularly appealing, since the nucleus contains other 3-phosphorylated inositol lipids and the enzyme which synthesize them is phosphoinositide 3-kinase (PI3 K). PI3 K/Akt axis is crucial to widely divergent physiological processes that include cell cycle progression, differentiation, transcription, translation, and apoptosis (Ye 2005). Constitutive activation of this axis, due to gene mutations, amplifications or rearrangements, results in altering these processes and leads to the growth of tumor cells. An impaired regulation of the PI3 K/Akt axis has been strongly implicated in carcinogenesis and is often associated with hematological malignancies, including acute and chronic human leukemias.

As for MDS, recent reports demonstrated a constitutive activation (phosphorylation) of the Akt pathway. Phosphorylation of Akt was highly present in bone marrow and peripheral blood mononuclear cells from high-risk MDS patients, while it was almost absent in low-risk and healthy donors (Nyakern et al. 2006). In high-risk MDS patients, this enzyme and its downstream targets, such as mTOR, are over-activated, thus leading to an imbalance in the apoptotic processes (Follo et al. 2007) (Fig. 11.2). Furthermore, there is a specific up-regulation of the mTOR pathway in the hematopoietic myeloid progenitors of high-risk MDS patients, since rapamycin, targeting mTOR, influenced the survival of $CD33^+$ cells and the clonogenity ability of $CD34^+$ MDS cells.

All in all, PI-PLC $\beta 1$ signaling, as well as Akt/mTOR pathways, play an important role in MDS cell proliferation and differentiation. In particular, an altered expression of nuclear PI-PLC $\beta 1$ and activated (phosphorylated) Akt could lead to a deregulation of cell cycle processes, therefore negatively influencing the apoptotic processes and affecting the survival of primary MDS cells (Follo et al. 2010) (Fig. 11.2). After azacitidine treatment in high-risk MDS patients, an increase in PI-PLC $\beta 1$ levels is followed by a reduction in activated Akt levels, thus indicating that PI-PLC $\beta 1$ and Akt could play opposite roles (Follo et al. 2008). The balance between these two molecules could be directly related to PIP_2 levels, which is both substrate for PI-PLC $\beta 1$ and a key player in the activation of the PI3 K/Akt axis (Follo et al. 2009b). Indeed, the combination therapy of azacitidine and VPA has demonstrated a potential increased activity in high-risk MDS patients by increasing PI-PLC $\beta 1$ and inhibiting Akt activation. Moreover, azacitidine and VPA together caused a higher dephosphorylation of p-Akt in patients who responded to the therapy, as compared with the effect of azacitidine alone, whereas non-responders did not show any significant Akt inhibition (Follo et al. 2011).

Interestingly, among the members of the PI-PLC β family, not only the isoform $\beta 1$ seems to be involved in leukemogenesis, but also the isoform $\beta 3$, which is expressed in both hematopoietic and non-hematopoietic cells.

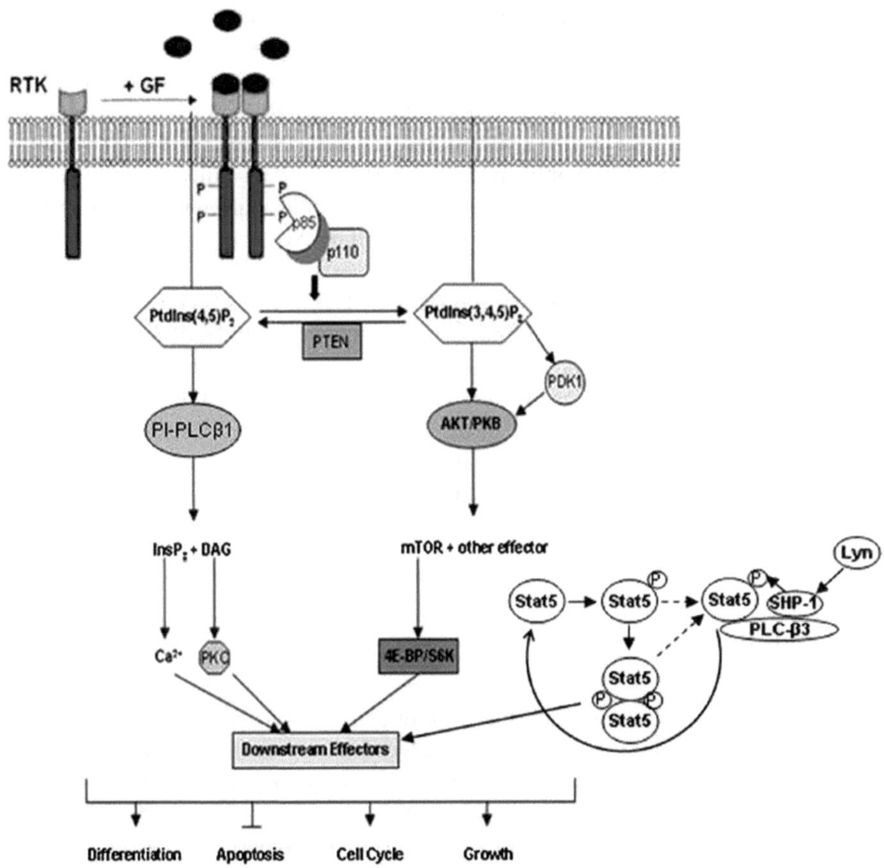

Fig. 11.2 *Possible correlation between PI-PLCs and Akt signaling in MDS.* PI-PLC β1, PI-PLC β3 and Akt signaling pathways can be inter-connected and play a role in the activation of cell cycle, differentiation, and apoptotic processes

Phospholipase C β3 (PI-PLC β3) is a member of the PI-PLC β family enzymes that can produce diacylglycerol and inositol 1,4,5-trisphosphate (IP3), downstream of heterotrimeric G proteins. Given its catalytic activity to generate diacylglycerol and inositol 1,4,5-trisphosphate, PI-PLC β3 is also involved in normal and pathological cell growth. In fact, it has been demonstrated that PI-PLC β3—deficient mice can develop myeloproliferative diseases, lymphoma, and other tumors.

Moreover, it has been established that in a form of human chronic myelomonocitic leukemia in mice, hyperactivation of transcription factor Stat5 is involved in cell proliferation and survival, and that this is due to a reduction of function of the phospatase SHP-1, which regulates Stat5 in a PI-PLC β3 dependent manner.

In particular, SHP-1 phosphatase activity is increased by PI-PLC β3 and Lyn, a Src family kinase, but it is the physical interaction among Stat5, SHP-1, and

PI-PLC $\beta 3$, which has an adaptor function, that is responsible for the suppression of Stat5 phosphorylation (Fig. 11.2). The dysregulation of this mechanism at the level of hematopoietic stem cells may lead to the development of myeloproliferative disorders (Xiao et al. 2009, 2010). In fact, PI-PLC $\beta 3$ augments SHP-1 mediated deactivation of Stat5 activity, and the loss of this regulation seems to lead to myeloproliferative neoplasms in aged PI-PLC $\beta 3$-/- mice.

Constitutive activation of Stat5 leads to hematologic malignancies, since Stat5 is activated by several oncoproteins, such as BCR-ABL, FLT3-ITD, and JAK2 V617F47 and is highly required for the pathogenesis of leukemia caused by these oncoproteins. Moreover, a reduced expression of Lyn, PI-PLC $\beta 3$, or SHP-1 might contribute to the development of other types of leukemia. Indeed, Stat5 activation was shown to be essential for myeloid leukemia induced by the activated oncogenes, such as TEL/JAK2 (Schwaller et al. 2000), TEL/PDGFRB (Cain et al. 2007), and FLT3 ITD (Choudhary et al. 2007). and by deficiencies of SHIP and Lyn (Xiao et al. 2008). All in all, the adaptor function of PI-PLC $\beta 3$ seems essential to protect the hematopoietic and non-hematopoietic systems from tumor development.

Conclusions

Nuclear PI-PLC $\beta 1$ plays an important role in cell proliferation and differentiation, in normal and pathological conditions. Indeed, recent findings indicate that the inositide signaling pathways might contribute to the further clarification of the therapeutic activity of some drugs currently used in high-risk MDS, such as azacitidine. In fact, not only PI-PLC $\beta 1$ promoter hypermethylation has been associated with the progression of high-risk MDS into AML, but also the activation of Cyclin D3 after successful epigenetic treatment strengthens the contention that a correct nuclear lipid signaling is essential for physiological processes such as cell growth and differentiation in MDS.

Moreover, the role of PI-PLC $\beta 3$ in the pathogenesis of hematological malignancies could represent an attractive new molecular target for testing the effect of epigenetic therapies. Further investigations are needed to fully understand the molecular mechanisms underlying the MDS progression into AML, but it is now clear that signal transduction pathways can be considered as innovative therapeutic targets in MDS treatments.

Acknowledgments This work was supported by Italian MIUR-FIRB (Human Proteome Net and Accordi di Programma 2010), Italian MIUR PRIN and Celgene Corp.

References

Bennett JM, Catovsky D, Daniel MT, Flandrin G, Galton DA, Gralnick HR, Sultan C (1982) Proposals for the classification of the myelodysplastic syndromes. Br J Haematol 51:189–199

Braiteh F, Soriano AO, Garcia-Manero G et al (2008) Phase I study of epigenetic modulation with 5-azacytidine and valproic acid in patients with advanced cancers. Clin Cancer Res 14:6296–6301

Cain JA, Xiang Z, O'Neal J et al (2007) Myeloproliferative disease induced by TEL-PDGFRB displays dynamic range sensitivity to Stat5 gene dosage. Blood 109:3906–3914

Choudhary C, Brandts C, Schwable J et al (2007) Activation mechanisms of STAT5 by oncogenic Flt3-ITD. Blood 110:370–374

Cooper AB, Sawai CM, Sicinska E, Powers SE, Sicinski P, Clark MR, Aifantis I (2006) A unique function for cyclin D3 in early B cell development. Nat Immunol 7:489–497

Daskalakis M, Nguyen TT, Nguyen C et al (2002) Demethylation of a hypermethylated P15/INK4B gene in patients with myelodysplastic syndrome by 5-Aza-2′-deoxycytidine (decitabine) treatment. Blood 100:2957–2964

Faenza I, Bregoli L, Ramazzotti G et al (2008) Nuclear phospholipase Cbeta1 and cellular differentiation. Front Biosci 13:2452–2463

Faenza I, Matteucci A, Manzoli L et al (2000) A role for nuclear phospholipase Cbeta1 in cell cycle control. J Biol Chem 275:30520–30524

Faenza I, Ramazzotti G, Bavelloni A et al (2007) Inositide-dependent phospholipase C signaling mimics insulin in skeletal muscle differentiation by affecting specific regions of the cyclin D3 promoter. Endocrinology 148:1108–1117

Fenaux P, Mufti GJ, Hellstrom-Lindberg E et al (2009) Efficacy of azacitidine compared with that of conventional care regimens in the treatment of higher-risk myelodysplastic syndromes: a randomised, open-label, phase III study. Lancet Oncol 10:223–232

Fenaux P, Raza A, Mufti GJ et al (2007) A multicenter phase 2 study of the farnesyltransferase inhibitor tipifarnib in intermediate—to high-risk myelodysplastic syndrome. Blood 109:4158–4163

Fiume R, Ramazzotti G, Teti G et al (2009) Involvement of nuclear PLCbeta1 in lamin B1 phosphorylation and G2/M cell cycle progression. FASEB J 23:957–966

Follo MY, Bosi C, Finelli C et al (2006) Real-time PCR as a tool for quantitative analysis of PI-PLCbeta1 gene expression in myelodysplastic syndrome. Int J Mol Med 18:267–271

Follo MY, Finelli C, Bosi C et al (2008) PI-PLCbeta-1 and activated Akt levels are linked to azacitidine responsiveness in high-risk myelodysplastic syndromes. Leukemia 22:198–200

Follo MY, Finelli C, Clissa C et al (2009a) Phosphoinositide-phospholipase Cbeta1 mono-allelic deletion is associated with myelodysplastic syndromes evolution into acute myeloid leukemia. J Clin Oncol 27:782–790

Follo MY, Finelli C, Mongiorgi S et al (2009b) Reduction of phosphoinositide-phospholipase Cbeta1 methylation predicts the responsiveness to azacitidine in high-risk MDS. Proc Natl Acad Sci U S A 106:16811–16816

Follo MY, Finelli C, Mongiorgi S et al (2011) Synergistic induction of PI-PLCbeta1 signaling by azacitidine and valproic acid in high-risk myelodysplastic syndromes. Leukemia 25:271–280

Follo MY, Mongiorgi S, Bosi C et al (2007) The Akt/mammalian target of rapamycin signal transduction pathway is activated in high-risk myelodysplastic syndromes and influences cell survival and proliferation. Cancer Res 67:4287–4294

Follo MY, Mongiorgi S, Finelli C et al (2010) Nuclear inositide signaling in myelodysplastic syndromes. J Cell Biochem 109:1065–1071

Furukawa Y (2002) Cell cycle control genes and hematopoietic cell differentiation. Leuk Lymphoma 43:225–231

Greenberg P, Cox C, LeBeau MM et al (1997) International scoring system for evaluating prognosis in myelodysplastic syndromes. Blood 89:2079–2088

Griffiths EA, Gore SD (2008) DNA methyltransferase and histone deacetylase inhibitors in the treatment of myelodysplastic syndromes. Semin Hematol 45:23–30

Jabbour E, Kantarjian HM, Koller C, Taher A (2008) Red blood cell transfusions and iron overload in the treatment of patients with myelodysplastic syndromes. Cancer 112:1089–1095

Kaminskas E, Farrell A, Abraham S et al (2005) Approval summary: azacitidine for treatment of myelodysplastic syndrome subtypes. Clin Cancer Res 11:3604–3608

Malcovati L, Germing U, Kuendgen A et al (2007) Time-dependent prognostic scoring system for predicting survival and leukemic evolution in myelodysplastic syndromes. J Clin Oncol 25:3503–3510

Martelli AM, Fiume R, Faenza I et al (2005) Nuclear phosphoinositide specific phospholipase C (PI-PLC)-beta1: a central intermediary in nuclear lipid-dependent signal transduction. Histol Histopathol 20:1251–1260

Martelli AM, Gilmour RS, Bertagnolo V, Neri LM, Manzoli L, Cocco L (1992) Nuclear localization and signalling activity of phosphoinositidase Cbeta in Swiss 3T3 cells. Nature 358:242–245

Mercurio C, Minucci S, Pelicci PG (2010) Histone deacetylases and epigenetic therapies of hematological malignancies. Pharmacol Res 62:18–34

Minucci S, Pelicci PG (2006) Histone deacetylase inhibitors and the promise of epigenetic (and more) treatments for cancer. Nat Rev Cancer 6:38–51

Morgan MA, Reuter CW (2006) Molecularly targeted therapies in myelodysplastic syndromes and acute myeloid leukemias. Ann Hematol 85:139–163

Neff T, Armstrong SA (2009) Chromatin maps, histone modifications and leukemia. Leukemia 23:1243–1251

Nyakern M, Tazzari PL, Finelli C et al (2006) Frequent elevation of Akt kinase phosphorylation in blood marrow and peripheral blood mononuclear cells from high-risk myelodysplastic syndrome patients. Leukemia 20:230–238

O'Carroll SJ, Mitchell MD, Faenza I, Cocco L, Gilmour RS (2009) Nuclear PLCbeta1 is required for 3T3-L1 adipocyte differentiation and regulates expression of the cyclin D3-cdk4 complex. Cell Signal 21:926–935

Park S, Chapuis N, Bardet V et al (2008) PI-103, a dual inhibitor of Class IA phosphatidylinositide 3-kinase and mTOR, has antileukemic activity in AML. Leukemia 22:1698–1706

Perl AE, Kasner MT, Tsai DE et al (2009) A phase I study of the mammalian target of rapamycin inhibitor sirolimus and MEC chemotherapy in relapsed and refractory acute myelogenous leukemia. Clin Cancer Res 15:6732–6739

Peruzzi D, Calabrese G, Faenza I et al (2000) Identification and chromosomal localisation by fluorescence in situ hybridisation of human gene of phosphoinositide-specific phospholipase Cbeta1. Biochim Biophys Acta 1484:175–182

Quintas-Cardama A, Tong W, Kantarjian H et al (2008) A phase II study of 5-azacitidine for patients with primary and post-essential thrombocythemia/polycythemia vera myelofibrosis. Leukemia 22:965–970

Raj K, John A, Ho A et al (2007) CDKN2B methylation status and isolated chromosome 7 abnormalities predict responses to treatment with 5-azacytidine. Leukemia 21:1937–1944

Schwaller J, Parganas E, Wang D et al (2000) Stat5 is essential for the myelo—and lymphoproliferative disease induced by TEL/JAK2. Mol Cell 6:693–704

Sekeres MA, Maciejewski JP, Giagounidis AA, Wride K, Knight R, Raza A, List AF (2008) Relationship of treatment-related cytopenias and response to lenalidomide in patients with lower-risk myelodysplastic syndromes. J Clin Oncol 26:5943–5949

Silverman LR, Mufti GJ (2005) Methylation inhibitor therapy in the treatment of myelodysplastic syndrome. Nat Clin Pract Oncol 2(Suppl 1):12–23

Srinivasan S, Schiffer CA (2008) Current treatment options and strategies for myelodysplastic syndromes. Expert Opin Pharmacother 9:1667–1678

Stresemann C, Lyko F (2008) Modes of action of the DNA methyltransferase inhibitors azacytidine and decitabine. Int J Cancer 123:8–13

Suh PG, Park JI, Manzoli L et al (2008) Multiple roles of phosphoinositide-specific phospholipase C isozymes. BMB Rep 41:415–434

Tefferi A, Vardiman JW (2009) Myelodysplastic syndromes. N Engl J Med 361:1872–1885

Vardiman JW, Thiele J, Arber DA et al (2009) The 2008 revision of the World Health Organization (WHO) classification of myeloid neoplasms and acute leukemia: rationale and important changes. Blood 114:937–951

Xiao W, Ando T, Wang HY, Kawakami Y, Kawakami T (2010) Lyn—and PLCbeta3-dependent regulation of SHP-1 phosphorylation controls Stat5 activity and myelomonocytic leukemia-like disease. Blood 116:6003–6013

Xiao W, Hong H, Kawakami Y et al (2009) Tumor suppression by phospholipase Cbeta3 via SHP-1-mediated dephosphorylation of Stat5. Cancer Cell 16:161–171

Xiao W, Hong H, Kawakami Y, Lowell CA, Kawakami T (2008) Regulation of myeloproliferation and M2 macrophage programming in mice by Lyn/Hck, SHIP, and Stat5. J Clin Invest 118:924–934

Ye K (2005) PIKE/nuclear PI 3-kinase signaling in preventing programmed cell death. J Cell Biochem 96:463–472

Chapter 12
Inositol Polyphosphate Phosphatases in Human Disease

Sandra Hakim, Micka C. Bertucci, Sarah E. Conduit, David L. Vuong and Christina A. Mitchell

Abstract Phosphoinositide signalling molecules interact with a plethora of effector proteins to regulate cell proliferation and survival, vesicular trafficking, metabolism, actin dynamics and many other cellular functions. The generation of specific phosphoinositide species is achieved by the activity of phosphoinositide kinases and phosphatases, which phosphorylate and dephosphorylate, respectively, the inositol headgroup of phosphoinositide molecules. The phosphoinositide phosphatases can be classified as 3-, 4- and 5-phosphatases based on their specificity for dephosphorylating phosphates from specific positions on the inositol head group. The SAC phosphatases show less specificity for the position of the phosphate on the inositol ring. The phosphoinositide phosphatases regulate PI3K/Akt signalling, insulin signalling, endocytosis, vesicle trafficking, cell migration, proliferation and apoptosis. Mouse knockout models of several of the phosphoinositide phosphatases have revealed significant physiological roles for these enzymes, including the regulation of embryonic development, fertility, neurological function, the immune system and insulin sensitivity. Importantly, several phosphoinositide phosphatases have been directly associated with a range of human diseases. Genetic mutations in the 5-phosphatase *INPP5E* are causative of the ciliopathy syndromes Joubert and MORM, and mutations in the 5-phosphatase *OCRL* result in Lowe's syndrome and Dent 2 disease. Additionally, polymorphisms in the 5-phosphatase SHIP2 confer diabetes susceptibility in specific populations, whereas reduced protein expression of SHIP1 is reported in several human leukaemias. The 4-phosphatase, INPP4B, has recently been identified as a tumour suppressor in human breast and prostate cancer. Mutations in one SAC phosphatase, *SAC3/FIG4,* results in the degenerative neuropathy, Charcot-Marie-Tooth disease. Indeed, an understanding of the precise

S. Hakim · M. C. Bertucci · S. E. Conduit · D. L. Vuong · C. A. Mitchell (✉)
Department of Biochemistry and Molecular Biology, Monash University,
Wellington Rd, Clayton 3800, Australia
e-mail: christina.mitchell@monash.edu

M. Falasca (ed.), *Phosphoinositides and Disease*, Current Topics in Microbiology and Immunology 362, DOI: 10.1007/978-94-007-5025-8_12,

functions of phosphoinositide phosphatases is not only important in the context of normal human physiology, but to reveal the mechanisms by which these enzyme families are implicated in an increasing repertoire of human diseases.

Abbreviations

ADAM	A disintegrin and a metalloprotease
ALL	Acute lymphoblastic leukaemia
AML	Acute myeloid leukaemia
AP	Adaptor protein
APPL1	Adaptor protein containing Pleckstrin homology domain, PTB domain and Leucine Zipper motif 1
AR	Androgen receptor
ARF-GAP	ADP-ribosylation factor GTPase activating protein
ARF-GEF	ADP-ribosylation factor Guanine nucleotide-exchange factor
ASH	Abnormal spindle-like microcephaly-associated protein/spindle pole body/hydrin
ASPM	Abnormal spindle-like microcephaly-associated protein
ATLL	Adult T cell leukaemia/lymphoma
AVP	Arginine vasopressin
BAFF	B cell activating factor belonging to the TNF family
BCR	B cell receptor
BMM	Bone marrow macrophage
BMMC	Bone marrow mast cell
Btk	Bruton's tyrosine kinase
C. elegans	*Caenorhabditis elegans*
C/EBPβ	CCAAT enhancer-binding protein β
CAP	Cbl interacting protein
CD2AP	CD2-associated protein
Cdk5	Cyclin-dependent kinase 5
CD-MPR	Cation-dependent mannose-6-phosphate receptor
CI-MPR	Cation-independent mannose-6-phosphate receptor
CK5/6	Cytokeratin 5/6
CLL	Chronic lymphocytic leukaemia
CML	Chronic myeloid leukaemia
CMT	Charcot-Marie-Tooth
CRMP2	Collapsin response mediator protein 2
DAP12	DNAX-activating protein of 12 kD
DC	Dendritic cell
DS	Down's syndrome
EGFR	Epidermal growth factor receptor
Epo	Erythropoietin
ER	Endoplasmic reticulum
ER	Oestrogen receptor
ER^-	Oestrogen receptor negative
ER^+	Oestrogen receptor positive

ERGIC	Endoplasmic reticulum-to-Golgi intermediate compartment
F. tularensis	*Francisella tularensis*
F-MuLV	Friend murine leukaemia virus
FYVE	Fab1, YOTB, Vac1, EEA1
G6Pase	Glucose-6-phosphatase
GAP	GTPase activating protein
G-CSF	Granulocyte colony-stimulating factor
GIPC	GAIP-interacting protein C terminus
GK	Goto Kakizaki
GSK3β	Glycogen synthase kinase-3 β
HDAC2	Histone deacetylase 2
HER-2	v-erb-b2 erythroblastic leukaemia viral oncogene homologue 2
IGF-1	Insulin-like growth factor
INPP4A	Inositol polyphosphate 4-phosphatase type I
INPP4B	Inositol polyphosphate 4-phosphatase type II
INPP5A	Inositol polyphosphate 5-phosphatase type I
INPP5B	Inositol polyphosphate 5-phosphatase type II
INPP5E	Inositol polyphosphate 5-phosphatase type IV
Ins(1,2,3,4,5)P_5	Inositol 1,2,3,4,5-pentakisphosphate
Ins(1,3,4)P_3	Inositol 1,3,4-trisphosphate
Ins(1,3,4,5)P_4	Inositol 1,3,4,5-tetrakisphosphate
Ins(1,4)P_2	Inositol 1,4-bisphosphate
Ins(1,4,5)P_3	Inositol 1,4,5-trisphosphate
Ins(1,4,5,6)P_4	Inositol 1,4,5,6-tetrakisphosphate
Ins(3,4)P_2	Inositol 3,4-bisphosphate
IRS	Insulin receptor substrate
ITIM	Immunoreceptor tyrosine-based inhibitory motif
LAT	Linker for activation of T cells
M-CSF	Monocyte colony-stimulating factor
MDS	Miller–Dieker syndrome
MEFs	Mouse embryonic fibroblasts
miR-155	MicroRNA-155
MNB/DYRK1A	Dual-specific tyrosine phosphorylation-regulated kinase 1A
NCA	Na^+/Ca^{2+} antiporter
NCoR	Nuclear corepressor
NF-κB	Nuclear factor-κB
NPF	Asparagine-proline-phenylalanine
NTAL	Non-T cell activation linker
OCRL	Oculocerebrorenal syndrome of Lowe
PAS	(PIKfyve–ArPIKfyve–Sac3)
PBMs	Peripheral blood monocytes
PDGF	Platelet-derived growth factor
PDK1	Phosphoinositide-dependent kinase-1
PEPCK	Phosphoenolpyruvate carboxykinase

PGN	Peptidoglycan
PH	Pleckstrin homology
PI	Phosphoinositide
PI(3)P	Phosphatidylinositol 3-phosphate
$PI(3,4)P_2$	Phosphatidylinositol 3,4-bisphosphate
$PI(3,4,5)P_3$	Phosphatidylinositol 3,4,5-trisphosphate
$PI(3,5)P_2$	Phosphatidylinositol 3,5-bisphosphate
PI(4)P	Phosphatidylinositol 4-phosphate
$PI(4,5)P_2$	Phosphatidylinositol 4,5-bisphosphate
PI(5)P	Phosphatidylinositol 5-phosphate
PI3K	Phosphatidylinositol 3-kinase
PIPP	Proline-rich inositol polyphosphate 5-phosphatase
PLCγ	Phospholipase C γ
PLD	Phospholipase D
PR	Progesterone receptor
PTB	Phosphotyrosine binding
PTEN	Phosphatase and tensin homolog
RANKL	Receptor activator of nuclear factor-κB ligand
Rho-GEFs	Rho-Guanine nucleotide exchange factors
S. flexneri	*Shigella flexneri*
SAC	Supressor of actin
SCC	Squamous cell carcinoma
SCF	Stem cell factor
SCIPs	SAC domain-containing inositol phosphatases
SCVs	*Salmonella* containing vacuoles
SF	Steel-factor
SHIP	SH2-containing inositol phosphatase
SKIP	Skeletal muscle and kidney inositol phosphatase
SNP	Single nucleotide polymorphism
SODD/BAG4	Silencer of death domain
SPD2	Spindle pole body 2
SYNJ1	Synaptojanin 1
SYNJ2	Synaptojanin 2
T-ALL	T cell acute lymphoblastic leukaemia
TGFβ	Transforming growth factor β
TGN	Trans-Golgi network
Tir	Translocated intimin receptor
TIRFM	Total internal reflection fluorescent microscopy
TLR-2	Toll-like receptor-2
TMEM55A	Transmembrane protein 55A
TMEM55B	Transmembrane protein 55A
TREM2	Triggering receptor expressed on myeloid cells-2
UTR	Untranslated region

Introduction

Phosphoinositides (PIs) are ubiquitous membrane associated signalling molecules, composed of an inositol headgroup which can be phosphorylated, a glycerol moiety and two fatty acid chains that enable insertion into lipid membranes. There are 7 different signalling species derived from phosphatidylinositol, which differ based on the number and position of phosphates on the inositol ring at the D-3, D-4 and/or D-5 positions. Specific phosphoinositides are restricted to and thus specify various subcellular compartments, recruit downstream effector proteins and initiate intracellular signalling networks to mediate a diverse range of cellular functions (Liu and Bankaitis 2010; Sasaki et al. 2009). The spatio-temporal production of phosphorylated phosphoinositides is critical for context-dependent cellular functions, and is achieved by an exquisitely orchestrated activity of phosphoinositide kinases and phosphatases (see Fig. 12.1).

Phosphoinositide phosphatases are highly conserved and comprise diverse classes of enzymes which hydrolyse phosphorylated phosphoinositide species. There are 35 mammalian phosphatases identified, which can be classified based on substrate specificity, into D-3, D-4 or D-5 position phosphate hydrolysing enzymes designated 3-, 4- and 5-phosphatases, respectively. The 3-phosphatases include PTEN and the myotubularins. The SAC phosphatases do not display phosphate positional phosphatase specificity, and along with the 4-phosphatases, share a Mg^{2+}-independent catalytic mechanism mediated by a conserved CX_5R motif (Guo et al. 1999; Norris et al. 1997a; Ungewickell et al. 2005). In contrast the 5-phosphatases contain a unique conserved catalytic domain and display Mg^{2+}-dependent phosphatase activity with a mechanism shared by base excision repair endonucleases (Tsujishita et al. 2001; Whisstock et al. 2000).

Phosphoinositide phosphatases are implicated in a large and diverse array of human diseases, which include cancer, diabetes and neurological diseases. Additionally, many knockout mouse models of specific phosphoinositide phosphatases have revealed the physiological role these enzymes play (see Table 12.1). The 3-phosphatases PTEN and the myotubularins will not be discussed here, rather, this chapter will focus on 4-, 5- and SAC phosphatases. Specifically, the cellular function of each enzyme will be illustrated, and the biological functions both in animal models, and particularly in the context of human disease will be discussed.

Inositol Polyphosphate 4-Phosphatases

The phosphoinositide 4-phosphatases comprise 4 mammalian enzymes. In addition, several bacterial homologues possessing 4-phosphatase catalytic activity exist that contribute to human diseases. The 4-phosphatases contain a conserved CX_5R catalytic motif, with Mg^{2+}-independent activity to hydrolyse the D-4 position phosphate from phosphoinositide species (Norris et al. 1997a, 1998; Ungewickell

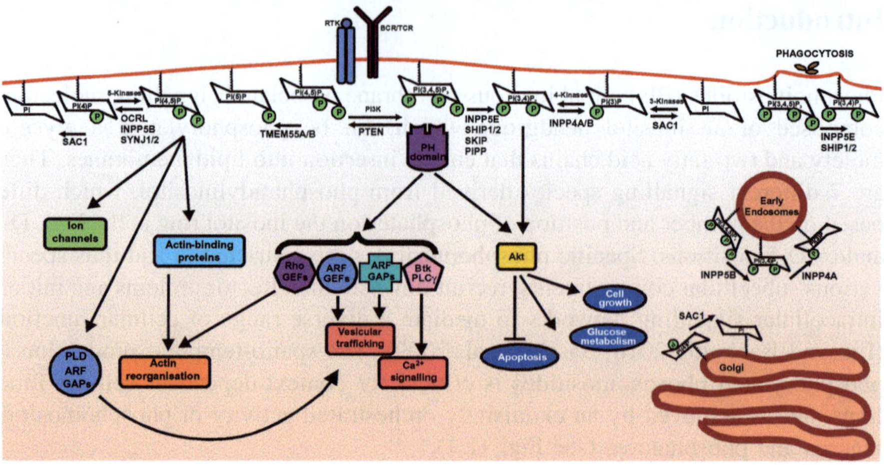

Fig. 12.1 *Phosphoinositide signalling is coordinated by phosphoinositide kinases and phosphatases*. The major PI phosphatases and kinases are depicted above, and highlight only the major substrates and functions of some PI phosphatase family members. Activation of cell surface receptors initiates PI3K activity, transiently phosphorylating PI(4,5)P_2 to the second messenger PI(3,4,5)P_3 on the inner leaflet of the plasma membrane. PI(3,4,5)P_3 recruits various signalling proteins via their PH domain, including Rho-Guanine nucleotide exchange factors (Rho-GEFs), ADP ribosylation factor-Guanine nucleotide exchange factors (ARF-GEFs), ARF-GTPase activating proteins (ARF-GAPs), as well as Bruton's tyrosine kinase (Btk), Phospholipase C γ (PLCγ) and Akt resulting in their allosteric activation and initiation of downstream signalling pathways to promote various cellular effects such as actin polymerisation, cell proliferation and cell growth. PI(3,4,5)P_3 signals are hydrolysed by 3-phosphatases such as PTEN to form PI(4,5)P_2, as well as 5-phosphatases (INPP5E, SHIP1/2, SKIP and PIPP) forming PI(3,4)P_2. PI(3,4)P_2 can be further hydrolysed to PI(3)P through the actions of 4-phosphatases (INPP4A/B). The 5-phosphatases OCRL, INPP5B and SYNJ1/2 dephosphorylate PI(4,5)P_2 to form PI(4)P, and therefore modulate cellular activities which are orchestrated by PI(4,5)P_2 signals, including vesicular trafficking via Phospholipase D (PLD) and ARF-GAPs, actin reorganisation via actin-binding proteins and ion channels. INPP5B and INPP4A hydrolyse PI(3,4,5)P_3 and PI(3,4)P_2, respectively, at forming/early endosomes, and the SAC phosphatases hydrolyse monophosphorylated PI species to form PI. PI(4,5)P_2 is also hydrolysed by the PI(4,5)P_2 4-phosphatases TMEM55A/B to form PI(5)P

et al. 2005). The 4-phosphatases are broadly classified based on their ability to hydrolyse PI(3,4)P_2 or PI(4,5)P_2.

PI(3,4)P_2 4-Phosphatases

The two mammalian PI(3,4)P_2 metabolising 4-phosphatase enzymes designated inositol polyphosphate 4-phosphatase I and II have been identified, encoded by the *INPP4A* and *INPP4B* genes, respectively. Although INPP4A and INPP4B are only

Table 12.1 Substrate specificity, human disease associations and animal model phenotypes of phosphoinositide 4-, 5- and SAC phosphatases

Protein name	Alias(es)	Gene name(s)	Substrate(s)	Human disease	Animal model
4-Phosphatases					
Inositol polyphosphate 4-phosphatase type I (INPP4A)		*INPP4A*	Ins(1,3,4)P_3, Ins(3,4)P_2, PI(3,4)P_2		*Inpp4a*^{wbl} mutant mouse—Purkinje cell loss, early onset ataxia, postnatal lethality 2–3 weeks of age (Nystuen et al. 2001). *Inpp4a*$^{-/-}$ mouse—neuronal death in striatum, involuntary movements, postnatal lethality by 2–3 weeks of age (Sasaki et al. 2010)
Inositol polyphosphate 4-phosphatase type II (INPP4B)		*INPP4B*	Ins(1,3,4)P_3, Ins(3,4)P_2, PI(3,4)P_2	Tumour suppressor in breast cancer (Fedele et al. 2010; Gewinner et al. 2009) and loss of expression in prostate cancer (Hodgson et al. 2011)	*Inpp4b*$^{-/-}$ mouse—reduced bone density and volume at 8 weeks of age, as well as decreased osteoclast number and size (Ferron et al. 2011)
IpgD (Bacteria)		*IpgD*	PI(4,5)P_2	Shigellosis (Bacillary dysentery) due to *Shigella flexineri* infection (Allaoui et al. 1993; Niebuhr et al. 2000)	Not applicable
SigD (Bacteria)		*SigD*	PI(4,5)P_2	Gastroeneteritis due to *Salmonella Typhimurium* infection (Marcus et al. 2001)	Not applicable
SopB (Bacteria)		*SopB*	PI(4,5)P_2	Gastroeneteritis due to *Salmonella Dublin* infection (Norris et al. 1998)	Not applicable
PI(4,5)P_2 4-phosphatase type I (TMEM55B)		*TMEM55B*	PI(4,5)P_2		Not reported
PI(4,5)P_2 4-phosphatase type II (TMEM55A)		*TMEM55A*	PI(4,5)P_2		Not reported
5-Phosphatases					
Inositol polyphosphate 5-phosphatase type I (INPP5A)	43 kDa inositol polyphosphate 5-phosphatase	*INPP5A*	Ins(1,3,4,5)P_4, Ins(1,4,5)P_3		Not reported

(continued)

Table 12.1 (continued)

Protein name	Alias(es)	Gene name(s)	Substrate(s)	Human disease	Animal model
Occulocerebrorenal of Lowe protein (OCRL)	Lowe's protein	*OCRL* *INPP5F*	Ins(1,3,4,5)P_4, Ins(1,4,5)P_3, PI(3,5)P_2, PI(3,4,5)P_3, PI(4,5)P_2	Mutations in *OCRL* occur in Lowe's syndrome and Dent 2 disease (Attree et al. 1992; Hoopes et al. 2005)	*Ocrl*$^{-/-}$ mouse—viable and does not display the characteristics of Lowe's syndrome (Jänne et al. 1998). *Ocrl:Inpp5b* double knockout mouse—embryonically lethal (Jänne et al. 1998)
Inositol polyphosphate 5-phosphatase type II (INPP5B)	75 kDa inositol polyphosphate 5-phosphatase, 5-phosphatase-II	*INPP5B*	cIns(1:2,4,5)P_3, Ins(1,3,4,5)P_4, Ins(1,4,5)P_3, PI(3,4,5)P_3, PI(4,5)P_2		*Inpp5b*$^{-/-}$ mouse—males testicular degeneration leading to infertility (Hellsten et al. 2001; Jänne et al. 1998)
Inositol polyphosphate 5-phosphatase type IV (INPP5E)	Pharbin (rat), 72 kDa inositol polyphosphate 5-phosphatase	*INPP5E*	PI(3,5)P_2, PI(3,4,5)P_3, PI(4,5)P_2	Mutations in *INPP5E* occur in ciliopathy syndromes Joubert and MORM (Bielas et al. 2009; Jacoby et al. 2009; Poretti et al. 2009)	*Inpp5e*$^{-/-}$ mouse—embryonic to early post natal lethality. Features recapitulate ciliopathy syndrome including anencephaly, exencephaly, ossification defects, kidney cysts, polydactyly, and bilateral anophthalmos (Jacoby et al. 2009).
SH2-containing inositol phosphatase-1 (SHIP1)	SHIP, SHIP-1	*INPP5D* *SHIP1*	Ins(1,3,4,5)P_4, PI(3,4,5)P_3	Mutations in *SHIP1* are detected in human acute myeloid leukaemia (AML), T cell acute lymphoblastic leukaemia (T-ALL) and acute lymphoblastic leukaemia (ALL) (Lo et al. 2009; Luo et al. 2004), and chronic myeloid leukaemia (CML) (Jiang et al. 2003; Satre et al. 1999)	*Ship1*$^{-/-}$ mouse—myeloid cell infiltration of the lung, haematopoietic perturbations and shortened life span (Helgason et al. 1998). bPten/Ship1$^{-/-}$ (B cell-specific double knockout) mouse—Spontaneous B cell lymphoma (Miletic et al. 2010). T cell-specific *Ship1*$^{-/-}$ mouse—$CD4^+$ skewing towardss a Th1 phenotype and altered cytokine production, and increased cytotoxic activity of $CD8^+$ T cells (Tarasenko et al. 2007)
SH2-containing inositol phosphatase-1 (SHIP2)	SHIP-2, 51C protein	*INPPL1* *SHIP2*	Ins(1,2,3,4,5)P_5, Ins(1,3,4,5)P_4, Ins(1,4,5,6)P_4, PI(3,5)P_2, PI(3,4,5)P_3, PI(4,5)P_2	Mutations in *SHIP2* which confer susceptibility to diabetes and the metabolic syndrome have been identified in population studies (Ishida et al. 2006; Kagawa et al. 2005; Kaisaki et al. 2004; Marion et al. 2002)	*Ship2*$^{-/-}$ mouse—obesity resistance and enhanced insulin sensitivity on a high fat diet (Sleeman et al. 2005). *Ship2*-AS rat (antisense oligonucleotide knockdown)—enhanced glucose tolerance on high fat diet (Buettner et al. 2007). Ship2-Tg (transgenic overexpression) mouse—mild body weight gain and elevated serum glucose (Kagawa et al. 2008)

(continued)

Table 12.1 (continued)

Protein name	Alias(es)	Gene name(s)	Substrate(s)	Human disease	Animal model
Skeletal muscle and kidney enriched inositol phosphatase (SKIP)		*SKIP* *INPP5K*	Ins(1,3,4,5)P_4, Ins(1,4,5)P_3, PI(3,4,5)P_3, PI(4,5)P_2		*Skip*$^{-/-}$ mouse—embryonically lethal at E10.5 (Ijuin et al. 2008). *Skip*$^{+/-}$ mouse—viable, insulin sensitivity and increased glucose tolerance (Ijuin et al. 2008). *Skip*-Tg (transgenic overexpression) mouse—reduced plasma osmolality and impaired water excretion in the kidney (Pernot et al. 2011)
Proline-rich inositol polyphosphate 5-phosphatase (PIPP)	Phosphatidylinositol (4,5) bisphosphate 5-phosphatase A	*PIB5PA* *INPP5* *INPP*	Ins(1,3,4,5)P_4, Ins(1,4,5)P_3, PI(3,4,5)P_3, PI(4,5)P_2		Not reported
Synaptojanin 1	SJ1, SYNJ1	*SYNJ1* *INPP5G*	Ins(1,3,4,5)P_4, Ins(1,4,5)P_3, PI(3)P, PI(3,5)P_2, PI(3,4,5)P_3, PI(4)P		*Synj1*$^{-/-}$ mouse—85 % die within 24 h of birth, with the remaining surviving up to 15 days after birth. Neurons exhibit an accumulation of clathrin-coated vesicles in nerve endings (Cremona et al. 1999)
Synaptojanin 2	SYNJ2	*SYNJ2* *INPP5H*	Ins(1,3,4,5)P_4, Ins(1,4,5)P_3, PI(3)P, PI(3,5)P_2, PI(3,4,5)P_3, PI(4)P		*Mozart* mouse (ENU-generated mutation)—hearing loss and severe deafness at 12 weeks of age (Manji et al. 2011)
SAC phosphatases					
SAC1		*SAC1*	PI(3)P, PI(4)P, PI(3,5)P_2		*Sac1*$^{-/-}$ mouse—preimplantation lethality (Liu et al. 2008)
SAC2	INPP5F	*INPP5F*	PI(4,5)P_2, PI(3,4,5)P_3		*Inpp5f*$^{-/-}$ mouse—increased stress-induced cardiac hypertrophy *Inpp5f*-Tg (transgenic overexpression) mouse—resistant to stress-induced cardiac hypertrophy (Zhu et al. 2009)
SAC3	FIG4	*FIG4*	PI(3,5)P_2, PI(4,5)P_2, PI(3,4,5)P_3	Mutations in *SAC3* are linked to Charcot-Marie-tooth (CMT) disease type 4J (Chow et al. 2007)	*Fig4*$^{-/-}$ "pale tremor" mouse—abnormal gait, severe tremor, neurodegeneration, juvenile lethality (Chow et al. 2007; Ferguson et al. 2009; Zhang et al. 2008)

37 % identical at the amino acid level, they share a highly conserved CX_5R catalytic phosphatase motif and catalytic domain (Norris et al. 1997a), with a similar substrate repertoire, hydrolysing the soluble inositol species $Ins(1,3,4)P_3$ and $Ins(3,4)P_2$, but preferentially hydrolysing membrane-bound $PI(3,4)P_2$ to form PI(3)P (Bansal et al. 1987, 1990; Norris et al. 1997b; Norris and Majerus 1994). Both enzymes contain N-terminal C2 domains which mediate interactions with lipid membranes and phosphoinositides (Ferron and Vacher 2006; Shearn and Norris 2007). Alternative splicing at the C-terminus generates α and β isoforms, with β forms lacking enzyme activity resulting from an additional C-terminal hydrophobic region (Norris et al. 1997a). Despite these similarities, INPP4A and INPP4B exhibit distinct cellular functions, tissue distributions and disease associations.

INPP4A

The human *INPP4A* gene is located on chromosome 2q11.2 (Joseph et al. 1999) and encodes inositol polyphosphate 4-phosphatase I, which was initially purified from rat brain extract (Norris et al. 1995). INPP4A undergoes alternative splicing of the 107 kDa α isoform to generate a 110 kDa isoform that contains an additional internal 40 amino acids encoding a PEST sequence. The α isoforms have different tissue distributions, with the 107 kDa form expressed almost exclusively in human brain, and the 110 kDa isoform predominantly expressed in human platelets. Expression of both α isoforms is evident in mouse heart, lung, uterus and spleen (Shearn et al. 2001).

INPP4A localises to the plasma membrane and early endosomes mediated by its C2 domain. INPP4A forms a complex with the 5-phosphatase, INPP5B and Rab5, to contribute to the production of PI(3)P from $PI(3,4)P_2$ on forming early endosomes (Ivetac et al. 2005; Shin et al. 2005). INPP4A generated PI(3)P promotes transferrin endocytosis (Shin et al. 2005). INPP4A negatively regulates cell proliferation in mouse primary megakaryocytes downstream of GATA-1 expression (Vyas et al. 2000), and in HeLa cells increases cell survival (MacKeigan et al. 2005). Loss of INPP4A expression leads to $PI(3,4)P_2$ accumulation in fibroblasts, enhanced Akt activation and cellular transformation. In addition, SV40 transformed *Inpp4a*$^{-/-}$ mouse embryonic fibroblasts (MEFs) form tumours in xenografted mouse models (Ivetac et al. 2009). INPP4A is a common site of viral insertion for murine leukaemia virus (Erkeland et al. 2004), and *INPP4A* gene expression is increased in cell culture models of B cell chronic lymphocytic leukaemia (CLL) (Edelmann et al. 2008). Furthermore *INPP4A* is upregulated in metastatic human prostate cancer (LaTulippe et al. 2002) and undergoes chromosomal translocations which result in production of a chimeric protein, INPP4A-HJURP, in metastatic cancer cell lines (Maher et al. 2009). However, whether INPP4A is functionally implicated in human cancers remains to be shown.

The in vivo function of INPP4A has been revealed by *Inpp4a*$^{-/-}$ mice, and the characterisation of a spontaneously occurring INPP4A mouse mutant designated 'Weeble' (*Inpp4a*^{wbl}). *Inpp4a*^{wbl} arises due to a single nucleotide deletion in exon 10 of *Inpp4a* which is predicted to be nonsense (Nystuen et al. 2001). *Inpp4a*^{wbl} mice exhibit a recessive but severe neurological phenotype resulting in death 2–3 weeks after birth. *Inpp4a*^{wbl} mice display severe locomotor instability (ataxia) and neuronal loss in the hippocampus and cerebellum as a consequence of degeneration of hippocampal pyramidal cells and cerebellar Purkinje cells (Nystuen et al. 2001). Targeted deletion of *Inpp4a* in mice results in a similar phenotype. *Inpp4a*$^{-/-}$ mice die by 4 weeks of age, are unable to walk and exhibit an involuntary movement disorder resembling Huntington's disease. Additionally, in contrast to the *Inpp4a*^{wbl} mouse, neurodegeneration in *Inpp4a*$^{-/-}$ mice is observed specifically in the striatum, with increased apoptosis of medium-sized spiny projection neurons (Sasaki et al. 2010).

The neurodegeneration in both *Inpp4a*$^{-/-}$ and *Inpp4a*^{wbl} mice may be a consequence of glutamate excitotoxicity. *Inpp4a*^{wbl} mice display neurodegeneration in specific Purkinje cell subsets dependent on the expression of specific glutamate receptor types (Sachs et al. 2009). In addition, *Inpp4a*$^{-/-}$ medium-sized spiny projection neurons display degeneration, characteristic of excitoxicity due to glutamate receptor hyperactivity. Furthermore, *Inpp4a*$^{-/-}$ mice exhibit PI(3,4)P_2 and glutamate receptor accumulation in the post synaptic density of neuronal cells (Sasaki et al. 2010). INPP4A loss is proposed to disrupt glutamate receptor internalisation leading to excitotoxicity in neuronal cells. Indeed, INPP4A regulates transferrin endocytosis in cell based systems (Shin et al. 2005). However, whether this mechanism is applicable to neuronal glutamate receptor internalisation remains to be explicitly demonstrated. Interestingly, the chromosomal locus 2q11.2 where *INPP4A* is located is frequently deleted in human neurological disorders such as schizophrenia and DiGeorge/velofacialcardio syndrome (Karayiorgou et al. 2010; Ou et al. 2008), and undergoes chromosomal translocation in FOXG1 syndrome, a congenital disorder related to Rett syndrome arising from mutations and chromosomal translocations involving the *FOXG1* gene (Kortüm et al. 2011). However, INPP4A is yet to be directly implicated in human neurological diseases.

INPP4A has been identified as a candidate gene in asthma pathogenesis. Microarray studies using platelet samples from human atopic asthmatic patients have identified *INPP4A* polymorphisms, which are likely to decrease INPP4A protein function (Sharma et al. 2008). Mouse models in which airway allergic inflammation is induced display a downregulation of *Inpp4a* expression (Agrawal et al. 2009). INPP4A associates with the p85 subunit of phosphoinositide-3 kinase (PI3K) in platelets (Munday et al. 1999). INPP4A is rapidly cleaved by calpain upon thrombin-mediated activation of human platelets and this may lead to PI(3,4)P_2 accumulation, which is required for platelet aggregation (Norris et al. 1997b). Emerging evidence suggests that platelets derived from *Inpp4a*^{wbl} mice may have an increased propensity to aggregate (Marjanovic et al. 2011) although this confers aberrant allergic responses in asthma remains to be shown.

INPP4B

The human *INPP4B* gene is located on chromosome 4q31.21 and encodes inositol polyphosphate 4-phosphatase II. INPP4B is more highly conserved across species than INPP4A, and is widely expressed in many human tissues, with highest expression in skeletal muscle and heart, and relatively low expression in the brain (Norris et al. 1997a). INPP4B also undergoes additional splicing of the α isoform to generate a shorter protein designated αs (Ferron and Vacher 2006).

Amplified PI3K signalling is a feature of many human cancers. Increased PI3K signalling leads to elevated $PI(3,4,5)_3$ signals, which initiates downstream signalling, including the phosphorylation and activation of Akt, to promote cancer cell proliferation and survival (Vara et al. 2004). Many recent studies support the contention that INPP4B is a tumour suppressor. The transgenic mouse spi-1/PU.1 develops erythroleukaemia characterised by erythropoietin (Epo) independent growth of proerythroblasts. *Inpp4b* expression is lost in the transformation of benign to malignant proerythroblast cells, and reintroduction of *Inpp4b* expression decreases Akt activation and cell transformation (Barnache et al. 2006). shRNA-mediated knockdown of INPP4B in human mammary epithelial cells enhances insulin-mediated Akt phosphorylation (Gewinner et al. 2009), and increases basal Akt phosphorylation, associated with increased cell proliferation and anchorage-independent growth of breast cancer cell lines (Fedele et al. 2010). Significantly, reconstitution of INPP4B expression in breast cancer cell lines which lack INPP4B expression, reduces Akt phosphorylation and anchorage-independent growth (Fedele et al. 2010). Furthermore, siRNA-mediated silencing of INPP4B induces cellular transformation in human mammary epithelial cells (Westbrook et al. 2005), and increases tumour formation in xenograft mouse models (Fedele et al. 2010). Significantly, overexpression of INPP4B in breast cancer cell lines reduces tumour formation in xenografted mouse models (Gewinner et al. 2009).

Loss of INPP4B expression has been recently defined in several human cancers. Loss of heterozygosity (LOH) at 4q27-31, which includes the *INPP4B* gene, is frequently observed in sporadic human primary breast tumours (Naylor et al. 2005). INPP4B is normally expressed in human mammary ductal luminal epithelial cells and a subset of lobular cells. Specifically, INPP4B is expressed in non-proliferative oestrogen receptor positive (ER^+) cells but not oestrogen receptor negative (ER^-) cells in normal breast tissue, and may function to inhibit proliferation in these cells (Fedele et al. 2010). INPP4B expression positively correlates with ER expression in human breast cancer cell lines and human primary breast tumours (Fedele et al. 2010; West et al. 2001), and INPP4B expression is frequently lost in ER^- cell lines and tumours.

The basal-like subtype of breast tumours is associated with increased metastatic potential and poor prognosis (Sørlie et al. 2003). Basal-like human breast cancers typically display loss of ER, progesterone receptor (PR) and v-erb-b2 erythroblastic leukaemia viral oncogene homologue 2 (HER-2), but express the epidermal growth factor receptor (EGFR) and/or cytokeratin 5/6 (CK5/6), and can be classified based on these molecular markers (Cheang et al. 2008). Reduced INPP4B protein expression occurs frequently in basal-like tumours, and rarely in the less aggressive hormone receptor positive luminal A and B subtypes (Fedele et al. 2010). INPP4B protein expression is lost in cancers with expression of basal marker CK5/6, and ER^- and PR^- status in both primary tumours and invasive ductal carcinoma (Fedele et al. 2010). *INPP4B* LOH occurs frequently in sporadic basal-like breast cancers which are triple negative (ER^-, PR^- and HER-2^-) and in *BRCA1* germline mutation tumours (Gewinner et al. 2009). *INPP4B* LOH correlates with reduced breast cancer patient survival (Gewinner et al. 2009).

The basis of the relationship between hormone receptor status and INPP4B expression levels still remains to be elucidated. Oestradiol and progesterone stimulation of MCF-7 cells does not modulate INPP4B protein expression (Fedele et al. 2010) and INPP4B gene expression is not induced in response to hormone treatment in breast cancer cell lines (Agoulnik et al. 2011). However, given the strong correlation between aggressive hormone receptor negative cancers and decreased INPP4B expression, INPP4B expression may be a useful prognostic marker to predict patient outcomes. Additionally, INPP4B loss correlates with high tumour grade, increased tumour size and proliferative potential (Fedele et al. 2010).

INPP4B LOH occurs frequently in ovarian cancer and melanoma, and *INPP4B* LOH in ovarian cancer correlates with increased lymph node metastases and decreased patient survival (Gewinner et al. 2009). Additionally, the tumour suppressor phosphatase and tensin homolog (PTEN) is frequently lost in many sporadic human cancers. Germline mutations in PTEN predispose individuals to developing a range of cancers (Chalhoub and Baker 2009; Hollander et al. 2011). INPP4B loss occurs in 49 % of human breast cancer samples with PTEN loss (Fedele et al. 2010). It is not understood how combined loss of PTEN and INPP4B cooperate in tumorigenesis, although p53 loss may be an additional event which promotes tumorigenesis in these cancers, since p53 is frequently lost in ovarian cancers with loss of PTEN and INPP4B (Gewinner et al. 2009).

INPP4B is also implicated in human prostate cancer. INPP4B is normally expressed in prostate epithelium, but is frequently lost in human primary prostate cancer (Hodgson et al. 2011), and metastatic prostate cancer (Taylor et al. 2010). Significantly, INPP4B loss correlates with reduced recurrence-free survival for prostate cancer patients (Hodgson et al. 2011). Interestingly, INPP4B expression is induced downstream of the transcriptional coactivator, nuclear corepressor (NCoR), in response to androgen receptor (AR) signalling in prostate cell lines (Hodgson et al. 2011), and may act to negatively modulate AR signalling (Agoulnik et al. 2011). However, it is not yet established if INPP4B expression is lost in aggressive AR independent tumours and how INPP4B loss contributes to AR independent cancer growth (Hodgson et al. 2011).

Additionally, a recent report has implicated INPP4B in osteoporosis. INPP4Bα is specifically expressed in osteoclasts (bone resorbing cells), and is upregulated during osteoclast differentiation (Ferron et al. 2011). Overexpression of INPP4B in the monocytic cell line RAW 264.7 decreases osteoclast differentiation, reduces bone resorption and increases apoptosis in vitro. INPP4B predominantly hydrolyses Ins(1,3,4)P_3 in osteoclasts, and overexpression of INPP4B in RAW 264.7

cells decreases intracellular calcium oscillations, correlating with decreased NFATc1 nuclear translocation and gene transcription. INPP4B expression is reduced in the bone tissue and osteoclasts of the osteoporotic mouse *gl/gl*. *Inpp4b*$^{-/-}$ mice display reduced bone density and volume at 8 weeks of age, along with decreased osteoclast number and size. Primary osteoclasts from *Inpp4b*$^{-/-}$ mice show increased intracellular calcium oscillations, as well as increased NFATc1 nuclear translocation and gene transcription. Interestingly, several single nucleotide polymorphisms (SNPs) occurring in the human *INPP4B* gene have been identified in healthy pre-menopausal subjects, which are associated with bone mineral density variability (Ferron et al. 2011). However, the role of INPP4B in human osteoclast function and osteoporosis remains to be fully characterised.

PI(4,5)P$_2$ 4-Phosphatases

Bacterial PI(4,5)P$_2$ 4-Phosphatases

Shigella flexneri (S. flexneri) is a bacterial pathogen which invades the colonic epithelium in humans to cause bacillary dysentery (Labrec et al. 1964), a severe form of diarrhoea also referred to as shigellosis. *Shigella* species reside and propagate in the cytoplasm of intestinal epithelial cells where following internalisation, bacterial cells escape from endosomes via lysis (Clerc et al. 1987). *Shigella* virulence is mediated by a plasmid which encodes the machinery for bacterial entry and pathogenesis in mammalian host cells (Sansonetti et al. 1982). *Shigella* virulence induces morphological changes in host cells to facilitate bacterial entry and invasion of neighbouring cells, and induces cell survival pathways to allow sustained propagation of bacteria (Ibarra and Steele-Mortimer 2009).

IpgD is one of several genes in the *Shigella* virulence plasmid and encodes a 60 kDa IpgD protein, secreted by the type III secretion system (Allaoui et al. 1993; Niebuhr et al. 2000). IpgD contains the highly conserved CX_5R catalytic motif (Norris et al. 1998), and acts as a potent PI(4,5)P$_2$ 4-phosphatase, with lesser activity towards other phosphoinositide and soluble inositol species (Niebuhr et al. 2002). IpgD hydrolysis of PI(4,5)P$_2$ leads to increased PI(5)P. Accumulation of PI(5)P is instrumental in *Shigella* pathogenesis. IpgD-mediated accumulation of PI(5)P induces Rac and Cdc42 dependent remodelling of the actin cytoskeleton to produce membrane blebbing, and reduced membrane adhesion of actin fibres (Niebuhr et al. 2002). Akt is recruited and activated at sites of PI(5)P production, which accumulates at *S. flexneri* entry points (Pendaries et al. 2006). Interestingly, *Shigella* infects T lymphocytes, and IpgD impairs T cell polarisation and chemotaxis via depletion of PI(4,5)P$_2$ (Konradt et al. 2011). IpgD also inhibits T cell receptor (TCR) activation and signalling via accumulation of PI(5)P (Guittard et al. 2010). However, the significance of *Shigella* infection of T cells in vivo is only emerging.

In addition to IpgD, other bacterial homologues have been identified that exhibit phosphoinositide phosphatase activity. SopB and SigD are highly homologous bacterial phosphatases required for virulence of *Salmonella Dublin* (Norris et al. 1998) and *Salmonella typhimurium* (Marcus et al. 2001), respectively. These *Salmonella* strains are also pathogenic in humans and cause gastroenteritis and severe diarrhoea. However, unlike *Shigella*, *Salmonella* species reside and propagate in intracellular vacuolar structures termed *Salmonella* containing vacuoles (SCVs), reviewed in (Knodler and Steele-Mortimer 2003).

SopB and SigD are encoded on virulence plasmids and are delivered into host cells via type III secretion systems. SopB and SigD both contain a region of homology with the synaptojanin 1 (SYNJ1) 5-phosphatase domain (Marcus et al. 2001) and share the highly conserved CX_5R catalytic motif (Norris et al. 1998). SopB and SigD exhibit broad substrate specificity towards soluble inositol species and phosphoinositides in vitro (Marcus et al. 2001; Norris et al. 1998) although $PI(4,5)P_2$ is the preferred substrate in cellular assays (Terebiznik et al. 2002).

SigD hydrolyses $PI(4,5)P_2$ at bacterial entry sites upon bacterial invasion to promote membrane fission in the formation of SCVs (Terebiznik et al. 2002). An interaction in the N-terminus of SopB with Cdc42 promotes the localisation of SopB to SCVs, and is required for efficient bacterial replication (Rodríguez-Escudero et al. 2011). SopB maintains PI(3)P pools on the outside of SCVs via depletion of $PI(4,5)P_2$, leading to Rab5 and Vps32 recruitment to SCVs to produce PI(3)P (Hernandez et al. 2004; Mallo et al. 2008). In addition, SopB is implicated in actin reorganisation of host cells, via IP_5 hydrolysis to induce Cdc42 and Rac-mediated reorganisation of the actin cytoskeleton (Zhou et al. 2001). SigD also promotes membrane translocation and activation of Akt (Marcus et al. 2001; Steele-Mortimer et al. 2000), however, Akt translocation to membrane ruffles is not dependent on SigD phosphatase activity (Steele-Mortimer et al. 2000). In addition, SopB-mediated Akt membrane translocation and activation is insensitive to the PI3K inhibitor, wortmannin, but requires the activity of Rictor and phosphoinositide-dependent kinase-1 (PDK1) (Cooper et al. 2011). However, the exact mechanisms by which SopB/SigD promotes Akt membrane translocation and activation remain to be fully elucidated.

TMEM55A and TMEM55B

The mammalian $PI(4,5)P_2$ 4-phosphatases were initially identified via screening the human genome for mammalian homologues of bacterial $PI(4,5)P_2$ 4-phosphatases. The mammalian $PI(4,5)P_2$ 4-phosphatase I and II are encoded by the *TMEM55B* and *TMEM55A* genes positioned at 14q11.2 and 8q21.2, respectively. TMEM55A and TMEM55B share 51 % amino acid identity, and contain the conserved CX_5R catalytic motif, however, show no homology to the $PI(3,4)P_2$ 4-phosphatases INPP4A and INPP4B outside of this motif.

TMEM55A and TMEM55B are both widely expressed in human tissues, with highest expression in the brain, heart and liver (Ungewickell et al. 2005).

Recombinant TMEM55A and TMEM55B enzymes hydrolyse PI(4,5)P_2, producing PI(5)P in vitro, and when expressed in cells localise to late endosomes and lysosomal membrane compartments (Ungewickell et al. 2005). The cellular functions of TMEM55A and TMEM55B have not been extensively investigated. TMEM55A/TMEM55B overexpression in EGF stimulated HeLa cells leads to the accumulation of PI(5)P, correlating with EGFR degradation (Ungewickell et al. 2005). Additionally, genotoxic treatment of HeLa cells causes redistribution of TMEM55B into the nucleus to generate nuclear pools of PI(5)P, which enhances p53 stability and increases stress-induced p53-dependent apoptosis (Zou et al. 2007). As yet there are no reports of TMEM55A/TMEM55B animal knockouts, or associations with human disease.

Inositol Polyphosphate 5-Phosphatases

There are 10 mammalian inositol polyphosphate 5-phosphatases. These enzymes are Mg^{2+} dependent and catalyse the hydrolysis of the D-5 position phosphate from both soluble inositol polyphosphates as well as membrane-bound phosphoinositides PI(3,5)P_2, PI(4,5)P_2 and PI(3,4,5)P_3. The 5-phosphatases contain a conserved 300 amino acid catalytic domain which folds similarly to the apurinic/apyramidic endonuclease family of enzymes (Tsujishita et al. 2001; Whisstock et al. 2000). The synaptojanins contain an additional catalytic suppressor of actin (SAC) domain containing a CX_5R motif, conferring catalytic activity to hydrolyse PI(3,4)P_2, PI(3)P and PI(4)P to PI (Guo et al. 1999). 5-phosphatases were initially classified based on in vitro substrate specificity, however, extensive characterisation of 5-phosphatases in vivo has revealed that the previous classifications do not hold true for all enzymes.

Many of the 5-phosphatases contain additional domains which mediate subcellular localisation and protein or substrate interactions. Although several 5-phosphatase enzymes have been subject to extensive analysis, others are far less characterised. The in vivo function of most 5-phosphatases has been explored by the generation of knockout mouse models, and furthermore, several human diseases have been directly linked to 5-phosphatase dysfunction. Each of the 10 mammalian 5-phosphatase enzymes will be discussed in the following sections.

INPP5A

Inositol polyphosphate 5-phosphatase type I is a 43 kDa enzyme encoded by the *INPP5A* gene located on human chromosome 10q26.3. INPP5A is the smallest 5-phosphatase and contains a central 5-phosphatase domain and a C-terminal CAAX motif (Laxminarayan et al. 1994). INPP5A hydrolyses only the soluble inositol species Ins(1,3,4,5)P_4 and Ins(1,4,5)P_3 forming Ins(1,3,4)P_3 and Ins(1,4)P_2, respectively (De Smedt et al. 1994; Laxminarayan et al. 1993; Speed and Mitchell 2000). INPP5A association with 14-3-3ζ increases its hydrolysis of Ins(1,4,5)P_3 (Campbell et al. 1997).

There are few studies which have identified the cellular functions of INPP5A, and there are no reports of an INPP5A knockout mouse. Antisense-mediated depletion of INPP5A increases Ins(1,4,5)P_3 and intracellular Ca^{2+} levels in stimulated and unstimulated cells (Speed et al. 1999), although the functional role INPP5A plays in calcium signalling in vivo remains to be determined. However, reduced INPP5A expression in normal rat kidney cells leads to cellular transformation and tumour formation in mouse xenograft models (Speed et al. 1996). Additionally, chromosomal deletions at 10q26.3 frequently occur in human brain cancers (Fults and Pedone 1993; Lee et al. 2001), and human cutaneous squamous cell carcinoma (SCC) tissue shows reduced INPP5A expression (Sekulic et al. 2010). The role INPP5A plays in human cancer is only emerging.

OCRL/INPP5F

OCRL (also known as *INPP5F*) is located on human chromosome Xq25, and encodes the occulocerebrorenal syndrome of Lowe protein (OCRL), also known as the Lowe's protein. OCRL is a 105 kDa 5-phosphatase with catalytic activity towards Ins(1,4,5)P_3, Ins(1,3,4,5)P_4, PI(3,4,5)P_3, PI(3,5)P_2 and PI(4,5)P_2, however, PI(4,5)P_2 is the preferred substrate in vivo (Olivos-Glander et al. 1995; Schmid et al. 2004; Zhang et al. 1995). OCRL shares 45 % amino acid identity and a similar domain structure with INPP5B. OCRL contains the conserved 5-phosphatase domain, an abnormal spindle-like microcephaly-associated protein (ASPM)/spindle pole body 2 (SPD2)/hydrin (ASH) domain, Rho-GTPase activating protein (Rho-GAP) domain and pleckstrin homology (PH) domain, all of which mediate multiple interactions with trafficking effectors and adaptor proteins (Hou et al. 2011; Mao et al. 2009). OCRL is widely expressed with high expression in the kidney and brain (Jänne et al. 1998; Olivos-Glander et al. 1995).

OCRL was first identified as the gene mutated in Lowe's syndrome (also known as oculocerebrorenal syndrome of Lowe) (Attree et al. 1992), an X-linked disorder characterised by renal Fanconi syndrome, congenital cataracts, mental retardation, maladaptive behaviour, rickets, hypotonia and areflexia (Kenworthy et al. 1993; Lowe et al. 1952). Disease causing mutations have been reported throughout the *OCRL* coding region (Addis et al. 2007; Peverall et al. 2000), and include nonsense mutations clustered in the 3' region (Attree et al. 1992; Leahey et al. 1993), and missense mutations which commonly occur in the 5-phosphatase, ASH, Rho-GAP and Rab binding domains, which disrupt catalytic activity, protein–protein interactions and/or protein stability (Addis et al. 2004; Erdmann et al. 2007; Hou et al. 2011; Kawano et al. 1998; Kubota et al. 1998; Lin et al. 1998; Pirruccello et al. 2011; Satre et al. 1999; Swan et al. 2010; Zhang et al. 1998). Recently, mutations in *OCRL* have been reported in the related rare X-linked genetic disorder, Dent 2 Disease (Hoopes et al. 2005). Dent 2 disease patients exhibit renal Fanconi syndrome including low molecular weight proteinuria, hypercalciuria and nephrocalcinosis and extrarenal manifestations such as mild mental impairment and rickets (Hoopes et al. 2005). Missense mutations in Dent 2 disease cluster in the *OCRL* 5-phosphatase domain and impair catalytic activity (Shrimpton et al. 2009). Nonsense and frameshift mutations are also reported in the 5' region of *OCRL* (Hoopes et al. 2005; Shrimpton et al. 2009). It is not clear how mutations in *OCRL* result in either Lowe's or comparatively mild Dent 2 disease, although the nature and position of *OCRL* mutations in Lowe's compared to Dent 2 disease indicates that differences in the nature of mutations may be instrumental. Dent 2 mutations generally result in the expression of a truncated protein which retains some catalytic activity (Shrimpton et al. 2009), whereas Lowe's mutations are frequently characterised by loss of protein expression and/or catalytic activity. Furthermore, the contribution from additional modifier genes may also contribute to disease severity (Bökenkamp et al. 2009).

Regardless of the *OCRL* mutation, Lowe's syndrome patient fibroblasts exhibit elevated PI(4,5)P_2 (Suchy et al. 1995; Zhang et al. 1995, 1998), and it is proposed that Lowe's and Dent 2 disease features arise from disruption of cellular processes via a PI(4,5)P_2 imbalance (Zhang et al. 1995, 1998). Lowe's syndrome fibroblasts exhibit abnormal actin cysoskeletal organisation including enhanced sensitivity to depolymerising agents (Suchy and Nussbaum 2002). In addition, the OCRL Rho-GAP domain, although generally classified as an inactive GTPase activating protein (GAP) domain, binds Rac and exhibits low level RacGAP activity (Faucherre et al. 2003). OCRL regulates lamellipodia formation and PI(4,5)P_2 at lamellipodia, in a Rac-dependent manner (Faucherre et al. 2003, 2005).

Several recent studies have demonstrated OCRL regulates cytokinesis, which may contribute to Lowe's syndrome pathogenesis. Depletion of dOCRL in Drosophila cells leads to increased and mislocalised PI(4,5)P_2 resulting in abscission defects during cytokinesis (Ben El Kadhi et al. 2011) and this defect is also evident in Lowe's affected renal cells (Dambournet et al. 2011). The function of OCRL in abscission occurs via an interaction with Rab35 at the intracellular bridge in the late stages of cytokinesis (Dambournet et al. 2011), although specifically how this defect affects Lowe's pathophysiology is not yet established.

OCRL binds to multiple regulators of vesicular trafficking localising the phosphatase to the *trans*-Golgi network (TGN), early endosomes and endocytic clathrin-coated pits (Dressman et al. 2000; Erdmann et al. 2007; Olivos-Glander et al. 1995; Suchy et al. 1995; Ungewickell et al. 2004; Zhang et al. 1998). OCRL interacts with several Rab GTPases including Rab1, Rab5, Rab6, Rab8, Rab14, Rab31 and Rab35 (Dambournet et al. 2011; Hou et al. 2011; Rodriguez-Gabin et al. 2010). Rabs are restricted to and define particular endomembrane compartments, and specifically Rab1, Rab6 and Rab31 target OCRL to the TGN (Hyvola et al. 2006; Rodriguez-Gabin et al. 2010). Interestingly these interactions stimulate 5-phosphatase activity. Rab binding defective OCRL disease mutations abrogate OCRL recruitment to the TGN or endosomes (Hou et al. 2011). OCRL therefore may function on endomembrane domains to restrict PI(4,5)P_2

accumulation, produce PI(4)P and thereby define membrane identity and regulate the recruitment of specific trafficking adaptors (Hyvola et al. 2006).

OCRL interacts with clathrin via its Rho-GAP and PH domains (Choudhury et al. 2005; Mao et al. 2009; Ungewickell et al. 2004) and assembles clathrin coats in vitro (Choudhury et al. 2005). OCRL regulates a range of clathrin-mediated trafficking processes. OCRL binds AP-2 and co-localises with transferrin and the cation-independent mannose-6-phosphate receptor (CI-MPR) on clathrin-coated vesicles (Ungewickell et al. 2004). Significantly, OCRL modulates transferrin receptor endocytosis via interactions with clathrin (Choudhury et al. 2009). Altered expression of OCRL blocks retrograde trafficking in vitro, resulting in CI-MPR and AP-1 redistribution to endosomes (Choudhury et al. 2005; Ungewickell et al. 2004). OCRL also regulates cation-dependent mannose-6-phosphate receptor (CD-MPR) TGN to endosome trafficking in a complex with Rab31 (Rodriguez-Gabin et al. 2010). The CI-MPR and CD-MPR deliver hydrolyases to the lysosome, and notably, lysosomal enzymes are elevated in Lowe's syndrome patient serum (Ungewickell and Majerus 1999). Hence, OCRL mutations may disrupt MPR recycling or endocytosis, resulting in exocytosis of lysosomal hydrolyase cargo via the default secretory pathway. Interestingly, disruption of the TGN to endosome pathway in Lowe's syndrome patient oligodendrocytes may perturb myelin synthesis and turnover, leading to the demyelination observed in the central nervous system of some affected individuals (Rodriguez-Gabin et al. 2010; Schneider et al. 2001).

Rab5 and its effector Adaptor protein containing PH domain, phosphotyrosine binding (PTB) domain and Leucine Zipper motif 1 (APPL1) and Ses1 and 2 interact with OCRL on subsets of early endosomes (Erdmann et al. 2007; Noakes et al. 2011; Swan et al. 2010). OCRL binding to APPL1 mediates an interaction with GAIP-interacting protein C terminus (GIPC), which together interact with and regulate the trafficking of multiple cell surface receptors, including megalin and TrkA (Erdmann et al. 2007). TrkA plays a critical role in neuronal signalling, and megalin is a scavenger receptor expressed on renal tubule epithelial cells that mediates the cellular uptake of proteins and solutes from the urine (Leheste et al. 1999). Similar to Lowe's syndrome and Dent 2 disease, GIPC and megalin knockout mice exhibit low molecular weight proteinuria (Leheste et al. 1999). Decreased megalin shedding is detected in Lowe's patient urine (Norden et al. 2002) which may arise as a consequence of reduced plasma membrane megalin levels on kidney tubule epithelial cells (Cui et al. 2010). Indeed, OCRL knockdown and Lowe's syndrome renal epithelial cells exhibit ectopic localisation of $PI(4,5)P_2$ on early endosomes, resulting in N-WASP-induced F-actin polymerisation (Vicinanza et al. 2011). This is proposed to block trafficking of multiple cell surface receptors, including megalin, at the early endodomes. However, the molecular mechanisms by which altered F-actin dynamics inhibits early endosomal trafficking remain unclear. The characterisation of the role of wild type and mutant OCRL play in megalin trafficking is still emerging and will provide useful insight into Dent 2 Disease and Lowe's syndrome pathogenesis.

Surprisingly, *Ocrl*$^{-/-}$ mice are viable, fertile and do not display the characteristic features of Lowe's syndrome (Jänne et al. 1998), possibly as a consequence of functional redundancy with other 5-phosphatase family members, such as INPP5B. Indeed, INPP5B and OCRL share many common binding partners, including various Rabs, APPL1, Ses1 and Ses2 (Erdmann et al. 2007; Noakes et al. 2011; Swan et al. 2010; Williams et al. 2007). Significantly, *Ocrl:Inpp5b* double knockout mice are embryonically lethal (Jänne et al. 1998). Blockade of both clathrin-dependent and independent trafficking pathways mediated by OCRL and INPP5B, respectively, may be the basis of lethality with loss of both enzymes (Erdmann et al. 2007; Williams et al. 2007), Interestingly, human *INPP5B* is unable to compensate for loss of mouse *Ocrl* and *Inpp5b*, as transgenic mice expressing human *INPP5B* which lack both mouse *Ocrl* and *Inpp5b* recapitulate the key features of Lowe's syndrome (Bothwell et al. 2011). Therefore, an intrinsic difference between human and mouse *INPP5B* in terms of 5-phosphatase expression, splicing and/or cellular function exists, resulting in the Lowe's syndrome phenotype in humans, but no phenotype in *Ocrl*$^{-/-}$ mice.

INPP5B

The *INPP5B* gene located on human chromosome 1p34 encodes INPP5B, also known as the inositol polyphosphate 5-phosphatase type II, 5-phosphatase-II or the 75 kDa 5-phosphatase. INPP5B hydrolyses the soluble inositol species Ins(1,4,5)P_3, Ins(1,3,4,5)P_4 and cIns(1:2,4,5)P_3, and membrane-bound PI(4,5)P_2 and PI(3,4,5)P_3 (Jackson et al. 1995; Jefferson and Majerus 1995; Matzaris et al. 1994, 1998; Mitchell et al. 1989; Ross et al. 1991). INPP5B contains a PH domain, 5-phosphatase catalytic domain, ASH domain, inactive Rho-GAP domain (Jefferson and Majerus 1995; Mao et al. 2009) and a C-terminal CAAX motif which contributes to its membrane localisation (Jefferson and Majerus 1995; Matzaris et al. 1994).

INPP5B is expressed during embryonic development, and is widely expressed in adults, with high levels in platelets (Hodgkin et al. 1994; Jänne et al. 1998; Speed et al. 1995; Takimoto et al. 1989). INPP5B binds to various Rabs, APPL1, Ses1 and Ses2 (Erdmann et al. 2007; Noakes et al. 2011; Swan et al. 2010; Williams et al. 2007). This indicates that similar to OCRL, INPP5B plays a role in the regulation of vesicular trafficking. However, INPP5B does not contain the clathrin or clathrin adaptor AP-2 binding sites present in OCRL (Williams et al. 2007). Significantly, INPP5B localises to components of the endomembrane network, particularly the *cis*-Golgi and the endoplasmic reticulum- (ER) to-Golgi intermediate compartment (ERGIC) (Shin et al. 2005; Williams et al. 2007). INPP5B overexpression leads to a block in Golgi-ER retrograde trafficking. However, there is evidence that INPP5B also regulates early endocytic events, where it is recruited to endosomes via ASH domain interactions with Rab5 (Shin et al. 2005; Williams et al. 2007), and cooperates in a cascade at the plasma membrane with Rab5 and INPP4A to produce PI(3)P and promote transferrin receptor endocytosis (Shin et al. 2005). In addition, INPP5B is recruited to complement receptor 3 (CR3)-induced phagosomes, where it is then rapidly displaced by the accumulation of PI(3)P, allowing $PI(3,4,5)P_3$ signals to increase and phagosome internalisation to occur (Bohdanowicz et al. 2010).

The in vivo function of INPP5B has been investigated with the generation of *Inpp5b*$^{-/-}$ mice. Male *Inpp5b*$^{-/-}$ mice display testicular degeneration and infertility, characterised by spermatocyte and spermatid defects (Hellsten et al. 2001; Jänne et al. 1998). *Inpp5b*$^{-/-}$ sperm exhibits reduced motility and oocyte adhesion leading to decreased fertilisation ability. A Disintegrin and A Metalloprotease (ADAM) 2 and ADAM3 processing during sperm maturation in the epididymis is disrupted with INPP5B loss, impeding sperm-oocyte interactions (Hellsten et al. 2001; Marcello and Evans 2010). Additionally, *Inpp5b*$^{-/-}$ mice exhibit progressive loss of spermatids and spermatocytes, resulting from sertoli cell dysfunction: *Inpp5b*$^{-/-}$ sertoli cells show enlarged actin coated endosome-like vacuoles in the cytosol which accumulate adherence junction components, resulting in disrupted adherence junction function and sloughing of cells (Hellsten et al. 2002).

There is a high degree of functional redundancy between OCRL and INPP5B, with similar domain structure, amino acid identity, common binding partners, overlap in subcellular localisation and regulation of similar trafficking events and *Inpp5b*$^{-/-}$ mice do not exhibit a Lowe's-like phenotype (Jänne et al. 1998). Interestingly, the human *INPP5B* chromosomal region, 1q34, is linked to multiple human diseases with characteristic features reminiscent of Lowe's syndrome including mental retardation, optic abnormalities and congenital cataracts (Bisgaard et al. 2007; Cormand et al. 1999; Shearman et al. 1996).

INPP5E

The *INPP5E* gene is located on human chromosome 9q34.3 and encodes INPP5E, also known as the 72 kDa 5-phosphatase, inositol polyphosphate 5-phosphtase IV or Pharbin. Unlike all other phosphoinositide 5-phosphatases, INPP5E has no activity towardss soluble inositol phosphates, but hydrolyses $PI(4,5)P_2$, $PI(3,5)P_2$ and has the highest activity towardss $PI(3,4,5)P_3$ of all 5-phosphatase family members (Kisseleva et al. 2000; Kong et al. 2000).

INPP5E is widely expressed, with high expression detected in the testis and brain (Kong et al. 2000). INPP5E exhibits a cytosolic distribution with perinuclear/TGN enrichment mediated via its N-terminal proline-rich domain (Kong et al. 2000). In quiescent cells, the CAAX motif is suggested to localise INPP5E to the primary cilium (Jacoby et al. 2009). In macrophages INPP5E is also recruited to the phagocytic cup in response to Fcγ receptor activation, where its hydrolysis of $PI(3,4,5)P_3$ regulates pseudopod extension and phagosome closure (Horan et al. 2007).

INPP5E inhibits Akt phosphorylation in response to platelet-derived growth factor (PDGF) and insulin-like growth factor (IGF-1) stimulation and sensitises cells to Fas-induced apoptosis (Kisseleva et al. 2002; Wang et al. 2011). INPP5E also regulates PI3K/Akt signalling downstream of insulin signalling in vivo:

INPP5E is recruited to the p85-PI3K subunit and insulin receptor substrates (IRS) in response to insulin stimulation degrading PI3K-generated PI(3,4,5)P_3 signals in the rat hypothalamus. siRNA-mediated depletion of INPP5E leads to hypothalamic accumulation of PI(3,4,5)P_3, associated with reduced food intake and body mass in rats (Bertelli et al. 2006). In addition, expression of INPP5E in adipocytes induces membrane translocation of the glucose transporter, GLUT4, via PI(3)P production at the plasma membrane (Kong et al. 2006), although, whether INPP5E mediates glucose homeostasis and insulin signalling in humans remains to be demonstrated.

An *Inpp5e*$^{-/-}$ mouse has recently been described that exhibits a lethal phenotype during late embryogenesis or early postnatal life (Jacoby et al. 2009). *Inpp5e*$^{-/-}$ embryos display multiple abnormalities including exencephaly, polydactyly, polycystic kidneys, bilateral anopthalmos and skeletal abnormalities. Tamoxifen induced knockout of *Inpp5e* in 4-week-old mice results in multiple defects including obesity, cystic kidneys and retinal degeneration (Jacoby et al. 2009). These features are characteristic of ciliopathy syndromes, developmental disorders resulting from defects in primary cilium structure or function (D'Angelo and Franco 2009). Primary cilia are microtubule-based organelles present on most quiescent cells in the human body that are critical for embryonic development and adult tissue homeostasis via regulation of diverse processes including signal transduction, mechanosensation, olfaction and photoreception. *Inpp5e*$^{-/-}$ MEFs display defects in cilia stability (Jacoby et al. 2009), although, how INPP5E regulates cilia to produce this profound phenotype remains to be explicitly demonstrated.

Significantly, mutations in *INPP5E* have been linked to the rare human ciliopathy syndromes Joubert and MORM (Bielas et al. 2009; Jacoby et al. 2009; Poretti et al. 2009). Joubert syndrome is characterised by midbrain–hindbrain malformation (the molar tooth sign), cognitive impairment, polydactyly, liver fibrosis, nephronophthisis and retinal dystrophy (Joubert et al. 1969). MORM syndrome affected individuals exhibit mental retardation, truncal obesity, retinal dystrophy and micropenis (Hampshire et al. 2006). All reported Joubert syndrome mutations cluster in the *INPP5E* 5-phosphatase domain and disrupt the phosphoinositide binding site, reducing INPP5E catalytic activity (Bielas et al. 2009). In contrast, MORM disease causing mutations arise from deletion of the C-terminal region of INPP5E including the CAAX motif and abrogate INPP5E cilia localisation, but do not impair catalytic activity (Jacoby et al. 2009). Both MORM and Joubert *INPP5E* mutations result in cilia instability phenotypes, indicating that INPP5E may regulate cilia phosphoinositide signalling, although, the phosphoinositide content of this organelle is yet to be defined (Bae et al. 2009; Jin et al. 2010). Indeed, INPP5E may regulate a variety of processes to maintain cilia function such as trafficking, signalling and/or cytoskeletal dynamics.

INPP5E is also implicated in human cancers, and is one of the top five genes upregulated (57-fold) in human cervical cancer specimens relative to non-cancerous tissue (Yoon et al. 2003). *INPP5E* is overexpressed 5.7-fold in leiomyosarcoma compared to the normal myometrium (Quade et al. 2004). *INPP5E* is also among the top 6 genes overexpressed in Non-Hodgkin's lymphoma following chemotherapy (Chow et al. 2006). In contrast, downregulation of *INPP5E* is detected in gastric cancer compared to normal stomach mucosa and in metastatic adenocarcinoma compared to primary tumours (Kim et al. 2003; Ramaswamy et al. 2003). Interestingly, primary cilia have recently been implicated in tumorigenesis via regulation of the cell cycle and signalling (Nigg and Raff 2009), however, the role INPP5E plays in tumorigenesis both dependent and independent of cilia regulation remains to be demonstrated.

SHIP1

The SH2-containing inositol phosphatase (SHIP) 5-phosphatases include two members, SHIP1 (also known as INPP5D) and SHIP2 (also called INPPL1) which have a similar domain structure comprising an N-terminal SH2 domain, central 5-phosphatase domain and divergent C-terminal proline-rich domains (Backers et al. 2003; Hejna et al. 1995; Liu and Dumont 1997). The SHIP1 C-terminus contains two NPXY motifs, which mediates the interaction following its phosphorylation with Shc, Dok 1, Dok 2 and four PxxP motifs that bind Grb2, Src, Lyn, Hck, Abl, PLCg1 and PIAS1, reviewed in (Hamilton et al. 2010; Rohrschneider et al. 2000). The human *SHIP1* gene is located on chromosome 2q37.1. Splicing of SHIP1 (SHIP1α) into three shorter isoforms generates SHIP1β, SHIP1δ and s-SHIP1 (Lucas and Rohrschneider 1999; Tu et al. 2001). SHIP1α and SHIP1δ are restricted to haematopoietic and spermatogenic cells (Liu et al. 1998b). s-SHIP is a stem cell-specific 104 kDa isoform, which lacks the SH2 domain that may be active in functioning mammary stem cells, the proposed precursor cells to basal-like human breast cancers (Bai and Rohrschneider 2010). s-SHIP1 may also contribute to the regulation of embryonic and haematopoietic stem cell growth and survival.

SHIP1 is a key negative regulator of PI3K/Akt signalling via hydrolysis of PI(3,4,5)P_3 in haematopoietic cells (Miletic et al. 2010) including B and T cells, dendritic cells, macrophages, mast cells, osteoclasts, platelets and neutrophils. In response to monocyte colony-stimulating factor (M-CSF), granulocyte colony-stimulating factor (G-CSF), interleukin-3 or stem cell factor (SCF) haematopoietic cells that lack SHIP1 exhibit enhanced proliferation and survival due to the activation of PI3K/Akt and mitogen activated protein kinase (MAPK) pathways, reviewed in (Hamilton et al. 2010). Many different groups have generated *Ship*$^{-/-}$ mice which exhibit a reproducible phenotype comprising splenomegaly and elevated white cell counts including macrophages and granulocytes, associated with myeloid cell infiltration of the lung, which leads to a shortened life span (Helgason et al. 1998; Liu et al. 1999). Interestingly, an ENU mutagenesis screen identified a mouse with concommittant loss of *Ship1* and *s-Ship* which leads to greater increases in myeloid cells, inflammatory markers and infiltration of the lungs by activated macrophages than that observed in *Ship1*$^{-/-}$ mice, suggesting that s-SHIP synergises with SHIP1 to suppress macrophage activation (Nguyen et al. 2011).

SHIP1 Regulation of B Cell Function

SHIP1 associates with immunoreceptor tyrosine-based inhibitory motif (ITIM) containing proteins in B cells and regulates B cell numbers and function. SHIP1 also regulates signalling downstream of the autonomous B cell receptor (BCR) (Brauweiler et al. 2000), B cell activating factor belonging to the TNF family (BAFF) receptor, FcγRIIb and the chemokine receptor CXCR4 (Brauweiler et al. 2007; Crowley et al. 2009). SHIP1 controls the size of the peripheral B cell compartment by regulating B cell proliferation (Brauweiler et al. 2000; Helgason et al. 2000; Liu et al. 1998a). Aged *Ship1*$^{-/-}$ mice exhibit a reduction in circulating B cells as a consequence of elevated IL-6 secretion by macrophages (Maeda et al. 2010). SHIP1 may also regulate B cell maturation. Mice subjected to irradiation to destroy the endogenous bone marrow and reconstituted with *Ship*$^{-/-}$ haematopoietic cells, display a reduction in both immature and mature forms of B cells (Helgason et al. 2000; Liu et al. 1998a).

MicroRNA-155 (miR-155) regulates immune cell development and function (Baltimore et al. 2008). miR-155 targets and suppresses *SHIP1* gene expression via direct 3'UTR interactions (O'Connell et al. 2009). Interestingly, increased miR-155 expression correlates with reduced SHIP1 expression in diffuse large B cell lymphoma (Pedersen et al. 2009). In addition, miR-155 transgenic mice develop B lymphoma, and leukaemic B cells display progressively reduced SHIP1 expression (Costinean et al. 2009). Retroviral delivery of a miR-155-formatted siRNA against SHIP1 induces a myeloproliferative-like syndrome reminiscent of miR-155 transgenic mice and *Ship1*$^{-/-}$ mice (O'Connell et al. 2009). miR-155 is maximally expressed in pre-B cells, and miR-155 overexpression results in the reduced expression of SHIP1 as well as CCAAT enhancer-binding protein β (C/EBPβ) which act to inhibit B cell differentiation and promote proliferation of pre-B cells (Costinean et al. 2009).

B cell-specific deletion of *Ship1* in mice does not lead to the development of lymphoma (Miletic et al. 2010). Both SHIP1 and PTEN act to negatively regulate PI3K-generated signals. Concomitant deletion of *Pten* and *Ship1* in B cells (b*Pten/Ship1*$^{-/-}$) is sufficient to promote the development of spontaneous B cell lymphomas. Significantly, b*Pten/Ship1*$^{-/-}$ mouse B cells display increased Akt phosphorylation, cell survival and an enhanced proliferative response to BAFF, which is not observed with either *Pten* or *Ship1* deletion alone (Miletic et al. 2010). Therefore, SHIP1 may act in cooperation with PTEN to suppress the development of B cell lymphoma (Miletic et al. 2010).

SHIP1 Activity in T Cells

Multiple in vitro studies suggest SHIP1 regulates TCR signalling. SHIP1 is tyrosine phosphorylated in response to CD3 or CD28 activation of T cells (Edmunds et al. 1999; Freeburn et al. 2002). Phosphorylation of SHIP1 NPXY motifs, via the tyrosine kinase Lck, increases SHIP1 5-phosphatase activity and promotes the association with PTB domains of the adaptor protein, Shc (Edmunds et al. 1999; Lamkin et al. 1997). This induces SHIP1 recruitment to the plasma membrane, placing SHIP1 in an ideal spatial and temporal localisation to negatively regulate PI3K-generated PI(3,4,5)P_3. SHIP1 deficient Jurkat leukaemic T cell lines exhibit elevated basal and CD3-stimulated PI(3,4,5)P_3 signals, associated with increased Akt phosphorylation and activity (Freeburn et al. 2002; Horn et al. 2004). Additionally in response to TCR activation SHIP1 forms a multi-protein complex with the inhibitory adaptors Dok1 and Dok2, Grb2 and LAT which regulate Akt and kinase Zap-70 activity (Dong et al. 2006). The complex is suggested to act via a negative feedback loop to regulate TCR signalling and therefore T cell tolerance.

Ship1$^{-/-}$ mice exhibit a reduced number of T lymphocytes in the circulation (Helgason et al. 1998). However, *Ship1*$^{-/-}$ mice display increased numbers of T_{reg} cells which have an increased immunosuppressive activity (Collazo et al. 2009). $CD8^+$ T lymphocytes from T cell-specific *Ship1* knockout mice exhibit increased cytotoxic activity, associated with high granzyme B enzyme activity (Tarasenko et al. 2007). The ratio of $CD4^+$:$CD8^+$ T cells is normal in the thymus and periphery of *Ship1*$^{-/-}$ mice, however, $CD4^+$ T cell proportions are increased in the spleen (Kashiwada et al. 2006). Interestingly, within the $CD4^+$ T cell subset, *Ship1*$^{-/-}$ mice are skewed towards Th2 phenotype (Kuroda et al. 2011). Hyperactivation of IL-3 and IgE signalling pathways in *Ship1*$^{-/-}$ basophils results in elevated IL-4 secretion, stimulating naïve Th cells to differentiate into Th2 cells. However, in contrast to global *Ship1*$^{-/-}$ mice, T cell-specific *Ship1* knockout mice exhibit skewing towards a Th1 phenotype, with reduced production of Th2 cytokines (Tarasenko et al. 2007) and exhibit a reduced Th2 response in vivo (Roongapinun et al. 2010). Therefore, SHIP1 regulates the intrinsic differentiation of T lymphocytes to Th2, and the pro-inflammatory cytokine environment of *Ship1*$^{-/-}$ mice may overcome signals regulated by SHIP1.

Dendritic cells (DCs) present antigen to T cells to stimulate the generation of an immune response (Guermonprez et al. 2002). *Ship1*$^{-/-}$ mice demonstrate a tolerance to allografts, and do not exhibit graft rejection (Ghansah et al. al. 2004; Wang et al. 2002). *Ship1*$^{-/-}$ mice display increased DC populations which are poorly differentiated (Antignano et al. 2010). Additionally, the elevated IL-6 levels in global *Ship1*$^{-/-}$ mice may impair the ability of dendritic cells to stimulate T cell proliferation (Neill et al. 2007). Hence, defective DC function may promote allograft tolerance in *Ship1*$^{-/-}$ mice. However, recent reports suggest *Ship1*$^{-/-}$ DCs have increased antigen presentation capacity to stimulate T cell activation, and consistent with this, *Ship1*$^{-/-}$ T cells have an impaired DC signal response (Roongapinun et al. 2010).

The Role of SHIP1 in Macrophage Functions

Macrophages undergo classical activation (M1) in response to inflammatory signals to release pro-inflammatory cytokines and effector molecules, or alternative activation (M2) in response to events such as tissue remodelling to produce anti-inflammatory cytokines and molecules (Mantovani et al. 2007). *Ship1*$^{-/-}$ mice exhibit high levels of the pro-inflammatory cytokine IL-6, which is predominantly secreted by peritoneal macrophages. Loss of SHIP1 in peritoneal macrophages, in addition to increased FcγR signalling due to elevated IgG secretion by *Ship1*$^{-/-}$ B cells, drives IL-6 production (Maeda et al. 2010). Despite elevated serum pro-inflammatory cytokines, *Ship1*$^{-/-}$ mice display decreased M1 and increased M2 populations, corresponding to a reduced inflammatory response with *Salmonella enterica* infection, and impaired infection clearance (Bishop et al. 2008; Rauh et al. 2005). However, skewing of *Ship1*$^{-/-}$ macrophages to an M2 phenotype may occur as a consequence of SHIP1 activity to inhibit M2 skewing. IL-4 promotes M2 macrophage skewing. SHIP1 expression is decreased in macrophages following IL-4 stimulation, and *Ship1*$^{-/-}$ macrophages exhibit enhanced sensitivity to IL-4 induced M2 skewing in vitro (Weisser et al. 2011). In addition, inhibition of PI3K signalling in *Ship1*$^{-/-}$ bone marrow macrophages suppresses expression of the M2 marker arginase 1 (Rauh et al. 2005).

SHIP1 directly regulates M1-associated macrophage phagocytic activity. Many 5-phosphatase family members play a role in macrophage phagocytosis including SHIP1, SHIP2, INPP5B and INPP5E. SHIP1 regulates macrophage phagocytosis by hydrolysing PI(3,4,5)P_3 at the phagocytic cup during phagosome formation to promote the extension of pseudopodia and phagosome closure (Horan et al. 2007; Kamen et al. 2007; Swanson and Hoppe 2004). Altered distribution of PI(3,4,5)P_3 during phagocytosis interferes with phagosome formation, and in vitro studies have demonstrated that reduction in SHIP1 expression inhibits phagocytosis (Horan et al. 2007). SHIP1 preferentially regulates CR3-mediated phagocytosis although displays activity in FcγR-mediated phagocytosis (Cox et al. 2001; Nakamura et al. 2002), whereas INPP5E regulates FcγR, and not CR3-mediated phagocytosis via regulation of PI(3,4,5)P_3 signals (Horan et al. 2007). The highly homologous SHIP2 also regulates FcγR-mediated phagocytosis independently of SHIP1, and is recruited to phagocytic cups, and inhibits Rac activity, thereby regulating actin dynamics during phagosome formation (Ai et al. 2006).

Francisella tularensis (*F. tularensis*) is a bacterial pathogen which causes the fatal disease tularemia in humans. Upon endocytosis, *F. tularensis* evades fusion with lysosomes and escapes into the cytosol of infected cells where it replicates and alters host cell signalling, reviewed in (Santic et al. 2010). In macrophages and monocytes, *Francisella* induces apoptosis which is associated with bacterial escape from endosomal compartments and failure to fuse with lysosomal compartments (Lai et al. 2001; Lai and Sjostedt 2003). Apoptosis induced by *Francisella* infection occurs via Fas signalling. Interestingly, Akt negatively regulates Fas expression and *Ship1*$^{-/-}$ macrophages display reduced Fas expression and increased phagosome-lysosomal fusion and bacterial clearance (Rajaram et al. 2009). Furthermore, SHIP1 is phosphorylated upon *Francisella novicida* infection and acts to negatively regulate PI3K/Akt signalling to inhibit nuclear factor-κB (NF-κB) gene transcription and pro-inflammatory cytokine production (Parsa et al. 2006). Human peripheral blood monocytes (PBMs) infected with the less virulent *F. tularensis* display increased miR-155 expression upon infection, leading to a reduction in SHIP1 expression.

However, this induction is not observed with the more virulent *F. tularensis* strain. The differential suppression of SHIP1 in host cells may therefore contribute to the pathogenicity of different *Francisella* strains (Cremer et al. 2009).

The Role of SHIP1 in Osteoclast, Mast cell, Neutrophil and Platelet Function

SHIP1 regulates the function of osteoclasts, bone marrow macrophage-derived cells which promote bone resorption. The development of osteoclasts from bone marrow macrophage (BMM) precursors requires M-CSF, which promotes their proliferation and survival, and receptor activator of nuclear factor-κB ligand (RANKL), which is required for osteoclast differentiation, reviewed in (Boyle et al. 2003; Pixley and Stanley 2004). *Ship1*$^{-/-}$ mice exhibit a severe osteoporotic phenotype, associated with a 2-fold increase in osteoclast numbers due to increased sensitivity to M-CSF, and also as a consequence of the increased bone resorptive activity of *Ship1*$^{-/-}$ osteoclasts (Takeshita et al. 2002). SHIP1 localises to the podosomes of osteoclasts to regulate bone resorption, and associates with Cas and c-Cbl (Yogo et al. 2006). *Ship1*$^{-/-}$ osteoclasts exhibit increased proliferation in response to M-CSF stimulation, without affecting differentiation or survival. SHIP1 inhibits M-CSF-mediated Akt activation and suppresses the expression of D-type cyclins, which are increased in *Ship1*$^{-/-}$ BMMs (Zhou et al. 2006). Although SHIP1 is not directly implicated in bone disease in humans, it is reported to interact with the Triggering receptor expressed on myeloid cells-2 (TREM2), and the associated adaptor protein DNAX-activating protein of 12 kD (DAP12) in osteoclasts and macrophages. TREM2 and DAP12 are mutated in Nasu-Hakola disease, a rare syndrome characterised by bone cysts and dementia. Specifically, SHIP1 inhibits TREM2/DAP12 intracellular signalling by binding to DAP12 and preventing the recruitment of PI3K to the activated receptor complex (Peng et al. 2010).

Mast cells express the high affinity IgE receptor (FcεRI), which aggregates upon IgE-antigen crosslinking. FcεRI clustering generates intracellular signalling leading to mast cell degranulation and the release of factors such as cytokines, chemokines and histamine, which contribute to allergic inflammation and anaphylaxis, reviewed in (Galli and Tsai 2010). SHIP1 directly binds FcεRI via its SH2 domain (Kimura et al. 1997; Osborne et al. 1996). The linker for activation of T cells (LAT) and the non-T cell activation linker (NTAL) mediate the organisation of FcεRI signalling complexes, with LAT promoting and NTAL inhibiting mast cell degranulation. SHIP1 also binds to LAT and is recruited to FcεRI, where it decreases Akt phosphorylation and inhibits mast cell survival (Roget et al. 2008). SHIP1 functions downstream of the FcεRI to regulate mast cell degranulation and pro-inflammatory cytokine release. *Ship1*$^{-/-}$ mast cells display degranulation in response to Steel-factor (SF, or mast cell growth factor) which is not observed in wildtype mast cells, associated with increased PI(3,4,5)P_3 signals and intracellular calcium (Huber et al. 1998b). In addition, *Ship1*$^{-/-}$ mast cells also show enhanced degranulation in response to antigen-loaded IgE and IgE alone. SHIP1 acts to restrict calcium influx via regulation of PI3K-generated signals, as PI3K inhibition inhibits calcium influx and subsequent degranulation in *Ship1*$^{-/-}$ mast cells (Huber et al. 1998a). Moreover, *Ship1*$^{-/-}$ mast cells display increased antigen-loaded IgE-induced IL-6 production downstream of NF-κB activation (Kalesnikoff et al. 2002).

Altered mast cell activity contributes to the phenotype observed in *Ship1*$^{-/-}$ mice. *Ship1*$^{-/-}$ mice exhibit mast cell hyperplasia, as well as increased serum pro-inflammatory cytokine release, and hypersensitivity to anaphylaxis (Haddon et al. 2009). *Ship1*$^{-/-}$ mice display infiltration of the lung with hyperactivated mast cells (degranulating) (Oh et al. 2007). SHIP1 binds to Allergin 1, an immunoglobulin-like receptor which opposes FcεRI signalling and suppresses mast cell degranulation and anaphylaxis (Hitomi et al. 2010). In addition, *11βHSD1*$^{-/-}$ mice which are susceptible to endotoxemia and cutaneous anaphylaxis display elevated SHIP1 expression (Zhang and Daynes 2007). LPS- and sMLA- stimulated BMMs and bone marrow mast cells (BMMCs) display TLR4/MyD88 dependent transforming growth factor β (TGFβ) production which has paracrine activity in inducing SHIP1 expression to mediate hyporesponsiveness with subsequent challenge (Cekic et al. 2011; Sly et al. 2003, 2009). Therefore, SHIP1 plays an important role in negatively regulating mast cell degranulation and allergic inflammation. Significantly, SHIP1 small molecule agonists reduce endotoxemia and cutaneous anaphylaxis in endotoxemia mouse models, via inhibition of macrophage and mast cell activation (Ong et al. 2007). Therefore, SHIP1 agonists may represent an attractive tool for potential therapeutic use in human inflammatory disease.

SHIP1 is the major phosphoinositide phosphatase that regulates neutrophil migration. *Ship1*$^{-/-}$ mice but not mice with granulocyte-specific deletion of *Pten*, exhibit granulocytic infiltration of many organs including the lung (Nishio et al. 2007). SHIP1 also regulates neutrophil polarisation and motility. In neutrophils SHIP1, but not PTEN, directs the spatial distribution of PI(3,4,5)P_3 which is required for normal neutrophil chemotaxis (Nishio et al. 2007). Bacterial peptidoglycan (PGN)-stimulated *Ship1*$^{-/-}$ neutrophils display enhanced Akt activation and pro-inflammatory cytokine release downstream of toll like receptor-2 (TLR-2) activation, and *Ship1*$^{-/-}$ mice develop acute lung injury in response to PGN (Trotta et al. 2005).

Platelet activation leading to aggregation is required for blood clot formation during normal haemostasis, reviewed in (Gratacap et al. 2008). PI(3,4,5)P_3 is rapidly produced upon platelet activation, followed by a more gradual and sustained increase of PI(3,4)P_2 correlating with platelet aggregation and integrin engagement, reviewed in (Gratacap et al. 2008). Both SHIP1 and SHIP2 are expressed in platelets, however, SHIP1 plays a more dominant role in regulating PI(3,4,5)P_3 signals in response to agonists (Giuriato et al. 2003). Upon thrombin-induced activation of platelet aggregation, SHIP1 is tyrosine phosphorylated in an integrin $\alpha_{IIb}\beta_3$-dependent mechanism, to increase PI(3,4)P_2 levels via PI(3,4,5)P_3 hydrolysis (Giuriato et al. 1997). *Ship1*$^{-/-}$ mouse platelets display increased PI(3,4,5)P_3 and oscillatory calcium influx, corresponding to enhanced adhesion and spreading on fibronectin, and an increased adhesion on fibronectin under in vitro blood flow conditions (Maxwell et al. 2004). However, platelets form *Ship1*$^{-/-}$ mice display decreased and disorganised aggregation in response to thrombin, collagen and thromboxane A_2 treatment, as well as decreased thrombus contraction. Functionally, *Ship1*$^{-/-}$ mice show an increased tail bleeding time in response to injury and exhibit smaller thrombus formation in vivo (Severin et al. 2007). Collectively, these studies reveal that SHIP1 plays a critical role in the initial thrombus formation and in clot retraction and is therefore an important regulator of haemostasis. However, the precise ways in which SHIP1 facilitates these activities remains to be fully elucidated.

SHIP1 Expression and Activity is Altered in Human Leukaemias

Alterations in SHIP1 expression or activity in solid human tumours has not been widely reported, although *Ship1*$^{-/-}$ mice demonstrate enhanced tumour growth following tumour xenograft implantation (Rauh et al. 2005). However, SHIP1 expression is lost in some human haematological cancers. SHIP1 is mutated and inactivated in T cell acute lymphoblastic leukaemia (T-ALL), as is the 3-phosphatase PTEN (Lo et al. 2009). In addition, mutation of the *SHIP1* gene has been reported in human acute myeloid leukaemia (AML) and acute lymphoblastic leukaemia (ALL) (Luo et al. 2004). Further characterisation of one somatic mutation V684E in AML revealed that this mutation occurs in the SHIP1 5-phosphatase domain, leading to reduced catalytic activity and increased Akt phosphorylation in leukaemic cells harbouring the mutation (Luo et al. 2003). Significantly, gene transfer of SHIP1 into AML patient leukaemic cells is sufficient to reduce proliferation following GM-CSF stimulation, and inhibit autonomous proliferation (Metzner et al. 2009).

The BCR/ABL fusion protein arises from chromosomal translocation, and causes chronic myeloid leukaemia (CML) (Sattler and Griffin 2003). SHIP1 protein expression is lost in human primary CML patient cells derived from the bone marrow, as well as CML cell lines expressing BCR/ABL (Sattler et al. 1999). Other studies indicate that SHIP1 expression is reduced in the chronic-blast stages of CML in some patients rather than in primitive leukaemic cells (Jiang et al. 2003), and is therefore associated with disease progression rather than initiation. Recent studies have revealed that SHIP1 is phosphorylated by BCR/ABL, leading to its ubiquitination and proteasomal degradation (Ruschmann et al. 2010).

HTLV-1 encodes the viral protein Tax, which causes human Adult T cell leukaemia/lymphoma (ATLL) (Yoshida 2001). Peripheral T cells from acute and chronic forms of ATLL also display reduced SHIP1 and PTEN expression associated with increased Akt phosphorylation (Fukuda et al. 2005). SHIP1 and PTEN expression is reduced as a result of Tax-mediated activation of NF-κB transcriptional suppression (Fukuda et al. 2009). Modulation of SHIP1 levels is also a feature of haematopoietic malignancies in some mouse models. The Friend murine leukaemia virus (F-MuLV) induces erythroleukaemia via activation of the Fli-1 transcription factor which promotes erythroblast transformation. Fli-1 binds to the

SHIP1 promoter leading to transcriptional repression of SHIP1, which contributes to malignant transformation (Lakhanpal et al. 2010). Interestingly, *Ship1*$^{-/-}$ mice demonstrate accelerated F-MuLV-induced erythroleukaemia progression (Lakhanpal et al. 2010), consistent with a tumour suppressive role of SHIP1 in leukaemia development.

SHIP2

SH2-containing inositol phosphatase 2 (SHIP2), encoded by the gene *INPPL1* on human chromosome 11q23, is a 142 kDa enzyme that contains an N-terminal SH2 domain, a central 5-phosphatase domain and a C-terminal proline-rich domain containing multiple WW, NPXY and SAM motifs, reviewed in (Backers et al. 2003; Dyson et al. 2005). SHIP2 shares 51% amino acid homology with SHIP1, although it is much more widely expressed, and is detected in the brain, heart and skeletal muscle, reviewed in (Muraille et al. 2001; Pesesse et al. 1997; Zhang et al. 2009). SHIP2 hydrolyses soluble inositol phosphate species Ins(1,2,3,4,5)P_5, Ins(1,3,4,5)P_4 and Ins(1,4,5,6)P_4 and membrane-bound PI(3,5)P_2, PI(4,5)P_2 and PI(3,4,5)P_3 in vitro, although its main substrate in vivo has not been extensively characterised (Chi et al. 2004; Taylor et al. 2000). SHIP2 is phosphorylated at tyrosine residues which include sites in the NXPY motif (Prasad et al. 2009). SHIP2 phosphorylation occurs in response to growth factor and insulin stimulation of cells (Blero et al. 2001; Habib et al. 1998) and increases its catalytic activity. Inhibition of protein tyrosine phosphatase activity using vanadate analogues increases SHIP2 phosphorylation and results in a 5–10-fold increase in its phosphatase activity in 1321N1 astrocytoma cells (Batty et al. 2007) and immunoprecipitated tyrosine phosphorylated SHIP2 displays 2.5-fold increased phosphatase activity (Prasad et al. 2009). Interestingly, it has recently been reported that SHIP2 is phosphorylated at additional sites including Ser132 which may alter SHIP2 stability and cellular localisation (Elong Edimo et al. 2011).

In quiescent cells, SHIP2 is localised to the cytoplasm, however, upon stimulation, SHIP2 translocates to the plasma membrane and extracellular matrix (ECM) adhesion points where it interacts with a number of proteins involved in actin cytoskeleton dynamics. SHIP2 binds to filamin, and adaptor protein p130^{cas} to regulate submembranous actin filament formation, cell spreading and adhesion (Dyson et al. 2001; Prasad et al. 2001). Other proteins which bind SHIP2 include vinexin (Paternotte et al. 2005), LL5β (Takabayashi et al. 2010), c-met (Koch et al. 2005) and Shc (Wisniewski et al. 1999). In addition, SHIP2 inhibits PI(3,4,5)P_3 accumulation and Rac1/cdc42 activation in PC12 neuronal cells to regulate neurite number and length (Aoki et al. 2007). SHIP2 also associates with c-Cbl and binds c-Cbl interacting protein (CAP) via the SHIP2 proline-rich domain (Vandenbroere et al. 2003). The SAM domain of SHIP2 binds the GAP Arap3 (Raaijmakers et al. 2007), and the Ephrin A2 receptor (Leone et al. 2008). SHIP2 also interacts with the bacterial translocated intimin receptor (Tir), a receptor which inserts into the membrane of epithelial cells and mediates epithelial cell adhesion via the induction of actin-rich pedestal formation. SHIP2 is recruited to Tir and interacts with Shc to induce $PI(3,4)P_2$-dependent actin remodelling (Smith et al. 2010). Recombinant SHIP2 has recently been shown to localise to endocytic clathrin-coated pits via interactions with intersectin, during early pit formation (Nakatsu et al. 2010).

PI3K/Akt signalling is critical in relaying insulin-stimulated signals to promote glucose uptake and glycogen synthesis (Taniguchi et al. 2006). There are several in vitro studies which have highlighted that SHIP2 negatively regulates PI3K-generated $PI(3,4,5)P_3$ in response to insulin signalling. SHIP2 translocates to the plasma membrane in response to insulin stimulation in 3T3-L1 adipocytes where it primarily regulates the phosphorylation of Akt2 (Sasaoka et al. 2004). Adenoviral-mediated gene transfer of SHIP2 in L6 myoblasts inhibits Akt and glycogen synthase kinase 3 β (GSK3β) phosphorylation, and decreases glycogen synthesis (Sasaoka et al. 2001), and in adipocytes depletes $PI(3,4,5)P_3$ levels associated with decreased Akt and GSK3β phosphorylation, as well as reduced glucose transporter GLUT4 translocation to the plasma membrane leading to decreased 2-deoxyglucose uptake (Wada et al. 2001). However, RNAi-mediated gene silencing of SHIP2 in adipoyctes does not significantly alter insulin signalling (Tang et al. 2005; Zhou et al. 2004), and although SHIP2 RNAi in C2C12 muscle cells enhances Akt phosphorylation, SHIP2 knockdown does not enhance glucose uptake (Gupta and Dey 2009).

Podocytes form the filtration barrier in the glomeruli of kidney nephrons, and podocyte damage occurs in diabetic nephropathy (Wolf and Ziyadeh 2007). SHIP2 is recruited to the plasma membrane in podocytes upon insulin stimulation, and binds to the actin-binding protein CD2-associated protein (CD2AP), although it is not yet established whether SHIP2 regulates the actin cytoskeleton in response to insulin. Overexpression of SHIP2 in podocytes reduces Akt activation and promotes apoptosis (Hyvonen et al. 2010). Interestingly, SHIP2 protein levels are increased in diabetic mouse and rat glomeruli (Hyvonen et al. 2010), however, the physiological relevance of SHIP2 in glomerular function and kidney disease remains to be defined.

SHIP2 Regulates Insulin Signalling in vivo: Studies Using Rodent Models

Two separate reports of SHIP2 knockout mice have emerged. Initially, *Ship2*$^{-/-}$ mice were reported to display enhanced insulin sensitivity as a consequence of increased membrane translocation of the glucose transporter GLUT4 in skeletal muscle, leading to hypoglycaemia and perinatal death (Clement et al. 2001). However, it was later revealed that the *Phox2a* gene was also inadvertently targeted in this mouse model. A more recent report of *Ship2*$^{-/-}$ mice has revealed that loss of SHIP2 results in viable mice. *Ship2*$^{-/-}$ mice exhibit normal insulin sensitivity, however, on a high fat diet, display obesity resistance and enhanced insulin sensitivity, corresponding to reduced serum insulin, glucose and lipid levels, compared to wildtype littermates (Sleeman et al. 2005). Insulin-stimulated Akt phosphorylation in the liver and skeletal muscle is enhanced in *Ship2*$^{-/-}$ mice (Sleeman et al. 2005). In rat models, administration of SHIP2 antisense oligonucleotides (*Ship2*-AS) leads to decreased SHIP2 expression in the liver and skeletal muscle, associated with a sustained Akt phosphorylation in skeletal muscle. *Ship2*-AS also does not affect insulin sensitivity in rats on a normal chow diet, however, as is seen in *Ship2*$^{-/-}$ mice, *Ship2*-AS rats on a high fat diet display significantly increased glucose tolerance in response to insulin (Buettner et al. 2007). Therefore, SHIP2 may play a role in the development of diet-induced obesity and insulin resistance.

Mutations in SHIP2 have been identified in the type 2 diabetic and hypertensive rat strain, Goto Kakizaki (GK). A mutation specifically in the SHIP2 proline-rich domain results in decreased insulin-stimulated Akt and MAPK activation in vitro (Marion et al. 2002). In addition, SHIP2 protein levels are increased in the insulin resistant and diabetic mouse strain *db/db*. SHIP2 expression is elevated in *db/db* mouse skeletal muscle and epididymal fat tissue, although not in the liver (Hori et al. 2002). Adenoviral-mediated and liver-specific overexpression of SHIP2 in heterozygote *db/+m* mice reduces Akt phosphorylation in the liver and decreases glucose tolerance in *db/+m* mice (Fukui et al. 2005). Additionally, overexpression of dominant negative ΔIP-*Ship2* in *db/db* mice restores the already diminished phospho-Akt levels in the liver, and rescues the decreased glucose tolerance (Fukui et al. 2005). Liver-overexpression of *Ship2* does not affect insulin signalling in other insulin responsive tissues such as skeletal muscle and adipose tissue (Fukui et al. 2005).

Interestingly, mice with *Ship2* overexpression (*Ship2*-transgenic/*Ship2*-tg) display only a mild phenotype, with 5% body weight gain and slightly elevated blood glucose and serum insulin levels on a normal chow diet compared to wildtype littermates. In addition, these effects are not observed on a high fat chow diet (Kagawa et al. 2008). However, *Ship2*-tg mice demonstrate decreased Akt activation in skeletal muscle, fat and the liver, and have increased liver expression of gluconeogenic enzymes glucose-6-phosphatase (G6Pase) and phosphoenolpyruvate carboxykinase (PEPCK), and decreased glycolytic enzyme glucokinase (Kagawa et al. 2008). Insulin signalling stimulates glycolytic and inhibits gluconeogenic hepatic gene expression, and these actions are disrupted in insulin resistance as a consequence of type 2 diabetes (Saltiel and Kahn 2001). Overexpression of ΔIP-*Ship2* in *db/db* mice decreases liver expression of G6Pase and PEPCK (Fukui et al. 2005). In addition, liver-specific inhibition of SHIP2 has been reported in a second type 2 diabetic mouse model, *KKA*^{Y} which is hyperglycaemic and insulin resistant. Adenoviral-mediated and liver-specific overexpression of ΔIP-*Ship2* in *KKA*^{Y} mice also increases Akt phosphorylation and decreases liver expression of G6Pase and PEPCK, in addition to increasing glucose tolerance (Grempler et al. 2007). Therefore, SHIP2 may negatively regulate insulin signalling to alter hepatic gene expression, and promote insulin resistance and systemic glucose homeostasis. Interestingly, administration of a small molecule

SHIP2 inhibitor, AS1949490, to *db/db* mice lowers plasma glucose and enhances glucose tolerance (Suwa et al. 2009) and may represent an attractive potential therapeutic tool for type two diabetes treatment in humans.

SHIP2 is Implicated in Human Diabetes

The metabolic syndrome is characterised by features such as type two diabetes, hypertension, abdominal obesity, dyslipidemia, impaired fibrolysis and insulin resistance, reviewed in (Palomo et al. 2006). A study which investigated type two diabetic patients in UK and French cohorts has identified several *SHIP2* SNPs which are associated with metabolic syndrome features such as hypertension and obesity (Kaisaki et al. 2004).

In addition, a deletion in the 3'UTR of the SHIP2 gene has been identified in human type two diabetic patients. The 3'UTR deletion disrupts a motif involved in protein synthesis, and the recombinant mutant SHIP2 displays increased mRNA and protein expression in vitro (Marion et al. 2002). Other reports of polymorphisms in Japanese cohorts have revealed SNPs in the promoter and 5'UTR regions of the *SHIP2* gene, which are associated with an impaired fasting glycaemia. In vitro studies using luciferase reporters have revealed increased promoter activity of mutant SHIP2 (Ishida et al. 2006). Interestingly, analysis of a Japanese cohort has revealed a number of polymorphisms associated with normal control rather than diabetic individuals, with one particular SNP (L632I) located in the SHIP2 5-phosphatase domain. In vitro studies using the recombinant L632I mutant SHIP2 results in enhanced insulin signalling, including reduced inhibition of $PI(3,4,5)P_3$ signals and Akt2 phosphorylation, suggesting that SNPs which suppress SHIP2 catalytic activity may be protective in the development of type two diabetes (Kagawa et al. 2005). Therefore, mutations which increase SHIP2 expression and catalytic activity are associated with type two diabetic patients in populations, and those which decrease SHIP2 expression or activity appear in apparently normal individuals, and may confer protection. Again, these data highlight the potential value of SHIP2 as a therapeutic target for the treatment of type two diabetes.

SHIP2 is Implicated in Human Cancer

SHIP2 has been reported to have both pro- and anti-tumorigenic roles in human cancer and murine cancer models. For example SHIP2 mRNA and protein levels are increased in human primary breast cancer samples and a number of breast cancer cell lines. High SHIP2 protein expression is reported in human breast primary tumours, ductal carcinoma and invasive carcinoma. SHIP2 expression in invasive carcinoma is significantly and positively associated with reduced disease-free and overall survival of patients. In addition, SHIP2 expression positively correlates with EGFR expression (Prasad et al. 2008b). Significantly, siRNA-mediated SHIP2 depletion in the human breast cancer cell line MDA-MB-231 enhances EGFR degradation leading to suppression of downstream signalling, reducing cell proliferation and suppressing tumour growth and lung metastases in xenografted nude mice (Prasad et al. 2008a). SHIP2 RNAi-treated HeLa cells demonstrate redistribution of EGFR-containing endocytic vesicles, accompanied by enhanced c-Cbl-mediated ubiquitination and degradation of EGFR (Prasad and Decker 2005). In MDA-MB-231 breast cancer cells, siRNA-depletion of SHIP2 decreases EGF-stimulated Akt phosphorylation, and the CXCR4 receptor which is implicated in cancer cell metastasis is also downregulated, associated with decreased cell adhesion and migration (Prasad et al. 2009). Interestingly, SHIP2 regulates the internalisation of another cell surface receptor which is amplified in human cancers, EphA2. Overexpression of SHIP2 in COS-7 cells inhibits EphA2 internalisation, whereas siRNA-mediated silencing of SHIP2 in MDA-MB-231 breast cancer cells promotes PI3K-generated PI(3,4,5)P_3 accumulation and Rac1 activation, associated with enhanced EphA2 endocytosis and degradation (Zhuang et al. 2007). However, an association between EphA2 and SHIP2 expression in human cancer has not yet been reported. These studies demonstrate that SHIP2 may possess proto-oncogenic activity to regulate growth factor receptor stability, endocytosis and/or signalling leading to proliferation and metastasis. Given the multiple associations of SHIP2 with the actin cytoskeleton, it is also possible that SHIP2 may regulate actin dynamics to regulate cancer progression and metastasis, although this is yet to be specifically demonstrated.

Contrasting roles of SHIP2 have also been described in cancer models in vitro. miRNA-205 supresses SHIP2 expression in squamous cell epithelium. Squamous cell carcinoma (SCC) epithelial cell lines demonstrate increased miRNA-205 correlating with reduced SHIP2 expression and enhanced Akt phosphorylation. siRNA-mediated depletion of miRNA-205 decreases Akt phosphorylation and promotes apoptosis in SCC cell lines (Yu et al. 2008). Although miRNA-205 upregulation has been reported in a number of human cancers, reviewed in (Sotiropoulou et al. 2009), the role of SHIP2 in these contexts remains to be determined. In addition, overexpression of SHIP2 in glioblastoma cell lines reduces PI(3,4,5)P_2 and Akt phosphorylation leading to cell cycle arrest (Taylor et al. 2000).

SKIP/INPP5K

Skeletal muscle and kidney inositol phosphatase (SKIP) is encoded by the *SKIP/INPP5K* gene located on human chromosome 17p13.3. SKIP is a 51 kDa 5-phosphatase which is alternatively spliced to form a 43 kDa variant. SKIP hydrolyses the soluble inositol species Ins(1,4,5)P_3 and Ins(1,3,4,5)P_4, and membrane-bound PI species PI(4,5)P_2 and PI(3,4,5)P_3 (Ijuin et al. 2000), with highest in vitro phosphatase activity towards PI(4,5)P_2 (Schmid et al. 2004). SKIP is widely expressed in mammalian cells and tissues, with highest levels observed in skeletal muscle and kidney (Ijuin et al. 2000). The smaller isoform is expressed highly in adipose tissue (Ijuin et al. 2008) and has a much lower 5-phosphatase activity to hydrolyse $PI(3,4,5)P_3$ and $PI(4,5)P_2$ compared to the full length enzyme (Ijuin et al. 2008). SKIP hydrolysis of $PI(3,4,5)P_3$ is inhibited by binding to the silencer of death domain (SODD/BAG4) (Rahman et al. 2011). SKIP localises to neurites in neuroblastoma cell lines, and has a perinuclear distribution in COS-7 cells (Ijuin et al. 2000). SKIP plays a major role in insulin signalling both in vitro and in vivo.

In response to insulin stimulation, a pool of SKIP redistributes from a perinuclear distribution to the plasma membrane, dependent on its C-terminal SKIP carboxyl homology (SKICH) domain, and acts to inhibit actin stress fibre formation and membrane ruffling (Gurung et al. 2003; Ijuin and Takenawa 2003). Interestingly, the SKICH domain of SKIP also interacts with the Hepatitis B virus core protein to inhibit HBV gene expression and viral replication (Hung et al. 2009). SKIP overexpression depletes $PI(3,4,5)P_3$ signals and decreases Akt phosphorylation in response to insulin stimulation, and decreases translocation of the glucose transporter GLUT4 to the cell membrane, leading to decreased glucose uptake (Ijuin and Takenawa 2003). RNAi-mediated SKIP depletion in C2C12 myoblast muscle cells enhances Akt and GSK3β phosphorylation in response to insulin, correlating with enhanced glycogen synthase dephosphorylation which promotes glycogen synthesis (Xiong et al. 2009). Therefore, in vitro, SKIP negatively regulates glucose uptake and glycogen synthesis in response to insulin.

Targeted deletion of the *Skip* gene in mice is embryonically lethal at E10.5 (Ijuin et al. 2008). *Skip*$^{+/-}$ mice are viable and physically indistinguishable from wildtype littermates, but display an insulin sensitivity phenotype, with enhanced serum glucose clearance following insulin administration, and enhanced glucose uptake in skeletal muscle. Skeletal muscle from *Skip*$^{+/-}$ mice exhibits increased insulin signalling, with enhanced Akt phosphorylation, leading to increased phosphorylation of downstream targets p70 S6 kinase and GSK3β. Although it has not been demonstrated in vivo, siRNA-mediated knockdown of SKIP in L6 myoblast cells leads to increased GLUT4 translocation to the cell membrane (Ijuin et al. 2008) and may be the basis of the enhanced glucose uptake in skeletal muscle of *Skip*$^{+/-}$ mice.

Significantly, *Skip*$^{+/-}$ mice display obesity resistance on a high fat diet, with lower serum leptin cholesterol and insulin levels compared to wildtype mice, correlating with enhanced Akt phosphorylation in skeletal muscle (Ijuin et al. 2008). SKIP therefore plays an important role in regulating diet-induced insulin resistance and obesity via regulation of glucose homeostasis and insulin signalling in skeletal muscle. However, the role SKIP plays in insulin signalling in brain and adipose tissue, and whether these effects have a regulatory role in human diabetes and in obesity remains to be demonstrated. Interestingly, *Skip*$^{+/-}$ mice also display increased muscle mass (Ijuin et al. 2008), and SKIP is upregulated during C2C12 myoblast differentiation dependent on MyoD regulation (Xiong et al. 2009).

Transgenic mice which overexpress *Skip* in all tissues have recently been generated. *Skip* transgenic mice display reduced plasma osmolality, and an impaired ability to excrete a water load. In addition, *Skip* transgenic mice have an increased response to arginine vasopressin (AVP) (Pernot et al. 2011). AVP induces an antidiuretic effect on the kidney and induces water retention. AVP precursor protein is released from the hypothalamus in response to dehydration and converted into active AVP in the bloodstream. *Skip* transgenic mice have an increased expression of the AVP receptor, AVPR2 and the water transporter AQP2 in the kidney, although they display decreased AVP expression in the brain. The altered AVP/AQP2 axis may be the basis of the altered plasma osmolality and urine excretion capacity of the transgenic mice (Pernot et al. 2011). However, the way in which SKIP regulates the expression and function of AQP2 and AVPR2 in the kidney, and the expression of AVP in the brain remains to be described.

The chromosomal locus in which the human *SKIP* gene is located is implicated in multiple human diseases. Miller–Dieker syndrome (MDS) arises from deletions in the 17p13.3 locus and is characterised by lissencephaly (lack of gyrations of the cerebellum) microcephaly, facial dysmorphism, cardiac malformation, mental retardation, seizures and a reduced lifespan (Dobyns et al. 1991). Critical regions which are commonly deleted include the *SKIP* gene (Bruno et al. 2010; Cardoso et al. 2003) although it has not been established what role SKIP may play in the development of MDS.

The *SKIP* gene locus, 17p13.3, is frequently deleted or hypermethylated in a diverse range of human cancers. For example, deletions at 17p13.3 occur frequently in human breast cancers (Ellsworth et al. 2005; Parrella et al. 2005), advanced colorectal tumours (Risio, Casorzo et al. 2003), lung cancers (Konishi et al. 1998), ovarian cancer, cervical cancer (Guoling et al. 1997), bladder cancer (Zheng et al. 2002) and childhood medulloblastoma (Lamont et al. 2004). Significantly, SKIP expression has been reported to be altered in human cancers and is decreased in lung adenocarcinoma (Stearman et al. 2005), but increased in renal cell cancer (Jones et al. 2005). In addition, SKIP expression is associated with lymph node metastasis and poor outcome in bladder cancer (Sanchez-Carbayo et al. 2006). Somatic mutation of the *SKIP* gene has been identified in human breast cancer samples (Sjöblom et al. 2006), although the frequency of this mutation in large data sets has not been evaluated. The precise roles that SKIP plays in human cancers are yet to be fully evaluated.

PIPP/INPP5J

The proline-rich inositol polyphosphate 5-phosphatase (PIPP) is encoded by the *PIPP/INPP5J* gene located on human chromosome 22q12.2. PIPP is a 107 kDa 5-phosphatase that is highly expressed in brain, heart, kidney lung, intestine and stomach. PIPP contains proline-rich N and C termini, a putative N-terminal SH3 domain, 6 putative RSXSXP protein binding domains, a SKICH domain and conserved 5-phosphatase catalytic domain (Mochizuki and Takenawa 1999).

PIPP hydrolyses both Ins(1,4,5)P_3, Ins(1,3,4,5)P_4, and membrane-bound PI(4,5)P_2 and PI(3,4,5)P_3 (Mochizuki and Takenawa 1999; Ooms et al. 2006).

Ectopic expression of PIPP in COS-7 cells results in the accumulation of PIPP at ruffling membranes via its SKICH domain (Gurung et al. 2003; Mochizuki and Takenawa 1999), which also mediates PIPP localisation to the neurite growth cone and shaft of growth factor differentiated neuronal cell lines (PC12) (Ooms et al. 2006). PIPP hydrolysis of PI(3,4,5)P_3 inhibits PC12 cell neurite elongation at the growth cone, but promotes polarised neurite formation in a complex with collapsin response mediator protein 2 (CRMP2) (Astle et al. 2011; Ooms et al. 2006). PIPP negatively regulates PI3K/Akt signalling via PI(3,4,5)P_3 hydrolysis, to inhibit PI3K/Akt induced cell transformation in fibroblasts (Denley et al. 2009), and mitotic and cleavage events in early embryogenesis (Deng et al. 2011). PIPP also promotes cell survival in HeLa cells (MacKeigan et al. 2005).

There have been few studies to investigate the role PIPP plays in vivo, and no knockout mouse has been reported. However, several studies have identified associations between PIPP expression and human cancers. The chromosomal region in which *PIPP* is located is frequently mutated in human breast cancer (Hartikainen et al. 2006), and LOH of 22q12 occurs in both the stroma and epithelium of breast cancers (Kurose et al. 2001). 22q12 deletions occur frequently in invasive lobular carcinoma (Weber-Mangal et al. 2003), and in ER^+ tumours, as well as tumours which are well differentiated (Richard et al. 2000). Significantly, *PIPP* is one of the top 10 predictive genes identified for breast cancer outcomes. PIPP expression positively correlates with increased time to metastasis, and increased disease-free survival (Gevaert et al. 2006; Takahashi et al. 2004; van't Veer et al. 2002). PIPP is more highly expressed in ER^+ than ER^- primary breast tumours (Gruvberger et al. 2001), and separate studies indicate that PIPP expression positively correlates with ER expression in human breast cancer samples (van't Veer et al. 2002). Additionally, PIPP expression negatively correlates with *BRCA1* germline mutation (van't Veer et al. 2002).

22q12.2. deletions are frequently observed in other human cancers including 35 % of sporadic colorectal carcinomas (Zhou et al. 2002), primary hepatocellular carcinoma (Zhu et al. 2004) and in ovarian cancer (Benetkiewicz et al. 2005; Englefield et al. 1994). PIPP is upregulated in response to the novel drug Ellagitannin, which inhibits cellular proliferation (Wen et al. 2009). However, the functional role PIPP plays in cancer remains unknown. Additionally, PIPP is implicated in rheumatoid arthritis where studies using microarray analysis to identify novel biomarkers for autoimmune diseases have revealed increased PIPP expression (Edwards et al. 2007; López-Pedrera et al. 2009).

SYNAPTOJANINS

There are two mammalian synaptojanin isoforms, designated synaptojanin 1 and synaptojanin 2, encoded by *SYNJ1/INPP5G* on human chromosome 21q22.11, and *SYNJ2/INPP5H* on chromosome 6q25.3, respectively. Both enzymes contain

N-terminal SAC domains, a central 5-phosphatase domain and a C-terminal proline-rich domain (McPherson et al. 1996), and share 53.8 and 57.2 % amino acid homology across their 5-phosphatase and SAC domains, respectively (Nemoto et al. 1997). The Sac domain hydrolyses PI(3)P, PI(4)P and PI(3,5)P_2 to PI, in addition to the 5-phosphatase domain which hydrolyses the 5-position phosphate from PI(4,5)P_2, PI(3,4,5)P_3, Ins(1,4,5)P_3 and Ins(1,3,4,5)P_4 (Guo et al. 1999; McPherson et al. 1996; Schmid et al. 2004).

SYNJ1 and SYNJ2 are alternatively spliced in the C-terminal proline-rich domain. The most widely expressed splice variants for SYNJ1 are 145 and 170 kDa isoforms (Nemoto et al. 1997, 2001; Ramjaun and McPherson 1996; Seet et al. 1998) which have broad expression in various tissues, with high levels in the brain for the 145 kDa isoform (Nemoto et al. 1997; Ramjaun and McPherson 1996; Sakisaka et al. 1997). Six splice variants have been identified for SYNJ2, with 2A, 2B1 and 2B2 forms being most widely expressed. Isoform 2A displays ubiquitous expression, and 2B1 and 2B2 are highly expressed in the brain and testis (Nemoto et al. 1997, 2001; Seet et al. 1998).

SYNJ1

SYNJ1 regulates synaptic vesicle recycling and endocytosis, forming complexes with multiple proteins involved in these processes. The C-terminal proline-rich region of SYNJ1 binds to the SH3 domains of endocytic proteins including endophilin, amphiphysin, syndapin/pacsin and intersectin/Dap160 (De Heuvel et al. 1997; McPherson et al. 1996; Qualmann et al. 1999; Ringstad et al. 1997; Roos and Kelly 1998). The 170 kDa isoform contains three asparagine-proline-phenylalanine (NPF) repeats which facilitate binding to the endocytic protein Eps15, the N-terminal domain of the clathrin heavy chain and to the ear domain of the α-adaptin component of the AP-2 complex (Haffner et al. 1997; Hughes et al. 2000). These interactions function to direct the subcellular localisation of SYNJ1 and/or increase hydrolysis of PI(4,5)P_2 (Lee et al. 2004; Ringstad et al. 1997). Total internal reflection fluorescent microscopy (TIRFM) studies have demonstrated differential temporal recruitment of the two SYNJ1 variants during clathrin-coated pit formation. The 170 kDa isoform is constitutively associated with forming endosomes during clathrin-coated pit formation, whilst the 145 kDa splice variant is rapidly recruited only at the late stages along with endophilin (Perera et al. 2006). Interestingly, the fatty acid composition of the lipid substrate PI(4,5)P_2 can influence the substrate specificity and subcellular targeting of SYNJ1, with preferential hydrolysis of long chain polyunsaturated fatty acids of PI(4,5)P_2 over short chain (Schmid et al. 2004). *Caenorhabditis elegans (C. elegans)* mutants which lack the preferred fatty acid groups on PI(4,5)P_2 display mislocalised SYNJ1 correlating with the accumulation of PI(4,5)P_2 at synapses and defective synaptic vesicle recycling (Marza et al. 2008).

SYNJ1 undergoes constitutive phosphorylation in unstimulated synapses, and in response to depolarisation is dephosphorylated by the serine threonine phosphatase, calcineurin. Several lines of evidence indicate SYNJ1 catalytic activity and protein interactions are modulated via specific phosphorylation events. Cyclin-dependent kinase 5 (Cdk5)-mediated phosphorylation of SYNJ1 at S1144, which is adjacent to the endophilin binding site, inhibits interactions with SYNJ1 binding partners endophilin and amphiphysin, and suppresses catalytic activity (Lee et al. 2004). Similarly, EphB2 tyrosine phosphorylation of SYNJ1 proline-rich domain, reduces 5-phosphatase activity and interaction with endophilin A1 (Irie et al. 2005). Conversely, calcineurin mediates dephosphorylation of SYNJ1 at S1144, following Cdk5-mediated phosphorylation that enhances interactions with endophilin and stimulates its 5-phosphatase activity (Lee et al. 2004). Phosphorylation of SYNJ1 by the dual-specific tyrosine phosphorylation-regulated kinase 1A (MNB/DYRK1A) at multiple sites regulates amphiphysin1 and intersectin1 binding and modestly increases catalytic activity towardss $PI(4,5)P_2$ (Adayev et al. 2006).

In vivo, SYNJ1 regulates clathrin-mediated endocytosis and neuronal function. The majority of *Synj1*$^{-/-}$ mice (85 %) die within 24 h of birth and the remaining mice that live for up to 15 days exhibit severe neurological defects such as progressive ataxia, weakness and convulsions (Cremona et al. 1999). *Synj1*$^{-/-}$ neurons exhibit increased $PI(4,5)P_2$, correlating with an accumulation of clathrin-coated vesicles (Cremona et al. 1999). Similar phenotypes are observed in both the SYNJ1 null *C. elegans* (*unc-26*) and Synj1 null *Drosophila melanogaster (D. melanogaster)* (Harris et al. 2000). ENU mutagenesis screens in *Danio rerio* (*D. rerio*) larvae with vestibular defects identified mutant SYNJ1 induces abnormalities in hair cell basal blebbing via disrupted synaptic vesicle production and release at hair cell ribbon synapses (Trapani et al. 2009). The SAC and 5-phosphatase domains of SYNJ1 are required for normal vesicle endocytosis and uncoating of clathrin in vitro (Mani et al. 2007). $PI(4,5)P_2$ regulates the interaction of synaptic vesicles with the plasma membrane by recruiting proteins which mediate vesicle docking (Cremona et al. 1999; Harris et al. 2000; Mani et al. 2007), as well as restricting a readily releasable pool of synaptic vesicles at nerve synapses (Milosevic et al. 2005; Paolo et al. 2004). Therefore $PI(4,5)P_2$ may affect several steps in the synaptic vesicle cycle which rely on clathrin-dependent endocytosis, functioning to supply new synaptic vesicles during nerve stimulation (Haucke 2003; Kasprowicz et al. 2008). $PI(4,5)P_2$ is essential for clathrin-coated pit formation via the recruitment of various endocytic proteins and clathrin adaptors, including AP-2 and AP-180, to the plasma membrane (Haucke 2003). By degrading $PI(4,5)P_2$, SYNJ1 directs the uncoating of clathrin and thereby postendocytic vesicle reavailability. The endocytic defects that occur as a consequence of SYNJ1 loss can be attributed to the accumulation of $PI(4,5)P_2$ on endocytic vesicles (Cremona et al. 1999; Mani et al. 2007). Recent studies describe that SYNJ1, along with membrane curvature sensors such as the BAR protein endophilin, mediate hydrolysis of $PI(4,5)P_2$ orientation on curved membranes opposed to comparatively flat ones, suggesting that the $PI(4,5)P_2$ to PI(4)P conversion may be spatially restricted during membrane internalisation (Chang-Ileto et al. 2011). Studies in *C. elegans* demonstrate disruption of genes encoding the NCA (Na^+/Ca^{2+} antiporter) ion channel, or *unc-80*, which encodes a novel neuronal protein required for ion channel subunit localisation, can partially suppress the defects observed in synaptojanin mutants (Jospin et al. 2007).

SYNJ1 maps to human chromosome 21q22.2. Trisomy of chromosome 21 causes Down's syndrome (DS), which is characterised by intellectual disability, congenital heart disease and susceptibility to several cancers, reviewed in (Patterson 2009). Significantly, increased expression of SYNJ1 is reported in DS brain tissue (Arai et al. 2002; Cheon et al. 2003). Furthermore, the DS mouse model, Ts65Dn, displays elevated Synj1 expression associated with decreased $PI(4,5)P_2$ levels, leading to cognitive defects. Transgenic mice overexpressing the *Synj1* gene recapitulate these features (Voronov et al. 2008). Remarkably, crossing *Synj1* heterozygous mice with Ts65Dn mice restores Synj1 expression to wildtype levels, associated with a reversal of the DS phenotype and normalisation of $PI(4,5)P_2$ levels (Voronov et al. 2008). Screens in *D. melanogaster*, suggest that SYNJ1 is likely to function in conjunction with other proteins located on human chromosome 21 to regulate synaptic vesicle recycling. Overexpression of *synj*, *dap160* (intersectin) and *nla* (nebula) individually results in abnormal synaptic morphology in *D. melanogaster*, although impaired vesicle recycling, locomotor defects and defective endocytosis are only observed with combined overexpression (Chang and Min 2009). Further details of how SYNJ1 mediates DS pathophysiology remain to be elucidated. Additionally, *C. elegans* with *unc-26* mutations display exacerbated polyglutamine toxicity implicating a possible protective potential of SYNJ1, similar to its binding partner endophilin, in Huntington's disease (Parker et al. 2007).

SYNJ2

SYNJ2 contains a similar domain structure to SYNJ1 however its C-terminal and proline-rich domains are unique (Nemoto et al. 1997). SYNJ2A contains a C-terminal PDZ domain which binds to the mitochondrial outer membrane protein OMP25, and may regulate the intracellular distribution of mitochondria (Nemoto and De Camilli 1999). In addition, SYNJ2B, SYNJ2B1 and SYNJ2B2, interact with SH3-containing endocytic proteins via their proline-rich domains (Nemoto et al. 2001), and are implicated in clathrin-mediated endocytosis. siRNA-mediated reduction of SYNJ2 expression in cell lines results in defective clathrin-mediated receptor internalisation as well as a reduction in clathrin-coated pits and vesicles (Rusk et al. 2003).

SYNJ2 acts as a Rac1 effector, and interacts with Rac1 in a GTP-dependent manner. In vitro, expression of constitutively active Rac1 promotes SYNJ2 translocation to the plasma membrane, and inhibits clathrin-mediated endocytosis (Malecz et al. 2000; Nemoto et al. 2001). Interestingly, SYNJ2 interactions with Rac1 are not restricted to endocytosis and have been implicated in Rac1-mediated malignant cell invasion and migration. siRNA-mediated depletion of either Rac1 or SYNJ2 inhibits the formation of lamellipodia and invadopodia, associated with reduced cell migration and invasion in matrigel and rat brain slices (Chuang et al. 2004).

Therefore, SYNJ2 may act as a downstream effector of Rac1 to regulate matrix degradation and cell migration.

Although no SYNJ2 knockout mouse has been reported, SYNJ2 has recently been identified as a possible regulator of adult hearing. The ENU-generated mutant mouse, *Mozart*, is characterised by recessively inherited, non-syndromic progressive hearing loss and severe deafness at 12 weeks of age, attributed to the degeneration of cochlear hair cells and stereocilia (Manji et al. 2011). This mouse carries a single mutation in a critical catalytic residue in the 5-phosphatase domain of Synj2 which impairs its catalytic activity towardss PI(4,5)P_2 and PI(3,4,5)P_3 (Manji et al. 2011). Cochlea expression of Synj2 is detected in both the inner and outer hair cells, but is absent from the spiral ganglion in wildtype mice. The mechanism by which SYNJ2 regulates hair cell maintenance and survival is yet to be established.

SAC Phosphatases

Members of the SAC phosphatase family contain a conserved SAC domain that was originally isolated in a genetic screen for modifiers of actin cytoskeleton defects in yeast (ySac1) (Novick et al. 1989). Several SAC domain-containing phosphoinositide phosphatases have since been identified which are evolutionarily conserved from yeast to mammals. The highly conserved SAC domain is comprised of $\sim$400 amino acids arranged into 7 motifs, with the conserved CX_5R active site located within the sixth motif. The yeast Sac domain crystal structure comprises two closely packed sub-domains, an N-terminal (SacN) and a catalytic sub-domain (Manford et al. 2010), which confers unique interactions with regulatory proteins and a differential dephosphorylation mechanism towardss phosphoinositide substrates compared with other phosphatases (Manford et al. 2010). The SAC domain displays broad substrate specificity efficiently dephosphorylating PI(3)P, PI(4)P and PI(3,5)P_2 to PI (Guo et al. 1999).

There are two classes of SAC family proteins, which are classified based on domain structure. Stand-alone SAC phosphatases contain a single N-terminal SAC domain and no other identifiable domains and include yeast ySac1 and yFig4, and human SAC1, SAC2/INPP5F and SAC3/FIG4. Enzymes containing a SAC domain in addition to a 5-phosphatase catalytic domain are termed SAC domain-containing inositol phosphatases (SCIPs) and include mammalian SYNJ1 and SYNJ2. SAC phosphatases exhibit varied cellular functions in mammalian cell systems and knockout mouse models, and are implicated in several human diseases including neurodegeneration and cardiac hypertrophy. The following section will outline the biological function of the mammalian stand-alone SAC phosphatases, highlighting their roles in human disease.

SAC1

SAC1 is a 67 kDa type II transmembrane protein, and was the first mammalian SAC phosphatase identified (Nemoto et al. 2000). *SAC1* is located on human chromosome 3p21.3, and displays an identical substrate specificity to ySac1, hydrolysing PI(3)P, PI(4)P and PI(3,5)P_2 both in vitro and in vivo (Nemoto et al. 2000). SAC1 displays the greatest catalytic activity towardss monophosphorylated phosphoinositides, and in yeast cells mediates the majority of PI(4)P degradation (Foti et al. 2001; Nemoto et al. 2000). SAC1 contains two C-terminal domains that anchor it to intracellular membranes such that the hydrophilic region of the C-terminus faces the cytosol (Konrad et al. 2002). The ySac1 crystal structure has revealed the presence of a linker sequence between the catalytic and the C-terminal transmembrane domain, which is predicted to enable access to its substrates on adjacent membranes, possibly allowing hydrolysis of multiple phosphoinositide pools (Manford et al. 2010).

SAC1 localises to the Golgi complex and ER membranes in both yeast and mammalian cells (Cleves et al. 1989; Nemoto et al. 2000; Rohde et al. 2003; Whitters et al. 1993). In yeast ySac1 regulates PI(3)P and PI(4)P hydrolysis at the ER, and PI(4)P levels at the Golgi (Foti et al. 2001; Konrad et al. 2002). In mammalian cells, SAC1 is shuttled to the Golgi via COP-II interactions where it degrades local PI(4)P pools upon serum deprivation (Blagoveshchenskaya et al. 2008; Cheong et al. 2010). Growth factor simulation of primary human fibroblasts promotes COP-I dependent localisation of SAC1 to the ER, leading to PI(4)P accumulation at the Golgi, which promotes the trafficking of signalling proteins to the cell periphery (Rohde et al. 2003; Schorr et al. 2001).

Cell culture studies and transgenic animal models have provided interesting insights to the functional role of SAC1. In yeast ySac1 depletion or mutation results in pleiotropic defects including disorganisation of the cytoskeleton (Cleves et al. 1989; Novick et al. 1989), inositol auxotrophy (Rivas et al. 1999), retardation of cell growth (Cheong et al. 2010; Liu et al. 2008), abnormal vacuole formation and trafficking (Foti et al. 2001; Tahirovic et al. 2005) and cell wall defects (Schorr et al. 2001). In mammalian cells, RNAi-mediated knockdown of SAC1 leads to disorganisation of the TGN and mitotic spindles, which is associated with inefficient cell cycle progression through G2/M (Liu et al. 2008). Additionally, SAC1 expression is required for the spatial regulation of Golgi PI(4)P and maintenance of normal Golgi organisation in mammalian cells (Cheong et al. 2010).

SAC1 is essential for embryonic development, with *Sac1*$^{-/-}$ mice displaying preimplantation lethality (Liu et al. 2008). In addition, *Drosophila Sac1* mutants display defective dorsal closure in embryogenesis (Wei et al. 2003). Mutant *Sac1 Drosophila* display elevated PI(4)P in addition to increased Hedgehog signalling (Jiang and Hui 2008; Yavari et al. 2010). Hedgehog signalling regulates many developmental processes, although whether phosphoinositide signalling directly influences hedgehog developmental programmes in *Drosophila* with loss of SAC1 remains to be elucidated.

SAC2/INPP5F

Mammalian *SAC2*, also known as *INPP5F*, is located on human chromosome 10q26.11. *SAC2* is ubiquitously expressed, with its highest expression detected in the heart, brain, kidney and skeletal muscle. Interestingly, mammalian SAC2 has no yeast homologue, and displays unique substrate specificity, dephosphorylating both $PI(4,5)P_2$ and $PI(3,4,5)P_3$ at the D-5 position (Minagawa et al. 2001). There has been little characterisation of SAC2, although a *Sac2*$^{-/-}$ mouse model has been reported which demonstrates increased sensitivity to stress-induced cardiac hypertrophy and hyperactivation of Akt and GSK3β by modulation of $PI(3,4,5)P_3$ (Zhu et al. 2009). Interestingly, histone deacetylase 2 (HDAC2) transgenic mice display increased hypertrophy associated with decreased *Sac2* gene expression and elevated Akt/GSK3β signalling (Trivedi et al. 2007). Therefore, *Sac2* is implicated in cardiac hypertrophic signalling in mouse models, however, a role for SAC2 in human cardiac hypertrophy remains to be demonstrated. Recently, the spatio-temporal pattern of a transcriptional variant of *Sac2/Inpp5f, Inpp5f-v3,* during mouse development revealed expression specifically in mouse brain, suggesting that *Inpp5f-v3* may be involved in brain development (Yan et al. 2011).

SAC3/FIG4

Yeast yFig4 was discovered via genetic screens as a pheromone-induced gene required for yeast mating (Erdman et al. 1998). The mammalian homologue, SAC3 also known as FIG4, was initially isolated in the rat (rSac3) (Sbrissa et al. 2007; Yuan et al. 2007). Human *SAC3/FIG4* is located on chromosome 6q21, and the SAC3 protein displays in vitro activity to hydrolyse $PI(3,5)P_2$, $PI(4,5)P_2$ and $PI(3,4,5)P_3$, with cell-based assays revealing a preference for $PI(3,5)P_2$ (Rudge et al. 2004; Sbrissa et al. 2007). SAC3 maintains $PI(3,5)P_2$ levels in a complex with the D5-kinase PIKfyve and PIKfyve regulator, ArPIKfyve termed the PAS (PIKfyve–ArPIKfyve–Sac3) complex (Ikonomov et al. 2009; Sbrissa et al. 2004, 2008; Sbrissa and Shisheva 2005). SAC3 is stabilised and protected from degradation in the PAS complex (Ikonomov et al. 2010; Sbrissa et al. 2007), and this association promotes $PI(3,5)P_2$ synthesis via PIKfyve and ArPIKfyve; and $PI(3,5)P_2$ hydrolysis via SAC3 activity (Sbrissa et al. 2004; Sbrissa and Shisheva 2005). SAC3 localises to the ER, Golgi, early and recycling endosomes (Sbrissa et al. 2007; Yuan et al. 2007).

SAC3 is widely expressed in human and mouse tissues, with highest expression in the brain, lung and white adipose tissue (Sbrissa et al. 2007). Insulin-stimulated Sac3 depleted mouse adipocytes display increased cellular $PI(3,5)P_2$ levels, increased GLUT4 membrane translocation and glucose uptake (Ikonomov et al. 2009) although whether SAC3 mediates insulin sensitivity in vivo and particularly in humans remains to be shown. In rats, *rSac3* is highly expressed in the central nervous system (Chow et al. 2007), and in vitro studies have revealed that *rSac3* regulates neurite elongation in PC12 cells (Yuan et al. 2007).

The 'pale tremor' mouse is a spontaneously occurring mouse mutant that exhibits an abnormal gait, peripheral neuropathy, selective neuronal degradation, severe tremor and juvenile lethality arising from a mutation due to a transposon insertion into intron 18 in *Sac3/Fig4* ($Fig4^{-/-}$) (Chow et al. 2007; Ferguson et al. 2009). $Fig4^{-/-}$ pale tremor mice display pathological features which are characteristic of neurodegenerative diseases such as increased levels of autophagy markers, inclusion bodies and protein aggregates in the brain (Ferguson et al. 2009). Pale tremor fibroblasts show significantly reduced levels of the Sac3 substrate, PI(3,5)P_2, due to disrupted PAS complex formation and decreased PIKfyve and ArPIKfyve activity (Duex et al. 2006a, b). In addition, $Fig4^{-/-}$ primary neuron and fibroblast cultures display extensive and enlarged LAMP-2 positive vacuoles (Chow et al. 2007; Ferguson et al. 2009; Zhang et al. 2008), although it is not yet clear how depleted PI(3,5)P_2 and loss of Sac3/Fig4 result in this phenotype.

The clinical and pathological features observed in the $Fig4^{-/-}$ mouse are consistent with the human neuropathy Charcot-Marie-Tooth (CMT) disease. Significantly, SAC3 mutations have been indentified in CMT and include an Ile-to-Thr substitution at amino acid 41 (I41T) predicted to disturb the folding or stability of the protein, and this subtype is designated CMT4J (Manford et al. 2010). In vitro studies have revealed that the CMT4J mutant SAC3 protein retains PI(3,5)P_2 hydrolysing capacity, but has disrupted ArPIKfyve-dependent stabilisation resulting in its rapid degradation (Ikonomov et al. 2010). A recent study has further elucidated the consequences of the CMT-associated mutation within SAC3. The I41T mutation inhibits the interaction of SAC3 with its binding partner and scaffold protein, VAC14. The importance of this interaction is observed in $Vac14^{-/-}$ mice, which also display neurodegeneration accompanied by loss of Sac3 expression and perinatal death (Zhang et al. 2007). Interestingly, $Fig4^{-/-}$ mice transgenically expressing moderate levels of the human FIG4-I41T mutated gene display low levels of FIG4-I41T protein, as is observed in human CMT4J patients, which is attributed to protein instability from impaired interactions of Vac14 with the mutant protein. Interestingly, when the FIG4-I41T mutant gene is highly overexpressed in the $Fig4^{-/-}$ mice, it rescues the $Fig4^{-/-}$ lethal phenotype. Therefore, it is possible that CMT4J patients could be treated by upregulating and/or stabilising the mutant phosphatase (Lenk et al. 2011).

Conclusion

Several new studies have revealed altered expression or function of the phosphoinositide phosphatases may contribute to human disease pathogenesis. However, despite two decades of investigation the functions of many phosphoinositide phosphatases remain to be fully elucidated. Animal studies, including emerging knockout mouse models, are proving to be a valuable tool for the characterisation of phosphatase function in vivo. In addition, gene array studies continue to report altered expression of phosphoinositide phosphatases in various human diseases, highlighting potential roles for these enzymes in specific diseases. Future studies will be required to determine if these phosphatases can be targeted therapeutically.

References

Adayev T, Chen-Hwang M-C, Murakami N, Wang R, Hwang Y-W (2006) MNB/DYRK1A phosphorylation regulates the interactions of synaptojanin 1 with endocytic accessory proteins. Biochem Biophys Res Commun 351:1060–1065

Addis M, Loi M, Lepiani C, Cau M, Melis MA (2004) OCRL mutation analysis in Italian patients with Lowe syndrome. Hum Mutat 23:524–525

Addis M, Meloni C, Congiu R, Santaniello S, Emma F, Zuffardi O, Ciccone R, Cao A, Melis MA, Cau M (2007) A novel interstitial deletion in Xq25, identified by array-CGH in a patient with Lowe syndrome. Eur J Med Genet 50:79–84

Agoulnik IU, Hodgson MC, Bowden WA, Ittmann MM (2011) INPP4B: the new kid on the PI3K block. Oncotarget 2:321–328

Agrawal A, Sinha A, Ahmad T, Aich J, Singh P, Sharma A, Ghosh B (2009) Maladaptation of critical cellular functions in asthma: bioinformatic analysis. Physiol Genomics 40:1–7

Ai J, Maturu A, Johnson W, Wang Y, Marsh CB, Tridandapani S (2006) The inositol phosphatase SHIP-2 down-regulates FcγR-mediated phagocytosis in murine macrophages independently of SHIP-1. Blood 107:813–820

Allaoui A, Menard R, Sansonetti PJ, Parsot C (1993) Characterization of the *Shigella flexneri* ipgD and ipgF genes, which are located in the proximal part of the mxi locus. Infect Immun 61:1707–1714

Antignano F, Ibaraki M, Kim C, Ruschmann J, Zhang A, Helgason CD, Krystal G (2010) SHIP is required for dendritic cell maturation. J Immunol 184:2805–2813

Aoki K, Nakamura T, Inoue T, Meyer T, Matsuda M (2007) An essential role for the SHIP2-dependent negative feedback loop in neuritogenesis of nerve growth factor-stimulated PC12 cells. J Cell Biol 177:817–827

Arai Y, Ijuin T, Takenawa T, Becker LE, Takashima S (2002) Excessive expression of synaptojanin in brains with down syndrome. Brain Dev 24:67–72

Astle MV, Ooms LM, Cole AR, Binge LC, Dyson JM, Layton MJ, Petratos S, Sutherland C, Mitchell CA (2011) Identification of a proline-rich inositol polyphosphate 5-phosphatase (PIPP): collapsin response mediator protein 2 (CRMP2) complex that regulates neurite elongation. J Biol Chem 286(26):23407–23418

Attree O, Olivos IM, Okabe I, Bailey LC, Nelson DL, Lewis RA, McLnnes RR, Nussbaum RL (1992) The Lowe's oculocerebrorenal syndrome gene encodes a protein highly homologous to inositol polyphosphate-5-phosphatase. Nature 358:239–242

Backers K, Blero D, Paternotte N, Zhang J, Erneux C (2003) The termination of PI3K signalling by SHIP1 and SHIP2 inositol 5-phosphatases. Adv Enzyme Regul 43:15–28

Bae Y-K, Kim E, L'Hernault SW, Barr MM (2009) The CIL-1 PI 5-phosphatase localizes TRP polycystins to cilia and activates sperm in *C. elegans*. Curr Biol 19:1599–1607

Bai L, Rohrschneider LR (2010) s-SHIP promoter expression marks activated stem cells in developing mouse mammary tissue. Genes Dev 24:1882–1892

Baltimore D, Boldin MP, O'Connell RM, Rao DS, Taganov KD (2008) MicroRNAs: new regulators of immune cell development and function. Nat Immunol 9:839–845

Bansal VS, Caldwell KK, Majerus PW (1990) The isolation and characterization of inositol polyphosphate 4-phosphatase. J Biol Chem 265:1806–1811

Bansal VS, Inhorn RC, Majerus PW (1987) The metabolism of inositol 1,3,4-trisphosphate to inositol 1,3-bisphosphate. J Biol Chem 262:9444–9447

Barnache S, Le Scolan E, Kosmider O, Denis N, Moreau-Gachelin F (2006) Phosphatidylinositol 4-phosphatase type II is an erythropoietin-responsive gene. Oncogene 25:1420–1423

Batty IH, vanÂ derÂ kaay J, Gray A, Telfer JF, Dixon MJ, Downes CP (2007) The control of phosphatidylinositol 3,4-bisphosphate concentrations by activation of the Src homology two domain containing inositol polyphosphate 5-phosphatase 2, SHIP2. Biochem J 407:255–266

Ben El Kadhi K, Roubinet C, Solinet S, Emery G, Carréno S (2011) The inositol 5-phosphatase dOCRL controls PI(4,5)P2 homeostasis and is necessary for cytokinesis. Curr Biol 21:1074–1079

Benetkiewicz M, Wang Y, Schaner M, Wang P, Mantripragada KK, Buckley PG, Kristensen G, Børresen-Dale A-L, Dumanski JP (2005) High-resolution gene copy number and expression profiling of human chromosome 22 in ovarian carcinomas. Genes Chromosom Cancer 42:228–237

Bertelli DF, Araujo EP, Cesquini M, Stoppa GR, Gasparotto-Contessotto M, Toyama MH, Felix JVC, Carvalheira JB, Michelini LC, Chiavegatto S, Boschero AC, Saad MJA, Lopes-Cendes I, Velloso LA (2006) Phosphoinositide-specific inositol polyphosphate 5-phosphatase IV inhibits inositide trisphosphate accumulation in hypothalamus and regulates food intake and body weight. Endocrinology 147:5385–5399

Bielas SL, Silhavy JL, Brancati F, Kisseleva MV, AL-Gazali L, Sztriha L, Bayoumi RA, Zaki MS, Abdel-Aleem A, Rosti RO, Kayserili H, Swistun D, Scott LC, Bertini E, Boltshauser E, Fazzi E, Travaglini L, Field SJ, Gayral S, Jacoby M, Schurmans S, Dallapiccola B, Majerus PW, Valente EM, Gleeson JG (2009) Mutations in INPP5E, encoding inositol polyphosphate-5-phosphatase E, link phosphatidyl inositol signaling to the ciliopathies. Nat Genet 41:1032–1036

Bisgaard A-M, Kirchhoff M, Nielsen JE, Brandt C, Hove H, Jepsen B, Jensen T, Ullmann R, Skovby F (2007) Transmitted cytogenetic abnormalities in patients with mental retardation: pathogenic or normal variants? Eur J Med Genet 50:243–255

Bishop JL, Sly LM, Krystal G, Finlay BB (2008) The inositol phosphatase SHIP controls *Salmonella enterica* serovar Typhimurium infection in vivo. Infect Immun 76:2913–2922

Blagoveshchenskaya A, Cheong FY, Rohde HM, Glover G, Knödler A, Nicolson T, Boehmelt G, Mayinger P (2008) Integration of Golgi trafficking and growth factor signaling by the lipid phosphatase SAC1. J Cell Biol 180:803–812

Blero D, De Smedt F, Pesesse X, Paternotte N, Moreau C, Payrastre B, Erneux C (2001) The SH2 domain containing inositol 5-phosphatase SHIP2 controls phosphatidylinositol 3,4,5-trisphosphate levels in CHO-IR cells stimulated by insulin. Biochem Biophys Res Commun 282:839–843

Bohdanowicz M, Cosío G, Backer JM, Grinstein S (2010) Class I and class III phosphoinositide 3-kinases are required for actin polymerization that propels phagosomes. J Cell Biol 191:999–1012

Bökenkamp A, Böckenhauer D, Cheong HI, Hoppe B, Tasic V, Unwin R, Ludwig M (2009) Dent-2 disease: a mild variant of Lowe syndrome. J Pediatr 155:94–99

Bothwell SP, Chan E, Bernardini IM, Kuo Y-M, Gahl WA, Nussbaum RL (2011) Mouse model for Lowe syndrome/dent disease two renal tubulopathy. J Am Soc Nephrol 22:443–448

Boyle WJ, Simonet WS, Lacey DL (2003) Osteoclast differentiation and activation. Nature 423:337–342

Brauweiler A, Merrell K, Gauld SB, Cambier JC (2007) Cutting edge: acute and chronic exposure of immature B cells to antigen leads to impaired homing and SHIP1-dependent reduction in stromal cell-derived factor-1 responsiveness. J Immunol 178:3353–3357

Brauweiler A, Tamir I, Dal Porto J, Benschop RJ, Helgason CD, Humphries RK, Freed JH, Cambier JC (2000) Differential regulation of B cell development, activation, and death by the Src homology 2 domain-containing 5' inositol phosphatase (ship). J Exp Med 191:1545–1554

Bruno DL, Anderlid B-M, Lindstrand A, van Ravenswaaij-Arts C, Ganesamoorthy D, Lundin J, Martin CL, Douglas J, Nowak C, Adam MP, Kooy RF, Van der Aa N, Reyniers E, Vandeweyer G, Stolte-Dijkstra I, Dijkhuizen T, Yeung A, Delatycki M, Borgström B, Thelin L, Cardoso C, van Bon B, Pfundt R, de Vries BBA, Wallin A, Amor DJ, James PA, Slater HR, Schoumans J (2010)

Further molecular and clinical delineation of co-locating 17p13.3 microdeletions and microduplications that show distinctive phenotypes. J Med Genet 47:299–311

Buettner R, Ottinger I, Gerhardt-Salbert C, Wrede CE, Scholmerich J, Bollheimer LC (2007) Antisense oligonucleotides against the lipid phosphatase SHIP2 improve muscle insulin sensitivity in a dietary rat model of the metabolic syndrome. Am J Physiol Endocrionol Metab 292:E1871–E1878

Campbell JK, Gurung R, Romero S, Speed CJ, Andrews RK, Berndt MC, Mitchell CA (1997) Activation of the 43 kDa inositol polyphosphate 5-phosphatase by 14–3-3ζ. Biochemistry 36:15363–15370

Cardoso C, Leventer RJ, Ward HL, Toyo-oka K, Chung J, Gross A, Martin CL, Allanson J, Pilz DT, Olney AH, Mutchinick OM, Hirotsune S, Wynshaw-Boris A, Dobyns WB, Ledbetter DH (2003) Refinement of a 400-kb critical region allows genotypic differentiation between isolated lissencephaly, Miller–Dieker syndrome, and other phenotypes secondary to deletions of 17p13.3. Am J Hum Genet 72:918–930

Cekic C, Casella CR, Sag D, Antignano F, Kolb J, Suttles J, Hughes MR, Krystal G, Mitchell TC (2011) MyD88-dependent SHIP1 regulates proinflammatory signaling pathways in dendritic cells after monophosphoryl lipid A stimulation of TLR4. J Immunol 186:3858–3865

Chalhoub N, Baker SJ (2009) PTEN and the PI3-kinase pathway in cancer. Annu Rev Pathol 4:127–150

Chang-Ileto B, Frere SG, Chan RB, Voronov SV, Roux AL, Di Paolo G (2011) Synaptojanin 1-mediated PI(4,5)P_2 hydrolysis is modulated by membrane curvature and facilitates membrane fission. Dev Cell 20:206–218

Chang KT, Min K-T (2009) Upregulation of three *Drosophila* homologs of human chromosome 21 genes alters synaptic function: implications for down syndrome. Proc Natl Acad Sci U S A 106:17117–17122

Cheang MCU, Voduc D, Bajdik C, Leung S, McKinney S, Chia SK, Perou CM, Nielsen TO (2008) Basal-like breast cancer defined by five biomarkers has superior prognostic value than triple-negative phenotype. Clin Cancer Res 14:1368–1376

Cheon MS, Kim SH, Ovod V, Kopitar Jerala N, Morgan JI, Hatefi Y, Ijuin T, Takenawa T, Lubec G (2003) Protein levels of genes encoded on chromosome 21 in fetal down syndrome brain: challenging the gene dosage effect hypothesis (Part III). Amino Acids 24:127–134

Cheong FY, Sharma V, Blagoveshchenskaya A, Oorschot VMJ, Brankatschk B, Klumperman J, Freeze HH, Mayinger P (2010) Spatial regulation of Golgi phosphatidylinositol-4-phosphate is required for enzyme localization and glycosylation fidelity. Traffic 11:1180–1190

Chi Y, Zhou B, Wang WQ, Chung SK, Kwon YU, Ahn YH, Chang YT, Tsujishita Y, Hurley JH, Zhang ZY (2004) Comparative mechanistic and substrate specificity study of inositol polyphosphate 5-phosphatase Schizosaccharomyces pombe synaptojanin and SHIP2. J Biol Chem 279:44987–44995

Choudhury R, Diao A, Zhang F, Eisenberg E, Saint-Pol A, Williams C, Konstantakopoulos A, Lucocq J, Johannes L, Rabouille C, Greene LE, Lowe M (2005) Lowe syndrome protein OCRL1 interacts with clathrin and regulates protein trafficking between endosomes and the trans-Golgi network. Mol Biol Cell 16:3467–3479

Choudhury R, Noakes CJ, McKenzie E, Kox C, Lowe M (2009) Differential clathrin binding and subcellular localization of OCRL1 splice isoforms. J Biol Chem 284:9965–9973

Chow CY, Zhang Y, Dowling JJ, Jin N, Adamska M, Shiga K, Szigeti K, Shy ME, Li J, Zhang X, Lupski JR, Weisman LS, Meisler MH (2007) Mutation of FIG4 causes neurodegeneration in the pale tremor mouse and patients with CMT4J. Nature 448:68–72

Chow KU, Nowak D, Kim S-Z, Schneider B, Komor M, Boehrer S, Mitrou PS, Hoelzer D, Weidmann E, Hofmann W-K (2006) In vivo drug-response in patients with leukemic non-Hodgkin's lymphomas is associated with in vitro chemosensitivity and gene expression profiling. Pharmacol Res 53:49–61

Chuang YY, Tran NL, Rusk N, Nakada M, Berens ME, Symons M (2004) Role of synaptojanin 2 in Glioma cell migration and invasion. Cancer Res 64:8271–8275

Clement S, Krause U, Desmedt F, Tanti JF, Behrends J, Pesesse X, Sasaki T, Penninger J, Doherty M, Malaisse W, Dumont JE, Le Marchand-Brustel Y, Erneux C, Hue L, Schurmans S (2001) The lipid phosphatase SHIP2 controls insulin sensitivity. Nature 409:92–97

Clerc PL, Ryter A, Mounier J, Sansonetti PJ (1987) Plasmid-mediated early killing of eucaryotic cells by *Shigella flexneri* as studied by infection of J774 macrophages. Infect Immun 55:521–527

Cleves AE, Novick PJ, Bankaitis VA (1989) Mutations in the SAC1 gene suppress defects in yeast Golgi and yeast actin function. J Cell Biol 109:2939–2950

Collazo MM, Wood D, Paraiso KHT, Lund E, Engelman RW, Le C-T, Stauch D, Kotsch K, Kerr WG (2009) SHIP limits immunoregulatory capacity in the T-cell compartment. Blood 113:2934–2944

Cooper KG, Winfree S, Malik-Kale P, Jolly C, Ireland R, Knodler LA, Steele-Mortimer O (2011) Activation of Akt by the bacterial inositol phosphatase, SopB, is Wortmannin insensitive. PLoS ONE 6:e22260

Cormand B, Avela K, Pihko H, Santavuori P, Talim B, Topaloglu H, de la Chapelle A, Lehesjoki A-E (1999) Assignment of the muscle–eye–brain disease gene to 1p32–p34 by linkage analysis and homozygosity mapping. Am J Hum Genet 64:126–135

Costinean S, Sandhu SK, Pedersen IM, Tili E, Trotta R, Perrotti D, Ciarlariello D, Neviani P, Harb J, Kauffman LR, Shidham A, Croce CM (2009) Src homology 2 domain-containing inositol-5-phosphatase and CCAAT enhancer-binding protein beta are targeted by miR-155 in B cells of Emicro-MiR-155 transgenic mice. Blood 114:1374–1382

Cox D, Dale BM, Kashiwada M, Helgason CD, Greenberg S (2001) A regulatory role for Src homology 2 domain-containing inositol 5'-phosphatase (SHIP) in phagocytosis mediated by Fcγ receptors and complement receptor 3 ($\alpha_M\beta_2$;CD11b/CD18). J Exp Med 193:61–71

Cremer TJ, Ravneberg DH, Clay CD, Piper-Hunter MG, Marsh CB, Elton TS, Gunn JS, Amer A, Kanneganti TD, Schlesinger LS, Butchar JP, Tridandapani S (2009) MiR-155 induction by *F. novicida* but not the virulent *F. tularensis* results in SHIP down-regulation and enhanced pro-inflammatory cytokine response. PLoS ONE 4:e8508

Cremona O, Di Paolo G, Wenk MR, Lüthi A, Kim WT, Takei K, Daniell L, Nemoto Y, Shears SB, Flavell RA, McCormick DA, De Camilli P (1999) Essential role of phosphoinositide metabolism in synaptic vesicle recycling. Cell 99:179–188

Crowley JE, Stadanlick JE, Cambier JC, Cancro MP (2009) FcγRIIβ signals inhibit BLyS signaling and BCR-mediated BLyS receptor up-regulation. Blood 113:1464–1473

Cui S, Guerriero CJ, Szalinski CM, Kinlough CL, Hughey RP, Weisz OA (2010) OCRL1 function in renal epithelial membrane traffic. Am J Physiol Ren Physiol 298:F335–F345

D'Angelo A, Franco B (2009) The dynamic cilium in human diseases. PathoGenetics 2:3

Dambournet D, Machicoane M, Chesneau L, Sachse M, Rocancourt M, El Marjou A, Formstecher E, Salomon R, Goud B, Echard A (2011) Rab35 GTPase and OCRL phosphatase remodel lipids and F-actin for successful cytokinesis. Nat Cell Biol 13:981–988

De Heuvel E, Bell AW, Ramjaun AR, Wong K, Sossin WS, McPherson PS (1997) Identification of the major synaptojanin-binding proteins in brain. J Biol Chem 272:8710–8716

De Smedt F, Verjans B, Mailleux P, Erneux C (1994) Cloning and expression of human brain type I inositol 1,4,5-trisphosphate 5-phosphatase. High levels of mRNA in cerebellar Purkinje cells. FEBS Lett 347:69–72

Deng X, Feng C, Wang EH, Zhu YQ, Cui C, Zong ZH, Li GS, Liu C, Meng J, Yu BZ (2011) Influence of proline-rich inositol polyphosphate 5-phosphatase, on early development of fertilized mouse eggs, via inhibition of phosphorylation of Akt. Cell Prolif 44:156–165

Denley A, Gymnopoulos M, Kang S, Mitchell C, Vogt PK (2009) Requirement of phosphatidylinositol(3,4,5)trisphosphate in phosphatidylinositol 3-kinase-induced oncogenic transformation. Mol Cancer Res 7:1132–1138

Dobyns WB, Curry CJR, Hoyme HE, Turlington L, Ledbetter DH (1991) Clinical and molecular diagnosis of Miller–Dieker syndrome. Cell Press, Cambridge

Dong S, Corre B, Foulon E, Dufour E, Veillette A, Acuto O, Michel F (2006) T cell receptor for antigen induces linker for activation of T cell-dependent activation of a negative signaling complex involving Dok-2, SHIP-1, and Grb-2. J Exp Med 203:2509–2518

Dressman MA, Olivos-Glander IM, Nussbaum RL, Suchy SF (2000) Ocrl1, a PtdIns(4,5)P_2 5-phosphatase, is localized to the trans-Golgi network of fibroblasts and epithelial cells. J Histochem Cytochem 48:179–189

Duex JE, Nau JJ, Kauffman EJ, Weisman LS (2006a) Phosphoinositide 5-phosphatase Fig4p is required for both acute rise and subsequent fall in stress-induced phosphatidylinositol 3,5-bisphosphate levels. Eukaryot Cell 5:723–731

Duex JE, Tang F, Weisman LS (2006b) The Vac14p-Fig4p complex acts independently of Vac7p and couples PI(3,5)P_2 synthesis and turnover. J Cell Biol 172:693–704

Dyson JM, Kong AM, Wiradjaja F, Astle MV, Gurung R, Mitchell CA (2005) The SH2 domain containing inositol polyphosphate 5-phosphatase-2: SHIP2. Int J Biochem Cell Biol 37:2260–2265

Dyson JM, O'Malley CJ, Becanovic J, Munday AD, Berndt MC, Coghill ID, Nandurkar HH, Ooms LM, Mitchell CA (2001) The SH2-containing inositol polyphosphate 5-phosphatase, SHIP-2, binds filamin and regulates submembraneous actin. J Cell Biol 155:1065–1079

Edelmann J, Klein-Hitpass L, Carpinteiro A, Führer A, Sellmann L, Stilgenbauer S, Dührsen U, Dürig J (2008) Bone marrow fibroblasts induce expression of PI3K/NF-κB pathway genes and a pro-angiogenic phenotype in CLL cells. Leuk Res 32:1565–1572

Edmunds C, Parry RV, Burgess SJ, Reaves B, Ward SG (1999) CD28 stimulates tyrosine phosphorylation, cellular redistribution and catalytic activity of the inositol lipid 5-phosphatase SHIP. Eur J Immunol 29:3507–3515

Edwards CJ, Feldman JL, Beech J, Shields KM, Stover JA, Trepicchio WL, Larsen G, Foxwell BMJ, Brennan FM, Feldmann M, Pittman DD (2007) Molecular profile of peripheral blood mononuclear cells from patients with rheumatoid arthritis. Mol Med 13:19

Ellsworth RE, Ellsworth DL, Neatrour DM, Deyarmin B, Lubert SM, Sarachine MJ, Brown P, Hooke JA, Shriver CD (2005) Allelic imbalance in primary breast carcinomas and metastatic tumors of the axillary lymph nodes. Mol Cancer Res 3:71–77

WS Elong Edimo, Derua R, Janssens V, Nakamura T, Vanderwinden JM, Waelkens E, Erneux C (2011) Evidence of SHIP2 Ser132 phosphorylation, its nuclear localization and stability. Biochem J 439:391–401

Englefield P, Foulkes WD, Campbell IG (1994) Loss of heterozygosity on chromosome 22 in ovarian carcinoma is distal to and is not accompanied by mutations in NF2 at 22q12. Br J Cancer 70:905–907

Erdman S, Lin L, Malczynski M, Snyder M (1998) Pheromone-regulated genes required for yeast mating differentiation. J Cell Biol 140:461–483

Erdmann KS, Mao Y, McCrea HJ, Zoncu R, Lee S, Paradise S, Modregger J, Biemesderfer D, Toomre D, De Camilli P (2007) A role of the Lowe syndrome protein OCRL in early steps of the endocytic pathway. Dev Cell 13:377–390

Erkeland SJ, Valkhof M, Heijmans-Antonissen C, van Hoven-Beijen A, Delwel R, Hermans MHA, Touw IP (2004) Large-scale identification of disease genes involved in acute myeloid leukemia. J Virol 78:1971–1980

Faucherre A, Desbois P, Nagano F, Satre V, Lunardi J, Gacon G, Dorseuil O (2005) Lowe syndrome protein Ocrl1 is translocated to membrane ruffles upon Rac GTPase activation: a new perspective on Lowe syndrome pathophysiology. Hum Mol Genet 14:1441–1448

Faucherre A, Desbois P, Satre V, Lunardi J, Dorseuil O, Gacon G (2003) Lowe syndrome protein OCRL1 interacts with Rac GTPase in the trans-Golgi network. Hum Mol Genet 12:2449–2456

Fedele CG, Ooms LM, Ho M, Vieusseux J, O'Toole SA, Millar EK, Lopez-Knowles E, Sriratana A, Gurung R, Baglietto L, Giles GG, Bailey CG, Rasko JEJ, Shields BJ, Price JT, Majerus PW, Sutherland RL, Tiganis T, McLean CA, Mitchell CA (2010) Inositol polyphosphate 4-phosphatase II regulates PI3K/Akt signaling and is lost in human basal-like breast cancers. Proc Natl Acad Sci U S A 107:22231–22236

Ferguson CJ, Lenk GM, Meisler MH (2009) Defective autophagy in neurons and astrocytes from mice deficient in PI(3,5)P_2. Hum Mol Genet 18:4868–4878

Ferron M, Boudiffa M, Arsenault M, Rached M, Pata M, Giroux S, Elfassihi L, Kisseleva M, Majerus Philip W, Rousseau F, Vacher J (2011) Inositol polyphosphate 4-phosphatase B as a regulator of bone mass in mice and humans. Cell Metab 14:466–477

Ferron M, Vacher J (2006) Characterization of the murine Inpp 4b gene and identification of a novel isoform. Gene 376:152–161

Foti M, Audhya A, Emr SD (2001) Sac1 lipid phosphatase and Stt4 phosphatidylinositol 4-kinase regulate a pool of phosphatidylinositol 4-phosphate that functions in the control of the actin cytoskeleton and vacuole morphology. Mol Biol Cell 12:2396–2411

Freeburn RW, Wright KL, Burgess SJ, Astoul E, Cantrell DA, Ward SG (2002) Evidence that SHIP-1 contributes to phosphatidylinositol 3,4,5-trisphosphate metabolism in T lymphocytes and can regulate novel phosphoinositide 3-kinase effectors. J Immunol 169:5441–5450

Fukuda R, Hayashi A, Utsunomiya A, Nukada Y, Fukui R, Itoh K, Tezuka K, Ohashi K, Mizuno K, Sakamoto M, Hamanoue M, Tsuji T (2005) Alteration of phosphatidylinositol 3-kinase cascade in the multilobulated nuclear formation of adult T cell leukemia/lymphoma (ATLL). Proc Natl Acad Sci U S A 102:15213–15218

Fukuda RI, Tsuchiya K, Suzuki K, Itoh K, Fujita J, Utsunomiya A, Tsuji T (2009) Human T-cell leukemia virus type I tax down-regulates the expression of phosphatidylinositol 3,4,5-trisphosphate inositol phosphatases via the NF-kappaB pathway. J Biol Chem 284:2680–2689

Fukui K, Wada T, Kagawa S, Nagira K, Ikubo M, Ishihara H, Kobayashi M, Sasaoka T (2005) Impact of the liver-specific expression of SHIP2 (SH2-containing inositol 5'-phosphatase 2) on insulin signaling and glucose metabolism in mice. Diabetes 54:1958–1967

Fults D, Pedone C (1993) Deletion mapping of the long arm of chromosome 10 in glioblastoma multiforme. Genes Chromosom Cancer 7:173–177

Galli SJ, Tsai M (2010) Mast cells in allergy and infection: versatile effector and regulatory cells in innate and adaptive immunity. Eur J Immunol 40:1843–1851

Gevaert O, Smet FD, Timmerman D, Moreau Y, Moor BD (2006) Predicting the prognosis of breast cancer by integrating clinical and microarray data with Bayesian networks. Bioinformatics 22:e184–e190

Gewinner C, Wang ZC, Richardson A, Teruya-Feldstein J, Etemadmoghadam D, Bowtell D, Barretina J, Lin WM, Rameh L, Salmena L, Pandolfi PP, Cantley LC (2009) Evidence that inositol polyphosphate 4-phosphatase type II is a tumor suppressor that inhibits PI3K signaling. Cancer Cell 16:115–125

Ghansah T, Paraiso KH, Highfill S, Desponts C, May S, McIntosh JK, Wang JW, Ninos J, Brayer J, Cheng F, Sotomayor E, Kerr WG (2004) Expansion of myeloid suppressor cells in SHIP-deficient mice represses allogeneic T cell responses. J Immunol 173:7324–7330

Giuriato S, Payrastre B, Drayer AL, Plantavid M, Woscholski R, Parker P, Erneux C, Chap H (1997) Tyrosine phosphorylation and relocation of SHIP are integrin-mediated in thrombin-stimulated human blood platelets. J Biol Chem 272:26857–26863

Giuriato S, Pesesse X, Bodin S, Sasaki T, Viala C, Marion E, Penninger J, Schurmans S, Erneux C, Payrastre B (2003) SH2-containing inositol 5-phosphatases 1 and 2 in blood platelets: their interactions and roles in the control of phosphatidylinositol 3,4,5-trisphosphate levels. Biochem J 376:199–207

Gratacap M-P, Séverin S, Chicanne G, Plantavid M, Payrastre B (2008) Different roles of SHIP1 according to the cell context: the example of blood platelets. Adv Enzyme Regul 48:240–252

Grempler R, Zibrova D, Schoelch C, van Marle A, Rippmann JF, Redemann N (2007) Normalization of prandial blood glucose and improvement of glucose tolerance by liver-specific inhibition of SH2 domain containing inositol phosphatase 2 (SHIP2) in diabetic KKAy mice: SHIP2 inhibition causes insulin-mimetic effects on glycogen metabolism, gluconeogenesis, and glycolysis. Diabetes 56:2235–2241

Gruvberger S, Ringnér M, Chen Y, Panavally S, Saal LH, Borg Å, Fernö M, Peterson C, Meltzer PS (2001) Estrogen receptor status in breast cancer is associated with remarkably distinct gene expression patterns. Cancer Res 61:5979–5984

Guermonprez P, Valladeau J, Zitvogel L, Théry C, Amigorena S (2002) Antigen presentation and T cell stimulation by dendritic cells. Annu Rev Immunol 20:621–667

Guittard G, Mortier E, Tronchère H, Firaguay G, Gérard A, Zimmermann P, Payrastre B, Nunès JA (2010) Evidence for a positive role of PtdIns5P in T-cell signal transduction pathways. FEBS Lett 584:2455–2460

Guo S, Stolz LE, Lemrow SM, York JD (1999) SAC1-like domains of yeast SAC1, INP52, and INP53 and of human synaptojanin encode polyphosphoinositide phosphatases. J Biol Chem 274:12990–12995
Guoling Z, Huijuan Y, Kaili X (1997) Loss of heterozygosity on chromosome 17p13.3 in ovarian cancer and cervical cancer. Chin J Oncol 19:401–403
Gupta A, Dey CS (2009) PTEN and SHIP2 regulates PI3K/Akt pathway through focal adhesion kinase. Mol Cell Endocrinol 309:55–62
Gurung R, Tan A, Ooms LM, McGrath MJ, Huysmans RD, Munday AD, Prescott M, Whisstock JC, Mitchell CA (2003) Identification of a novel domain in two mammalian inositol-polyphosphate 5-phosphatases that mediates membrane ruffle localization: the inositol 5-phosphatase SKIP localizes to the endoplasmic reticulum and translocates to membrane ruffles following epidermal growth factor stimulation. J Biol Chem 278:11376–11385
Habib T, Hejna JA, Moses RE, Decker SJ (1998) Growth factors and insulin stimulate tyrosine phosphorylation of the 51C/SHIP2 protein. J Biol Chem 273:18605–18609
Haddon DJ, Antignano F, Hughes MR, Blanchet M-R, Zbytnuik L, Krystal G, McNagny KM (2009) SHIP1 is a repressor of mast cell hyperplasia, cytokine production, and allergic inflammation in vivo. J Immunol 183:228–236
Haffner C, Takei K, Chen H, Ringstad N, Hudson A, Butler MH, Salcini AE, Di Fiore PP, De Camilli P (1997) Synaptojanin 1: localization on coated endocytic intermediates in nerve terminals and interaction of its 170 kDa isoform with Eps15. FEBS Lett 419:175–180
Hamilton MJ, Ho VW, Kuroda E, Ruschmann J, Antignano F, Lam V, Krystal G (2010) Role of SHIP in cancer. Exp Hematol 39(1):2–13
Hampshire DJ, Ayub M, Springell K, Roberts E, Jafri H, Rashid Y, Bond J, Riley JH, Woods CG (2006) MORM syndrome (mental retardation, truncal obesity, retinal dystrophy and micropenis), a new autosomal recessive disorder, links to 9q34. Eur J Hum Genet 14:543–548
Harris TW, Hartwieg E, Horvitz HR, Jorgensen EM (2000) Mutations in synaptojanin disrupt synaptic vesicle recycling. J Cell Biol 150:589–600
Hartikainen JM, Tuhkanen H, Kataja V, Eskelinen M, Uusitupa M, Kosma V-M, Mannermaa A (2006) Refinement of the 22q12–q13 breast cancer-associated region: evidence of TMPRSS6 as a candidate gene in an Eastern Finnish population. Clin Cancer Res 12:1454–1462
Haucke V (2003) Where proteins and lipids meet: membrane trafficking on the move. Dev Cell 4:153–157
Hejna JA, Saito H, Merkens LS, Tittle TV, Jakobs PM, Whitney MA, Grompe M, Friedberg AS, Moses RE (1995) Cloning and characterization of a human cDNA (INPPL1) sharing homology with inositol polyphosphate phosphatases. Genomics 29:285–287
Helgason CD, Damen JE, Rosten P, Grewal R, Sorensen P, Chappel SM, Borowski A, Jirik F, Krystal G, Humphries RK (1998) Targeted disruption of SHIP leads to hemopoietic perturbations, lung pathology, and a shortened life span. Genes Dev 12:1610–1620
Helgason CD, Kalberer CP, Damen JE, Chappel SM, Pineault N, Krystal G, Humphries RK (2000) A dual role for Src homology 2 domain-containing inositol-5-phosphatase (Ship) in immunity. J Exp Med 191:781–794
Hellsten E, Bernard DJ, Owens JW, Eckhaus M, Suchy SF, Nussbaum RL (2002) Sertoli cell vacuolization and abnormal germ cell adhesion in mice deficient in an inositol polyphosphate 5-phosphatase. Biol Reprod 66:1522–1530
Hellsten E, Evans JP, Bernard DJ, Jänne PA, Nussbaum RL (2001) Disrupted sperm function and fertilin β processing in mice deficient in the inositol polyphosphate 5-phosphatase Inpp 5b. Dev Biol 240:641–653
Hernandez LD, Hueffer K, Wenk MR, Galán JE (2004) *Salmonella* modulates vesicular traffic by altering phosphoinositide metabolism. Science 304:1805–1807
Hitomi K, Tahara-Hanaoka S, Someya S, Fujiki A, Tada H, Sugiyama T, Shibayama S, Shibuya K, Shibuya A (2010) An immunoglobulin-like receptor, Allergin-1, inhibits immunoglobulin E-mediated immediate hypersensitivity reactions. Nat Immunol 11:601–607

Hodgkin M, Craxton A, Parry J, Hughes P, Potter B, Michell R, Kirk C (1994) Bovine testis and human erythrocytes contain different subtypes of membrane-associated Ins(1,4,5)P3/ Ins(1,3,4,5)P4 5-phosphomonoesterases. Biochem J 297:637

Hodgson MC, Shao LJ, Frolov A, Li R, Peterson LE, Ayala G, Ittmann MM, Weigel NL, Agoulnik IU (2011) Decreased expression and androgen regulation of the tumor suppressor gene INPP4B in prostate cancer. Cancer Res 71:572–582

Hollander MC, Blumenthal GM, Dennis PA (2011) PTEN loss in the continuum of common cancers, rare syndromes and mouse models. Nat Rev Cancer 11:289–301

Hoopes RR Jr, Shrimpton AE, Knohl SJ, Hueber P, Hoppe B, Matyus J, Simckes A, Tasic V, Toenshoff B, Suchy SF, Nussbaum RL, Scheinman SJ (2005) Dent disease with mutations in OCRL1. Am J Hum Genet 76:260–267

Horan KA, Watanabe K-I, Kong AM, Bailey CG, Rasko JEJ, Sasaki T, Mitchell CA (2007) Regulation of FcγR-stimulated phagocytosis by the 72-kDa inositol polyphosphate 5-phosphatase: SHIP1, but not the 72-kDa 5-phosphatase, regulates complement receptor 3 mediated phagocytosis by differential recruitment of these 5-phosphatases to the phagocytic cup. Blood 110:4480–4491

Hori H, Sasaoka T, Ishihara H, Wada T, Murakami S, Ishiki M, Kobayashi M (2002) Association of SH2-containing inositol phosphatase 2 with the insulin resistance of diabetic db/db Mice. Diabetes 51:2387–2394

Horn S, Endl E, Fehse B, Weck MM, Mayr GW, Jucker M (2004) Restoration of SHIP activity in a human leukemia cell line downregulates constitutively activated phosphatidylinositol 3-kinase/Akt/GSK-3β signaling and leads to an increased transit time through the G1 phase of the cell cycle. Leukemia 18:1839–1849

Hou X, Hagemann N, Schoebel S, Blankenfeldt W, Goody RS, Erdmann KS, Itzen A (2011) A structural basis for Lowe syndrome caused by mutations in the Rab-binding domain of OCRL1. EMBO J 30:1659–1670

Huber M, Helgason CD, Damen JE, Liu L, Humphries RK, Krystal G (1998a) The src homology 2-containing inositol phosphatase (SHIP) is the gatekeeper of mast cell degranulation. Proc Natl Acad Sci U S A 95:11330–11335

Huber M, Helgason CD, Scheid MP, Duronio V, Humphries RK, Krystal G (1998b) Targeted disruption of SHIP leads to steel factor-induced degranulation of mast cells. EMBO J 17:7311–7319

Hughes WE, Cooke FT, Parker PJ (2000) Sac phosphatase domain proteins. Biochem J 350:337–352

Hung CS, Lin YL, Wu CI, Huang CJ, Ting LP (2009) Suppression of hepatitis B viral gene expression by phosphoinositide 5-phosphatase SKIP. Cell Microbiol 11:37–50

Hyvola N, Diao A, McKenzie E, Skippen A, Cockcroft S, Lowe M (2006) Membrane targeting and activation of the Lowe syndrome protein OCRL1 by rab GTPases. EMBO J 25:3750–3761

Hyvonen ME, Saurus P, Wasik A, Heikkila E, Havana M, Trokovic R, Saleem M, Holthofer H, Lehtonen S (2010) Lipid phosphatase SHIP2 downregulates insulin signalling in podocytes. Mol Cell Endocrinol 328:70–79

Ibarra JA, Steele-Mortimer O (2009) *Salmonella*—the ultimate insider. *Salmonella* virulence factors that modulate intracellular survival. Cell Microbiol 11:1579–1586

Ijuin T, Mochizuki Y, Fukami K, Funaki M, Asano T, Takenawa T (2000) Identification and characterization of a novel inositol polyphosphate 5- phosphatase. J Biol Chem 275: 10870–10875

Ijuin T, Takenawa T (2003) SKIP negatively regulates insulin-induced GLUT4 translocation and membrane ruffle formation. Mol Cell Biol 23:1209–1220

Ijuin T, Yu YE, Mizutani K, Pao A, Tateya S, Tamori Y, Bradley A, Takenawa T (2008) Increased insulin action in SKIP heterozygous knockout mice. Mol Cell Biol 28:5184–5195

Ikonomov OC, Sbrissa D, Fligger J, Delvecchio K, Shisheva A (2010) ArPIKfyve regulates Sac3 protein abundance and turnover. J Biol Chem 285:26760–26764

Ikonomov OC, Sbrissa D, Ijuin T, Takenawa T, Shisheva A (2009) Sac3 is an insulin-regulated PtdIns(3,5)P_2 phosphatase: gain in insulin responsiveness through Sac3 downregulation in adipocytes. J Biol Chem 284:23961–23971

Irie F, Okuno M, Pasquale EB, Yamaguchi Y (2005) EphrinB-EphB signalling regulates clathrin-mediated endocytosis through tyrosine phosphorylation of synaptojanin 1. Nat Cell Biol 7:501–509

Ishida S, Funakoshi A, Miyasaka K, Shimokata H, Ando F, Takiguchi S (2006) Association of SH-2 containing inositol 5'-phosphatase 2 gene polymorphisms and hyperglycemia. Pancreas 33:63–67

Ivetac I, Gurung R, Hakim S, Horan KA, Sheffield DA, Binge LC, Majerus PW, Tiganis T, Mitchell CA (2009) Regulation of PI(3)K/Akt signalling and cellular transformation by inositol polyphosphate 4-phosphatase-1. EMBO Rep 10:487–493

Ivetac I, Munday AD, Kisseleva MV, Zhang XM, Luff S, Tiganis T, Whisstock JC, Rowe T, Majerus PW, Mitchell CA (2005) The type Iα inositol polyphosphate 4-phosphatase generates and terminates phosphoinositide 3-kinase signals on endosomes and the plasma membrane. Mol Biol Cell 16:2218–2233

Jackson S, Schoenwaelder S, Matzaris M, Brown S, CA M (1995) Phosphatidylinositol 3,4,5-trisphosphate is a substrate for the 75 kDa inositol polyphosphate 5-phosphatase and a novel 5-phosphatase which forms a complex with the p85/p110 form of phosphoinositide 3-kinase. EMBO J 14:4490–4500

Jacoby M, Cox JJ, Gayral S, Hampshire DJ, Ayub M, Blockmans M, Pernot E, Kisseleva MV, Compère P, Schiffmann SN, Gergely F, Riley JH, Pérez-Morga D, Woods CG, Schurmans S (2009) INPP5E mutations cause primary cilium signaling defects, ciliary instability and ciliopathies in human and mouse. Nat Genet 41:1027–1031

Jänne PA, Suchy SF, Bernard D, MacDonald M, Crawley J, Grinberg A, Wynshaw-Boris A, Westphal H, Nussbaum RL (1998) Functional overlap between murine Inpp 5b and Ocrl1 may explain why deficiency of the murine ortholog for OCRL1 does not cause Lowe syndrome in mice. J Clin Investig 101:2042–2053

Jefferson AB, Majerus PW (1995) Properties of type II inositol polyphosphate 5-phosphatase. J Biol Chem 270:9370–9377

Jiang J, Hui CC (2008) Hedgehog signaling in development and cancer. Dev Cell 15:801–812

Jiang X, Stuible M, Chalandon Y, Li A, Chan WY, Eisterer W, Krystal G, Eaves A, Eaves C (2003) Evidence for a positive role of SHIP in the BCR-ABL-mediated transformation of primitive murine hematopoietic cells and in human chronic myeloid leukemia. Blood 102:2976–2984

Jin H, White SR, Shida T, Schulz S, Aguiar M, Gygi SP, Bazan JF, Nachury MV (2010) The conserved Bardet–Biedl syndrome proteins assemble a coat that traffics membrane proteins to cilia. Cell 141:1208–1219

Jones J, Otu H, Spentzos D, Kolia S, Inan M, Beecken WD, Fellbaum C, Gu X, Joseph M, Pantuck AJ, Jonas D, Libermann TA (2005) Gene signatures of progression and metastasis in renal cell cancer. Clin Cancer Res 11:5730–5739

Joseph RE, Walker J, Norris FA (1999) Assignment of the inositol polyphosphate 4-phosphatase type I gene (INPP4A) to human chromosome band 2q11.2 by in situ hybridization. Cytogenet Cell Genet 87:276–277

Jospin M, Watanabe S, Joshi D, Young S, Hamming K, Thacker C, Snutch TP, Jorgensen EM, Schuske K (2007) UNC-80 and the NCA ion channels contribute to endocytosis defects in synaptojanin mutants. Curr Biol 17:1595–1600

Joubert M, Eisenring J, Robb J, Andermann F (1969) Familial agenesis of the cerebellar vermis. A syndrome of episodic hyperpnea, abnormal eye movements, ataxia, and retardation. Neurology 19:813–825

Kagawa S, Sasaoka T, Yaguchi S, Ishihara H, Tsuneki H, Murakami S, Fukui K, Wada T, Kobayashi S, Kimura I, Kobayashi M (2005) Impact of SRC homology 2-containing inositol 5'-phosphatase 2 gene polymorphisms detected in a Japanese population on insulin signaling. J Clin Endocrinol Metab 90:2911–2919

Kagawa S, Soeda Y, Ishihara H, Oya T, Sasahara M, Yaguchi S, Oshita R, Wada T, Tsuneki H, Sasaoka T (2008) Impact of transgenic overexpression of SH2-containing inositol 5'-phosphatase 2 on glucose metabolism and insulin signaling in mice. Endocrinology 149:642–650

Kaisaki PJ, Delpine M, Woon PY, Sebag-Montefiore L, Wilder SP, Menzel S, Vionnet N, Marion E, Riveline JP, Charpentier G, Schurmans S, Levy JC, Lathrop M, Farrall M, Gauguier D (2004) Polymorphisms in type II SH2 domain-containing inositol 5-phosphatase (INPPL1, SHIP2) are associated with physiological abnormalities of the metabolic syndrome. Diabetes 53:1900–1904

Kalesnikoff J, Lam V, Krystal G (2002) SHIP represses mast cell activation and reveals that IgE alone triggers signaling pathways which enhance normal mast cell survival. Mol Immunol 38:1201–1206

Kamen LA, Levinsohn J, Swanson JA (2007) Differential association of phosphatidylinositol 3-kinase, SHIP-1, and PTEN with forming phagosomes. Mol Biol Cell 18:2463–2472

Karayiorgou M, Simon TJ, Gogos JA (2010) 22q11.2 microdeletions: linking DNA structural variation to brain dysfunction and schizophrenia. Nat Rev Neurosci 11:402–416

Kashiwada M, Cattoretti G, McKeag L, Rouse T, Showalter BM, Al-Alem U, Niki M, Pandolfi PP, Field EH, Rothman PB (2006) Downstream of tyrosine kinases-1 and Src homology 2-containing inositol 5'-phosphatase are required for regulation of CD4 + CD25+ T Cell development. J Immunol 176:3958–3965

Kasprowicz J, Kuenen S, Miskiewicz K, Habets RLP, Smitz L, Verstreken P (2008) Inactivation of clathrin heavy chain inhibits synaptic recycling but allows bulk membrane uptake. J Cell Biol 182:1007–1016

Kawano T, Indo Y, Nakazato H, Shimadzu M, Matsuda I (1998) Oculocerebrorenal syndrome of Lowe: three mutations in the OCRL1 gene derived from three patients with different phenotypes. Am J Med Genet 77:348–355

Kenworthy L, Park T, Charnas LR (1993) Cognitive and behavioral profile of the oculocerebrorenal syndrome of Lowe. Am J Med Genet 46:297–303

Kim B, Bang S, Lee S, Kim S, Jung Y, Lee C, Choi K, Lee S-G, Lee K, Lee Y, Kim S-S, Yeom Y-I, Kim Y-S, Yoo H-S, Song K, Lee I (2003) Expression profiling and subtype-specific expression of stomach cancer. Cancer Res 63:8248–8255

Kimura T, Sakamoto H, Appella E, Siraganian RP (1997) The negative signaling molecule SH2 domain-containing inositol-polyphosphate 5-phosphatase (SHIP) binds to the tyrosine-phosphorylated β subunit of the high affinity IgE receptor. J Biol Chem 272:13991–13996

Kisseleva MV, Cao L, Majerus PW (2002) Phosphoinositide-specific inositol polyphosphate 5-phosphatase IV inhibits Akt/protein kinase B phosphorylation and leads to apoptotic cell death. J Biol Chem 277:6266–6272

Kisseleva MV, Wilson MP, Majerus PW (2000) The isolation and characterization of a cDNA encoding phospholipid-specific inositol polyphosphate 5-phosphatase. J Biol Chem 275:20110–20116

Knodler LA, Steele-Mortimer O (2003) Taking possession: biogenesis of the *Salmonella*-containing vacuole. Traffic 4:587–599

Koch A, Mancini A, El Bounkari O, Tamura T (2005) The SH2-domian-containing inositol 5-phosphatase (SHIP)-2 binds to c-Met directly via tyrosine residue 1356 and involves hepatocyte growth factor (HGF)-induced lamellipodium formation, cell scattering and cell spreading. Oncogene 24:3436–3447

Kong AM, Horan KA, Sriratana A, Bailey CG, Collyer LJ, Nandurkar HH, Shisheva A, Layton MJ, Rasko JEJ, Rowe T, Mitchell CA (2006) Phosphatidylinositol 3-phosphate [PtdIns(3)P] is generated at the plasma membrane by an inositol polyphosphate 5-phosphatase: endogenous PtdIns(3)P can promote GLUT4 translocation to the plasma membrane. Mol Cell Biol 26:6065–6081

Kong AM, Speed CJ, O'Malley CJ, Layton MJ, Meehan T, Loveland KL, Cheema S, Ooms LM, Mitchell CA (2000) Cloning and characterization of a 72-kDa inositol-polyphosphate 5-phosphatase localized to the Golgi network. J Biol Chem 275:24052–24064

Konishi H, Takahashi T, Kozaki KI, Yatabe Y, Mitsudomi T, Fujii Y, Sugiura T, Matsuda H (1998) Detailed deletion mapping suggests the involvement of a tumor suppressor gene at 17p13.3, distal to p53, in the pathogenesis of lung cancers. Oncogene 17:2095–2100

Konrad G, Schlecker T, Faulhammer F, Mayinger P (2002) Retention of the yeast Sac1p phosphatase in the endoplasmic reticulum causes distinct changes in cellular phosphoinositide levels and stimulates microsomal ATP transport. J Biol Chem 277:10547–10554

Konradt C, Frigimelica E, Nothelfer K, Puhar A, Salgado-Pabon W, di Bartolo V, Scott-Algara D, Rodrigues Cristina D, Sansonetti Philippe J, Phalipon A (2011) The *Shigella flexneri* type three secretion system effector IpgD inhibits T cell migration by manipulating host phosphoinositide metabolism. Cell Host Microbe 9:263–272

Kortüm F, Das S, Flindt M, Morris-Rosendahl DJ, Stefanova I, Goldstein A, Horn D, Klopocki E, Kluger G, Martin P, Rauch A, Roumer A, Saitta S, Walsh LE, Wieczorek D, Uyanik G, Kutsche K, Dobyns WB (2011) The core FOXG1 syndrome phenotype consists of postnatal microcephaly, severe mental retardation, absent language, dyskinesia, and corpus callosum hypogenesis. J Med Genet 48:396–406

Kubota T, Sakurai A, Arakawa K, Shirnazu M, Wakui K, Furihata K, Fukushirna Y (1998) Identification of two novel mutations in the OCRLI gene in Japanese families with Lowe syndrome. Clin Genet 54:199–202

Kuroda E, Antignano F, Ho VW, Hughes MR, Ruschmann J, Lam V, Kawakami T, Kerr WG, McNagny KM, Sly LM, Krystal G (2011) SHIP represses Th2 skewing by inhibiting IL-4 production from basophils. J Immunol 186:323–332

Kurose K, Hoshaw-Woodard S, Adeyinka A, Lemeshow S, Watson HP, Eng C (2001) Genetic model of multi-step breast carcinogenesis involving the epithelium and stroma: clues to tumour–microenvironment interactions. Hum Mol Genet 10:1907–1913

Labrec EH, Schneider H, Magnani TJ, Formal SB (1964) Epithelial cell penetration as an essential step in the pathogenesis of bacillary dysentery. J Bacteriol 88:1503–1518

Lai X-H, Golovliov I, Sjostedt A (2001) *Francisella tularensis* induces cytopathogenicity and apoptosis in murine macrophages via a mechanism that requires intracellular bacterial multiplication. Infect Immun 69:4691–4694

Lai X-H, Sjostedt A (2003) Delineation of the molecular mechanisms of *Francisella tularensis*-induced apoptosis in murine macrophages. Infect Immun 71:4642–4646

Lakhanpal GK, Vecchiarelli-Federico LM, Li YJ, Cui JW, Bailey ML, Spaner DE, Dumont DJ, Barber DL, Ben-David Y (2010) The inositol phosphatase SHIP-1 is negatively regulated by Fli-1 and its loss accelerates leukemogenesis. Blood 116:428–436

Lamkin TD, Walk SF, Liu L, Damen JE, Krystal G, Ravichandran KS (1997) Shc interaction with Src homology two domain containing inositol phosphatase (SHIP) in vivo requires the Shc-phosphotyrosine binding domain and two specific phosphotyrosines on SHIP. J Biol Chem 272:10396–10401

Lamont JM, McManamy CS, Pearson AD, Clifford SC, Ellison DW (2004) Combined histopathological and molecular cytogenetic stratification of medulloblastoma patients. Clin Cancer Res 10:5482–5493

LaTulippe E, Satagopan J, Smith A, Scher H, Scardino P, Reuter V, Gerald WL (2002) Comprehensive gene expression analysis of prostate cancer reveals distinct transcriptional programs associated with metastatic disease. Cancer Res 62:4499–4506

Laxminarayan KM, Chan BK, Tetaz T, Bird PI, Mitchell CA (1994) Characterization of a cDNA encoding the 43-kDa membrane-associated inositol-polyphosphate 5-phosphatase. J Biol Chem 269:17305–17310

Laxminarayan KM, Matzaris M, Speed CJ, Mitchell CA (1993) Purification and characterization of a 43-kDa membrane-associated inositol polyphosphate 5-phosphatase from human placenta. J Biol Chem 269:17305–17310

Leahey A-M, Charnas LR, Nussbaum RL (1993) Nonsense mutations in the OCRL-1 gene in patients with the oculocerebrorenal syndrome of Lowe. Hum Mol Genet 2:461–463

Lee SH, Davison JA, Vidal SM, Belouchi A (2001) Cloning, expression and chromosomal location of NKX6B to 10q26, a region frequently deleted in brain tumors. Mamm Genome 12:157–162

Lee SY, Wenk MR, Kim Y, Nairn AC, De Camilli P (2004) Regulation of synaptojanin 1 by cyclin-dependent kinase five at synapses. Proc Natl Acad Sci U S A 101:546–551

Leheste J-R, Rolinski B, Vorum H, Hilpert J, Nykjaer A, Jacobsen C, Aucouturier P, Moskaug JØ, Otto A, Christensen EI, Willnow TE (1999) Megalin knockout mice as an animal model of low molecular weight proteinuria. Am J Pathol 155:1361–1370

Lenk GM, Ferguson CJ, Chow CY, Jin N, Jones JM, Grant AE, Zolov SN, Winters JJ, Giger RJ, Dowling JJ, Weisman LS, Meisler MH (2011) Pathogenic Mechanism of the FIG4 mutation responsible for charcot-marie-tooth disease CMT4J. PLoS Genet 7:e1002104

Leone M, Cellitti J, Pellecchia M (2008) NMR studies of a heterotypic Sam–Sam domain association: the interaction between the lipid phosphatase Ship2 and the EphA2 receptor. Biochemistry 47:12721–12728

Lin T, Orrison BM, Suchy SF, Lewis RA, Nussbaum RL (1998) Mutations are not uniformly distributed throughout the OCRL1 gene in Lowe syndrome patients. Mol Genet Metab 64:58–61

Liu Q, Dumont DJ (1997) Molecular cloning and chromosomal localization in human and mouse of the SH2-containing inositol phosphatase, INPP5D (SHIP). Genomics 39:109–112

Liu Q, Oliveira-Dos-Santos AJ, Mariathasan S, Bouchard D, Jones J, Sarao R, Kozieradzki I, Ohashi PS, Penninger JM, Dumont DJ (1998a) The inositol polyphosphate 5-phosphatase Ship is a crucial negative regulator of B cell antigen receptor signaling. J Exp Med 188:1333–1342

Liu Q, Sasaki T, Kozieradzki I, Wakeham A, Itie A, Dumont DJ, Penninger JM (1999) SHIP is a negative regulator of growth factor receptor-mediated PKB/Akt activation and myeloid cell survival. Genes Dev 13:786–791

Liu Q, Shalaby F, Jones J, Bouchard D, Dumont DJ (1998b) The SH2-containing inositol polyphosphate 5-phosphatase, ship, is expressed during hematopoiesis and spermatogenesis. Blood 91:2753–2759

Liu Y, Bankaitis VA (2010) Phosphoinositide phosphatases in cell biology and disease. Prog Lipid Res 49:201–217

Liu Y, Boukhelifa M, Tribble E, Morin-Kensicki E, Uetrecht A, Bear JE, Bankaitis VA (2008) The Sac1 phosphoinositide phosphatase regulates Golgi membrane morphology and mitotic spindle organization in mammals. Mol Biol Cell 19:3080–3096

Lo TC, Barnhill LM, Kim Y, Nakae EA, Yu AL, Diccianni MB (2009) Inactivation of SHIP1 in T-cell acute lymphoblastic leukemia due to mutation and extensive alternative splicing. Leuk Res 33:1562–1566

López-Pedrera C, Barbarroja N, Villalba JM (2009) Novel biomarkers of atherosclerosis and cardiovascular risk in autoimmune diseases: genomics and proteomics approaches poteomics. Clin Appl 3:213–225

Lowe C, Terrey M, Maclachlan EA (1952) Organic-aciduria, decreased renal ammonia production, hydrophthalmos, and mental retardation: a clinical entity. AMA Am J Dis Child 83:164–184

Lucas DM, Rohrschneider LR (1999) A novel spliced form of SH2-containing inositol phosphatase is expressed during myeloid development. Blood 93:1922–1933

Luo JM, Liu ZL, Hao HL, Wang FX, Dong ZR, Ohno R (2004) Mutation analysis of SHIP gene in acute leukemia. Zhongguo Shi Yan Xue Ye Xue Za Zhi 12:420–426

Luo JM, Yoshida H, Komura S, Ohishi N, Pan L, Shigeno K, Hanamura I, Miura K, Iida S, Ueda R, Naoe T, Akao Y, Ohno R, Ohnishi K (2003) Possible dominant-negative mutation of the SHIP gene in acute myeloid leukemia. Leukemia 17:1–8

MacKeigan JP, Murphy LO, Blenis J (2005) Sensitized RNAi screen of human kinases and phosphatases identifies new regulators of apoptosis and chemoresistance. Nat Cell Biol 7:591–600

Maeda K, Mehta H, Drevets DA, Coggeshall KM (2010) IL-6 increases B-cell IgG production in a feed-forward proinflammatory mechanism to skew hematopoiesis and elevate myeloid production. Blood 115:4699–4706

Maher CA, Kumar-Sinha C, Cao X, Kalyana-Sundaram S, Han B, Jing X, Sam L, Barrette T, Palanisamy N, Chinnaiyan AM (2009) Transcriptome sequencing to detect gene fusions in cancer. Nature 458:97–101

Malecz N, McCabe PC, Spaargaren C, Qiu R-G, Chuang Y, Symons M (2000) Synaptojanin 2, a novel Rac1 effector that regulates clathrin-mediated endocytosis. Curr Biol 10:1383–1386

Mallo GV, Espina M, Smith AC, Terebiznik MR, Alemán A, Finlay BB, Rameh LE, Grinstein S, Brumell JH (2008) SopB promotes phosphatidylinositol 3-phosphate formation on *Salmonella* vacuoles by recruiting Rab5 and Vps34. J Cell Biol 182:741–752

Manford A, Xia T, Saxena AK, Stefan C, Hu F, Emr SD, Mao Y (2010) Crystal structure of the yeast Sac1: implications for its phosphoinositide phosphatase function. EMBO J 29:1489–1498

Mani M, Lee SY, Lucast L, Cremona O, Di Paolo G, De Camilli P, Ryan TA (2007) The dual phosphatase activity of synaptojanin1 is required for both efficient synaptic vesicle endocytosis and reavailability at nerve terminals. Neuron 56:1004–1018

Manji SSM, Williams LH, Miller KA, Ooms LM, Bahlo M, Mitchell CA, Dahl H-HM (2011) A mutation in synaptojanin 2 causes progressive hearing loss in the ENU-mutagenised mouse strain *Mozart*. PLoS ONE 6:e17607

Mantovani A, Sica A, Locati M (2007) New vistas on macrophage differentiation and activation. Eur J Immunol 37:14–16

Mao Y, Balkin DM, Zoncu R, Erdmann KS, Tomasini L, Hu F, Jin MM, Hodsdon ME, De Camilli P (2009) A PH domain within OCRL bridges clathrin-mediated membrane trafficking to phosphoinositide metabolism. EMBO J 28:1831–1842

Marcello MR, Evans JP (2010) Multivariate analysis of male reproductive function in Inpp 5b−/− mice reveals heterogeneity in defects in fertility, sperm–egg membrane interaction and proteolytic cleavage of sperm ADAMs. Mol Hum Reprod 16:492–505

Marcus SL, Wenk MR, Steele-Mortimer O, Finlay BB (2001) A synaptojanin-homologous region of *Salmonella typhimurium* SigD is essential for inositol phosphatase activity and Akt activation. FEBS Lett 494:201–207

Marion E, Kaisaki PJ, Pouillon V, Gueydan C, Levy JC, Bodson A, Krzentowski G, Daubresse JC, Mockel J, Behrends J, Servais G, Szpirer C, Kruys V, Gauguier D, Schurmans S (2002) The gene INPPL1, encoding the lipid phosphatase SHIP2, is a candidate for type 2 diabetes in rat and man. Diabetes 51:2012–2017

Marjanovic J, Wilson MP, Zhang C, Zou J, Nicholas P, Majerus PW (2011) The role of inositol polyphosphate 4-phosphatase 1 in platelet function using a weeble mouse model. Adv Enzyme Regul 51:101–105

Marza E, Long T, Saiardi A, Sumakovic M, Eimer S, Hall DH, Lesa GM (2008) Polyunsaturated fatty acids influence synaptojanin localization to regulate synaptic vesicle recycling. Mol Biol Cell 19:833–842

Matzaris M, Jackson SP, Laxminarayan KM, Speed CJ, Mitchell CA (1994) Identification and characterization of the phosphatidylinositol-(4, 5)-bisphosphate 5-phosphatase in human platelets. J Biol Chem 269:3397–3402

Matzaris M, O'Malley CJ, Badger A, Speed CJ, Bird PI, Mitchell CA (1998) Distinct membrane and cytosolic forms of inositol polyphosphate 5-phosphatase II. J Biol Chem 273:8256–8267

Maxwell MJ, Yuan Y, Anderson KE, Hibbs ML, Salem HH, Jackson SP (2004) SHIP1 and Lyn kinase negatively regulate integrin αIIbβ3 signaling in platelets. J Biol Chem 279:32196–32204

McPherson PS, Garcia EP, Slepnev VI, David C, Zhang X, Grabs D, Sossini WS, Bauerfeind R, Nemoto Y, De Camilli P (1996) A presynaptic inositol-5-phosphatase. Nature 379:353–357

Metzner A, Precht C, Fehse B, Fiedler W, Stocking C, Günther A, Mayr GW, Jücker M (2009) Reduced proliferation of CD34(+) cells from patients with acute myeloid leukemia after gene transfer of INPP5D. Gene Ther 16:570–573

Miletic AV, Anzelon-Mills AN, Mills DM, Omori SA, Pedersen IM, Shin D-M, Ravetch JV, Bolland S, Morse HC, Rickert RC (2010) Coordinate suppression of B cell lymphoma by PTEN and SHIP phosphatases. J Exp Med 207:2407–2420

Milosevic I, Sorensen JB, Lang T, Krauss M, Nagy GB, Haucke V, Jahn R, Neher E (2005) Plasmalemmal phosphatidylinositol-4,5-bisphosphate level regulates the releasable vesicle pool size in chromaffin cells. J Neurosci 25:2557–2565

Minagawa T, Ijuin T, Mochizuki Y, Takenawa T (2001) Identification and characterization of a Sac domain-containing phosphoinositide 5-phosphatase. J Biol Chem 276:22011–22015

Mitchell CA, Connolly TM, Majerus PW (1989) Identification and isolation of a 75-kDa inositol polyphosphate-5-phosphatase from human platelets. J Biol Chem 264:8873–8877

Mochizuki Y, Takenawa T (1999) Novel inositol polyphosphate 5-phosphatase localizes at membrane ruffles. J Biol Chem 274:36790–36795

Munday AD, Norris FA, Caldwell KK, Brown S, Majerus PW, Mitchell CA (1999) The inositol polyphosphate 4-phosphatase forms a complex with phosphatidylinositol 3-kinase in human platelet cytosol. Proc Natl Acad Sci U S A 96:3640–3645

Muraille E, Dassesse D, Vanderwinden JM, Cremer H, Rogister B, Erneux C, Schiffmann SN (2001) The SH2 domain-containing 5-phosphatase SHIP2 is expressed in the germinal layers of embryo and adult mouse brain: increased expression in N-CAM-deficient mice. Neuroscience 105:1019–1030

Nakamura K, Malykhin A, Coggeshall KM (2002) The Src homology two domain-containing inositol 5-phosphatase negatively regulates Fcγ receptor-mediated phagocytosis through immunoreceptor tyrosine-based activation motif-bearing phagocytic receptors. Blood 100:3374–3382

Nakatsu F, Perera RM, Lucast L, Zoncu R, Domin J, Gertler FB, Toomre D, De Camilli P (2010) The inositol 5-phosphatase SHIP2 regulates endocytic clathrin-coated pit dynamics. J Cell Biol 190:307–315

Naylor T, Greshock J, Wang Y, Colligon T, Yu QC, Clemmer V, Zaks T, Weber B (2005) High resolution genomic analysis of sporadic breast cancer using array-based comparative genomic hybridization. Breast Cancer Res 7:1–13

Neill L, Tien AH, Rey-Ladino J, Helgason CD (2007) SHIP-deficient mice provide insights into the regulation of dendritic cell development and function. Exp Hematol 35:627–639

Nemoto Y, Arribas M, Haffner C, DeCamilli P (1997) Synaptojanin 2, a novel synaptojanin isoform with a distinct targeting domain and expression pattern. J Biol Chem 272:30817–30821

Nemoto Y, De Camilli P (1999) Recruitment of an alternatively spliced form of synaptojanin 2 to mitochondria by the interaction with the PDZ domain of a mitochondrial outer membrane protein. EMBO J 18:2991–3006

Nemoto Y, Kearns BG, Wenk MR, Chen H, Mori K, Alb JG, De Camilli P, Bankaitis VA (2000) Functional characterization of a mammalian Sac1 and mutants exhibiting substrate-specific defects in phosphoinositide phosphatase activity. J Biol Chem 275:34293–34305

Nemoto Y, Wenk MR, Watanabe M, Daniell L, Murakami T, Ringstad N, Yamada H, Takei K, De Camilli P (2001) Identification and characterization of a synaptojanin 2 splice isoform predominantly expressed in nerve terminals. J Biol Chem 276:41133–41142

Nguyen N-YN, Maxwell MJ, Ooms LM, Davies EM, Hilton AA, Collinge JE, Hilton DJ, Kile BT, Mitchell CA, Hibbs ML, Jane SM, Curtis DJ (2011) An ENU-induced mouse mutant of SHIP1 reveals a critical role of the stem cell isoform for suppression of macrophage activation. Blood 117:5362–5371

Niebuhr K, Giuriato S, Pedron T, Philpott DJ, Gaits F, Sable J, Sheetz MP, Parsot C, Sansonetti PJ, Payrastre B (2002) Conversion of PtdIns(4,5)P_2 into PtdIns(5)P by the S.flexneri effector IpgD reorganizes host cell morphology. EMBO J 21:5069–5078

Niebuhr K, Jouihri N, Allaoui A, Gounon P, Sansonetti PJ, Parsot C (2000) IpgD, a protein secreted by the type III secretion machinery of *Shigella flexneri*, is chaperoned by IpgE and implicated in entry focus formation. Mol Microbiol 38:8–19

Nigg EA, Raff JW (2009) Centrioles, centrosomes, and cilia in health and disease. Cell 139:663–678

Nishio M, Watanabe K, Sasaki J, Taya C, Takasuga S, Iizuka R, Balla T, Yamazaki M, Watanabe H, Itoh R, Kuroda S, Horie Y, Förster I, Mak TW, Yonekawa H, Penninger JM, Kanaho Y, Suzuki A, Sasaki T (2007) Control of cell polarity and motility by the PtdIns(3,4,5)P_3 phosphatase SHIP1. Nat Cell Biol 9:36–44

Noakes CJ, Lee G, Lowe M (2011) The PH domain proteins IPIP27A and B link OCRL1 to receptor recycling in the endocytic pathway. Mol Biol Cell 22:606–623

Norden AGW, Lapsley M, Igarashi T, Kelleher CL, Lee PJ, Matsuyama T, Scheinman SJ, Shiraga H, Sundin DP, Thakker RV, Unwin RJ, Verroust P, Moestrup SK (2002) Urinary megalin deficiency implicates abnormal tubular endocytic function in fanconi syndrome. J Am Soc Nephrol 13:125–133

Norris FA, Atkins RC, Majerus PW (1997a) The cDNA cloning and characterization of inositol polyphosphate 4- phosphatase type II. Evidence for conserved alternative splicing in the 4-phosphatase family. J Biol Chem 272:23859–23864

Norris FA, Atkins RC, Majerus PW (1997b) Inositol polyphosphate 4-phosphatase is inactivated by calpain-mediated proteolysis in stimulated human platelets. J Biol Chem 272:10987–10989

Norris FA, Auethavekiat V, Majerus PW (1995) The isolation and characterization of cDNA encoding human and rat brain inositol polyphosphate 4-phosphatase. J Biol Chem 270:16128–16133

Norris FA, Majerus PW (1994) Hydrolysis of phosphatidylinositol 3,4-bisphosphate by inositol polyphosphate 4-phosphatase isolated by affinity elution chromatography. J Biol Chem 269:8716–8720

Norris FA, Wilson MP, Wallis TS, Galyov EE, Majerus PW (1998) SopB, a protein required for virulence of *Salmonella Dublin*, is an inositol phosphate phosphatase. Proc Natl Acad Sci U S A 95:14057–14059

Novick P, Osmond BC, Botstein D (1989) Suppressors of yeast actin mutations. Genetics 121:659–674

Nystuen A, Legare ME, Shultz LD, Frankel WN (2001) A null mutation in inositol polyphosphate 4-phosphatase type I causes selective neuronal loss in weeble mutant mice. Neuron 32:203–212

O'Connell RM, Chaudhuri AA, Rao DS, Baltimore D (2009) Inositol phosphatase SHIP1 is a primary target of miR-155. Proc Natl Acad Sci U S A 106:7113–7118

Oh S-Y, Zheng T, Bailey ML, Barber DL, Schroeder JT, Kim Y-K, Zhu Z (2007) Src homology 2 domain-containing inositol 5-phosphatase 1 deficiency leads to a spontaneous allergic inflammation in the murine lung. J Allergy Clin Immunol 119:123–131

Olivos-Glander I, Jänne P, Nussbaum R (1995) The oculocerebrorenal syndrome gene product is a 105-kD protein localized to the Golgi complex. Am J Hum Genet 57:817–823

Ong CJ, Ming-Lum A, Nodwell M, Ghanipour A, Yang L, Williams DE, Kim J, Demirjian L, Qasimi P, Ruschmann J, Cao LP, Ma K, Chung SW, Duronio V, Andersen RJ, Krystal G, Mui AL (2007) Small-molecule agonists of SHIP1 inhibit the phosphoinositide 3-kinase pathway in hematopoietic cells. Blood 110:1942–1949

Ooms LM, Fedele CG, Astle MV, Ivetac I, Cheung V, Pearson RB, Layton MJ, Forrai A, Nandurkar HH, Mitchell CA (2006) The inositol polyphosphate 5-phosphatase, PIPP, is a novel regulator of phosphoinositide 3-kinase-dependent neurite elongation. Mol Biol Cell 17:607–622

Osborne MA, Zenner G, Lubinus M, Zhang X, Songyang Z, Cantley LC, Majerus P, Burn P, Kochan JP (1996) The inositol 5′-phosphatase SHIP binds to immunoreceptor signaling motifs and responds to high affinity IgE receptor aggregation. J Biol Chem 271:29271–29278

Ou Z, Berg JS, Yonath H, Enciso VB, Miller DT, Picker J, Lenzi T, Keegan CE, Sutton VR, Belmont J, Chinault AC, Lupski JR, Cheung SW, Roeder E, Patel A (2008) Microduplications of 22q11.2 are frequently inherited and are associated with variable phenotypes. Genet Med 10:267–277

Palomo I, Alarcon M, Moore-Carrasco R, Argiles JM (2006) Hemostasis alterations in metabolic syndrome. Int J Mol Med 18:969–974

Paolo GD, Moskowitz HS, Gipson K, Wenk MR, Voronov S, Obayashi M, Flavell R, Fitzsimonds RM, Ryan TA, Camilli PD (2004) Impaired PtdIns(4,5)P_2 synthesis in nerve terminals produces defects in synaptic vesicle trafficking. Nature 431:415–422

Parker JA, Metzler M, Georgiou J, Mage M, Roder JC, Rose AM, Hayden MR, Neri C (2007) Huntingtin-interacting protein 1 influences worm and mouse presynaptic function and protects caenorhabditis elegans neurons against mutant polyglutamine toxicity. J Neurosci 27:11056–11064

Parrella P, Scintu M, Prencipe M, Poeta ML, Gallo AP, Rabitti C, Rinaldi M, Tommasi S, Paradiso A, Schittulli F, Valori VM, Toma S, Altomare V, Fazio VM (2005) HIC1 promoter methylation and 17p13.3 allelic loss in invasive ductal carcinoma of the breast. Cancer Lett 222:75–81

Parsa KV, Ganesan LP, Rajaram MV, Gavrilin MA, Balagopal A, Mohapatra NP, Wewers MD, Schlesinger LS, Gunn JS, Tridandapani S (2006) Macrophage pro-inflammatory response to *Francisella novicida* infection is regulated by SHIP. PLoS Pathog 2:e71

Paternotte N, Zhang J, Vandenbroere I, Backers K, Blero D, Kioka N, Vanderwinden JM, Pirson I, Erneux C (2005) SHIP2 interaction with the cytoskeletal protein Vinexin. FEBS J 272:6052–6066

Patterson D (2009) Molecular genetic analysis of Down syndrome. Hum Genet 126:195–214

Pedersen IM, Otero D, Kao E, Miletic AV, Hother C, Ralfkiaer E, Rickert RC, Gronbaek K, David M (2009) Onco-miR-155 targets SHIP1 to promote TNFα-dependent growth of B cell lymphomas. EMBO Mol Med 1:288–295

Pendaries C, Tronchère H, Arbibe L, Mounier J, Gozani O, Cantley L, Fry MJ, Gaits-Iacovoni F, Sansonetti PJ, Payrastre B (2006) PtdIns(5)P activates the host cell PI3-kinase/Akt pathway during *Shigella flexneri* infection. EMBO J 25:1024–1034

Peng Q, Malhotra S, Torchia JA, Kerr WG, Coggeshall KM, Humphrey MB (2010) TREM2- and DAP12-dependent activation of PI3 K requires DAP10 and is inhibited by SHIP1. Sci Signal 3:ra38

Perera RM, Zoncu R, Lucast L, De Camilli P, Toomre D (2006) Two synaptojanin 1 isoforms are recruited to clathrin-coated pits at different stages. Proc Natl Acad Sci U S A 103:19332–19337

Pernot E, Terryn S, Cheong S, Markadieu N, Janas S, Blockmans M, Jacoby M, Pouillon V, Gayral S, Rossier B, Beauwens R, Erneux C, Devuyst O, Schurmans S (2011) The inositol Inpp5k 5-phosphatase affects osmoregulation through the vasopressin-aquaporin 2 pathway in the collecting system. Pflügers Archiv Eur J Physiol 462(6):871-883

Pesesse X, Deleu S, De Smedt F, Drayer L, Erneux C (1997) Identification of a second SH2-domain-containing protein closely related to the phosphatidylinositol polyphosphate 5-phosphatase SHIP. Biochem Biophys Res Commun 239:697–700

Peverall J, Edkins E, Goldblatt J, Murch A (2000) Identification of a novel deletion of the entire OCRL1 gene detected by FISH analysis in a family with Lowe syndrome. Clin Genet 58:479–482

Pirruccello M, Swan LE, Folta-Stogniew E, De Camilli P (2011) Recognition of the F&H motif by the Lowe syndrome protein OCRL. Nat Struct Mol Biol 18:789–795

Pixley FJ, Stanley ER (2004) CSF-1 regulation of the wandering macrophage: complexity in action. Trends Cell Biol 14:628–638

Poretti A, Dietrich-Alber F, Brancati F, Dallapiccola B, Valente EM, Boltshauser E (2009) Normal cognitive functions in Joubert syndrome. Neuropediatrics 40:287–290

Prasad N, Topping RS, Decker SJ (2001) SH2-containing inositol 5'-phosphatase SHIP2 associates with the p130(Cas) adapter protein and regulates cellular adhesion and spreading. Mol Cell Biol 21:1416–1428

Prasad NK, Decker SJ (2005) SH2-containing 5'-inositol phosphatase, SHIP2, regulates cytoskeleton organization and ligand-dependent down-regulation of the epidermal growth factor receptor. J Biol Chem 280:13129–13136

Prasad NK, Tandon M, Badve S, Snyder PW, Nakshatri H (2008a) Phosphoinositol phosphatase SHIP2 promotes cancer development and metastasis coupled with alterations in EGF receptor turnover. Carcinogenesis 29:25–34

Prasad NK, Tandon M, Handa A, Moore GE, Babbs CF, Snyder PW, Bose S (2008b) High expression of obesity-linked phosphatase SHIP2 in invasive breast cancer correlates with reduced disease-free survival. Tumour Biol 29:330–341

Prasad NK, Werner ME, Decker SJ (2009) Specific tyrosine phosphorylations mediate signal-dependent stimulation of SHIP2 inositol phosphatase activity, while the SH2 domain confers an inhibitory effect to maintain the basal activity. Biochemistry 48:6285–6287

Quade BJ, Wang T-Y, Sornberger K, Cin PD, Mutter GL, Morton CC (2004) Molecular pathogenesis of uterine smooth muscle tumors from transcriptional profiling. Genes Chromosom Cancer 40:97–108

Qualmann B, Roos J, DiGregorio PJ, Kelly RB (1999) Syndapin I, a synaptic dynamin-binding protein that associates with the neural Wiskott–Aldrich syndrome protein. Mol Biol Cell 10:501–513

Raaijmakers JH, Deneubourg L, Rehmann H, de Koning J, Zhang Z, Krugmann S, Erneux C, Bos JL (2007) The PI3K effector Arap3 interacts with the PI(3,4,5)P_3 phosphatase SHIP2 in a SAM domain-dependent manner. Cell Signal 19:1249–1257

Rahman P, Huysmans RD, Wiradjaja F, Gurung R, Ooms LM, Sheffield DA, Dyson JM, Layton MJ, Sriratana A, Takada H, Tiganis T, Mitchell CA (2011) Silencer of death domains (SODD) inhibits skeletal muscle and kidney enriched inositol phosphatase (SKIP) and regulates phosphoinositide 3-kinase (PI3K)/Akt signalling to the actin cytoskeleton. J Biol Chem 286(34):29758–29770

Rajaram MVS, Butchar JP, Parsa KVL, Cremer TJ, Amer A, Schlesinger LS, Tridandapani S (2009) Akt and SHIP modulate *Francisella* escape from the phagosome and induction of the Fas-mediated death pathway. PLoS ONE 4:e7919

Ramaswamy S, Ross KN, Lander ES, Golub TR (2003) A molecular signature of metastasis in primary solid tumors. Nat Genet 33:49–54

Ramjaun AR, McPherson PS (1996) Tissue-specific alternative splicing generates two synaptojanin isoforms with differential membrane binding properties. J Biol Chem 271:24856–24861

Rauh MJ, Ho V, Pereira C, Sham A, Sly LM, Lam V, Huxham L, Minchinton AI, Mui A, Krystal G (2005) SHIP represses the generation of alternatively activated macrophages. Immunity 23:361–374

Richard F, Pacyna-Gengelbach M, Schlüns K, Fleige B, Winzer KJ, Szymas J, Dietel M, Petersen I, Schwendel A (2000) Patterns of chromosomal imbalances in invasive breast cancer. Int J Cancer 89:305–310

Ringstad N, Nemoto Y, De Camilli P (1997) The SH3p4/Sh3p8/SH3p13 protein family: binding partners for synaptojanin and dynamin via a Grb2-like Src homology 3 domain. Proc Natl Acad Sci U S A 94:8569–8574

Risio M, Casorzo L, et al (2003) Deletions of 17p are associated with transition from early to advanced colorectal cancer. Cancer Gene Cytogenet 147(1):44–49

Rivas MP, Kearns BG, Xie Z, Guo S, Sekar MC, Hosaka K, Kagiwada S, York JD, Bankaitis VA (1999) Pleiotropic alterations in lipid metabolism in yeast sac1 mutants: relationship to "Bypass Sec14p" and inositol auxotrophy. Mol Biol Cell 10:2235–2250

Rodríguez-Escudero I, Ferrer NL, Rotger R, Cid VJ, Molina M (2011) Interaction of the *Salmonella* Typhimurium effector protein SopB with host cell Cdc42 is involved in intracellular replication. Mol Microbiol 80:1220–1240

Rodriguez-Gabin AG, Ortiz E, Demoliner K, Si Q, Almazan G, Larocca JN (2010) Interaction of Rab31 and OCRL-1 in oligodendrocytes: its role in transport of mannose 6-phosphate receptors. J Neurosci Res 88:589–604

Roget K, Malissen M, Malbec O, Malissen B, Daëron M (2008) Non-T cell activation linker promotes mast cell survival by dampening the recruitment of SHIP1 by linker for activation of T cells. J Immunol 180:3689–3698

Rohde HM, Cheong FY, Konrad G, Paiha K, Mayinger P, Boehmelt G (2003) The human phosphatidylinositol phosphatase SAC1 interacts with the coatomer I complex. J Biol Chem 278:52689–52699

Rohrschneider LR, Fuller JF, Wolf I, Liu Y, Lucas DM (2000) Structure, function, and biology of SHIP proteins. Genes Dev 14:505–520

Roongapinun S, Oh S-Y, Wu F, Panthong A, Zheng T, Zhu Z (2010) Role of SHIP-1 in the adaptive immune responses to aeroallergen in the airway. PLoS ONE 5:e14174

Roos J, Kelly RB (1998) Dap160, a neural-specific Eps15 homology and multiple SH3 domain-containing protein that interacts with Drosophila Dynamin. J Biol Chem 273:19108–19119

Ross TS, Jefferson AB, Mitchell CA, Majerus PW (1991) Cloning and expression of human 75-kDa inositol polyphosphate-5-phosphatase. J Biol Chem 266:20283–20289

Rudge SA, Anderson DM, Emr SD (2004) Vacuole size control: regulation of PtdIns(3,5)P_2 levels by the vacuole-associated Vac14–Fig4 complex, a PtdIns(3,5)P_2-specific phosphatase. Mol Biol Cell 15:24–36

Ruschmann J, Ho V, Antignano F, Kuroda E, Lam V, Ibaraki M, Snyder K, Kim C, Flavell RA, Kawakami T, Sly L, Turhan AG, Krystal G (2010) Tyrosine phosphorylation of SHIP promotes its proteasomal degradation. Exp Hematol 38(392–402):402 e1

Rusk N, Le PU, Mariggio S, Guay G, Lurisci C, Nabi IR, Corda D, Symons M (2003) Synaptojanin 2 functions at an early step of clathrin-mediated endocytosis. Curr Biol 13:659–663

Sachs AJ, David SA, Haider NB, Nystuen AM (2009) Patterned neuroprotection in the Inpp4a^{wbl} mutant mouse cerebellum correlates with the expression of Eaat4. PLoS ONE 4

Sakisaka T, Itoh T, Miura K, Takenawa T (1997) Phosphatidylinositol 4,5-bisphosphate phosphatase regulates the rearrangement of actin filaments. Mol Cell Biol 17:3841–3849

Saltiel AR, Kahn CR (2001) Insulin signalling and the regulation of glucose and lipid metabolism. Nature 414:799–806

Sanchez-Carbayo M, Socci ND, Lozano J, Saint F, Cordon-Cardo C (2006) Defining molecular profiles of poor outcome in patients with invasive bladder cancer using oligonucleotide microarrays. J Clin Oncol 24:778–789

Sansonetti PJ, Kopecko DJ, Formal SB (1982) Involvement of a plasmid in the invasive ability of *Shigella flexineri.* Infect Immun 35:852–860

Santic M, Al-Khodor S, Abu Kwaik Y (2010) Cell biology and molecular ecology of *Francisella tularensis.* Cell Microbiol 12:129–139

Sasaki J, Kofuji S, Itoh R, Momiyama T, Takayama K, Murakami H, Chida S, Tsuya Y, Takasuga S, Eguchi S, Asanuma K, Horie Y, Miura K, Davies EM, Mitchell C, Yamazaki M, Hirai H, Takenawa T, Suzuki A, Sasaki T (2010) The PtdIns(3,4)P_2 phosphatase INPP4A is a suppressor of excitotoxic neuronal death. Nature 465:497–501

Sasaki T, Takasuga S, Sasaki J, Kofuji S, Eguchi S, Yamazaki M, Suzuki A (2009) Mammalian phosphoinositide kinases and phosphatases. Prog Lipid Res 48:307–343

Sasaoka T, Hori H, Wada T, Ishiki M, Haruta T, Ishihara H, Kobayashi M (2001) SH2-containing inositol phosphatase 2 negatively regulates insulin-induced glycogen synthesis in L6 myotubes. Diabetologia 44:1258–1267

Sasaoka T, Wada T, Fukui K, Murakami S, Ishihara H, Suzuki R, Tobe K, Kadowaki T, Kobayashi M (2004) SH2-containing inositol phosphatase 2 predominantly regulates Akt2, and not Akt1, phosphorylation at the plasma membrane in response to insulin in 3T3-L1 adipocytes. J Biol Chem 279:14835–14843

Satre V, Monnier N, Berthoin F, Ayuso C, Joannard A, Jouk P-S, Lopez-Pajares I, Megabarne A, Philippe HJ, Plauchu H, Torres ML, Lunardi J (1999) Characterization of a Germline Mosaicism in families with Lowe syndrome, and identification of seven novel mutations in the OCRL1 gene. Am J Hum Genet 65:68–76

Sattler M, Griffin JD (2003) Molecular mechanisms of transformation by the BCR-ABL oncogene. Seminars in hematology 40. Supplement 2:4–10

Sattler M, Verma S, Byrne CH, Shrikhande G, Winkler T, Algate PA, Rohrschneider LR, Griffin JD (1999) BCR/ABL directly inhibits expression of SHIP, an SH2-containing polyinositol-5-phosphatase involved in the regulation of hematopoiesis. Mol Cell Biol 19:7473–7480

Sbrissa D, Ikonomov OC, Fenner H, Shisheva A (2008) ArPIKfyve homomeric and heteromeric interactions Scaffold PIKfyve and Sac3 in a complex to promote PIKfyve activity and functionality. J Mol Biol 384:766–779

Sbrissa D, Ikonomov OC, Fu Z, Ijuin T, Gruenberg J, Takenawa T, Shisheva A (2007) Core protein machinery for mammalian phosphatidylinositol 3,5-bisphosphate synthesis and turnover that regulates the progression of endosomal transport. J Biol Chem 282:23878–23891

Sbrissa D, Ikonomov OC, Strakova J, Dondapati R, Mlak K, Deeb R, Silver R, Shisheva A (2004) A mammalian ortholog of saccharomyces cerevisiae Vac14 that associates with and up-regulates PIKfyve phosphoinositide 5-kinase activity. Mol Biol Cell 24:10437–10447

Sbrissa D, Shisheva A (2005) Acquisition of unprecedented phosphatidylinositol 3,5-bisphosphate rise in hyperosmotically stressed 3T3-L1 adipocytes, mediated by ArPIKfyve-PIKfyve pathway. J Biol Chem 280:7883–7889

Schmid AC, Wise HM, Mitchell CA, Nussbaum R, Woscholski R (2004) Type II phosphoinositide 5-phosphatases have unique sensitivities towardss fatty acid composition and head group phosphorylation. FEBS Lett 576:9–13

Schneider JF, Boltshauser E, Neuhaus TJ, Rauscher C, Martin E (2001) MRI and proton spectroscopy in Lowe syndrome. Neuropediatrics 32(45):48

Schorr M, Then A, Tahirovic S, Hug N, Mayinger P (2001) The phosphoinositide phosphatase Sac1p controls trafficking of the yeast Chs3p chitin synthase. Curr Biol 11:1421–1426

Seet L-F, Cho S, Hessel A, Dumont DJ (1998) Molecular cloning of multiple isoforms of synaptojanin 2 and assignment of the gene to mouse chromosome 17A2-3.1. Biochem Biophys Res Commun 247:116–122

Sekulic A, Kim SY, Hostetter G, Savage S, Einspahr JG, Prasad A, Sagerman P, Curiel-Lewandrowski C, Krouse R, Bowden GT, Warneke J, Alberts DS, Pittelkow MR, DiCaudo D, Nickoloff BJ, Trent JM, Bittner M (2010) Loss of inositol polyphosphate 5-phosphatase is an early event in development of cutaneous squamous cell carcinoma. Cancer Prev Res 3:1277–1283

Severin S, Gratacap MP, Lenain N, Alvarez L, Hollande E, Penninger JM, Gachet C, Plantavid M, Payrastre B (2007) Deficiency of Src homology 2 domain-containing inositol 5-phosphatase 1 affects platelet responses and thrombus growth. J Clin Investig 117:944–952

Sharma M, Batra J, Mabalirajan U, Sharma S, Nagarkatti R, Aich J, Sharma SK, Niphadkar PV, Ghosh B (2008) A genetic variation in inositol polyphosphate 4 phosphatase A enhances susceptibility to asthma. Am J Respir Crit Care Med 177:712–719

Shearman AM, Hudson TJ, Andresen JM, Wu X, Sohn RL, Haluska F, Housman DE, Weiss JS (1996) The gene for Schnyder's crystalline corneal dystrophy maps to human chromosome 1p34.1–p36. Hum Mol Genet 5:1667–1672

Shearn CT, Norris FA (2007) Biochemical characterization of the type I inositol polyphosphate 4-phosphatase C2 domain. Biochem Biophys Res Commun 356:255–259

Shearn CT, Walker J, Norris FA (2001) Identification of a novel spliceoform of inositol polyphosphate 4-phosphatase type Iα expressed in human platelets: structure of human inositol polyphosphate 4-phosphatase type I gene. Biochem Biophys Res Commun 286:119–125

Shin H-W, Hayashi M, Christoforidis S, Lacas-Gervais S, Hoepfner S, Wenk MR, Modregger J, Uttenweiler-Joseph S, Wilm M, Nystuen A, Frankel WN, Solimena M, De Camilli P, Zerial M (2005) An enzymatic cascade of Rab5 effectors regulates phosphoinositide turnover in the endocytic pathway. J Cell Biol 170:607–618

Shrimpton A, Hoopes RJ, Knohl S, Hueber P, Reed A, Christie P, Igarashi T, Lee P, Lehman A, White C, Milford D, Sanchez M, Unwin R, Wrong O, Thakker R, Scheinman S (2009) OCRL1 mutations in Dent 2 patients suggest a mechanism for phenotypic variability. Nephron Physiol 112:27–36

Sjöblom T, Jones S, Wood LD, Parsons DW, Lin J, Barber TD, Mandelker D, Leary RJ, Ptak J, Silliman N, Szabo S, Buckhaults P, Farrell C, Meeh P, Markowitz SD, Willis J, Dawson D, Willson JKV, Gazdar AF, Hartigan J, Wu L, Liu C, Parmigiani G, Park BH, Bachman KE, Papadopoulos N, Vogelstein B, Kinzler KW, Velculescu VE (2006) The consensus coding sequences of human breast and colorectal cancers. Science 314:268–274

Sleeman MW, Wortley KE, Lai KM, Gowen LC, Kintner J, Kline WO, Garcia K, Stitt TN, Yancopoulos GD, Wiegand SJ, Glass DJ (2005) Absence of the lipid phosphatase SHIP2 confers resistance to dietary obesity. Nat Med 11:199–205

Sly LM, Hamilton MJ, Kuroda E, Ho VW, Antignano FL, Omeis SL, van Netten-Thomas CJ, Wong D, Brugger HK, Williams O, Feldman ME, Houseman BT, Fiedler D, Shokat KM, Krystal G (2009) SHIP prevents lipopolysaccharide from triggering an antiviral response in mice. Blood 113:2945–2954

Sly LM, Rauh MJ, Kalesnikoff J, Büchse T, Krystal G (2003) SHIP, SHIP2, and PTEN activities are regulated in vivo by modulation of their protein levels: SHIP is up-regulated in macrophages and mast cells by lipopolysaccharide. Exp Hematol 31:1170–1181

Smith K, Humphreys D, Hume PJ, Koronakis V (2010) Enteropathogenic *Escherichia coli* recruits the cellular inositol phosphatase SHIP2 to regulate actin-pedestal formation. Cell Host Microbe 7:13–24

Sørlie T, Tibshirani R, Parker J, Hastie T, Marron JS, Nobel A, Deng S, Johnsen H, Pesich R, Geisler S, Demeter J, Perou CM, Lønning PE, Brown PO, Børresen-Dale A-L, Botstein D (2003) Repeated observation of breast tumor subtypes in independent gene expression data sets. Proc Natl Acad Sci U S A 100:8418–8423

Sotiropoulou G, Pampalakis G, Lianidou E, Mourelatos Z (2009) Emerging roles of microRNAs as molecular switches in the integrated circuit of the cancer cell. RNA 15:1443–1461

Speed CJ, Little PJ, Hayman JA, Mitchell CA (1996) Underexpression of the 43 kDa inositol polyphosphate 5-phosphatase is associated with cellular transformation. EMBO J 15:4852–4861

Speed CJ, Matzaris M, Bird PI, Mitchell CA (1995) Tissue distribution and intracellular localisation of the 75-kDa inositol polyphosphate 5-phosphatase. Eur J Biochem 234:216–224

Speed CJ, Mitchell CA (2000) Sustained elevation in inositol 1,4,5-trisphosphate results in inhibition of phosphatidylinositol transfer protein activity and chronic depletion of the agonist-sensitive phosphoinositide pool. J Cell Sci 113:2631–2638

Speed CJ, Neylon CB, Little PJ, Mitchell CA (1999) Underexpression of the 43 kDa inositol polyphosphate 5-phosphatase is associated with spontaneous calcium oscillations and enhanced calcium responses following endothelin-1 stimulation. J Cell Sci 112:669–679

Stearman RS, Dwyer-Nield L, Zerbe L, Blaine SA, Chan Z, Bunn PA Jr, Johnson GL, Hirsch FR, Merrick DT, Franklin WA, Baron AE, Keith RL, Nemenoff RA, Malkinson AM, Geraci MW (2005) Analysis of orthologous gene expression between human pulmonary adenocarcinoma and a carcinogen-induced murine model. Am J Pathol 167:1763–1775

Steele-Mortimer O, Knodler LA, Marcus SL, Scheid MP, Goh B, Pfeifer CG, Duronio V, Finlay BB (2000) Activation of Akt/protein kinase B in epithelial cells by the *Salmonella typhimurium* effector SigD. J Biol Chem 275:37718–37724

Suchy SF, Nussbaum RL (2002) The deficiency of PIP2 5-phosphatase in Lowe syndrome affects actin polymerization. Am J Hum Genet 71:1420–1427

Suchy SF, Olivos-Glander IM, Nussbaum RL (1995) Lowe syndrome, a deficiency of a phosphatidyl-inositol 4,5-bisphosphate 5-phosphatase in the Golgi apparatus. Hum Mol Genet 4:2245–2250

Suwa A, Yamamoto T, Sawada A, Minoura K, Hosogai N, Tahara A, Kurama T, Shimokawa T, Aramori I (2009) Discovery and functional characterization of a novel small molecule inhibitor of the intracellular phosphatase, SHIP2. Br J Pharmacol 158:879–887

Swan LE, Tomasini L, Pirruccello M, Lunardi J, De Camilli P (2010) Two closely related endocytic proteins that share a common OCRL-binding motif with APPL1. Proc Natl Acad Sci U S A 107:3511–3516

Swanson JA, Hoppe AD (2004) The coordination of signaling during Fc receptor-mediated phagocytosis. J Leukoc Biol 76:1093–1103

Tahirovic S, Schorr M, Mayinger P (2005) Regulation of intracellular phosphatidylinositol-4-phosphate by the Sac1 lipid phosphatase. Traffic 6:116–130

Takabayashi T, Xie MJ, Takeuchi S, Kawasaki M, Yagi H, Okamoto M, Tariqur RM, Malik F, Kuroda K, Kubota C, Fujieda S, Nagano T, Sato M (2010) LL5β directs the translocation of filamin A and SHIP2 to sites of phosphatidylinositol 3,4,5-triphosphate (PtdIns(3,4,5)P_3) accumulation, and PtdIns(3,4,5)P_3 localization is mutually modified by co-recruited SHIP2. J Biol Chem 285:16155–16165

Takahashi H, Masuda K, Ando T, Kobayashi T, Honda H (2004) Prognostic predictor with multiple fuzzy neural models using expression profiles from DNA microarray for metastases of breast cancer. J Biosci Bioeng 98:193–199

Takeshita S, Namba N, Zhao JJ, Jiang Y, Genant HK, Silva MJ, Brodt MD, Helgason CD, Kalesnikoff J, Rauh MJ, Humphries RK, Krystal G, Teitelbaum SL, Ross FP (2002) SHIP-deficient mice are severely osteoporotic due to increased numbers of hyper-resorptive osteoclasts. Nat Med 8:943–949

Takimoto K, Okada M, Nakagawa H (1989) Purification and characterization of membrane-bound inositolpolyphosphate 5-phosphatase. J Biochem 106:684–690

Tang X, Powelka AM, Soriano NA, Czech MP, Guilherme A (2005) PTEN, but Not SHIP2, suppresses insulin signaling through the phosphatidylinositol 3-kinase/Akt pathway in 3T3-L1 adipocytes. J Biol Chem 280:22523–22529

Taniguchi CM, Emanuelli B, Kahn CR (2006) Critical nodes in signalling pathways: insights into insulin action. Nat Rev Mol Cell Biol 7:85–96

Tarasenko T, Kole HK, Chi AW, Mentink-Kane MM, Wynn TA, Bolland S (2007) T cell-specific deletion of the inositol phosphatase SHIP reveals its role in regulating Th1/Th2 and cytotoxic responses. Proc Natl Acad Sci U S A 104:11382–11387

Taylor BS, Schultz N, Hieronymus H, Gopalan A, Xiao Y, Carver BS, Arora VK, Kaushik P, Cerami E, Reva B, Antipin Y, Mitsiades N, Landers T, Dolgalev I, Major JE, Wilson M, Socci ND, Lash AE, Heguy A, Eastham JA, Scher HI, Reuter VE, Scardino PT, Sander C, Sawyers CL, Gerald WL (2010) Integrative genomic profiling of human prostate cancer. Cancer Cell 18:11–22

Taylor V, Wong M, Brandts C, Reilly L, Dean NM, Cowsert LM, Moodie S, Stokoe D (2000) 5' phospholipid phosphatase SHIP-2 causes protein kinase B inactivation and cell cycle arrest in glioblastoma cells. Mol Cell Biol 20:6860–6871

Terebiznik MR, Vieira OV, Marcus SL, Slade A, Yip CM, Trimble WS, Meyer T, Finlay BB, Grinstein S (2002) Elimination of host cell PtdIns(4,5)P_2 by bacterial SigD promotes membrane fission during invasion by *Salmonella*. Nat Cell Biol 4:766–773

Trapani JG, Obholzer N, Mo W, Brockerhoff SE, Nicolson T (2009) Synaptojanin1 is required for temporal fidelity of synaptic transmission in hair cells. PLoS Genet 5:e1000480

Trivedi CM, Luo Y, Yin Z, Zhang M, Zhu W, Wang T, Floss T, Goettlicher M, Noppinger PR, Wurst W, Ferrari VA, Abrams CS, Gruber PJ, Epstein JA (2007) Hdac2 regulates the cardiac hypertrophic response by modulating Gsk3β activity. Nat Med 13:324–331

Trotta R, Parihar R, Yu J, Becknell B, Allard JN, Wen J, Ding W, Mao H, Tridandapani S, Carson WE, Caligiuri MA (2005) Differential expression of SHIP1 in CD56bright and CD56dim NK cells provides a molecular basis for distinct functional responses to monokine costimulation. Blood 105:3011–3018

Tsujishita Y, Guo S, Stolz LE, York JD, Hurley JH (2001) Specificity determinants in phosphoinositide dephosphorylation: crystal structure of an archetypal inositol polyphosphate 5-phosphatase. Cell 105:379–389

Tu Z, Ninos JM, Ma Z, Wang J-W, Lemos MP, Desponts C, Ghansah T, Howson JM, Kerr WG (2001) Embryonic and hematopoietic stem cells express a novel SH2-containing inositol 5′-phosphatase isoform that partners with the Grb2 adapter protein. Blood 98:2028–2038

Ungewickell A, Hugge C, Kisseleva M, Chang SC, Zou J, Feng Y, Galyov EE, Wilson M, Majerus PW (2005) The identification and characterization of two phosphatidylinositol-4,5-bisphosphate 4-phosphatases. Proc Natl Acad Sci U S A 102:18854–18859

Ungewickell A, Ward ME, Ungewickell E, Majerus PW (2004) The inositol polyphosphate 5-phosphatase Ocrl associates with endosomes that are partially coated with clathrin. Proc Natl Acad Sci U S A 101:13501–13506

Ungewickell AJ, Majerus PW (1999) Increased levels of plasma lysosomal enzymes in patients with Lowe syndrome. Proc Natl Acad Sci U S A 96:13342–13344

van't Veer LJ, Dai H, van de Vijver MJ, He YD, Hart AAM, Mao M, Peterse HL, van der Kooy K, Marton MJ, Witteveen AT, Schreiber GJ, Kerkhoven RM, Roberts C, Linsley PS, Bernards R, Friend SH (2002) Gene expression profiling predicts clinical outcome of breast cancer. Nature 415:530–536

Vandenbroere I, Paternotte N, Dumont JE, Erneux C, Pirson I (2003) The c-Cbl-associated protein and c-Cbl are two new partners of the SH2-containing inositol polyphosphate 5-phosphatase SHIP2. Biochem Biophys Res Commun 300:494–500

Vara JÁF, Casado E, de Castro J, Cejas P, Belda-Iniesta C, González-Barón M (2004) PI3K/Akt signalling pathway and cancer. Cancer Treat Rev 30:193–204

Vicinanza M, Di Campli A, Polishchuk E, Santoro M, Di Tullio G, Godi A, Levtchenko E, De Leo MG, Polishchuk R, Sandoval L, Marzolo M, De Matteis MA (2011) OCRL controls trafficking through early endosomes via PtdIns4,5P_2-dependent regulation of endosomal actin. EMBO J 30(24):4970–4985

Voronov SV, Frere SG, Giovedi S, Pollina EA, Borel C, Zhang H, Schmidt C, Akeson EC, Wenk MR, Cimasoni L, Arancio O, Davisson MT, Antonarakis SE, Gardiner K, De Camilli P, Di Paolo G (2008) Synaptojanin 1-linked phosphoinositide dyshomeostasis and cognitive deficits in mouse models of Down's syndrome. Proc Natl Acad Sci U S A 105:9415–9420

Vyas P, Norris FA, Joseph R, Majerus PW, Orkin SH (2000) Inositol polyphosphate 4-phosphatase type I regulates cell growth downstream of transcription factor GATA-1. Proc Natl Acad Sci U S A 97:13696–13701

Wada T, Sasaoka T, Funaki M, Hori H, Murakami S, Ishiki M, Haruta T, Asano T, Ogawa W, Ishihara H, Kobayashi M (2001) Overexpression of SH2-containing inositol phosphatase 2 results in negative regulation of insulin-induced metabolic actions in 3T3-L1 adipocytes via its 5'-phosphatase catalytic activity. Mol Cell Biol 21:1633–1646

Wang F, Ijuin T, Itoh T, Takenawa T (2011) Regulation of IGF-1/PI3K/Akt signaling by the phosphoinositide phosphatase pharbin. J Biochem 150:83–93

Wang JW, Howson JM, Ghansah T, Desponts C, Ninos JM, May SL, Nguyen KH, Toyama-Sorimachi N, Kerr WG (2002) Influence of SHIP on the NK repertoire and allogeneic bone marrow transplantation. Science 295:2094–2097

Weber-Mangal S, Sinn H-P, Popp S, Klaes R, Emig R, Bentz M, Mansmann U, Bastert G, Bartram CR, Jauch A (2003) Breast cancer in young women ($\leq$35 years): genomic aberrations detected by comparative genomic hybridization. Int J Cancer 107:583–592

Wei H-C, Sanny J, Shu H, Baillie DL, Brill JA, Price JV, Harden N (2003) The Sac1 lipid phosphatase regulates cell shape change and the JNK cascade during dorsal closure in drosophila. Curr Biol 13:1882–1887

Weisser SB, McLarren KW, Voglmaier N, van Netten-Thomas CJ, Antov A, Flavell RA, Sly LM (2011) Alternative activation of macrophages by IL-4 requires SHIP degradation. Eur J Immunol 41:1742–1753

Wen X-Y, Wu S-Y, Li Z-Q, Liu Z-Q, Zhang J-J, Wang G-F, Jiang Z-H, Wu; S-G (2009) Ellagitannin (BJA3121), an anti-proliferative natural polyphenol compound, can regulate the expression of MiRNAs in HepG2 cancer cells. Phytotherapy Res 23:778–784

West M, Blanchette C, Dressman H, Huang E, Ishida S, Spang R, Zuzan H, Olson JA, Marks JR, Nevins JR (2001) Predicting the clinical status of human breast cancer by using gene expression profiles. Proc Natl Acad Sci U S A 98:11462–11467

Westbrook TF, Martin ES, Schlabach MR, Leng Y, Liang AC, Feng B, Zhao JJ, Roberts TM, Mandel G, Hannon GJ, DePinho RA, Chin L, Elledge SJ (2005) A genetic screen for candidate tumor suppressors identifies REST. Cell 121:837–848

Whisstock JC, Romero S, Gurung R, Nandurkar H, Ooms LM, Bottomley SP, Mitchell CA (2000) The inositol polyphosphate 5-phosphatases and the apurinic/apyrimidinic base excision repair endonucleases share a common mechanism for catalysis. J Biol Chem 275:37055–37061

Whitters E, Cleves A, McGee T, Skinner H, Bankaitis V (1993) SAC1p is an integral membrane protein that influences the cellular requirement for phospholipid transfer protein function and inositol in yeast. J Cell Biol 122:79–94

Williams C, Choudhury R, McKenzie E, Lowe M (2007) Targeting of the type II inositol polyphosphate 5-phosphatase INPP5B to the early secretory pathway. J Cell Sci 120: 3941–3951

Wisniewski D, Strife A, Swendeman S, Erdjument-Bromage H, Geromanos S, Kavanaugh WM, Tempst P, Clarkson B (1999) A novel SH2-containing phosphatidylinositol 3,4,5-trisphosphate 5-phosphatase (SHIP2) is constitutively tyrosine phosphorylated and associated with src homologous and collagen gene (SHC) in chronic myelogenous leukemia progenitor cells. Blood 93:2707–2720

Wolf G, Ziyadeh FN (2007) Cellular and molecular mechanisms of proteinuria in diabetic nephropathy. Nephron Physiol 106:p26–p31

Xiong Q, Deng CY, Chai J, Jiang SW, Xiong YZ, Li FE, Zheng R (2009) Knockdown of endogenous SKIP gene enhanced insulin-induced glycogen synthesis signaling in differentiating C2C12 myoblasts. BMB Reports 42:119–124

Yan C, Hongjuan H, Yanjiang X, Zhengbin H, Kai L, Fengwei Z, Jing H, Qiong W (2011) Expression patterns of imprinted gene Inpp 5f-v3 during mouse brain development. J Mol Histol 42:167–173

Yavari A, Nagaraj R, Owusu-Ansah E, Folick A, Ngo K, Hillman T, Call G, Rohatgi R, Scott MP, Banerjee U (2010) Role of lipid metabolism in smoothened derepression in hedgehog signaling. Dev Cell 19:54–65

Yogo K, Mizutamari M, Mishima K, Takenouchi H, Ishida-Kitagawa N, Sasaki T, Takeya T (2006) Src homology 2 (SH2)-containing 5'-inositol phosphatase localizes to podosomes, and the SH2 domain is implicated in the attenuation of bone resorption in osteoclasts. Endocrinology 147:3307–3317

Yoon J, Lee J, Namkoong S, Bae S, Kim Y, Han S, Cho Y, Nam G, Kim C, Seo J, Ahn W (2003) cDNA microarray analysis of gene expression profiles associated with cervical cancer. Cancer Res Treat 35:451–459

Yoshida M (2001) Multiple viral strategies of HTLV-1 for dysregulation of cell growth control. Annu Rev Immunol 19:475–496

Yu J, Ryan DG, Getsios S, Oliveira-Fernandes M, Fatima A, Lavker RM (2008) MicroRNA-184 antagonizes microRNA-205 to maintain SHIP2 levels in epithelia. Proc Natl Acad Sci U S A 105:19300–19305

Yuan Y, Gao X, Guo N, Zhang H, Xie Z, Jin M, Li B, Yu L, Jing N (2007) rSac3, a novel Sac domain phosphoinositide phosphatase, promotes neurite outgrowth in PC12 cells. Cell Res 17:919–932

Zhang TY, Daynes RA (2007) Macrophages from 11β-hydroxysteroid dehydrogenase type 1-deficient mice exhibit an increased sensitivity to lipopolysaccharide stimulation due to TGF-β-mediated up-regulation of SHIP1 expression. J Immunol 179:6325–6335

Zhang X, Chow CY, Sahenk Z, Shy ME, Meisler MH, Li J (2008) Mutation of FIG4 causes a rapidly progressive, asymmetric neuronal degeneration. Brain 131:1990–2001

Zhang X, Hartz PA, Philip E, Racusen LC, Majerus PW (1998) Cell lines from kidney proximal tubules of a patient with Lowe syndrome lack OCRL inositol polyphosphate 5-phosphatase and accumulate phosphatidylinositol 4,5-bisphosphate. J Biol Chem 273:1574–1582

Zhang X, Jefferson AB, Auethavekiat V, Majerus PW (1995) The protein deficient in Lowe syndrome is a phosphatidylinositol-4,5-bisphosphate 5-phosphatase. Proc Natl Acad Sci U S A 92:4853–4856

Zhang Y, Wavreille AS, Kunys AR, Pei D (2009) The SH2 domains of inositol polyphosphate 5-phosphatases SHIP1 and SHIP2 have similar ligand specificity but different binding kinetics. Biochemistry 48:11075–11083

Zhang Y, Zolov SN, Chow CY, Slutsky SG, Richardson SC, Piper RC, Yang B, Nau JJ, Westrick RJ, Morrison SJ, Meisler MH, Weisman LS (2007) Loss of Vac14, a regulator of the signaling lipid phosphatidylinositol 3,5-bisphosphate, results in neurodegeneration in mice. Proc Natl Acad Sci U S A 104:17518–17523

Zheng S, Zhang J, Di X, Xiao Z, Wang D, Li C, He Z, Han N, Guo S, Cheng S, Gao Y (2002) Loss of heterozygosity fine mapping of chromosome 17p13 in transitional cell carcinoma of human urinary bladder. Zhonghua yi xue za zhi 82:161–163

Zhou CZ, Peng ZH, Zhang F, Qiu GQ, He L (2002) Loss of heterozygosity on long arm of chromosome 22 in sporadic colorectal carcinoma. World J Gastroenterol 8:668–673

Zhou D, Chen L-M, Hernandez L, Shears SB, Galán JE (2001) A *Salmonella* inositol polyphosphatase acts in conjunction with other bacterial effectors to promote host cell actin cytoskeleton rearrangements and bacterial internalization. Mol Microbiol 39:248–260

Zhou P, Kitaura H, Teitelbaum SL, Krystal G, Ross FP, Takeshita S (2006) SHIP1 negatively regulates proliferation of osteoclast precursors via Akt-dependent alterations in D-type cyclins and p27. J Immunol 177:8777–8784

Zhou QL, Park JG, Jiang ZY, Holik JJ, Mitra P, Semiz S, Guilherme A, Powelka AM, Tang X, Virbasius J, Czech MP (2004) Analysis of insulin signalling by RNAi-based gene silencing. Biochem Soc Trans 32:817–821

Zhu GN, Zuo L, Zhou Q, Zhang SM, Zhu HQ, Gui SY, Wang Y (2004) Loss of heterozygosity on chromosome 10q22–10q23 and 22q11.2–22q12.1 and p53 gene in primary hepatocellular carcinoma. World J Gastroenterol 10:1975–1978

Zhu W, Trivedi CM, Zhou D, Yuan L, Lu MM, Epstein JA (2009) Inpp 5f is a polyphosphoinositide phosphatase that regulates cardiac hypertrophic responsiveness. Circ Res 105:1240–1247

Zhuang G, Hunter S, Hwang Y, Chen J (2007) Regulation of EphA2 receptor endocytosis by SHIP2 lipid phosphatase via phosphatidylinositol 3-kinase-dependent Rac1 activation. J Biol Chem 282:2683–2694

Zou J, Marjanovic J, Kisseleva MV, Wilson M, Majerus PW (2007) Type I phosphatidylinositol-4,5-bisphosphate 4-phosphatase regulates stress-induced apoptosis. Proc Natl Acad Sci U S A 104:16834–16839

Index

M. Falasca (ed.), *Phosphoinositides and Disease*, Current Topics in Microbiology and Immunology 362, DOI: 10.1007/978-94-007-5025-8,

MIX
Papier aus verantwortungsvollen Quellen
Paper from responsible sources
FSC® C105338

If you have any concerns about our products,
you can contact us on
ProductSafety@springernature.com

In case Publisher is established outside the EU,
the EU authorized representative is:
Springer Nature Customer Service Center GmbH
Europaplatz 3, 69115 Heidelberg, Germany

Printed by Libri Plureos GmbH
in Hamburg, Germany